Dynamical Models of Biology and Medicine

Dynamical Models of Biology and Medicine

Special Issue Editors

Yang Kuang
Meng Fan
Shengqiang Liu
Wanbiao Ma

MDPI • Basel • Beijing • Wuhan • Barcelona • Belgrade

Special Issue Editors
Yang Kuang
Arizona State University
USA

Meng Fan
Northeast Normal University
China

Shengqiang Liu
Tianjin Polytechnic University
China

Wanbiao Ma
University of Science and
Technology Beijing
China

Editorial Office
MDPI
St. Alban-Anlage 66
4052 Basel, Switzerland

This is a reprint of articles from the Special Issue published online in the open access journal *Applied Sciences* (ISSN 2076-3417) in 2016 (available at: https://www.mdpi.com/journal/applsci/special_issues/dynamical_models)

For citation purposes, cite each article independently as indicated on the article page online and as indicated below:

LastName, A.A.; LastName, B.B.; LastName, C.C. Article Title. *Journal Name* **Year**, *Article Number*, Page Range.

ISBN 978-3-03921-217-0 (Pbk)
ISBN 978-3-03921-218-7 (PDF)

Contents

About the Special Issue Editors . . . . **vii**

Yang Kuang, Meng Fan, Shengqiang Liu and Wanbiao Ma
Preface for the Special Issue on Dynamical Models of Biology and Medicine
Reprinted from: *Appl. Sci.* **2019**, *9*, 2380, doi:10.3390/app9112380 . . . . **1**

Javier Baez and Yang Kuang
Mathematical Models of Androgen Resistance in Prostate Cancer Patients under Intermittent Androgen Suppression Therapy
Reprinted from: *Appl. Sci.* **2016**, *6*, 352, doi:10.3390/app6110352 . . . . **4**

Urszula Ledzewicz and Helen Moore
Dynamical Systems Properties of a Mathematical Model for the Treatment of CML
Reprinted from: *Appl. Sci.* **2016**, *6*, 291, doi:10.3390/app6100291 . . . . **20**

Shinji Nakaoka, Sota Kuwahara, Chang Hyeong Lee, Hyejin Jeon, Junho Lee, Yasuhiro Takeuchi and Yangjin Kim
Chronic Inflammation in the Epidermis: A Mathematical Model
Reprinted from: *Appl. Sci.* **2016**, *6*, 252, doi:10.3390/app6090252 . . . . **42**

Wei Wang, Wanbiao Ma and Hai Yan
Global Dynamics of Modeling Flocculation of Microorganism
Reprinted from: *Appl. Sci.* **2016**, *6*, 221, doi:10.3390/app6080221 . . . . **77**

Jonathan E. Forde, Stanca M. Ciupe, Ariel Cintron-Arias and Suzanne Lenhart
Optimal Control of Drug Therapy in a Hepatitis B Model
Reprinted from: *Appl. Sci.* **2016**, *6*, 219, doi:10.3390/app6080219 . . . . **101**

Sara Manzano, Manuel Doblaré and Mohamed Hamdy Doweidar
Altered Mechano-Electrochemical Behavior of Articular Cartilage in Populations with Obesity
Reprinted from: *Appl. Sci.* **2016**, *6*, 186, doi:10.3390/app6070186 . . . . **119**

Jonathan Martin and Thomas Hillen
The Spotting Distribution of Wildfires
Reprinted from: *Appl. Sci.* **2016**, *6*, 177, doi:10.3390/app6060177 . . . . **132**

Michael Stemkovski, Robert Baraldi, Kevin B. Flores and H.T. Banks
Validation of a Mathematical Model for Green Algae (*Raphidocelis Subcapitata*) Growth and Implications for a Coupled Dynamical System with *Daphnia Magna*
Reprinted from: *Appl. Sci.* **2016**, *6*, 155, doi:10.3390/app6050155 . . . . **166**

Bing Li, Shengqiang Liu, Jing'an Cui and Jia Li
A Simple Predator-Prey Population Model with Rich Dynamics
Reprinted from: *Appl. Sci.* **2016**, *6*, 151, doi:10.3390/app6050151 . . . . **184**

Maria Vittoria Barbarossa, Christina Kuttler
Mathematical Modeling of Bacteria Communication in Continuous Cultures
Reprinted from: *Appl. Sci.* **2016**, *6*, 149, doi:10.3390/app6050149 . . . . **202**

Zejing Xing, Hongtao Cui and Jimin Zhang
Dynamics of a Stochastic Intraguild Predation Model
Reprinted from: *Appl. Sci.* **2016**, *6*, 118, doi:10.3390/app6040118 . . . . **219**

Chun Li, Wenchao Fei, Yan Zhao and Xiaoqing Yu
Novel Graphical Representation and Numerical Characterization of DNA Sequences
Reprinted from: *Appl. Sci.* **2016**, *6*, 63, doi:10.3390/app6030063 . . . . **236**

Chun Li, Xueqin Li and Yan-Xia Lin
Numerical Characterization of Protein Sequences Based on the Generalized Chou's Pseudo Amino Acid Composition
Reprinted from: *Appl. Sci.* **2016**, *6*, 406, doi:10.3390/app6120406 . . . . **251**

Bai Li and Xiaoyang Li
A Liquid-Solid Coupling Hemodynamic Model with Microcirculation Load
Reprinted from: *Appl. Sci.* **2016**, *6*, 28, doi:10.3390/app6010028 . . . . **268**

About the Special Issue Editors

Yang Kuang has been a professor of mathematics at Arizona State University (ASU) since 1988. He received his BSc degree from the University of Science and Technology of China in 1984 and a PhD in mathematics in 1988 from the University of Alberta. Dr. Kuang is the author or editor of 176 refereed journal publications and 11 books, and is the founder and editor of *Mathematical Biosciences and Engineering*. He has directed 21 PhD dissertations in mathematical and computational biology and several large scale multi-disciplinary research projects in the US. He is well-known for his efforts toward developing practical theories for the study of delay differential equation models and models incorporating resource quality in biology and medicine. His recent research focuses on the formulation and validation of scientifically well-grounded and computationally tractable mathematical models to describe the rich and intriguing dynamics of various within-host diseases and their treatments.

Meng Fan was a full professor of mathematics at Northeast Normal University (NENU) of PR China starting in 2003. He is now the dean of the School of Mathematics and Statistics and the director of the Center for Mathematical Biosciences at NENU. He received his MS degree in pure mathematics and his PhD in ecology from Northeast Normal University in 1998 and 2001, respectively. Dr. Fan is the author or co-editor of more than 130 refereed journal publications and 7 books/Special Issues. He has directed 13 PhD dissertations. Dr. Fan works on dynamical systems and mathematical biology. His research is motivated by both pure mathematics and mathmatical applications in bioscience. His recent research focuses on the dynamical modeling of aquatic ecosystems, zoonotic diseases, and grazing systems.

Shengqiang Liu has been a professor of mathematics at Harbin Institute of Technology (HIT) since 2007. He received his PhD in mathematics in 2002 from the Chinese Academy of Sciences. Dr. Liu has authored more than 60 refereed journal publications. He has supervised 10 PhD dissertations in mathematical biology and applied dynamical systems. His recent research focuses on the formulation of mathematical models to describe the impacts of heterogeneity/random noises on the spreading dynamics of infectious diseases, and their control strategies.

Wanbiao Ma received a BS and MS degree in mathematics in 1979 and 1983, respectively, from Inner Mongolia Normal University (Huhhot, China), and his PhD in engineering in 1997 from Shizuoka University (Hamamatsu, Japan). He is a professor of mathematics of the Department of Applied Mathematics of the University of Science and Technology, Beijing, China. He also served as the dean of the department from 2011–2016. Before he became a professor in 2003 he worked at Inner Mongolia Normal University, Osaka Prefecture University, and Shizuoka University as an assistant and associate professor. His current research focuses on stability theory of functional differential equations, dynamic models in biology, epidemiology, and immunology.

applied sciences

MDPI

Editorial

Preface for the Special Issue on Dynamical Models of Biology and Medicine

Yang Kuang [1,*], Meng Fan [2], Shengqiang Liu [3] and Wanbiao Ma [4]

1 School of Mathematical and Statistical Sciences, Arizona State University, Tempe, AZ 85287, USA
2 School of Mathematics and Statistics, Northeast Normal University, 5268 Renmin Street, Changchun 130024, China; mfan@nenu.edu.cn
3 The Academy of Fundamental and Interdisciplinary Science, Harbin Institute of Technology, 3026#, 2 Yi-Kuang Street, Nan-Gang District, Harbin 150080, China; sqliu@hit.edu.cn
4 Department of Applied Mathematics, School of Mathematics and Physics, University of Science and Technology Beijing, 30 Xue Yuan Road, Beijing 100083, China; wanbiao_ma@ustb.edu.cn
* Correspondence: atyxk@asu.edu; Tel.: +1-480-965-6915

Received: 5 June 2019; Accepted: 10 June 2019; Published: 11 June 2019

Mathematical and computational modeling approaches in biological and medical research are experiencing rapid growth globally. This special issue intends to catch a glimpse of this exciting phenomenon. Areas covered include general mathematical methods and their applications in biology and medicine, with an emphasis on work related to mathematical and computational modeling of the complex dynamics observed in biological and medical research. Specifically, there are fourteen rigorously reviewed papers included in this special issue. These papers cover several timely topics in classical population biology, fundamental biology and modern medicine.

There are four papers in the general area of computational biology dealing with modeling liquid-solid-porous media seepage coupling, bacterial cell-to-cell communication, representation and characterization of DNA sequences and protein sequences, respectively. The work of Bai Li and Xiaoyang Li [1] demonstrates the importance of microcirculation load in a hemodynamic model and their model offers a possibility for the simulation of the dynamic adjustment process of the human circulation system, which may also generate clinical applications. The work of Chun Li et al. [2] presented a cell-based descriptor vector based on the idea of "piecewise function" to numerically characterize the DNA sequence. The utility of their approach was fully illustrated by the examination of phylogenetic analysis on four datasets. In another paper by Chun Li et al. [3], the authors constructed a high dimensional vector to characterize protein sequences. The application of their method on two datasets and the identification of DNA-binding proteins suggested the potential for their user-friendly method. Most noteworthy is the data-validated delay differential equation modeling work of Maria Barbarossa and Christina Kuttler [4] on bacteria communication in continuous cultures. They observed that for a certain choice of parameter values, the model system presented stability switches with respect to the delay. On the other hand, when the delay was set to zero, a Hopf bifurcation might occur with respect to one of the negative feedback parameters. This delay differential model system is capable of explaining and predicting the biological observations.

There are also four papers in the general area of mathematical ecology. The work of Zejing Xing et al. [5] deals with the coexistence of multiple populations species in the context of intraguild predation (IPG). IPG is an ecological phenomenon, which occurs when one predator species attacks another predator species with which it competes for a shared prey species. Their study shows that it is possible for the coexistence of three species aided by the influence of environmental noise. The other three papers involve deterministic differential equation models. The paper by Bing Li et al. [6] studies a simple but non-smooth switched harvest model. The authors established that when the net reproductive number for the predator was greater than unity, the system was capable of generating rich dynamics. In addition to positive equilibrium due to the effects of the switched harvest, the model generated a saddle-node bifurcation, a limit cycle, and the coexistence of a stable equilibrium and an unstable circled inside limit cycle and a stable circled outside limit cycle. When the net reproductive number was less than unity, a backward bifurcation from a positive equilibrium occurred. In another paper, Wei Wang et al. [7] proposed a dynamic model describing the cultivation and flocculation of a microorganism that used two distinct nutrients (carbon and nitrogen). Their model also exhibited rich dynamics, including the existence of possibly five positive equilibria and the possibility of backward and forward bifurcations. In addition, the authors obtained some interesting global stability results of the positive equilibrium. While the aforementioned ecological modeling papers are theoretical, the paper by Michael Stemkovski et al. [8] focused on the validation of a model for green algae (Raphidocelis Subcapitata) growth and the implications for a coupled dynamical system with Daphnia Magna. They collected longitudinal data from three replicate population experiments of R. subcapitata. These data together with statistical model comparison tests and uncertainty quantification techniques allowed the authors to compare the performance of four models: The Logistic model, the Bernoulli model, the Gompertz model, and a discretization of the Logistic model.

There are five papers in the general area of mathematical medicine. In the paper by Urszula Ledzewicz and Helen Moore [9], a mathematical model for the treatment of chronic myeloid leukemia (CML) through a combination of tyrosine kinase inhibitors and immunomodulatory therapies was analyzed as a dynamical system for the case of constant drug concentrations. The model exhibited a variety of behaviors which resembled the chronic, accelerated and blast phases typical of the disease. This work provided qualitative insights into the system which should be useful for understanding the interaction between CML and the therapies considered here. In the paper by Sara Manzano et al. [10], the authors extended an existing mechano-electrochemical computational model and employed the extended model to analyze and quantify the effects of obesity on the articular cartilage of the femoral hip. Their results suggested that people with obesity should undergo preventive treatments for osteoarthritis to avoid homeostatic alterations and, subsequent, tissue deterioration. Combination antiviral drug therapy improves the survival rates of patients chronically infected with hepatitis B virus by controlling viral replication and enhancing immune responses. To address the trade-off between the positive and negative effects of the combination therapy, Jonathan Forde et al. [11] investigated an optimal control problem for a delay differential equation model of immune responses to hepatitis virus B infection. Their results indicated that the high drug levels that induced immune modulation rather than suppression of virological factors were essential for the clearance of hepatitis B virus. In the paper by Shinji Nakaoka et al. [12], the authors developed some mathematical models for the inflammation process using ordinary differential equations and delay differential equations. They investigated the complex microbial community dynamics via transcription factors, protease and extracellular cytokines. They found that large time delays in the activation of immune responses on the dynamics of those bacterial populations led to the onset of oscillations in harmful bacteria and immune activities. The mathematical model suggested the possible annihilation of time-delay-driven oscillations by therapeutic drugs. The paper by Javier Baez and Yang Kuang [13] was motivated and based on clinical data. They proposed and validated a novel type of mathematical model of androgen resistance development in prostate cancer patients under intermittent androgen suppression therapy. More specifically, they formulated and analyzed two mathematical models that aimed to forecast future levels of prostate-specific antigen (PSA). While these models were simplifications of an existing model, they fit data with similar accuracy and improved forecasting results. Their findings suggested that including more realistic mechanisms of androgen dynamics in a two-population model may improve androgen resistance timing prediction.

Last but not the least; this special issue also included a paper on modeling the distribution of wildfires by Jonathan Martin and Thomas Hillen [14]. Their model was based on detailed physical processes. They systematically discussed the use and measurement of their model in fire spread, fire management and fire breaching.

While authors of these papers deal with very different modeling questions, they are all well motivated by specific applications in biology and medicine and employ innovative mathematical and computational methods to study their complex model dynamics. We hope that these papers provide timely case studies that will inspire many more additional mathematical modeling efforts in biology and medicine.

Funding: This research received no external funding.

Conflicts of Interest: The authors declare no conflicts of interest.

References

1. Li, B.; Li, X. A Liquid-Solid Coupling Hemodynamic Model with Microcirculation Load. *Appl. Sci.* **2016**, *6*, 28. [CrossRef]
2. Li, C.; Fei, W.; Zhao, Y.; Yu, X. Novel Graphical Representation and Numerical Characterization of DNA Sequences. *Appl. Sci.* **2016**, *6*, 63. [CrossRef]
3. Li, C.; Li, X.; Lin, Y. Numerical Characterization of Protein Sequences Based on the Generalized Chou's Pseudo Amino Acid Composition. *Appl. Sci.* **2016**, *6*, 406. [CrossRef]
4. Barbarossa, M.; Kuttler, C. Mathematical Modeling of Bacteria Communication in Continuous Cultures. *Appl. Sci.* **2016**, *6*, 149. [CrossRef]
5. Xing, Z.; Cui, H.; Zhang, J. Dynamics of a Stochastic Intraguild Predation Model. *Appl. Sci.* **2016**, *6*, 118. [CrossRef]
6. Li, B.; Liu, S.; Cui, J.; Li, J. A Simple Predator-Prey Population Model with Rich Dynamics. *Appl. Sci.* **2016**, *6*, 151. [CrossRef]
7. Wang, W.; Ma, W.; Yan, H. Global Dynamics of Modeling Flocculation of Microorganism. *Appl. Sci.* **2016**, *6*, 221. [CrossRef]
8. Stemkovski, M.; Baraldi, R.; Flores, K.; Banks, H. Validation of a Mathematical Model for Green Algae (Raphidocelis Subcapitata) Growth and Implications for a Coupled Dynamical System with Daphnia Magna. *Appl. Sci.* **2016**, *6*, 155. [CrossRef]
9. Ledzewicz, U.; Moore, H. Dynamical Systems Properties of a Mathematical Model for the Treatment of CML. *Appl. Sci.* **2016**, *6*, 291. [CrossRef]
10. Manzano, S.; Doblaré, M.; Hamdy Doweidar, M. Altered Mechano-Electrochemical Behavior of Articular Cartilage in Populations with Obesity. *Appl. Sci.* **2016**, *6*, 186. [CrossRef]
11. Forde, J.; Ciupe, S.; Cintron-Arias, A.; Lenhart, S. Optimal Control of Drug Therapy in a Hepatitis B Model. *Appl. Sci.* **2016**, *6*, 219. [CrossRef]
12. Nakaoka, S.; Kuwahara, S.; Lee, C.; Jeon, H.; Lee, J.; Takeuchi, Y.; Kim, Y. Chronic Inflammation in the Epidermis: A Mathematical Model. *Appl. Sci.* **2016**, *6*, 252. [CrossRef]
13. Baez, J.; Kuang, Y. Mathematical Models of Androgen Resistance in Prostate Cancer Patients under Intermittent Androgen Suppression Therapy. *Appl. Sci.* **2016**, *6*, 352. [CrossRef]
14. Martin, J.; Hillen, T. The Spotting Distribution of Wildfires. *Appl. Sci.* **2016**, *6*, 177. [CrossRef]

applied sciences

MDPI

Article

Mathematical Models of Androgen Resistance in Prostate Cancer Patients under Intermittent Androgen Suppression Therapy

Javier Baez and Yang Kuang *

School of Mathematical and Statistical Sciences, Arizona State University, Tempe, AZ 85287, USA; jbaez2@asu.edu
* Correspondence: kuang@asu.edu; Tel.: +1-480-965-6915

Academic Editor: Serafim Kalliadasis
Received: 15 August 2016; Accepted: 5 November 2016; Published: 16 November 2016

Abstract: Predicting the timing of a castrate resistant prostate cancer is critical to lowering medical costs and improving the quality of life of advanced prostate cancer patients. We formulate, compare and analyze two mathematical models that aim to forecast future levels of prostate-specific antigen (PSA). We accomplish these tasks by employing clinical data of locally advanced prostate cancer patients undergoing androgen deprivation therapy (ADT). While these models are simplifications of a previously published model, they fit data with similar accuracy and improve forecasting results. Both models describe the progression of androgen resistance. Although Model 1 is simpler than the more realistic Model 2, it can fit clinical data to a greater precision. However, we found that Model 2 can forecast future PSA levels more accurately. These findings suggest that including more realistic mechanisms of androgen dynamics in a two population model may help androgen resistance timing prediction.

Keywords: mathematical modeling; prostate cancer; androgen deprivation therapy; data fitting

1. Introduction

Ever since the discovery of androgen dependency of prostate cells, androgen deprivation therapy (ADT) has played a vital role in the treatment of metastatic and locally advanced prostate cancer [1–3]. However, controversy remains regarding its best application. Although this treatment will regress tumors in over 90% of patients [4], after prolonged androgen depletion, patients will eventually develop castration-resistant prostate cancer (CRPC) [5]. The development of CRPC can take from a few months to more than ten years [3,6], after which there is a very limited number of effective treatments and patients suffer high mortality [7]. ADT is expensive and its side effects include sexual dysfunction, hot flashes, and fatigue [8]. Based on some preclinical studies, intermittent androgen suppression (IAS) is suggested as a sensible alternative to ADT [9]. During off-treatment periods, patients enjoy a "vacation" from the severe side effects of ADT [8], and studies have suggested that IAS may not negatively affect the time to resistance progression or survival in comparison to ADT [10]. Consequently, IAS is selected by some patients to improve the quality of life and also hopefully to delay the progression to CRPC [4].

Many mathematical models have studied the dynamics of prostate cancer during ADT or IAS [11–18]. A detailed review of some of these models are presented in the recent book of Kuang et al. [19]. Ideta et al. are pioneers of mathematically modeling and analyzing the dynamics of IAS [12]. They formulated a system of ordinary differential equations to study the mechanics of ADT and IAS. They considered castrate-resistant (CR) and castrate-sensitive (CS) cell populations as well as androgen levels. Their model included mutations from CS to CR cells, and their focus was on comparing continuous and intermittent therapy and the development of resistance. Hirata et al. [14] introduced a piece-wise linear model of three cancer cell populations. Their model included CS cells, CR cells that may mutate into CS cells, and CR cells that will not mutate. Several investigators using Hirata et al.'s model [14] have studied estimation of parameters [20,21], optimal switching times and control in IAS [20,22,23], and forecasting CRPC progression [24,25].

Built on the works of Ideta et al. [12] and Jackson [26], Portz, Kuang, and Nagy (PKN) [13] developed a novel mathematical model to study the dynamics of IAS by using the cell quota model [27] from mathematical ecology, which relates growth rate to an intracellular nutrient, to modeling the growth of both the CS and CR cell populations. The cell quota in [13] is defined as the intracellular androgen concentrations for each cell population. This model is carefully fitted with clinical prostate-specific antigen (PSA) data, where androgen data was used to model the cell quota and other growth parameters. Everett et al. [28] compared the models of Hirata et al. [14] and PKN [13] regarding their accuracy of fitting clinical data and predicting future PSA levels. They concluded that while a biologically-based model is important to reveal the underlying processes and my present more robust and better predictions, a simpler model such as that of Hirata et al. might also be practical for fitting clinical data and predicting future PSA outcomes of individual patients.

In this paper, we present a simplified model to the final model in PKN [13]. Several key terms in our model will be mechanistically formulated. This model is concise and amenable to systematical mathematical analysis of its dynamics. For simplicity, we shall use serum androgen concentration to approximate intracellular androgen. This is reasonable since androgen passively and quickly diffuses through the prostate membrane via concentration gradient [29]. This approach is practical for a typical clinical setting, where the data collected can be applied directly to the model. Most importantly, our model can fit PSA and androgen values simultaneously, which enables us to be more accurate in making future PSA value predictions.

2. Clinical Trial Data

We use data from Bruchovsky et al. [9] in our analysis and model calibration. This clinical trial admitted patients who demonstrated a rising serum PSA level after they received radiotherapy and had no evidence of metastasis [9]. The treatment in each cycle consisted of administering cyproterone acetate for four weeks, followed by a combination of leuprolide acetate and cyproterone acetate, for an average of 36 weeks. If serum PSA is less than 4 $\frac{\mu g}{L}$ by the end of this period, the androgen suppression therapy is stopped. If a patient's serum PSA stays above the threshold, the patient will be taken off of the study. After treatment is interrupted, PSA and androgen are monitored every four weeks. The therapy is restarted when patient's serum PSA increases to ≥10 μg/L [9]. The data set is available at [30]. Figure 1 shows a typical patient that undergoes IAS.

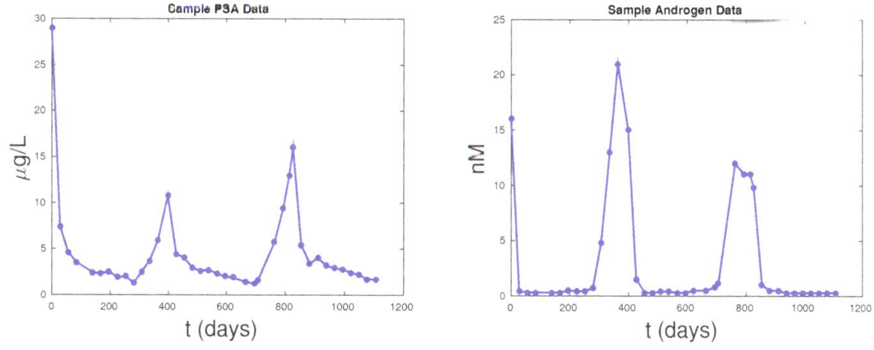

Figure 1. Sample data for prostate-specific antigen (PSA) and androgen data for a patient in a clinical trial.

3. Formulation of Mathematical Models

We develop two plausible mathematical models to study the temporal dynamics of prostate cancer progression to CRPR. In Model 1, we do not distinguish CS from CR cells. In this model, tumor cells' death rate is assumed to be a monotonically decreasing exponential function to implicitly account for the resistance development in cancer cells. Then, we propose a two cell population model where we separate CS from CR cells explicitly. To be more biologically relevant and consistent with the PKN model formulation, we assume in Model 2 that the development of cancer cell resistance to IAS is a decreasing function of androgen levels.

In both models, the cell growth rate is determined by the androgen *cell quota*. Specifically, as in the PKN model [13], we model the growth rate by a two parameter function of androgen cell quota,

$$G(Q) = \mu(1 - \frac{q}{Q}), \tag{1}$$

where Q is the androgen cell quota. Equation (1) is known as Droop equation or a Droop growth rate model [19]. It assumes that Q is the concentration of the most limiting resource or nutrient, and q is the minimum level of Q required to prevent cell death [27].

To be biologically relevant, for both models, we assume that the initial values for all variables are positive. This shall ensure that all components of their solutions are positive. Accordingly, we are only interested in studying the stabilities of nonnegative steady states and their biological and clinical implications.

3.1. Model 1: Single Population Model

In the following model, tumor cell volume is denoted by x (mm^3), and we assume that the total volume is a combination of CS and CR cells. Intracellular androgen cell levels are denoted by Q (nM), and PSA levels by P ($\frac{\mu g}{L}$). Droop's equations govern the growth rate of cancer cells [27], where μ represents the maximum cell growth rate and q the minimum concentration of androgen to sustain the tumor. Similar to [28], we assume an androgen-dependent death rate, where R denotes the half saturation level. However, we also assume a time dependent maximum baseline death rate ν, which decreases exponentially at rate d to reflect the cell castration-resistance development due to the decreasing death rate. We also include a density-independent death rate δ that constrains the total volume of cancer cells to be within realistic ranges [31]:

$$\frac{dx}{dt} = \underbrace{\mu(1 - \frac{q}{Q})x}_{\text{growth}} - \underbrace{(\nu\frac{R}{Q + R} + \delta x)x}_{\text{death}}, \tag{2}$$

$$\frac{d\nu}{dt} = -d\nu, \tag{3}$$

$$\frac{dQ}{dt} = \underbrace{\gamma}_{\text{production}} \underbrace{(Q_m - Q)}_{\text{diffusion}} - \underbrace{\mu(Q - q)}_{\text{uptake}}, \tag{4}$$

$$\frac{dP}{dt} = \underbrace{bQ}_{\text{baseline}} + \underbrace{\sigma x Q}_{\text{tumor production}} - \underbrace{\epsilon P}_{\text{clearance}}, \tag{5}$$

$$\gamma = \gamma_1 u(t) + \gamma_2, \quad u(t) = \begin{cases} 1, \text{ on treatment,} \\ 0, \text{ off treatment.} \end{cases} \tag{6}$$

In this model, androgen is assumed to be the most limiting nutrient. We assume that the androgen concentration in cancer cells is approximately the same as the androgen concentration in serum [29]. Parameter γ_1 denotes the constant production of androgen by the testes, and γ_2 denotes the production of androgen by the adrenal gland and kidneys. As over 95% of androgen is produced in the testes, we have that $\gamma_1 >> \gamma_2$. Parameter $u(t)$ is a switch between on and off treatment cycles. Luteinizing hormone releasing hormone agonists only stop testes production of androgen during treatment. During treatment, γ_2 will be the only production of androgen. $Q_m > q$ denotes the maximum androgen level in serum. The androgen uptake by prostate cells is assumed to be proportional to the difference of the maximum possible and the current androgen levels in serum. Androgen in cells is depleted for growth at a rate of $\mu(Q - q)$. PSA is produced by both the regular cells in the prostate at the rate bQ and by the cancer cells at the rate $\sigma x Q$. Notice that we have assumed that cell production of PSA is assumed to be dependent on levels of androgen. Finally, PSA is cleared from serum at rate ϵ.

3.2. Model 2: Two Population Model

Now, we present a two cell population model. In this model, we explicitly differentiate between CS and CR cells. x_1 (mm^3) and x_2 (mm^3) denote the CS and CR cell populations, respectively. The proliferation of each cancer cell population is denoted by

$$G_i(Q) = \mu(1 - \frac{q_i}{Q}), i = 1, 2,$$

for x_1 and x_2 respectively. Since CR cell populations proliferate at lower levels of androgen, we assume that $q_2 < q_1$. Death rates are denoted by:

$$D_i(Q) = d_i \frac{R_i}{Q + R_i}, i = 1, 2,$$

for their respective cell populations. We shall assume that $d_1 > d_2$, as CR cells are less susceptible to apoptosis by androgen deprivation than CS cells. Parameters $\delta_i, i = 1, 2$ denote the density dependent death rates, and we use these parameters to keep the maximum tumor volume in biological ranges.

Mutation between cell populations is assumed to take the form of a Hill equation of coefficient 1, given by:

$$\underbrace{\lambda(Q) = c\frac{K}{Q + K}}_{\text{CS to CR}}.$$

The CS to CR rate, $\lambda(Q)$, is a decreasing function of the androgen levels. We assume that when cells are experiencing androgen depletion, they have higher selective pressure to develop resistance. Likewise, in an androgen rich environment, CS cells are more likely to stay sensitive. IAS started under this assumption, with the intention to delay resistance [10]. c is the maximum rate of mutation between cells and K is the cell concentration for achieving half of the maximum rate of mutation. In this model, d_is are held constant and are not time dependent, as the mechanism of the development of resistance is due to mutations from x_1 to x_2 via $\lambda(Q)$ and not by a decreasing androgen dependent death rate.

The increase of intracellular androgen levels by diffusion from the serum level is modeled by $\gamma(Q_m - Q)$. For simplicity, and in contrast to the PKN model [13] and the model in Morken et al. [32], we assume the same PSA production rate σ for both cell populations:

$$\frac{dx_1}{dt} = \underbrace{\mu(1-\frac{q_1}{Q})x_1}_{\text{growth}} - \underbrace{(D_1(Q)+\delta_1 x_1)x_1}_{\text{death}} - \underbrace{\lambda(Q)x_1}_{\text{CS to CR}}, \tag{7}$$

$$\frac{dx_2}{dt} = \underbrace{\mu(1-\frac{q_2}{Q})x_2}_{\text{growth}} - \underbrace{(D_2(Q)+\delta_2 x_2)x_2}_{\text{death}} + \underbrace{\lambda(Q)x_1}_{\text{CS to CR}}, \tag{8}$$

$$\frac{dQ}{dt} = \underbrace{\gamma}_{\text{production}}\ \underbrace{(Q_m - Q)}_{\text{diffusion}} - \underbrace{\frac{\mu(Q-q_1)x_1 + \mu(Q-q_2)x_2}{x_1+x_2}}_{\text{uptake}}, \tag{9}$$

$$\frac{dP}{dt} = \underbrace{bQ}_{\text{baseline}} + \underbrace{\sigma(Qx_1 + Qx_2)}_{\text{tumor production}} - \underbrace{\epsilon P.}_{\text{clearence}}\ . \tag{10}$$

In a biologically realistic situation, one expects that $Q_m > \max\{q_1, q_2\}$.

3.3. Derivation of dQ/dt

Now, we provide a conservation law based derivation for the cell quota Q Equations (4) and (9). Specifically, we derive Equation (4) in detail and leave to the readers the straightforward task of its extension to (9). Our formulation comes from the conservation of androgen as it moves in and out of the tumor. Let Q_x be the total androgen inside tumor x (mm^3). We assume that Q (nM)is uniformly distributed in x, and

$$Q_x = Q(t)x(t) \quad \text{nmol}.$$

The inflow of androgen to the tumor comes from the serum which can be approximated by

$$\gamma(Q_m - Q(t))x(t).$$

The outflow of androgen from the tumor is due to death, which is

$$(\nu\frac{R}{Q+R} + \delta x(t))Q(t)x(t).$$

Then, the rate of change of androgen inside the tumor is:

$$(Q(t)x(t))' = \gamma(Q_m - Q(t))x(t) - (\nu\frac{R}{Q(t)+R} + \delta x(t))Q(t)x(t).$$

However,

$$\begin{aligned} (Q(t)x(t))' &= Q'(t)x(t) + Q(t)x'(t) \\ &= Q'(t)x(t) + \mu(Q(t)-q)x(t) - (\nu\frac{R}{Q(t)+R} + \delta x(t))Q(t)x(t), \end{aligned}$$

which implies that

$$Q'(t) = \gamma(Q_m - Q(t)) - \mu(Q(t) - q).$$

A similar approach can be applied to derive $Q'(t)$ for Model 2.

3.4. Portz, Kuang, and Nagy (PKN) Model

In this section, we briefly review the PKN model. For a more detailed explanation of this model, the reader is referred to [13]. The PKN model assumes constant death rates for cancer cells (d_1, d_2). CS and CR cells have androgen cell quota Q_1, Q_2 respectively. A denotes the serum androgen concentration, which is interpolated and used in the model:

$$\frac{dx_1}{dt} = \underbrace{\mu_m(1-\frac{q_1}{Q_1})x_1}_{\text{growth}} - \underbrace{d_1x_1}_{\text{death}} - \underbrace{\lambda_1(Q_1)x_1}_{\text{CS to CR}} + \underbrace{\lambda_2(Q_2)x_2}_{\text{CR to CS}}, \tag{11}$$

$$\frac{dx_2}{dt} = \underbrace{\mu_m(1-\frac{q_2}{Q_2})x_2}_{\text{growth}} - \underbrace{d_2x_2}_{\text{death}} - \underbrace{\lambda_2(Q_2)x_2}_{\text{CR to CS}} + \underbrace{\lambda_1(Q_1)x_1}_{\text{CS to CR}}, \tag{12}$$

$$\frac{dQ_1}{dt} = \underbrace{v_m\frac{q_m-Q_1}{q_m-q_1}\frac{A}{A+v_h}}_{\text{Androgen influx to CS cells}} - \underbrace{\mu(Q_1-q_1)}_{\text{uptake}} - \underbrace{bQ_1}_{\text{degradation}}, \tag{13}$$

$$\frac{dQ_2}{dt} = \underbrace{v_m\frac{q_m-Q_2}{q_m-q_2}\frac{A}{A+v_h}}_{\text{Androgen influx to CR cells}} - \underbrace{\mu(Q_2-q_2)}_{\text{uptake}} - \underbrace{bQ_2}_{\text{degradation}}, \tag{14}$$

$$\frac{dP}{dt} = \underbrace{\sigma_0(x_1+x_2)}_{\text{baseline production}} + \underbrace{\sigma_1x_1\frac{Q_1^m}{Q_1^m+\rho_1^m}}_{\text{tumor production}} + \underbrace{\sigma_2x_2\frac{Q_2^m}{Q_2^m+\rho_2^m}}_{\text{tumor production}} - \underbrace{\delta P.}_{\text{clearence}}. \tag{15}$$

4. Model Dynamics

Now, we study the mathematical properties and dynamics of our two models. For Model 1, we shall state the results without providing proofs as they are routine. The detailed mathematical analysis for Model 2 will be presented. Proposition 1 summarizes the mathematical dynamics of Model 1. Since P is decoupled from the system, we shall refer only to the dynamics of Equations (2)–(4). This proposition reveals that there is no cure for cancer. Since ADT is non-curative, this property is biologically reasonable.

Proposition 1. *Solutions of the system Equations (2)–(4) are positive and bounded. The system Equations (2)–(4) has a cancer free steady state* $E_0 = (0, 0, \frac{\gamma Q_m+\mu q}{\mu+\gamma})$ *that is unstable, and a steady state* $E_1 = (\frac{\mu\gamma}{\delta}\frac{Q_m-q}{\gamma Q_m+\mu q}, 0, \frac{\gamma Q_m+\mu q}{\mu+\gamma})$ *that is globally stable.*

Next, we do a thorough mathematical analysis of Model 2. First, we study boundedness and positivity of the system. Followed by the number and existence of steady states. Finally, we analyze the local stability of the steady states. Observe that P is also decoupled from Equations (2)–(4) and we do not include it in the analysis.

Proposition 2. *Assume* $q_2 \le q_1 < Q_m$ *and* $\delta_1 \ge \delta_2$. *Then, solutions of Equations (7)–(9) with initial conditions* $x_1(0) > 0$, $x_2(0) > 0$, *and* $q_2 \le Q(0) \le Q_m$ *stay in the region* $\{(x_1, x_2, Q) : x_1 \ge 0, x_2 \ge 0, x_1 + x_2 \le \frac{G_2(Q_m)-D_m(q_2)}{\delta_2}, q_2 \le Q \le Q_m\}$, *where* $D_m = \min\{D_1(q_2), D_2(q_2)\}$.

Proof. We note that in Equation (7), x_1 appears in every term ensuring its positivity. Since x_2 appears in the first two terms of (8) and x_1 appears in the last term, the positivity of x_2 is also guaranteed.

In addition, $q_2 \leq q_1 < Q_m$, and

$$Q' = \gamma(Q_m - Q) - \frac{\mu(Q - q_1)x_1 + \mu(Q - q_2)x_2}{x_1 + x_2}.$$

We see that $Q'(q_2) > 0$ and $Q'(Q_m) < 0$. It is thus easy to see that $q_2 \leq Q(t) \leq Q_m$ for $t > 0$ with initial conditions $q_2 \leq Q(0) \leq Q_m$.

For boundedness of x_1 and x_2, we let $N = x_1 + x_2$. Since we have that $\delta_1 \geq \delta_2$, and the growth rate $G_i(Q)$, $i = 1, 2$ are increasing functions of Q, we have

$$\begin{aligned} N' &\leq (G_2(Q) - D_m)N - \delta_2 N^2, & (16)\\ &\leq (G_2(Q_m) - D_m)N - \delta_2 N^2, & (17) \end{aligned}$$

which implies that $\limsup_{t\to\infty} N(t) \leq \frac{G_2(Q_m) - D_m}{\delta_2}$. □

Now, we study the steady states of Model 2. We seek to understand the conditions under which one population will overtake the other, and the circumstances under which they may coexist.

Proposition 3. *Assume $q_2 \leq q_1 < Q_m$ and $\delta_1 \geq \delta_2$. The system Equations (7)–(9) have a CR cell only steady state $E_1 = (0, \frac{G_2(Q^1) - D_2(Q^1)}{\delta_2}, Q^1)$, and a coexistence steady state $E_2 = (\frac{G_1(Q^*) - D_1(Q^*) - \lambda_1(Q^*)}{\delta_1}, x_2^*, Q^*)$, where $Q^1 = \frac{\gamma Q_m + \mu q_2}{\gamma + \mu}$ and $Q^* > Q^1$.*

Proof. Let $E = (x_1^*, x_2^*, Q^*)$ be a steady state of the system Equations (7)–(9). We have two mutually exclusive cases: $x_1^* = 0$ and $x_1^* > 0$.

If $x_1^* = 0$, then we have two possibilities: (i) $x_2^* = 0$ or (ii) $x_2^* > 0$. In the case of (i), we see that $E = E_0$. In the case of (ii), we see that $E = E_1$.

If $x_1^* > 0$, we see that $x_2^* > 0$ from the equation of dx_2/dt. In this case, $E = E_2$. In addition, we have the following:

$$\begin{aligned} 0 &= \gamma(Q_m - Q^*) - \frac{\mu(Q^* - q_1)x_1^* + \mu(Q^* - q_2)x_2^*}{x_1^* + x_2^*} & (18)\\ &\geq \gamma(Q_m - Q^*) - \mu(Q^* - q_2) \\ Q^* &\geq \frac{\gamma Q_m + \mu q_2}{\gamma + \mu} = Q^1. \end{aligned}$$

This proves the proposition. □

Proposition 3 demonstrates that if the CS cell population survives, then the CR must also survive. Biologically, this makes sense, as the CR will always receive new mutated CR cells as ADT continues.

Next, we study the extinction of cancer cell populations and stability conditions for each of these steady states when feasible. Observe that we can not linearize at the steady state E_0 since the last term of dQ/dt is not differentiable at E_0. This prevents us from carrying out a routine local stability analysis of E_0.

Proposition 4 below simply confirms the intuition that if both cancer cell populations growth rates are too low, they will die out eventually. For ease of computations in the following propositions, we shall define $S_1(Q) = G_1(Q) - D_1(Q) - \lambda(Q)$ and $S_2(Q) = G_2(Q) - D_2(Q)$.

Proposition 4. *Assume that $S_1(Q_m) < 0$, then CS population will die out. If, in addition, $S_2(Q_m) < 0$, then both cancer populations will die out.*

Proof. Observe that both $S_1(Q)$ and $S_2(Q)$ are strictly increasing with respect to positive values of Q. Since,

$$\frac{x_1'(t)}{x_1(t)} = G_1(Q) - D_1(Q) - \lambda(Q) - \delta_2 x_1,$$

and $S_1(Q_m) < 0$, we know that $G_1(Q) - D_1(Q) - \lambda(Q) \leq S_1(Q_m) < 0$ for any Q. Let $m = -S_1(Q_m)$, and, since $x_1(t) > 0$, we have that

$$\begin{aligned} \frac{x_1'(t)}{x_1(t)} &\leq -m \\ x_1(t) &\leq ce^{-mt}. \end{aligned}$$

Therefore $\lim_{t\to\infty} x_1(t) = 0$. Applying a similar but slightly more delicate comparison argument to $x_2(t)$ with $\lim_{t\to\infty} x_1(t) = 0$ yields $\lim_{t\to\infty} x_2(t) = 0$. This completes the proof of this proposition. □

The following proposition provides a simple set of conditions that yields the biologically realistic final outcome when sensitive cells are overtaken by resistant cells.

Proposition 5. *The CR only steady state E_1 is locally asymptotically stable when $S_1(Q^1) < 0$ and $S_2(Q^1) > 0$.*

Proof. The Jacobian matrix evaluated at E_1 is given by:

$$J(E_1) = \begin{pmatrix} S_1(Q^1) & 0 & 0 \\ \lambda(Q^1) & -S_2(Q^1) & \left(\frac{\mu q_2}{Q^2} + \frac{d_2}{(R_2+Q)^2}\right)\frac{G_2(Q^1)-D_2(Q^1)}{\delta_2} \\ \frac{\mu\delta_2(q_1-q_2)}{G_2(Q^1)-D_2(Q^1)} & 0 & -\gamma-\mu \end{pmatrix}.$$

The eigenvalues are the diagonal elements. We see that when $G_1(Q^1) - D_1(Q^1) - \lambda_1(Q^1) < 0$ and $G_2(Q^1) - D_2(Q^1) > 0$, all diagonal elements are negative. Hence, E_1 is locally asymptotically stable. □

If both CS and CI cells can proliferate under treatment, then the coexistence equilibrium may be stable. Figure 2 displays the regions where this could happen. If CS cells have a high growth rate μ, they may survive under relatively low levels of androgen. Alternatively, if these cells have a very low death rate d_1, they may persist as well.

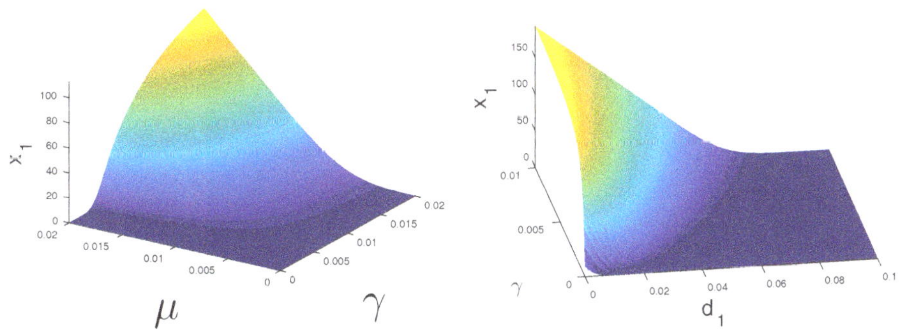

Figure 2. Bifurcation diagram displaying x_1 cell population vs. parameters μ and γ (**left**) and γ and d_1 (**right**). This figure depicts the regions in which x_1 can go extinct. This happens when androgen levels γ are very low, or cancer cells' proliferation rate μ is very low, or cancer cells' death rate d_1 is very high.

5. Parameter Estimation

In order to perform realistic model simulations, we need to obtain reasonable parameter values and their ranges. We start by estimating the realistic ranges for each of them. Parameters μ, d_1, d_2 are taken from [33], where they assess the growth and death rates of prostate cells under different concentrations of androgen. In [12], it was shown that, under continuous treatment, the fastest resistance rate is $c \approx 0.0001$. The approximate levels at which sensitive and resistant cells proliferate was studied in [34], from which we approximated q, q_1, and q_2.

In patients with no prostate cancer, PSA levels are usually less than 5 $\frac{\mu g}{L}$, accounting for benign tumor hyperplasia [8]. This implies that when tumor volume is near zero, the steady state of PSA given by: $\frac{bQ}{\epsilon}$ shall be approximately 5 $\frac{\mu g}{L}$. Prostate tumor volumes are normally bounded by 80 mm in length and, on average, they are about 13.4 mm [31]. Since all of our patients have advanced prostate cancer, we assumed a maximum length of 40 mm, and we compute the corresponding tumor volume assuming that tumors are spherical. Under complete androgen independence, tumor volume should not exceed 700 (mm^3). Thus, $\frac{\mu}{\delta} \approx \frac{\mu}{\delta_2} + \frac{\mu}{\delta_2} \approx 700$ (mm^3). Parameter Q_m is patient specific and is taken from the maximum androgen serum concentration of each patient during the first 1.5 cycles of treatment. Parameter γ_1 is held constant among every patient and γ_2 has a range of 0–0.01 $\frac{nmol}{Lday}$. The half-saturation variables $K, R, R1$, and $R2$ are estimated from [28]. Table 1 shows definitions, ranges, units, and sources for each of the parameters in our models.

Table 1. Parameter definitions, units, and ranges.

Parameters	Definition	Range	Units	Source
μ	Maximum proliferation rate	0.001–0.09	day^{-1}	[33]
q	Minimum cell quota	0.1–0.5	nM	[34]
q_1	Minimum CS cell quota	0.1–0.5	nM	[34]
q_2	Minimum CR cell quota	0.1–0.3	nM	[34]
b	Prostate baseline PSA	0.1–2.5	10^{-3} μg/L/nM/day	[9]
σ	Tumor PSA production rate	0.001–0.9	μg/L/nM/mm^3/day	[28]
ϵ	PSA clearance rate	0.001–0.01	day^{-1}	[28]
d	Maximum cell death rate	0.0001–0.09	day^{-1}	[33]
d_1	Maximum CS cell death rate	0.001–0.09	day^{-1}	[33]
d_2	Maximum CR cell death rate	0.0001–0.001	day^{-1}	[33]
δ_1	Density death rate	0.1–9 × 10^{-5}	1/day/mm^3	[31]
δ_2	Density death rate	0.01–4.5 × 10^{-4}	1/day/mm^3	[31]
R	Cell death rate half-saturation level	0–3	nM	[28]
R_1	CS cell death rate half-saturation level	0–3	nM	[28]
R_2	CR cell death rate half-saturation level	0–3	nM	[28]
c_1	Maximum CS to CR rate	10^{-5}–10^{-4}	day^{-1}	[12]
K	CS to CR half-saturation level	0–1	nM	[28]
γ_1	Testes androgen production	20	day^{-1}	ad hoc
γ_2	Secondary androgen production	0.001–0.01	day^{-1}	ad hoc
Q_m	Maximum androgen	15–30	nM	[9]
v	death rate decay rate	0.01	unitless	ad hoc

5.1. Sensitivity Analysis

Sensitivity analysis can be used to show which parameters play a bigger role in a model. The normalized sensitivity, S_p, for parameter p and state variable x is given by:

$$S_p = \frac{\partial x}{\partial p}\frac{p}{x}.$$

Figure 3 shows the sensitivities of every parameter and state variable for Model 1. We observe that, with the exception of d, no parameter has a much larger sensitivity than the rest. d is the only parameter that can dramatically affect cancer cell death rate, and we see a spike in the sensitivity figures. Cancer cell growth rate μ has the greatest effect on androgen and cancer cells, whereas σ and ϵ play the greatest roles in PSA production. Figure 4 shows the sensitivity of each parameter in Model 2.

We see similarities with Model 1 in that μ affects the production androgen and CR cells the most. In addition, PSA production is affected by the same parameters as in Model 1. Notice that CS cells are affected by d_1 the most since they are the most susceptible to changes in the androgen concentration.

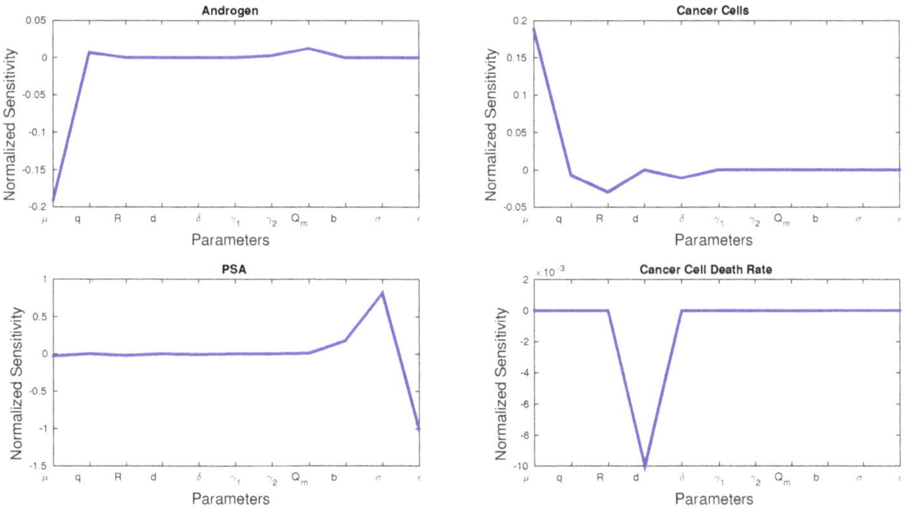

Figure 3. Normalized sensitivities of Model 1.

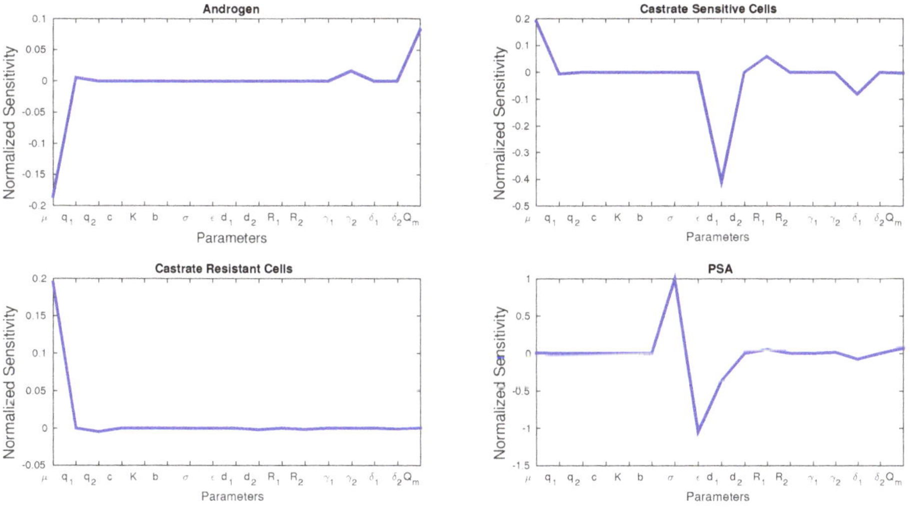

Figure 4. Normalized sensitivities of Model 2.

6. Comparison of Models

We use data from the Vancouver Prostate Center (Vancouver, BC, Canada) to validate and compare the accuracy of each model. From the 109 patients registered, 103 were eligible for interruption of treatment, with a PSA response rate of 95% [9]. Using the criteria of having at least 20 data points for both androgen and PSA in the initial 1.5 cycles, we select 62 from those 109 patients. The individual PSA and androgen mean square error (MSE) are provided in Table 2 from these 62 selected patients.

Notice that the PKN model did not include an androgen equation, and thus we cannot compare the fittings of androgen with the PKN model.

For the PKN model, we interpolated androgen serum data using a cubic spline interpolation between every androgen data point. This created a function in terms of time that was utilized as A in (13) and (14). We implemented the method used by Portz et al. [13] for generating future androgen levels by generating a rectangular function based on the average off and on-treatment serum androgen values. Parameter ranges were taken from PKN [13] and Everett et al. [28], and the reader is referred to these papers for more details on forecasting serum PSA levels and parameter values of the PKN model. For every patient selected, we fitted 1.5 cycles of treatment and performed parameter estimation. Then, to measure the forecasting ability of every model, we ran the models for one more cycle of data using the parameters estimated from the initial 1.5 cycles.

Table 2. Comparison of Mean Squared Error (MSE) for Androgen and prostate-specific antigen (PSA) for the first 1.5 cycles.

Model	PSA			Androgen		
	Min	Mean	Max	Min	Mean	Max
PKN Model	0.5119	9.4463	93.1587	N/A	N/A	N/A
Model 1	0.9735	8.6763	71.8471	5.0351	100.1071	710.2604
Model 2	0.2461	10.3993	137.4345	5.1283	101.4763	710.4412

To compare models, we conduct simulations with MATLAB's (MATLAB 9.1, The MathWorks, Inc., Natick, MA, USA) built in function fmincon, which uses the Interior Point Algorithm, to find the optimum parameters for each patient. The algorithm searches for a minimum value in a range of pre-specified parameter ranges, which we estimated from various literature sources. We use this algorithm to minimize the MSE for PSA and androgen data. The MSE is calculated with the following equations:

$$P_{error} = \frac{\sum_{i=1}^{N}(P_i - \hat{P}_i)^2}{N},$$

$$Q_{error} = \frac{\sum_{i=1}^{N}(Q_i - \hat{Q}_i)^2}{N},$$

where N represents the total number of data points, P_i represents the PSA data value, and $\hat{P}_i$ the value from the model. Likewise, Q_i represents the androgen data value, and $\hat{Q}_i$ the value from the model. We then use an equally weighted combination of both errors

$$error = P_{error} + Q_{error},$$

as our objective function, which is then minimized with fmincon.

Figure 5 shows PSA fitting and forecasting simulations for patients 1, 15, 17, and 63. We selected these patients to display the typical behavior shown in all 62 patients. Patient 1 shows that Models 1, 2, and PKN fit data with about the same accuracy. However, PKN overshoots in forecasting and Model 2 outperforms Model 1 in forecasting. Patient 17 shows that PKN underestimates future PSA levels, but Models 1 and 2 both perform well. Patients 15 and 63 provide the cases where PKN does a better forecast while Models 1 and 2 still do better. The rest of the patients can be classified similarly.

Table 2 documents the error of fitting 1.5 cycles of treatment and Table 3 displays the errors in forecasting one more cycle of treatment. On average, PKN and Model 1 perform prediction at the same level of accuracy. However, Model 2 performs prediction on average about three times better than the PKN model and Model 1.

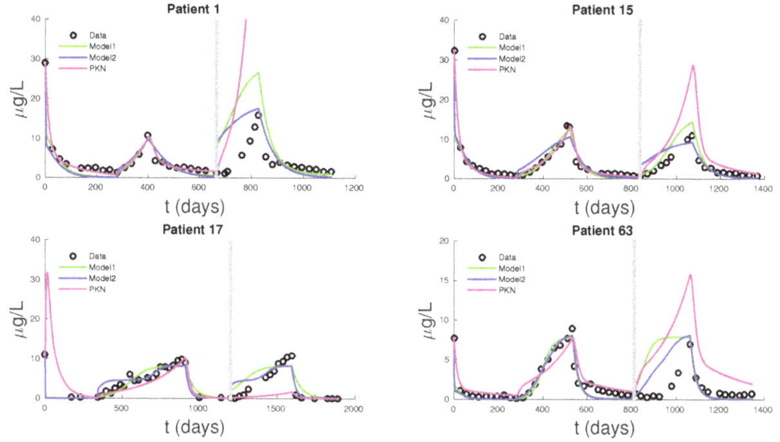

Figure 5. Simulations of fittings for every model for 1.5 cycles of treatment (**left** of gray line), and one cycle of forecast (**right** of gray line). For these four patients, we can see that models fit data at comparable accuracy but Model 2 perform much better in PSA forecasting.

Table 3. Comparison of forecast Mean Square Error for PSA.

Model	Min	Mean	Max
PKN Model	12.234	162.5494	1868.6394
Model 1	11.3935	141.9280	1663.0218
Model 2	2.2727	56.3478	278.4050

7. Conclusions

The main goal of this research is to produce a basic model capable of describing prostate cancer cell growth subject to IAS. Such a model may be amendable to detailed and systematical mathematical and computational study aimed at revealing the near term and intermediate term growth dynamics, including cancer cells treatment resistance development. Ultimately, such models may be helpful in establishing user-friendly treatment tools for both patients and physicians. To this end, we presented two models that can accurately fit clinical PSA and androgen data simultaneously. Existing models can only fit the PSA data. While these models are simplifications of PKN, they are just as accurate in data fitting and even better at forecasting future PSA levels. Model 1 had the lowest mean MSE for data fitting of all the models, followed by PKN and Model 2—not surprisingly, due to its more biologically realistic model assumptions, Model 2 had the lowest forecast MSE, with PKN doing the worst. The unreliability of PKN's forecasts stems from its dependence on androgen data and hence lacks the ability to predict androgen dynamics. Androgen cell quota values, which are not directly measurable from data, represent a significant source of uncertainty for the PKN model. Figure 6, shows how the new models can fit clinical androgen data and reduce uncertainty. For Models 1 and 2, Q is directly computable from clinical data.

Predicting the timing of resistance is a highly desirable objective of this modeling work. Mathematical models alone can not make practical predictions. However, we can use these models and apply statistical methods to produce reliable forecasts with confidence intervals [35,36]. In Pell et al. [35] and Chowell et al. [36], the authors have used these methods successfully in an epidemiological setting to make predictions. In a future paper, the authors plan to use the models presented in this paper to produce predictions using such statistical methods. Therefore, the main biological contribution of this work is the development of a clear and basic androgen cell-quota based model to aid our understanding of the resistance development dynamics of prostate cancer cells.

The dynamics of Model 1 is characterized by globally stable steady states. The mathematical analysis of Model 2 is only partially tractable. In fact, the stability and global stability of the cancer cell coexistence steady state remain unsettled. However, with our bifurcation analysis, we observe that under ADT or IAS, x_2 cells may drive x_1 cells to extinction. In Figure 2, we see that with lower and realistic levels of androgen production when the patient is under ADT, x_1 cells will eventually become extinct even at a higher level of proliferation compared to x_2 cells. Thus, we concluded that, under continuous treatment, almost all patients will eventually become androgen resistant. However, it is still not clear if IAS delays the speed at which this occurs. With the models presented in this work, we have moved closer to the ultimate goal of modeling the androgen resistance of prostate cancer.

Our work is limited by the small number of patients considered. We selected 62 patients that had at least 20 data points in the first 1.5 cycles of treatment. Using a larger time interval and more patients to calibrate models might reveal more subtle differences in the models' ability to fit data. Additionally, tumor volume data will allow us to validate the model more naturally. Figures 7 and 8 show the cancer populations for Models 1 and 2 in resistant and non-resistant patients. Tumor volume data will allow us to verify the results in these figures even if only a few data points are available.

In addition, identifiability analysis to determine if our parameter values can be represented uniquely by clinical data is essential if these models are to be used in a clinical setting to reliably and accurately predict PSA dynamics for individual patients. Allowing parameters to vary as treatment progresses and studying the changes in key parameter values such as proliferation and death rates as functions of time might be useful to describe and predict resistance mechanisms as suggested in the work of Morken et al. [32].

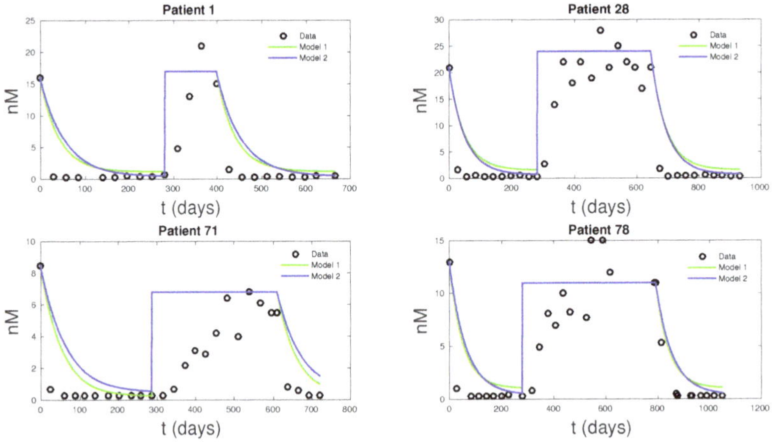

Figure 6. Simulations of fittings of androgen levels for Models 1 and 2. These two models have comparable goodness in fitting androgen data as their derivations are very similar.

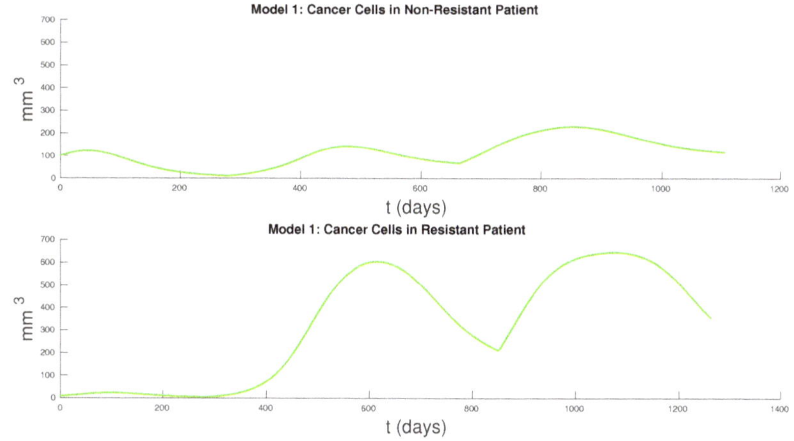

Figure 7. Cancer cells in resistant and non-resistant patient for Model 1. For the non-resistant patient we see a slight increase in volume over the course of several cycles. In the resistant patient we see that cancer volume has grown substantially.

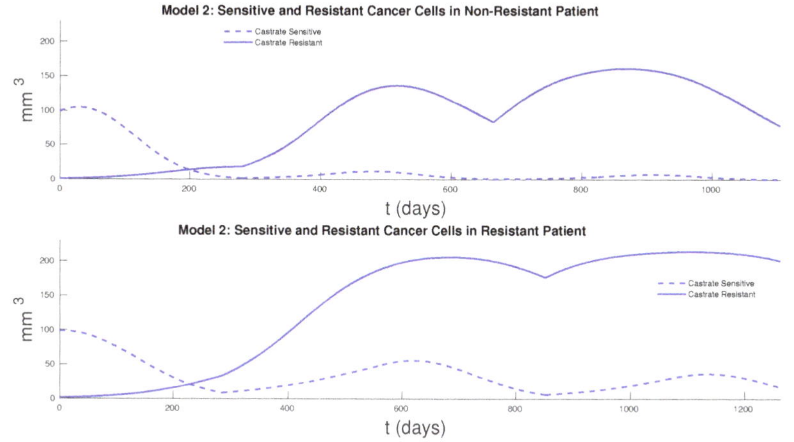

Figure 8. Cancer cells in resistant and non-resistant patients for Model 2. For the non-resistant patient, we see an increase in the volume of CR cells, but the original volume is about the same. In the resistant patient, we see that cancer volume has grown to double the volume compared to the non resistant patient.

Acknowledgments: The authors are partially supported by a US NSF grant DMS-1518529 and are profoundly grateful to Nicholas Bruchovsky for the clinical data set and his encouragement. In addition, the authors would like to thank Tin Phan and the reviewers for many helpful comments that improved the presentation of this manuscript.

Author Contributions: J.B. and Y.K. conceived and designed the models; J.B. created the code and ran simulations; J.B. and Y.K. wrote the paper.

Conflicts of Interest: The authors declare no conflict of interest.

References

1. Heinlein, C.; Chang, C. Androgen receptor in prostate cancer. *Endocr. Rev.* **2004**, *25*, 276–308.
2. Shafi, A.; Yen, A.; Weigel, N. Androgen receptors in hormone-dependent and castration-resistant prostate cancer. *Pharmacol. Ther.* **2013**, *140*, 223–238.
3. Tsao, C.K.; Small, A.; Galsky, M.; Oh, W. Overcoming castration resistance in prostate cancer. *Curr. Opin. Urol.* **2012**, *22*, 167–174.
4. Bruchovsky, N.; Klotz, L.; Crook, J.; Phillips, N.; Abersbach, J.; Goldenberg, S. Quality of life, morbidity, and mortality results of a prospective phase II study of intermittent androgen suppression for men with evidence of prostate-specific antigen relapse after radiation therapy for locally advanced prostate cancer. *Clin. Genitourin. Cancer* **2008**, *6*, 46–52.
5. Feldman, B.; Feldman, D. The development of androgen-independent prostate cancer. *Nat. Rev. Cancer* **2001**, *1*, 34–45.
6. Hussain, M.; Tangen, C.; Berry, D.; Higano, C.; Crawford, E.; Liu, G.; Wilding, G.; Prescott, S.; Sundaram, S.K.; Small, E.J.; et al. Intermittent versus continuous androgen deprivation in prostate cancer. *N. Engl. J. Med.* **2013**, *368*, 1314–1325.
7. Karantanos, T.; Evans, C.P.; Tombal, B.; Thompson, T.C.; Montironi, R.; Isaacs, W.B. Understanding the mechanisms of androgen deprivation resistance in prostate cancer at the molecular level. *Eur. Urol.* **2015**, *67*, 470–479.
8. Klotz, L.; Toren, P. Androgen deprivation therapy in advanced prostate cancer: Is intermittent therapy the new standard of care? *Curr. Oncol.* **2012**, *19*, S13–S21.
9. Bruchovsky, N.; Klotz, L.; Crook, J.; Malone, S.; Ludgate, C.; Morris, W.J.; Gleave, M.E.; Goldenberg, S.L. Final results of the Canadian prospective phase II trial of intermittent androgen suppression for men in biochemical recurrence after radiotherapy for locally advanced prostate cancer. *Cancer* **2006**, *107*, 389–395.
10. Gleave, M. Prime time for intermittent androgen suppression. *Eur. Urol.* **2014**, *66*, 240–242.
11. Jackson, T. A Mathematical Investigation of the Multiple Pathways to Recurrent Prostate Cancer: Comparison with Experimental Data. *Neoplasia* **2004**, *6*, 697–704.
12. Ideta, A.; Tanaka, G.; Takeuchi, T.; Aihara, K. A mathematical model of intermittent androgen suppression for prostate cancer. *J. Nonlinear Sci.* **2008**, *18*, 593–614.
13. Portz, T.; Kuang, Y.; Nagy, J. A clinical data validated mathematical model of prostate cancer growth under intermittent androgen suppression therapy. *AIP Adv.* **2012**, *2*, 1–14.
14. Hirata, Y.; Bruchovsky, N.; Aihara, K. Development of a mathematical model that predicts the outcome of hormone therapy for prostate cancer. *J. Theor. Biol.* **2010**, *264*, 517–527.
15. Swanson, K.; True, L.; Lin, D.; Buhler, K.; Vessella, R.; Murray, J. A quantitative model for the dynamics of serum prostate-specific antigen as a marker for cancerous growth: An explanation for a medical anomaly. *Am. J. Pathol.* **2001**, *158*, 2195–2199.
16. Jain, H.; Clinton, S.; Bhinder, A.; Friedman, A. Mathematical modeling of prostate cancer progression in response to androgen ablation therapy. *Proc. Natl. Acad. Sci. USA* **2011**, *108*, 19701–19706.
17. Jain, H.; Friedman, A. Modeling prostate cancer response to continuous versus intermittent androgen ablation therapy. *Discret. Contin. Dyn. Syst. B* **2013**, *18*, 945–967.
18. Jain, H.; Friedman, A. A partial differential equation model of metastasized prostatic cancer. *Math. Biosci. Eng.* **2013**, *10*, 591–608.
19. Kuang, Y.; Nagy, J.; Eikenberry, S. *Introduction to Mathematical Oncology*; CRC Press: Boca Raton, FL, USA, 2016.
20. Guo, Q.; Lu, Z.; Hirata, Y.; Aihara, K. Parameter estimation and optimal scheduling algorithm for a mathematical model of intermittent androgen suppression therapy for prostate cancer. *Chaos* **2013**, *23*, 43125.
21. Tao, Y.; Guo, Q.; Aihara, K. A partial differential equation model and its reduction to an ordinary differential equation model for prostate tumor growth under intermittent hormone therapy. *J. Math. Biol.* **2014**, *69*, 817–838.
22. Suzuki, Y.; Sakai, D.; Nomura, T.; Hirata, Y.; Aihara, K. A new protocol for intermittent androgen suppression therapy of prostate cancer with unstable saddle-point dynamics. *J. Theor. Biol.* **2014**, *350*, 1–16.

23. Hirata, Y.; Akakura, K.; Higano, C.; Bruchovsky, N.; Aihara, K. Quantitative mathematical modeling of PSA dynamics of prostate cancer patients treated with intermittent androgen suppression. *J. Mol. Cell Biol.* **2012**, *4*, 127–132.
24. Hirata, Y.; Tanaka, G.; Bruchovsky, N.; Aihara, K. Mathematically modelling and controlling prostate cancer under intermittent hormone therapy. *Asian J. Androl.* **2012**, *14*, 270–277.
25. Hirata, Y.; Azuma, S.; Aihara, K. Model predictive control for optimally scheduling intermittent androgen suppression of prostate cancer. *Methods* **2014**, *67*, 278–281.
26. Jackson, T. A mathematical model of prostate tumor growth and androgen-independent relapse. *Discret. Contin. Dyn. Syst. B* **2004**, *4*, 187–202.
27. Droop, M. Some thoughts on nutrient limitation in algae1. *J. Phycol.* **1973**, *9*, 264–272.
28. Everett, R.; Packer, A.; Kuang, Y. Can Mathematical Models Predict the Outcomes of Prostate Cancer Patients Undergoing Intermittent Androgen Deprivation Therapy? *Biophys. Rev. Lett.* **2014**, *9*, 173–191.
29. Roy, A.; Chatterjee, B. Androgen action. *Crit. Rev. Eukaryot. Gene Expr.* **1995**, *5*, 157–176.
30. Bruchovsky, N. Clinical Research. 2006. Available online: http://www.nicholasbruchovsky.com/clinicalResearch.html (accessed on 11 November 2016).
31. Vollmer, R. Tumor Length in Prostate Cancer. *Am. J. Clin. Pathol.* **2008**, *130*, 77–82.
32. Morken, J.; Packer, A.; Everett, R.; Nagy, J.; Kuang, Y. Mechanisms of resistance to intermittent androgen deprivation in patients with prostate cancer identified by a novel computational method. *Cancer Res.* **2014**, *74*, 3673–3683.
33. Berges, R.R.; Vukanovic, J.; Epstein, J.I.; CarMichel, M.; Cisek, L.; Johnson, D.E.; Veltri, R.W.; Walsh, P.C.; Isaacs, J.T. Implication of cell kinetic changes during the progression of human prostatic cancer. *Clin. Cancer Res.* **1995**, *1*, 473–480.
34. Nishiyama, T. Serum testosterone levels after medical or surgical androgen deprivation: A comprehensive review of the literature. *Urol. Oncol.* **2013**, *32*, 38.e17–38.e28.
35. Pell, B.; Baez, J.; Phan, T.; Gao, D.; C, G.; Kuang, Y. *Patch Models of EVD Transmission Dynamics*; Springer: Cham, Switzerland, 2016.
36. Chowell, G.; Simonsen, L.; Kuang, Y.; Sciences, S. Is West Africa Approaching a Catastrophic Phase or is the 2014 Ebola Epidemic Slowing Down? Different Models Yield Different Answers for Liberia. *PLOS Curr. Outbreaks* **2014**, doi:10.1371/currents.outbreaks.b4690859d91684da963dc40e00f3da81.

applied sciences

MDPI

Article

Dynamical Systems Properties of a Mathematical Model for the Treatment of CML

Urszula Ledzewicz [1,2,*] and Helen Moore [3]

[1] Deppartment of Mathematics and Statistics, Southern Illinois University Edwardsville, Edwardsville, IL 62026, USA
[2] Institute of Mathematics, Lodz University of Technology, Lodz 90-924, Poland
[3] Bristol-Myers Squibb, Quantitative Clinical Pharmacology, Princeton, NJ 08543, USA; helen.moore@bms.com
* Correspondence: uledzew@siue.edu; Tel.: +1-618-650-2361

Academic Editor: Yang Kuang
Received: 1 September 2016; Accepted: 28 September 2016; Published: 12 October 2016

Abstract: A mathematical model for the treatment of chronic myeloid leukemia (CML) through a combination of tyrosine kinase inhibitors and immunomodulatory therapies is analyzed as a dynamical system for the case of constant drug concentrations. Equilibria and their stability are determined and it is shown that, depending on the parameter values, the model exhibits a variety of behaviors which resemble the chronic, accelerated and blast phases typical of the disease. This work provides qualitative insights into the system which should be useful for understanding the interaction between CML and the therapies considered here.

Keywords: chronic myeloid leukemia; tyrosine kinase inhibitors; immunomodulatory therapies; combination therapy; equilibrium points

1. Introduction

Chronic myeloid leukemia (CML) is a hematologic cancer that accounts for about 15% of all leukemias in adults and is characterized by uncontrolled expansion of myeloid cells in the bone marrow and their accumulation in the blood [1]. The progression of the disease can be divided into three phases denoted *chronic*, *accelerated* and *blast* [1]. The chronic phase can last several years with levels of immature white blood cells (blasts) growing steadily but at a low rate. Once the disease enters the accelerated or blast phase, cells proliferate rapidly and the disease can be lethal within a few months if not treated. Current standard of care includes targeted tyrosine kinase inhibitors (TKIs), which have significantly improved long-term survival rates [2].

Responses to certain treatments have offered evidence of an immune component in the disease [3]. Early indications were provided by a correlation between incidence of graft-vs-host disease and improved leukemia-free survival in CML patients who had received allogeneic stem cell transplants [4]. Additionally, treatment with interferons (which are known to be immunomodulatory) has led to complete or partial responses in some fraction of CML patients [5]. More recently, studies that include immunomodulatory therapies such as nivolumab have been initiated [6].

Mathematical modeling of CML dynamics has a history dating back to the late 1960s with early work of Rubinow and Lebowitz [7,8]. Models by Fokas et al. [9] in the 1990s focused on maturation and proliferation of T-cell precursors. In 2004, Moore and Li [10] published a model of CML dynamics, which accounts for the actions of naive and effector T-cells separately. In [11], this model was analyzed as an optimal control problem. The model presented here first appeared in [12] and models the immune system effects with one compartment, and separates the CML cells into quiescent and proliferating classes. The rationale behind this new model is the ability to represent certain types of therapies for use in combination treatment. These therapies are: a BCR-ABL1 tyrosine kinase inhibitor (e.g., a therapy such as imatinib), an immunomodulatory therapy (e.g., a therapy such as nivolumab), and a therapy that combines both actions (e.g., a therapy such as dasatinib).

The model introduced in [12] is reviewed in Section 2 and then analyzed as a dynamical system with constant drug concentrations in Section 3. The analysis is carried out theoretically for values of parameters covering a range of dynamic possibilities. As will be seen, there are parameter values for which the model can have an asymptotically stable equilibrium point in which all the state variables are positive. This could be interpreted as disease control through continuous therapy. As parameters change, the system can become unstable and undergo exponential growth, representing the accelerated or blast phases of the disease. Our analysis incorporates constant drug concentrations, and thus provide insights into the dynamics both without and with treatment. In particular, we analyze how an increase in the levels of each of the three treatments affects the values of all three populations, the two types of leukemia cells and the strength of the immune effect. The combination of theoretical analysis and simulations is intended to shed some light on understanding the long-term dynamics of this disease under treatment.

2. A Mathematical Model for the Treatment of CML with BCR-ABL1-Targeted and Immunomodulatory Drugs

The mathematical model below was originally published in [12] in 2015.

2.1. A Brief Review of the Mathematical Model

Let Q be the concentration of quiescent leukemic cells, P the concentration of proliferating leukemic cells, and E the strength of immune system effects. We will consider E to represent effector T cell concentration levels, and will refer to E in the remainder as a concentration of effector T cells. The model contains three controls u_1, u_2 and u_3 that all denote normalized levels of different therapies. The roles of the specific drugs are illustrated in Figure 1 taken from [12] with arrows indicating amplification of effects and vertical bars indicating inhibition. The control u_1 represents the normalized concentration of a BCR-ABL1 inhibitor (such as imatinib) that mainly has an inhibitory effect on the highly-proliferating leukemic cells; u_2 is a BCR-ABL1 inhibitor that inhibits BCR-ABL1 that also has immune effects (such as dasatinib); while u_3 represents an immunomodulatory compound (such as nivolumab).

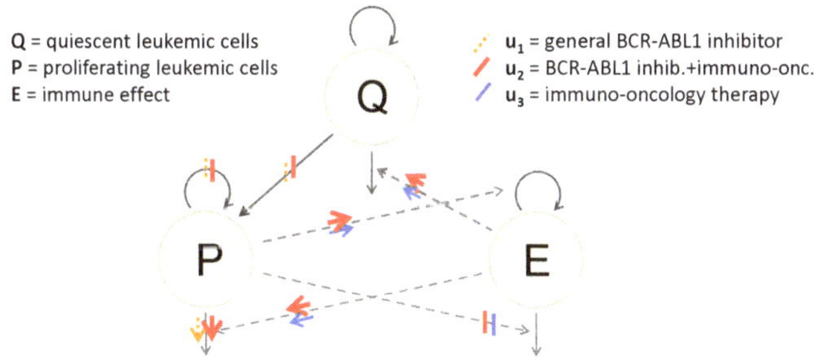

Figure 1. Diagram of the dynamical system. The green circular areas represent the "populations" included in the model. Solid arrows extending from or to the populations represent changes in numbers, with inward-pointing arrows representing increases and outward-pointing arrows decreases. Dashed arrows indicate indirect effects on those increases or decreases. Bars represent inhibition of a production or an indirect effect, due to the represented treatment; arrows represent amplification of a rate or an indirect effect. The effects of the general BCR-ABL1 inhibitor u_1 are shown using orange dashed bars and arrows, the effects of the BCR-ABL1 inhibitor u_2 which also has immune effects are shown using wide red solid bars and arrows and the effects of the immunomodulatory compound u_3 are shown using blue solid bars and arrows.

Representing the pharmacodynamic effects of the drugs using Michaelis-Menten terms results in the following equations:

$$\frac{dQ}{dt} = r_Q Q - \delta_Q \left[1 + \left(1 + \frac{U2_{\max,1}u_2}{U2C_{50} + u_2}\right)\left(1 + \frac{U3_{\max,1}u_3}{U3C_{50} + u_3}\right)\frac{E_{\max,1}E}{EC_{50} + E}\right] Q, \quad (1)$$

$$\begin{aligned} \frac{dP}{dt} = &\left(1 - \frac{U1_{\max,1}u_1}{U1C_{50} + u_1}\right)\left(1 - \frac{U2_{\max,2}u_2}{U2C_{50} + u_2}\right)\left[k_P Q + r_P P \ln\left(\frac{P_{ss}}{P}\right)\right] \\ &- \delta_P \left(1 + \frac{U1_{\max,2}u_1}{U1C_{50} + u_1}\right)\left(1 + \frac{U2_{\max,3}u_2}{U2C_{50} + u_2}\right) P \\ &- \delta_P \left(1 + \frac{U2_{\max,1}u_2}{U2C_{50} + u_2}\right)\left(1 + \frac{U3_{\max,1}u_3}{U3C_{50} + u_3}\right)\frac{E_{\max,2}E}{EC_{50} + E} P, \end{aligned} \quad (2)$$

$$\begin{aligned} \frac{dE}{dt} = &s_E \left[1 + \left(1 + \frac{U2_{\max,4}u_2}{U2C_{50} + u_2}\right)\left(1 + \frac{U3_{\max,2}u_3}{U3C_{50} + u_3}\right)\frac{P_{\max,1}P}{PC_{50} + P}\right] E \ln\left(\frac{E_{ss}}{E}\right) \\ &- \delta_E \left[1 + \left(1 - \frac{U2_{\max,5}u_2}{U2C_{50} + u_2}\right)\left(1 - \frac{U3_{\max,3}u_3}{U3C_{50} + u_3}\right)\frac{P_{\max,2}P}{PC_{50} + P}\right] E. \end{aligned} \quad (3)$$

In this system, all parameters are non-negative. For $P = 0$ or $E = 0$, we extend the system by defining it using the limits as $P \to 0$ or $E \to 0$, respectively. The cell count numbers for Q are relatively small and are therefore modeled by an exponential function with growth coefficient r_Q. For the proliferating cells P we model growth with a Gompertz function, as Afenya and Calderón state that this is best for describing CML growth [13]. The immune effect E (effector T cells) also has its rate of increase modeled by a Gompertz function, so as to have approximately exponential growth when numbers are very small, but still be bounded above. In the populations P and E, replication rate constants are represented by r_P and s_E, and carrying capacities (or steady states) by P_{ss} and E_{ss}, respectively. The natural death rate constants of the respective populations are denoted by δ_Q, δ_P and δ_E. The population Q consists of leukemic cells that are quiescent. Some or all of quiescent leukemic cells may be stem cells [14]. When quiescent cells divide, one copy is assumed to be the same kind as the original cell while the second copy may differentiate further into a proliferating type. For this reason, the transition term $k_P Q$ is not subtracted from the quiescent cell population in (1). This term represents the rate at which quiescent cells produce differentiated proliferating cancer cells, with the population Q the source for the population P.

The control variables represent the concentrations of the respective drugs, and their effects (pharmacodynamics) are modeled by Michaelis-Menten terms with different maximum effectiveness on the various populations. In modeling the combined drug actions it is assumed that any two drugs act independently of each other. Thus the term

$$\left(1 - \frac{U1_{\max,1}u_1}{U1C_{50} + u_1}\right)\left(1 - \frac{U2_{\max,2}u_2}{U2C_{50} + u_2}\right)\left[k_P Q + r_P P \ln\left(\frac{P_{ss}}{P}\right)\right]$$

represents the effects that drugs 1 and 2 have on decreasing the proliferation of the population P. A term of the type

$$-\delta_Q \left(1 + \frac{U2_{\max,1}u_2}{U2C_{50} + u_2}\right)\left(1 + \frac{U3_{\max,1}u_3}{U3C_{50} + u_3}\right)\frac{E_{\max,1}E}{EC_{50} + E} Q$$

represents the enhancement of the actions of the effector T cells E on the quiescent cells Q as a consequence of the activities of drugs 2 and 3. In each of the equations, the enhancement and inhibition effects of the drugs by means of the immune system are modeled additively.

The "C_{50}" parameters $U1C_{50}$, $U2C_{50}$, and $U3C_{50}$ represent the concentrations required to achieve half of the maximal effects of u_1, u_2, and u_3, respectively. These and EC_{50} and PC_{50} are assumed to be fixed across effects being modeled. These represent "potency" levels depending intrinsically on the particular therapy or population, and not on the setting of the effect. The maximum possible effect size is allowed to depend on the setting.

The equations above represent a semi-mechanistic, fit-for-purpose, minimal model. It is minimal in the sense that it only includes the levels of cell interactions needed to allow the controls to have their expected effects. Some of the terms are based on models validated with data, but other terms take forms that are more heuristic. For example, all of the control effect terms take a Michaelis-Menten or "Emax" form. This is because we wish to model very small effect at low levels of drug, as well as a limiting or asymptotic maximal effect at high levels of drugs. We chose the simplest among the models with this behavior that are typically used in drug development [15].

The states, controls, and related parameters are listed in Tables 1 and 2. Table 1 gives those parameters that are unrelated to the drug actions and make up the untreated, or uncontrolled, system; Table 2 lists the treatment-specific parameters in the model. In this paper, we do not fit or fix specific parameter values, and instead analyze the dynamic properties of the system (1)–(3) for large ranges of possible values. We include in the tables below two different sets of numerical values that we use to illustrate the dynamic properties of the system. These parameter values are purely for numerical illustration and do not reflect specific model fits or therapies. The focus of this paper is the mathematical analysis of the entire system rather than an analysis for particular parameter values.

Table 1. States and parameters for the dynamical system.

Symbol	Interpretation	Units	Values Used in Figure 2	Values Used in Figures 3 and 4
Q	concentration of quiescent leukemic cells	10^2 cells/mL		
P	concentration of proliferating leukemic cells	10^7 cells/mL		
P_{ss}	carrying capacity of proliferating leukemic cells	10^7 cells/mL	10	15
E	effector T cells	2×10^3 cells/mL		
E_{ss}	carrying capacity of effector T cells	2×10^3 cells/mL	1.75	2.25
r_Q	replication rate constant of quiescent cells	1/day		0.02
δ_Q	natural death rate constant of quiescent cells	1/day		0.005
k_P	rate constant for quiescent cells Q differentiating into proliferating cells P	1/day		0.10
r_P	replication rate constant of proliferating leukemic cells	1/day	8	0.30
δ_P	natural death rate constant of proliferating leukemic cells	1/day	0.75	0.02
s_E	growth rate constant for effector T cells	1/day	0.25	0.01
δ_E	natural death rate constant of effector T cells	1/day	0.25	0.005
$P_{max,1}$	maximum stimulation effect of proliferating leukemic cells P on effector T cells E		2	0.50
$P_{max,2}$	maximum inhibition effect of proliferating leukemic cells P on effector T cells E		5	0.20
PC_{50}	size of P with half the maximum effect	1/mL	10^7	10^7
$E_{max,1}$	maximum effect of effector T cells E on quiescent leukemic cells Q			5
$E_{max,2}$	maximum effect of effector T cells E on proliferating leukemic cells P		1	5
EC_{50}	size of E with half the maximum effect	1/mL	2000	2000

Table 2. Controls and pharmacodynamic parameters.

Symbol	Interpretation	Values Used in Figure 2	Values Used in Figures 3 and 4
u_1	normalized concentration of a general BCR-ABL1 inhibitor (e.g., imatinib)		
$U1_{\mathrm{max},1}$	maximum possible effect of u_1 on slowing transfer of quiescent cells Q into P and inhibiting growth of proliferating cells P	0.8	0.8
$U1_{\mathrm{max},2}$	maximum possible effect of u_1 on death of proliferating cells P	10	2
$U1C_{50}$	concentration of u_1 that gives half the maximum effect	1	1
u_2	normalized concentration of a BCR-ABL1 inhibitor which also has immunomodulatory effects (e.g., dasatinib)		
$U2_{\mathrm{max},1}$	maximum possible effect of u_2 on death of leukemic cells (the same for P and Q)	2	0.01
$U2_{\mathrm{max},2}$	maximum possible effect of u_2 slowing new P from Q and inhibiting growth of proliferating cells P	0.6	0.03
$U2_{\mathrm{max},3}$	maximum possible effect of u_2 on death of proliferating cells P	10	0.01
$U2_{\mathrm{max},4}$	maximum possible effect of u_2 on stimulating proliferation of effector T cells	10	0.025
$U2_{\mathrm{max},5}$	maximum possible effect of u_2 on prevention of the death of effector T cells	0.4	0.02
$U2C_{50}$	concentration of u_2 that gives half the maximum effect	0.6	0.8
u_3	normalized concentration of an immunomodulatory d agent (e.g., nivolumab)		
$U3_{\mathrm{max},1}$	maximum possible effect of u_3 on death of leukemic cells (the same for P and Q)	5	0.02
$U3_{\mathrm{max},2}$	maximum possible effect of u_3 on stimulating proliferation of effector T cells	5	0.05
$U3_{\mathrm{max},3}$	maximum possible effect of u_3 on prevention of the death of effector T cells	0.7	0.07
$U3C_{50}$	concentration of u_3 that gives half the maximum effect	0.7	0.7

2.2. Scaling of Parameters

We note that the dynamical system has various groups of symmetries that can be used to scale the variables and controls. Here we normalize all the "C_{50}" parameter values to 1 by rescaling the corresponding variables in terms of these quantities. This simply minimizes the number of parameters to be considered in the analysis of the system. For example, let Q_{ref} be a constant to be determined later, and define

$$\widetilde{Q} = \frac{Q}{Q_{\mathrm{ref}}}, \qquad \widetilde{P} = \frac{P}{PC_{50}}, \qquad \widetilde{E} = \frac{E}{EC_{50}},$$

and

$$\widetilde{u}_1 = \frac{u_1}{U1C_{50}}, \qquad \widetilde{u}_2 = \frac{u_2}{U2C_{50}}, \qquad \widetilde{u}_3 = \frac{u_3}{U3C_{50}}.$$

Then we have that

$$\frac{E}{EC_{50} + E} = \frac{\widetilde{E}}{1 + \widetilde{E}}$$

and analogously for the other terms.

For the differential equations, we obtain

$$\begin{aligned}\frac{d\widetilde{Q}}{dt} &= \frac{1}{Q_{\text{ref}}}\frac{dQ}{dt} \\ &= \frac{1}{Q_{\text{ref}}}\left\{r_Q Q - \delta_Q\left[1+\left(1+\frac{U2_{\max,1}\widetilde{u}_2}{1+\widetilde{u}_2}\right)\left(1+\frac{U3_{\max,1}\widetilde{u}_3}{1+\widetilde{u}_3}\right)\frac{E_{\max,1}\widetilde{E}}{1+\widetilde{E}}\right]Q\right\} \\ &= r_Q\widetilde{Q} - \delta_Q\left[1+\left(1+\frac{U2_{\max,1}\widetilde{u}_2}{1+\widetilde{u}_2}\right)\left(1+\frac{U3_{\max,1}\widetilde{u}_3}{1+\widetilde{u}_3}\right)\frac{E_{\max,1}\widetilde{E}}{1+\widetilde{E}}\right]\widetilde{Q}.\end{aligned}$$

Under this scaling all remaining parameters in this equation are invariant and need not be changed. Similarly,

$$\begin{aligned}\frac{d\widetilde{P}}{dt} &= \frac{1}{PC_{50}}\frac{dP}{dt} \\ &= \frac{1}{PC_{50}}\left\{\left(1-\frac{U1_{\max,1}\widetilde{u}_1}{1+\widetilde{u}_1}\right)\left(1-\frac{U2_{\max,2}\widetilde{u}_2}{1+\widetilde{u}_2}\right)\left[k_P Q + r_P P\ln\left(\frac{P_{ss}}{P}\right)\right]\right. \\ &\quad - \delta_P\left(1+\frac{U1_{\max,2}\widetilde{u}_1}{1+\widetilde{u}_1}\right)\left(1+\frac{U2_{\max,3}\widetilde{u}_2}{1+\widetilde{u}_2}\right)P \\ &\quad \left. -\delta_P\left(1+\frac{U2_{\max,1}\widetilde{u}_2}{1+\widetilde{u}_2}\right)\left(1+\frac{U3_{\max,1}\widetilde{u}_3}{1+\widetilde{u}_3}\right)\frac{E_{\max,2}\widetilde{E}}{1+\widetilde{E}}P\right\} \\ &= \left(1-\frac{U1_{\max,1}\widetilde{u}_1}{1+\widetilde{u}_1}\right)\left(1-\frac{U2_{\max,2}\widetilde{u}_2}{1+\widetilde{u}_2}\right)\left[\left(k_P\frac{Q_{\text{ref}}}{PC_{50}}\right)\widetilde{Q} + r_P\widetilde{P}\ln\left(\frac{P_{ss}}{\widetilde{P}\cdot PC_{50}}\right)\right] \\ &\quad - \delta_P\left(1+\frac{U1_{\max,2}\widetilde{u}_1}{1+\widetilde{u}_1}\right)\left(1+\frac{U2_{\max,3}\widetilde{u}_2}{1+\widetilde{u}_2}\right)\widetilde{P} \\ &\quad - \delta_P\left(1+\frac{U2_{\max,1}\widetilde{u}_2}{1+\widetilde{u}_2}\right)\left(1+\frac{U3_{\max,1}\widetilde{u}_3}{1+\widetilde{u}_3}\right)\frac{E_{\max,2}\widetilde{E}}{1+\widetilde{E}}\widetilde{P},\end{aligned}$$

and

$$\begin{aligned}\frac{d\widetilde{E}}{dt} &= \frac{1}{EC_{50}}\frac{dE}{dt} \\ &= s_E\left[1+\left(1+\frac{U2_{\max,4}\widetilde{u}_2}{1+\widetilde{u}_2}\right)\left(1+\frac{U3_{\max,2}\widetilde{u}_3}{1+\widetilde{u}_3}\right)\frac{P_{\max,1}\widetilde{P}}{1+\widetilde{P}}\right]\widetilde{E}\ln\left(\frac{E_{ss}}{\widetilde{E}\cdot EC_{50}}\right) \\ &\quad - \delta_E\left[1+\left(1-\frac{U2_{\max,5}\widetilde{u}_2}{1+\widetilde{u}_2}\right)\left(1-\frac{U3_{\max,3}\widetilde{u}_3}{1+\widetilde{u}_3}\right)\frac{P_{\max,2}\widetilde{P}}{1+\widetilde{P}}\right]\widetilde{E}.\end{aligned}$$

Thus, if we re-scale k_P as

$$\widetilde{k}_P = \frac{Q_{\text{ref}}}{PC_{50}}k_P \tag{4}$$

and the steady-state values as

$$\widetilde{P}_{ss} = \frac{P_{ss}}{PC_{50}} \quad \text{and} \quad \widetilde{E}_{ss} = \frac{E_{ss}}{EC_{50}}, \tag{5}$$

then formally the equations are the same as before with all "C_{50}" values in the Michaelis-Menten expressions normalized to 1. All other parameters remain unchanged and even their interpretation is the same as before. For the theoretical analysis and numerical computations this eliminates five parameters and introduces a favorable scaling to the variables. Naturally, the original parameters are still calculated for an interpretation of the results.

3. System Properties for Constant Concentrations

CML has three distinct phases, a *chronic* one that can last from three to five years, during which leukemic cell counts are low but may grow steadily, and *accelerated* and *blast* phases that may last for a only a few months and are characterized by higher cell counts or a rapid increase in cell counts followed by death of the patient [10]. Here we analyze the dynamical system to determine if it can capture such features.

3.1. Reduction to the Uncontrolled System and Basic Dynamical System Properties

We carry out the dynamical systems analysis for constant controls, i.e., concentrations. We do not explicitly include pharmacokinetics (fluctuations in concentrations that depend on doses). The treatments considered are either administered daily or have long half-lives, and such pharmacokinetics are not expected to be significant for the treatment periods we consider here (five years or longer). We also mention the 2009 paper by Shudo et al. [16] that supports this assumption in the setting of hepatitis C.

Keeping the "C_{50}" parameters in their original formulation in the controls, we define new drug-dependent parameters as

$$
\begin{aligned}
\hat{k}_P &= \left(1 - \frac{U1_{\max,1}u_1}{U1C_{50}+u_1}\right)\left(1 - \frac{U2_{\max,2}u_2}{U2C_{50}+u_2}\right) k_P, \\
\hat{r}_P &= \left(1 - \frac{U1_{\max,1}u_1}{U1C_{50}+u_1}\right)\left(1 - \frac{U2_{\max,2}u_2}{U2C_{50}+u_2}\right) r_P, \\
\hat{\delta}_P &= \left(1 + \frac{U1_{\max,2}u_1}{U1C_{50}+u_1}\right)\left(1 + \frac{U2_{\max,3}u_2}{U2C_{50}+u_2}\right) \delta_P, \\
\hat{E}_{\max,1} &= \left(1 + \frac{U2_{\max,1}u_2}{U2C_{50}+u_2}\right)\left(1 + \frac{U3_{\max,1}u_3}{U3C_{50}+u_3}\right) E_{\max,1}, \\
\hat{E}_{\max,2} &= \frac{\left(1 + \frac{U2_{\max,1}u_2}{U2C_{50}+u_2}\right)\left(1 + \frac{U3_{\max,1}u_3}{U3C_{50}+u_3}\right)}{\left(1 + \frac{U1_{\max,2}u_1}{U1C_{50}+u_1}\right)\left(1 + \frac{U2_{\max,3}u_2}{U2C_{50}+u_2}\right)} E_{\max,2}, \\
\hat{P}_{\max,1} &= \left(1 + \frac{U2_{\max,4}u_2}{U2C_{50}+u_2}\right)\left(1 + \frac{U3_{\max,2}u_3}{U3C_{50}+u_3}\right) P_{\max,1}, \\
\hat{P}_{\max,2} &= \left(1 - \frac{U2_{\max,5}u_2}{U2C_{50}+u_2}\right)\left(1 - \frac{U3_{\max,3}u_3}{U3C_{50}+u_3}\right) P_{\max,2}.
\end{aligned}
$$

With these identifications, the dynamical system with constant controls is identical with the uncontrolled system and therefore, without loss of generality, the analysis can be done on the uncontrolled system. Returning to the original notation without the carets, we thus consider the following equations:

$$\frac{dQ}{dt} = \left[r_Q - \delta_Q\left(1 + \frac{E_{\max,1}E}{1+E}\right)\right] Q, \tag{6}$$

$$\frac{dP}{dt} = k_P Q + \left[r_P \ln\left(\frac{P_{ss}}{P}\right) - \delta_P\left(1 + \frac{E_{\max,2}E}{1+E}\right)\right] P, \tag{7}$$

$$\frac{dE}{dt} = \left[s_E\left(1 + \frac{P_{\max,1}P}{1+P}\right)\ln\left(\frac{E_{ss}}{E}\right) - \delta_E\left(1 + \frac{P_{\max,2}P}{1+P}\right)\right] E. \tag{8}$$

The model with an exponential growth term on Q has various long-term behaviors. These include the extremes in which Q decays exponentially to zero or grows exponentially beyond limits, but there also is the possibility that nontrivial equilibrium points (Q_*, P_*, E_*) exist for which all three populations are positive. The first case corresponds to a scenario in which the patient goes into a stable deep molecular response. For the uncontrolled system, this may not seem to be of interest, but since the model includes the case with controls, this gives us information about which combinations of constant concentrations of the drugs would lead to an eradication of Q. The case of exponential growth may characterize the *accelerated* or *blast phase* as these phases have short doubling times [17]. The conditions under which this is the long-term behavior of the system give information about what controls are needed for successful treatment. An asymptotically stable equilibrium point (Q_*, P_*, E_*) with positive values could be interpreted as describing a subset of the *chronic* phase where net growth rate is zero, controlled by therapy or immune effects. Depending on the values of the parameters, this equilibrium point may be stable or unstable. Since in real life parameters may not be constant, bifurcation phenomena would be a mathematical description of the transition from chronic to the accelerated or blast phases. Knowing the parameter values when this may occur would be of interest. Our aim in the following is thus to determine the asymptotic behavior of the trajectories of the system.

We start with some basic properties. The positive orthant

$$\mathbb{P} = \{(Q, P, E) : Q > 0, P > 0, E > 0\}$$

is positively invariant for the dynamics. This is because the planes $Q = 0$ and $E = 0$ are invariant under Equations (6) and (8) and $\dot{P} \geq k_P Q$ whenever $P = 0$. Thus, starting at a positive initial condition (Q_0, P_0, E_0), it follows that the solutions remain positive for all times. For the long-term behavior of the system, the equilibrium solutions in the closure of $\mathbb{P}$, $\bar{\mathbb{P}} = \text{clos}(\mathbb{P})$, also matter. Recall that the system is defined and continuous on $\bar{\mathbb{P}}$ due to the use of the limits as $P \to 0$ and $E \to 0$ in place of $P = 0$ and $E = 0$, respectively. The vector field defining the P and E dynamics is not continuously differentiable at $P = 0$ or $E = 0$, but these values are repelling and thus this does not become an issue.

Lemma 1. *The equilibrium solution $E_* = 0$ is repelling: there exists a positive threshold $E_\Delta < E_{ss}$ such that $\frac{dE}{dt}$ is positive on $(0, E_\Delta]$. In particular, once $E_{ss} > E(\tau) \geq E_\Delta$, then $E(t) \geq E_\Delta$ for all $t \geq \tau$. Furthermore, for $E(0) < E_{ss}$, E will remain below E_{ss}.*

Proof. The terms in the last parentheses in Equation (8) are bounded between 1 and $1 + P_{\max,2}$ and thus, as $E \to 0$, the Gompertzian growth dominates the dynamics. Specifically, let

$$E_\Delta = E_{ss} \exp\left(-\frac{\delta_E}{s_E}\left(1 + P_{\max,2}\right)\right);$$

then $E_\Delta \leq E_{ss}$ and for $E < E_\Delta$ we have that $\frac{dE}{dt} > 0$. Furthermore, for $E = E_{ss}$, Equation (8) reduces to $\frac{dE}{dt} = -\delta_E\left(1 + \frac{P_{\max,2}P}{1+P}\right)E < 0$ and thus the values of E cannot reach the value E_{ss} if they start below E_{ss}.

Lemma 2. *The equilibrium solution $P_* = 0$ is repelling: there exists a positive threshold $P_\Delta < P_{ss}$ such that $\frac{dP}{dt}$ is positive on $(0, P_\Delta]$. In particular, once $P_{ss} > P(\tau) \geq P_\Delta$, we have $P(t) \geq P_\Delta$ for all $t \geq \tau$.*

Proof. For values of E less than E_{ss}, we have that

$$\delta_P\left(1 + \frac{E_{\max,2}E}{1+E}\right) < \delta_P\left(1 + \frac{E_{\max,2}E_{ss}}{1+E_{ss}}\right)$$

for all times. Choosing P_Δ as

$$P_\Delta = P_{ss} \exp\left(-\frac{\delta_P}{r_P}\left(1 + \frac{E_{\max,2}E_{ss}}{1+E_{ss}}\right)\right)$$

the result follows: for $P < P_\Delta$ we have that

$$r_P \ln\left(\frac{P_{ss}}{P}\right) - \delta_P\left(1 + \frac{E_{\max,2}E}{1+E}\right) > r_P \ln\left(\frac{P_{ss}}{P_\Delta}\right) - \delta_P\left(1 + \frac{E_{\max,2}E_{ss}}{1+E_{ss}}\right) = 0.$$

This proves the result.

Corollary 1. *The equilibrium solutions $E_* \equiv 0$ and $P_* \equiv 0$ are unstable.*

Note, however, that P is not necessarily bounded. For, with $P = P_{ss}$, Equation (7) becomes

$$\frac{dP}{dt} = k_P Q - \delta_P\left(1 + \frac{E_{\max,2}E}{1+E}\right) P_{ss}$$

and thus, if Q is large enough, this term will be positive. Hence, if Q grows exponentially, P will diverge to $+\infty$.

Lemma 3. *If Q increases exponentially with time, then $\lim_{t\to\infty} P(t) = +\infty$.*

Proof. We need to show that for every positive value $\hat{P}$ there exists a time $\hat{T}$ so that $P(t) \geq \hat{P}$ for all $t \geq \hat{T}$.

We first remark that P is unbounded. For, if there exists a value $\bar{P}$ with $P_{ss} < \bar{P} < \infty$ so that $P(t) \leq \bar{P}$ for all times t, then the term $\left[r_P \ln\left(\frac{P_{ss}}{P}\right) - \delta_P\left(1 + \frac{E_{\max,2}E}{1+E}\right)\right] P$ is bounded below. By assumption, there exist positive constants α and β so that $Q(t) \geq \alpha e^{\beta t}$ for all t. Hence, for t sufficiently large we have that

$$\frac{dP}{dt}(t) = \left[r_P \ln\left(\frac{P_{ss}}{P(t)}\right) - \delta_P\left(1 + \frac{E_{\max,2}E(t)}{1+E(t)}\right)\right] P(t) + k_P Q(t) > 1.$$

Contradiction.

Given $\hat{P} \geq P_{ss}$, choose $\check{T}$ so that

$$\alpha e^{\beta \check{T}} = \frac{1}{k_P}\left[\delta_P\left(1 + \frac{E_{\max,2}E_{ss}}{1+E_{ss}}\right) - r_P \ln\left(\frac{P_{ss}}{\hat{P}}\right)\right]\hat{P}.$$

Since P is not bounded, there exists a first time $\hat{T} > \check{T}$ so that $P(\hat{T}) = \hat{P} + 1$. We claim that $P(t) > \hat{P}$ for all $t \geq \hat{T}$. For, if there exists a time $\tau > \hat{T}$ such that $P(\tau) = \hat{P}$, then

$$\begin{aligned}\frac{dP}{dt}(\tau) &= \left[r_P \ln\left(\frac{P_{ss}}{\hat{P}}\right) - \delta_P\left(1 + \frac{E_{\max,2}E(\tau)}{1+E(\tau)}\right)\right]\hat{P} + k_P Q(\tau) \\ &> \left[r_P \ln\left(\frac{P_{ss}}{\hat{P}}\right) - \delta_P\left(1 + \frac{E_{\max,2}E_{ss}}{1+E_{ss}}\right)\right]\hat{P} + k_P Q(\tau) \\ &\geq k_P \alpha\left(e^{\beta\tau} - e^{\beta\check{T}}\right) > 0.\end{aligned}$$

Contradiction. Thus P diverges to $+\infty$.

3.2. Dynamics on the Plane $Q = 0$

The plane $Q = 0$ is invariant under the dynamics and can have regions that are repelling or attractive. We first analyze the reduced dynamical system in this boundary stratum of $\mathbb{P}$, i.e., consider the equations

$$\frac{dP}{dt} = \left[r_P \ln\left(\frac{P_{ss}}{P}\right) - \delta_P \left(1 + \frac{E_{\max,2}E}{1+E}\right)\right] P, \tag{9}$$

$$\frac{dE}{dt} = \left[s_E \left(1 + \frac{P_{\max,1}P}{1+P}\right) \ln\left(\frac{E_{ss}}{E}\right) - \delta_E \left(1 + \frac{P_{\max,2}P}{1+P}\right)\right] E \tag{10}$$

$$= \left[s_E \ln\left(\frac{E_{ss}}{E}\right) - \delta_E \frac{1+P+P_{\max,2}P}{1+P+P_{\max,1}P}\right] \left(1 + \frac{P_{\max,1}P}{1+P}\right) E. \tag{11}$$

Let $\mathbb{P}_0$ denote the open rectangle

$$\mathbb{P}_0 = \{(P,E) : 0 < P < P_{ss},\ 0 < E < E_{ss}\}$$

and denote by $\bar{\mathbb{P}}_0$ its closure, $\bar{\mathbb{P}}_0 = \{(P,E) : 0 \le P \le P_{ss},\ 0 \le E \le E_{ss}\}$. For $Q \equiv 0$ the variable P is bounded above by P_{ss} and therefore the compact set $\bar{\mathbb{P}}_0$ is positively invariant under Equations (9) and (11). The dynamical system has the following *trivial equilibrium solutions* in the boundary of $\bar{\mathbb{P}}_0$: $(0,0)$, $(P_*,0)$ with

$$P_* = P_{ss} \exp\left(-\frac{\delta_P}{r_P}\right) < P_{ss}$$

and $(0,E_*)$ with E_* given by

$$E_* = E_{ss} \exp\left(-\frac{\delta_E}{s_E}\right) < E_{ss}.$$

In view of Lemmas 1 and 2 these solutions are unstable. While the origin has two unstable modes, the equilibrium points $(P_*,0)$ and $(0,E_*)$ are saddles with the respective axes forming the stable manifolds and the unstable modes entering the interior of $\mathbb{P}_0$. It is clear from this that there needs to exist at least one more equilibrium point (P_*,E_*) in $\mathbb{P}_0$.

Lemma 4. *There are no periodic orbits in* $\mathbb{P}_0$.

Proof. Changing variables to $\tilde{P} = \ln P$ and $\tilde{E} = \ln E$, the dynamics transforms into

$$\frac{d\tilde{P}}{dt} = r_P \left(\ln P_{ss} - \tilde{P}\right) - \delta_P \left(1 + E_{\max,2}\frac{e^{\tilde{E}}}{1+e^{\tilde{E}}}\right),$$

$$\frac{d\tilde{E}}{dt} = s_E \left(1 + P_{\max,1}\frac{e^{\tilde{P}}}{1+e^{\tilde{P}}}\right)\left(\ln E_{ss} - \tilde{E}\right) - \delta_E \left(1 + P_{\max,2}\frac{e^{\tilde{P}}}{1+e^{\tilde{P}}}\right).$$

The divergence of this vector field is given by

$$-r_P - s_E \left(1 + P_{\max,1}\frac{e^{\tilde{P}}}{1+e^{\tilde{P}}}\right) < 0$$

and thus the result follows from Bendixson's negative criterion because of the monotonicity of the logarithm function.

The relations defining equilibrium points inside $\mathbb{P}_0$ are

$$P_* = P_{ss} \exp\left(-\frac{\delta_P}{r_P}\left(1 + E_{\max,2}\frac{E_*}{1+E_*}\right)\right) \tag{12}$$

and

$$\frac{s_E}{\delta_E} \ln\left(\frac{E_{ss}}{E_*}\right) = \frac{1+P_*+P_{\max,2}P_*}{1+P_*+P_{\max,1}P_*}$$

or, equivalently,

$$E_* = E_{ss} \exp\left(-\frac{\delta_E}{s_E} \times \frac{1 + P_* + P_{max,2}P_*}{1 + P_* + P_{max,1}P_*}\right). \tag{13}$$

Define

$$\Xi(P) = \frac{1 + P + P_{max,2}P}{1 + P + P_{max,1}P}, \qquad \Psi(\Xi) = \frac{E_{ss} \exp\left(-\frac{\delta_E}{s_E}\Xi\right)}{1 + E_{ss} \exp\left(-\frac{\delta_E}{s_E}\Xi\right)},$$

and

$$\Phi(P) = P_{ss} \exp\left(-\frac{\delta_P}{r_P}\left(1 + E_{max,2}\Psi(\Xi(P))\right)\right).$$

Then equilibrium values P_* are fixed points of the function Φ, $P = \Phi(P)$, in the interval $[0, P_{ss}]$. Since $\Phi(0) > 0$, $\Phi(P_{ss}) < P_{ss}$, and Φ is continuous in P, it follows that there exists at least one solution. The derivative Φ' of Φ is given by

$$\Phi'(P) = \Phi(P)\left(-\frac{\delta_P}{r_P}E_{max,2}\right) \frac{E_{ss} \exp\left(-\frac{\delta_E}{s_E}\Xi(P)\right)}{\left(1 + E_{ss} \exp\left(-\frac{\delta_E}{s_E}\Xi(P)\right)\right)^2} \left(-\frac{\delta_E}{s_E}\right) \Xi'(P)$$

and thus has the same sign as $\Xi'(P)$. Now

$$\Xi'(P) = \frac{P_{max,2} - P_{max,1}}{(1 + P + P_{max,1}P)^2}.$$

Thus Φ is strictly increasing for $P_{max,2} > P_{max,1}$ and strictly decreasing for $P_{max,2} < P_{max,1}$. If $P_{max,2} = P_{max,1}$, then

$$\Phi(P) = P_{ss} \exp\left(-\frac{\delta_P}{r_P}\left(1 + E_{max,2}\frac{E_{ss} \exp\left(-\frac{\delta_E}{s_E}\right)}{1 + E_{ss} \exp\left(-\frac{\delta_E}{s_E}\right)}\right)\right) = \text{const.}$$

Equilibria are intersections of the graph of Φ with the diagonal and thus there exists a unique equilibrium point $(P_*, E_*) \in \mathbb{P}_0$ if $P_{max,1} \geq P_{max,2}$, but multiple solutions are possible if $P_{max,1} < P_{max,2}$.

We determine the stability of (P_*, E_*) for the reduced system, i.e., within the invariant plane $Q = 0$. The Jacobian matrix at the equilibrium point is given by

$$\begin{pmatrix} -r_P & -\delta_P \frac{E_{max,2}}{(1+E_*)^2} P_* \\ \frac{s_E P_{max,1} \ln\left(\frac{E_{ss}}{E_*}\right) - \delta_E P_{max,2}}{(1+P_*)^2} E_* & -s_E\left(1 + \frac{P_{max,1}P_*}{1+P_*}\right) \end{pmatrix}.$$

Using the equilibrium relations we can write the $(2,1)$-term as

$$\begin{aligned} \frac{\partial}{\partial P}_{|(P_*,E_*)}\left(\frac{dE}{dt}\right) &= \frac{\delta_E P_{max,1} E_*}{(1 + P_*)^2}\left\{\frac{s_E}{\delta_E} \ln\left(\frac{E_{ss}}{E_*}\right) - \frac{P_{max,2}}{P_{max,1}}\right\} \\ &= \frac{\delta_E P_{max,1} E_*}{(1 + P_*)^2}\left\{\frac{1 + P_* + P_{max,2}P_*}{1 + P_* + P_{max,1}P_*} - \frac{P_{max,2}}{P_{max,1}}\right\} \\ &= \delta_E E_* \frac{P_{max,1} - P_{max,2}}{(1 + P_*)(1 + P_* + P_{max,1}P_*)}. \end{aligned}$$

The characteristic polynomial of this 2×2 matrix is given by

$$\chi(t) = \begin{vmatrix} t + r_P & \delta_P \frac{E_{\max,2}}{(1+E_*)^2} P_* \\ -\frac{(P_{\max,1} - P_{\max,2})\delta_E E_*}{(1+P_*)(1+P_*+P_{\max,1}P_*)} & t + s_E \left(1 + \frac{P_{\max,1}P_*}{1+P_*}\right) \end{vmatrix}$$

$$= t^2 + \left(r_P + s_E \left(1 + \frac{P_{\max,1}P_*}{1+P_*}\right)\right) t + r_P s_E \left(1 + \frac{P_{\max,1}P_*}{1+P_*}\right)$$

$$+ \delta_E \delta_P \frac{E_{\max,2} E_*}{(1+E_*)^2} \frac{(P_{\max,1} - P_{\max,2}) P_*}{(1+P_*)(1+P_*+P_{\max,1}P_*)}.$$

If we write $\chi(t) = t^2 + a_1 t + a_0$, then a_1 is positive and thus the equilibrium point is locally asymptotically stable if a_0 is positive while it is unstable if a_0 is negative. A saddle node bifurcation occurs as $a_0 = 0$. It immediately follows that (P_*, E_*) is locally asymptotically stable if $P_{\max,1} \geq P_{\max,2}$, i.e., if the stimulating effect of the tumor on the effector cells is larger than the inhibiting effect of the tumor on the effector cells. We have the following result:

Proposition 1. *If $P_{\max,1} \geq P_{\max,2}$, then there exists a unique equilibrium point (P_*, E_*) in $\mathbb{P}_0$ and it is globally asymptotically stable in the sense that its region of attraction is the full rectangle $\mathbb{P}_0$.*

Proof. The set $\mathbb{P}_0$ is positively invariant and every trajectory γ starting in $\mathbb{P}_0$ has a non-empty ω-limit set $\Omega(\gamma)$. Because of the stability properties of the equilibria in the boundary of $\mathbb{P}_0$, this ω-limit set $\Omega(\gamma)$ lies in $\mathbb{P}_0$. Since there exist no periodic orbits and since (P_*, E_*) is the only equilibrium point, it follows from Poincaré-Bendixson theory that $\Omega(\gamma) = \{(P_*, E_*)\}$, i.e., all trajectories starting in $\mathbb{P}_0$ converge to (P_*, E_*) as $t \to \infty$. □

It is clear from Poincaré-Bendixson theory that even if $P_{\max,1} < P_{\max,2}$, the equilibrium point (P_*, E_*) is globally asymptotically stable (in the sense that its region of attraction contains the set $\mathbb{P}_0$, and only this region is relevant for the problem) as long as it is the only equilibrium point in $\mathbb{P}_0$. This is shown in the phase portraits for the uncontrolled system in Figure 2; Figure 3 shows a case where $P_{\max,1} > P_{\max,2}$. (The values of the parameters are given in Tables 1 and 2.) We also show the phase-portraits for the systems when one of the controls is set to be equal to 1 and all others are zero. The two sets of figures illustrate two different scenarios, one where the control parameters are such that the equilibrium can be effectively controlled by all the drugs (Figure 2), the other where it is essentially only the control u_1 that is able to move the equilibrium point. However, this behavior depends on the fact that $Q = 0$.

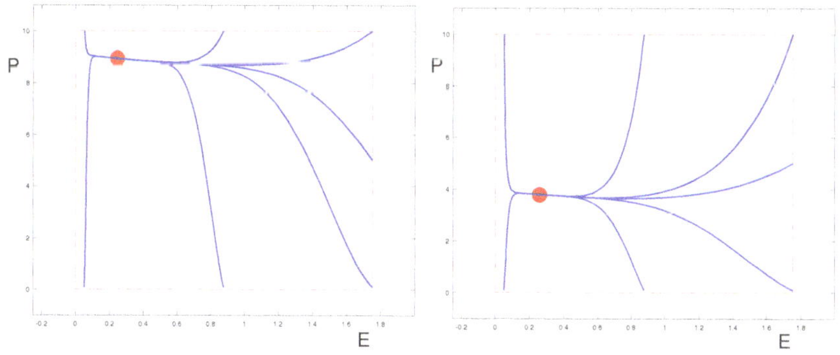

Figure 2. *Cont.*

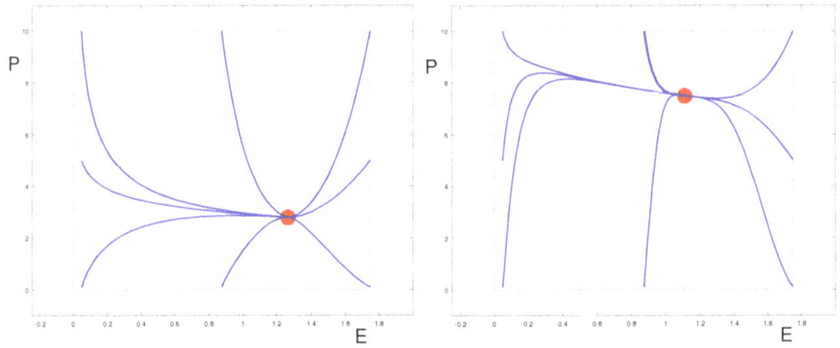

Figure 2. Phase portraits of the reduced dynamics for $Q = 0$ and $P_{\max,1} < P_{\max,2}$ for the uncontrolled system (**top, left**) and for constant controls $u_1 \equiv 1$ (**top, right**), $u_2 \equiv 1$ (**bottom, left**) and $u_3 \equiv 1$ (**bottom, right**). The numerical values for these phase portraits are given in Tables 1 and 2.

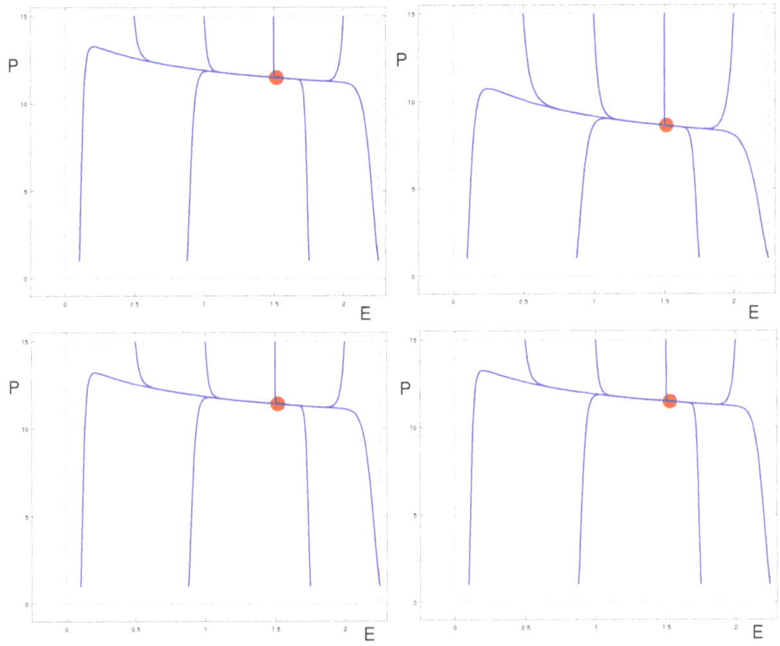

Figure 3. Phase portraits of the reduced dynamics for $Q = 0$ and $P_{\max,1} > P_{\max,2}$ for the uncontrolled system (**top, left**) and for constant controls $u_1 \equiv 1$ (**top, right**), $u_2 \equiv 1$ (**bottom, left**) and $u_3 \equiv 1$ (**bottom, right**). The numerical values for these phase portraits are given in Tables 1 and 2.

Since the coefficient a_1 is always positive, as a_0 vanishes the Jacobian matrix has the eigenvalue 0 and the other eigenvalue is negative. At such a point *saddle-node bifurcations* arise and two new equilibria, one stable, the other unstable, are born.

Proposition 2. *If* $P_{\max,2} > P_{\max,1}$*, then multiple equilibria* (P_*, E_*) *inside* $\mathbb{P}_0$ *can exist. At points* (P_*, E_*) *where*

$$\frac{\delta_P}{r_P} \frac{(P_{\max,2} - P_{\max,1})\, P_*}{(1 + P_* + P_{\max,1} P_*)^2} \frac{\delta_E}{s_E} \frac{E_{\max,2} E_*}{(1 + E_*)^2} = 1 \tag{14}$$

saddle-node bifurcations occur in which a stable and an unstable equilibrium point merge.

Proof. The coefficient a_0 vanishes if and only if

$$r_P s_E \frac{1 + P_* + P_{\max,1} P_*}{1 + P_*} = \delta_P \frac{E_{\max,2} P_*}{(1 + E_*)^2} \frac{(P_{\max,2} - P_{\max,1})\, \delta_E E_*}{(1 + P_*)\,(1 + P_* + P_{\max,1} P_*)}.$$

This condition is equivalent to (14).

For the underlying biological problem it is natural that an inhibition effect would be smaller than a stimulation effect. Also, the denominators are quadratic in the respective variables E and P, but these variables are scaled. In principle it is possible to satisfy (14), but we did not come across this in our simulations.

3.3. *Dynamic Behavior for Positive Q-Values*

For the behavior of the overall system, the Q dynamics are essential. If one considers the above equilibria in the plane $Q = 0$ now in the full three-dimensional space, then the first row of the Jacobian matrix at $(0, P_*, E_*)$ takes the form

$$\left(r_Q - \delta_Q \left(1 + \frac{E_{\max,1} E_*}{1 + E_*} \right), 0, 0 \right)$$

and thus $(0, P_*, E_*)$ is unstable if $r_Q > \delta_Q \left(1 + \frac{E_{\max,1}E_*}{1+E_*}\right)$ while the local stability properties for the overall system are the same as in the (P, E)-plane if $r_Q < \delta_Q \left(1 + \frac{E_{\max,1}E_*}{1+E_*}\right)$. If $r_Q = \delta_Q \left(1 + \frac{E_{\max,1}E_*}{1+E_*}\right)$, then there exists a 1-dimensional center manifold (corresponding to the 0 eigenvalue). In this case we have $\dot{Q} = 0$ and there exists a curve of equilibria emerging from $(0, P_*, E_*)$ parameterized by Q or P (also see below).

Generally, (1)–(3) is a time-varying linear system dominated by exponential growth and decay, depending on the parameter values. If

$$r_Q \geq \delta_Q \left(1 + E_{\max,1} \frac{E_{ss}}{1 + E_{ss}} \right),$$

then Q grows exponentially and no steady state exists. In this case, the influx $k_P Q$ eventually becomes the dominant term in Equation (2) and P also grows beyond limits (Lemma 3). This represents the *malignant* scenario in the model which corresponds to a highly-aggressive form of the disease or the *accelerated* or *blast* phase. The other extreme arises if $r_Q < \delta_Q$. In this case Q exponentially decays to 0 for the uncontrolled system and overall trajectories converge to one of the equilibria $(0, P_*, E_*)$ in the plane $Q = 0$. If there exist multiple such equilibria, there exists a stable manifold for the unstable one that separates the regions of attraction for the stable equilibria. This would reflect a scenario when Q initiates the disease, but eventually dies off and the remaining P population determines the outcome of the disease. This could be benign if P_* is small (a form of successful *immune surveillance*) or malignant if this value is larger. In such a case, however, one only needs to deal with the proliferating cells as far as treatment is concerned. This appears less likely (unless it could be induced by the drugs) and in the uncontrolled case of the disease we would have $r_Q > \delta_Q$.

The interesting and most difficult case arises when the uncontrolled system has a chronic steady state or undergoes exponential growth without treatment, but has a negative net growth rate for Q with treatment. This is the case if the parameters satisfy the following condition (A):

$$E_{\max,1} \frac{E_{ss}}{1 + E_{ss}} > \frac{r_Q}{\delta_Q} - 1 > 0, \tag{15}$$

or, with the controls in the original form,

$$E_{\max,1}\frac{E_{ss}}{1+E_{ss}}\left(1+\frac{U2_{\max,1}u_2}{U2C_{50}+u_2}\right)\left(1+\frac{U3_{\max,1}u_3}{U3C_{50}+u_3}\right)>\frac{r_Q}{\delta_Q}-1>0.$$

Thus the replication rate constant r_Q needs to be greater than the death rate constant δ_Q (this naturally will be satisfied for parameters in a disease state), but at the same time, the drugs need to be able to raise the maximum effectiveness $\hat{E}_{\max,1} = E_{\max,1}\left(1+\frac{U2_{\max,1}u_2}{U2C_{50}+u_2}\right)\left(1+\frac{U3_{\max,1}u_3}{U3C_{50}+u_3}\right)$ high enough that the magnitude of the immune system effect can overcome the difference. These appear to be natural conditions. Assuming that (15) holds, there exists a unique value $E_* \in (0, E_{ss})$ for which $\dot{Q} = 0$, namely

$$r_Q = \delta_Q\left(1+E_{\max,1}\frac{E_*}{1+E_*}\right) \iff E_* = \frac{\frac{r_Q}{\delta_Q}-1}{E_{\max,1}-\left(\frac{r_Q}{\delta_Q}-1\right)} \tag{16}$$

with Q increasing for $E < E_*$ and decreasing for $E > E_*$. In this case, the interplay between the variables allows for a steady state (Q_*, P_*, E_*) to exist with all values positive. We call such an equilibrium point (Q_*, P_*, E_*) *positive.*

3.4. Special Case: $P_{\max,1} = P_{\max,2}$

We first discuss the dynamical behavior of the system for the case $P_{\max,1} = P_{\max,2}$ which is quite different from the cases $P_{\max,1} \neq P_{\max,2}$. If these effective rates are equal, we have that

$$\frac{dE}{dt} = \left(s_E \ln\left(\frac{E_{ss}}{E}\right) - \delta_E\right)\left(1+\frac{P_{\max,1}P}{1+P}\right)E$$

and it follows that E is strictly increasing for $E < E_* = E_{ss}\exp\left(-\frac{\delta_E}{s_E}\right)$ and strictly decreasing for $E > E_*$. Therefore, as $t \to \infty$, the E-dynamics approach E_*, monotonically increasing if the initial condition is smaller, monotonically decreasing if it is higher. Consequently also the Q-dynamics approach the steady-state behavior

$$\frac{dQ}{dt} = \left[r_Q - \delta_Q\left(1+\frac{E_{\max,1}E_*}{1+E_*}\right)\right]Q$$

and Q will increase exponentially if

$$r_Q > \delta_Q\left(1+\frac{E_{\max,1}E_*}{1+E_*}\right)$$

and decrease exponentially if

$$r_Q < \delta_Q\left(1+\frac{E_{\max,1}E_*}{1+E_*}\right).$$

In the first case this also generates unbounded growth in P (Lemma 3) leading to behavior consistent with the *blast* phase of the system. In the second case, Q decays exponentially to 0 and P converges to the unique and asymptotically stable equilibrium point P_* on $Q = 0$. Overall, and writing in the constant controls (the respective concentrations u_i) we have the following result:

Proposition 3. *Suppose* $P_{\max,1} = P_{\max,2}$ *and let*

$$\hat{E} = E_{ss}\exp\left(-\frac{\delta_E}{s_E}\right)$$

and

$$\hat{P} = P_{ss} \exp\left(-\frac{\delta_P}{r_P}\left(1 + E_{\max,2}\frac{\hat{E}}{1+\hat{E}}\right)\right).$$

If

$$r_Q < \delta_Q \left(1 + \left(1 + \frac{U2_{\max,1}u_2}{U2C_{50}+u_2}\right)\left(1 + \frac{U3_{\max,1}u_3}{U3C_{50}+u_3}\right)\frac{E_{\max,1}\hat{E}}{1+\hat{E}}\right),$$

then all trajectories $(Q(t), P(t), E(t))$ *converge to the unique and asymptotically stable equilibrium point* $(0, \hat{P}, \hat{E})$ *in the boundary of* $\mathbb{P}_0$*, whereas if*

$$r_Q > \delta_Q \left(1 + \left(1 + \frac{U2_{\max,1}u_2}{U2C_{50}+u_2}\right)\left(1 + \frac{U3_{\max,1}u_3}{U3C_{50}+u_3}\right)\frac{E_{\max,1}\hat{E}}{1+\hat{E}}\right),$$

then Q *grows exponentially and* $\lim_{t\to\infty} P(t) = +\infty$ *and* $\lim_{t\to\infty} E(t) = \hat{E}$.

If

$$r_Q = \delta_Q \left(1 + \left(1 + \frac{U2_{\max,1}u_2}{U2C_{50}+u_2}\right)\left(1 + \frac{U3_{\max,1}u_3}{U3C_{50}+u_3}\right)\frac{E_{\max,1}\hat{E}}{1+\hat{E}}\right),$$

then a positive equilibrium point (Q_*, P_*, E_*) *exists, but this relation is non-generic and generally will not be satisfied for a given set of parameters.*

3.5. Existence and Stability of a Positive Equilibrium Point (Q_*, P_*, E_*) for $P_{\max,1} \neq P_{\max,2}$

We analyze whether positive equilibrium points (Q_*, P_*, E_*) exist. Throughout this section we assume that condition (15) is satisfied, i.e., that

$$E_{\max,1}\frac{E_{ss}}{1+E_{ss}} > \frac{r_Q}{\delta_Q} - 1 > 0,$$

since otherwise Q grows exponentially.

Lemma 5. *For* $P_{\max,1} \neq P_{\max,2}$*, there exists at most one positive equilibrium point* (Q_*, P_*, E_*).

Proof. The equilibrium relation for Equation (6) uniquely determines E_*:

$$r_Q = \delta_Q\left(1 + E_{\max,1}\frac{E_*}{1+E_*}\right) \iff E_* = \frac{\frac{r_Q}{\delta_Q} - 1}{E_{\max,1} - \left(\frac{r_Q}{\delta_Q} - 1\right)} > 0.$$

Given E_*, the equilibrium condition on the effector cells, $\dot{E} = 0$, is equivalent to

$$\frac{s_E}{\delta_E}\ln\left(\frac{E_{ss}}{E_*}\right) = \frac{1 + P_* + P_{\max,2}P_*}{1 + P_* + P_{\max,1}P_*}. \tag{17}$$

The quantity $\frac{s_E}{\delta_E}\ln\left(\frac{E_{ss}}{E_*}\right)$ is already determined. If $\frac{s_E}{\delta_E}\ln\left(\frac{E_{ss}}{E_*}\right) = 1$, then (17) only has the solution $P_* = 0$; otherwise there exists a unique solution $P_* = P_*(E_*)$ given by

$$P_* = \frac{1}{1+P_{\max,1}} \times \frac{1 - \frac{s_E}{\delta_E}\ln\left(\frac{E_{ss}}{E_*}\right)}{\frac{s_E}{\delta_E}\ln\left(\frac{E_{ss}}{E_*}\right) - \frac{1+P_{\max,2}}{1+P_{\max,1}}}. \tag{18}$$

If $P_{\max,1} < P_{\max,2}$, this solution is positive if and only if

$$1 < \frac{s_E}{\delta_E}\ln\left(\frac{E_{ss}}{E_*}\right) < \frac{1+P_{\max,2}}{1+P_{\max,1}}$$

and if $P_{\max,1} > P_{\max,2}$, the solution is positive if and only if

$$1 > \frac{s_E}{\delta_E} \ln\left(\frac{E_{ss}}{E_*}\right) > \frac{1+P_{\max,2}}{1+P_{\max,1}}.$$

If one of these inequalities is violated, no positive equilibrium solution $P_* = P_*(E_*)$ exists and the overall dynamics are determined either by exponential growth or decay of Q. If $P_* = P_*(E_*)$ exists and is positive, then Equation (7) defines Q_* as

$$k_P Q_* = \left[\delta_P \left(1 + E_{\max,2}\frac{E_*}{1+E_*}\right) - r_P \ln\left(\frac{P_{ss}}{P_*}\right)\right] P_*. \tag{19}$$

Using the equilibrium relation for E_*, this can equivalently be expressed in the form

$$k_P Q_* = \left[1 + \frac{E_{\max,2}}{E_{\max,1}}\left(\frac{r_Q}{\delta_Q} - 1\right) - \frac{r_P}{\delta_P} \ln\left(\frac{P_{ss}}{P_*}\right)\right] \delta_P P_*. \tag{20}$$

Note that Q_* is positive if $P_* \geq P_{ss}$ while otherwise this becomes a requirement on the equilibrium value $P_* = P_*(E_*)$, namely

$$P_* > P_{ss} \exp\left(-\frac{\delta_P}{r_P}\left[1 + \frac{E_{\max,2}}{E_{\max,1}}\left(\frac{r_Q}{\delta_Q} - 1\right)\right]\right).$$

If $E_{\max,2} = E_{\max,1}$, then this simply becomes $P_* > P_{ss} \exp\left(-\frac{\delta_P}{r_P}\frac{r_Q}{\delta_Q}\right)$. In either case, there exists at most one positive equilibrium point given by Equations (16), (18) and (20).

Remark 1. *As $P_{\max,1} \to P_{\max,2}$, condition (15) implies that along a positive solution $P_*(E_*)$ we must have*

$$\frac{s_E}{\delta_E} \ln\left(\frac{E_{ss}}{E_*}\right) \to 1$$

and thus the limit taken along these positive solutions only exists if $E_* \to \hat{E} = E_{ss} \exp\left(-\frac{\delta_E}{s_E}\right)$ and if

$$r_Q = \delta_Q \left(1 + E_{\max,1}\frac{\hat{E}}{1+\hat{E}}\right).$$

In this degenerate case, the equilibrium conditions $\dot{Q} = 0$ and $\dot{E} = 0$ are automatically satisfied and there exists a one-dimensional equilibrium manifold, namely $M = \{(Q_*(P), P, \hat{E}) : P > 0\}$ with the P value arbitrary and $Q_*(P)$ given by

$$Q_*(P) = \frac{1}{k_P}\left[\delta_P \left(1 + E_{\max,2}\frac{\hat{E}}{1+\hat{E}}\right) - r_P \ln\left(\frac{P_{ss}}{P}\right)\right] P.$$

We now investigate the *stability of the positive equilibrium point*. The partial derivatives of the equations defining the dynamics at the equilibrium point are given by

$$\frac{\partial f_1}{\partial Q}_{|(Q_*,P_*,E_*)} = 0, \quad \frac{\partial f_1}{\partial P}_{|(Q_*,P_*,E_*)} = 0, \quad \frac{\partial f_1}{\partial E}_{|(Q_*,P_*,E_*)} = -\delta_Q Q_* \frac{E_{\max,1}}{(1+E_*)^2},$$
$$\frac{\partial f_2}{\partial Q}_{|(Q_*,P_*,E_*)} = k_P, \quad \frac{\partial f_2}{\partial P}_{|(Q_*,P_*,E_*)} = -k_P \frac{Q_*}{P_*} - r_P, \quad \frac{\partial f_2}{\partial E}_{|(Q_*,P_*,E_*)} = -\delta_P P_* \frac{E_{\max,2}}{(1+E_*)^2},$$
$$\frac{\partial f_3}{\partial Q}_{|(Q_*,P_*,E_*)} = 0, \quad \frac{\partial f_2}{\partial E}_{|(Q_*,P_*,E_*)} = -s_E\left(1+\frac{P_{\max,1}P_*}{1+P_*}\right)$$
$$\frac{\partial f_3}{\partial P}_{|(Q_*,P_*,E_*)} = \frac{s_E P_{\max,1}\ln\left(\frac{E_{ss}}{E_*}\right) - \delta_E P_{\max,2}}{(1+P_*)^2} E_* = \delta_E E_* \frac{(P_{\max,1}-P_{\max,2})}{(1+P_*)(1+P_*+P_{\max,1}P_*)}.$$

Note that the equilibrium condition for P brings in Q_* in $\frac{\partial f_2}{\partial P}_{|(Q_*,P_*,E_*)}$. The characteristic polynomial for the Jacobian matrix is given by

$$\chi(t) = \begin{vmatrix} t & 0 & \delta_Q Q_* \frac{E_{\max,1}}{(1+E_*)^2} \\ -k_P & t + k_P \frac{Q_*}{P_*} + r_P & \delta_P P_* \frac{E_{\max,2}}{(1+E_*)^2} \\ 0 & \delta_E E_* \frac{P_{\max,2}-P_{\max,1}}{(1+P_*)(1+P_*+P_{\max,1}P_*)} & t + s_E\left(1+\frac{P_{\max,1}P_*}{1+P_*}\right) \end{vmatrix}$$
$$= t^3 + a_2 t^2 + a_1 t + a_0.$$

By the Routh-Hurwitz criterion, all eigenvalues have negative real parts if and only if $a_0 > 0$, $a_1 > 0$ and $a_1 a_2 > a_0$. These coefficients are given by

$$a_2 = \left(k_P \frac{Q_*}{P_*} + r_P\right) + s_E\left(1 + \frac{P_{\max,1}P_*}{1+P_*}\right) > 0,$$
$$a_1 = \left(k_P \frac{Q_*}{P_*} + r_P\right) s_E \left(1 + \frac{P_{\max,1}P_*}{1+P_*}\right) + \frac{E_{\max,2}\delta_E E_*}{(1+E_*)^2} \frac{(P_{\max,1}-P_{\max,2})\,\delta_P P_*}{(1+P_*)(1+P_*+P_{\max,1}P_*)}$$
$$a_0 = \delta_Q Q_* \frac{E_{\max,1}\delta_E E_*}{(1+E_*)^2} k_P \frac{P_{\max,1}-P_{\max,2}}{(1+P_*)(1+P_*+P_{\max,1}P_*)}.$$

If $P_{\max,1} < P_{\max,2}$, then a_0 is negative and the positive equilibrium point is *unstable*, i.e., once the maximal inhibiting effect of the tumor on the effector cells is larger than the maximal stimulating effect, no steady-state positive solution exists. Note further that for $a_0 < 0$ the characteristic polynomial $\chi(t) = t^3 + a_2 t^2 + a_1 t + a_0$ has exactly one change of sign in its coefficients and thus there exists a unique positive root. So the equilibrium point has a two-dimensional stable manifold that separates the regions where Q and P diverge to infinity from the region where Q converges to 0. Thus we have the following result:

Theorem 1. *If $P_{\max,1} < P_{\max,2}$, then the positive equilibrium point (Q_*, P_*, E_*) is unstable with a two-dimensional stable manifold in parameter space.*

If $P_{\max,1} = P_{\max,2}$, then the equilibrium point has the eigenvalue 0 and two negative eigenvalues. Thus there exists a one-dimensional center manifold which in this case consists of all equilibria, namely the equilibrium manifold M defined earlier.

For $P_{\max,1} > P_{\max,2}$, the coefficients a_0, a_1 and a_2 are all positive. Furthermore

$$\begin{aligned}
a_1 a_2 - a_0 = & \left[\left(k_P \frac{Q_*}{P_*} + r_P\right) + s_E\left(1 + \frac{P_{\max,1}P_*}{1+P_*}\right)\right] \times \\
& \times \left[\left(k_P \frac{Q_*}{P_*} + r_P\right) s_E \left(1 + \frac{P_{\max,1}P_*}{1+P_*}\right) + \frac{E_{\max,2}\delta_E E_*}{(1+E_*)^2} \frac{(P_{\max,1} - P_{\max,2})\,\delta_P P_*}{(1+P_*)(1+P_*+P_{\max,1}P_*)}\right] \\
& - \delta_Q Q_* \frac{E_{\max,1}\delta_E E_*}{(1+E_*)^2} \frac{k_P (P_{\max,1} - P_{\max,2})}{(1+P_*)(1+P_*+P_{\max,1}P_*)} \\
& > \frac{\delta_E E_*}{(1+E_*)^2} \frac{P_{\max,1} - P_{\max,2}}{(1+P_*)(1+P_*+P_{\max,1}P_*)} k_P Q_* \left(\delta_P E_{\max,2} - \delta_Q E_{\max,1}\right).
\end{aligned}$$

This expression is positive if we make the following assumption (B):

$$\delta_P E_{\max,2} \geq \delta_Q E_{\max,1}. \tag{21}$$

Note from Equations (2) and (3) that $\delta_P E_{\max,2}$ represents the maximal size of the immune effect E on P while $\delta_Q E_{\max,1}$ represents the maximal size of the immune effect E on Q. This effect is assumed to be stronger on the proliferating class of cells than on the quiescent class of cells. Thus assumption (21) is a natural one to make. This assumption is invariant under the actions of the drugs:

$$\begin{aligned}
\hat{\delta}_P \hat{E}_{\max,2} &= \left(1 + \frac{U1_{\max,2}u_1}{U1C_{50} + u_1}\right)\left(1 + \frac{U2_{\max,3}u_2}{U2C_{50} + u_2}\right)\delta_P \cdot \frac{\left(1 + \frac{U2_{\max,1}u_2}{U2C_{50}+u_2}\right)\left(1 + \frac{U3_{\max,1}u_3}{U3C_{50}+u_3}\right)}{\left(1 + \frac{U1_{\max,2}u_1}{U1C_{50}+u_1}\right)\left(1 + \frac{U2_{\max,3}u_2}{U2C_{50}+u_2}\right)} E_{\max,2} \\
&= \left(1 + \frac{U2_{\max,1}u_2}{U2C_{50} + u_2}\right)\left(1 + \frac{U3_{\max,1}u_3}{U3C_{50} + u_3}\right)\delta_P E_{\max,2}
\end{aligned}$$

while, letting $\hat{\delta}_Q = \delta_Q$,

$$\begin{aligned}
\hat{\delta}_Q \hat{E}_{\max,1} &= \delta_Q \cdot \left(1 + \frac{U2_{\max,1}u_2}{U2C_{50} + u_2}\right)\left(1 + \frac{U3_{\max,1}u_3}{U3C_{50} + u_3}\right) E_{\max,1} \\
&= \left(1 + \frac{U2_{\max,1}u_2}{U2C_{50} + u_2}\right)\left(1 + \frac{U3_{\max,1}u_3}{U3C_{50} + u_3}\right)\delta_Q E_{\max,1}
\end{aligned}$$

so that these terms are multiplied by the same coefficients. Hence we also have the following result:

Theorem 2. *If $P_{\max,1} > P_{\max,2}$ and $\delta_P E_{\max,2} \geq \delta_Q E_{\max,1}$, then the positive equilibrium point (Q_*, P_*, E_*) is locally asymptotically stable.*

The limiting case $P_{\max,1} = P_{\max,2}$ represents a degenerate scenario. In many cases no positive equilibrium exists. For example, if $\frac{s_E}{\delta_E} \ln\left(\frac{E_{ss}}{E_*}\right) \neq 1$, then it follows from (18) that

$$\lim_{P_{\max,1} \to P_{\max,2}} P_* = -\frac{1}{1 + P_{\max,1}} < 0.$$

In such a case equilibria will cease to exist, as $P_{\max,1} \searrow P_{\max,2}$, once the parameter values satisfy

$$1 > \frac{s_E}{\delta_E} \ln\left(\frac{E_{ss}}{E_*}\right) = \frac{1 + P_{\max,2}}{1 + P_{\max,1}}, \qquad E_* = \frac{\frac{r_Q}{\delta_Q} - 1}{E_{\max,1} - \left(\frac{r_Q}{\delta_Q} - 1\right)}.$$

Also, although the positive equilibrium point in Theorem 2 is stable, the value can be very high. In fact, P_* diverges to $+\infty$ as these parameter relations are reached (c.f. (18)):

$$P_* = \frac{1}{1+P_{\max,1}} \frac{1 - \frac{s_E}{\delta_E} \ln\left(\frac{E_{ss}}{E_*}\right)}{\frac{s_E}{\delta_E} \ln\left(\frac{E_{ss}}{E_*}\right) - \frac{1+P_{\max,2}}{1+P_{\max,1}}}.$$

For the equilibrium values to be relatively small ('chronic'), we see that $P_{\max,1}$ must be significantly larger than $P_{\max,2}$. In terms of the parameter values with drug actions, this can be achieved using the drugs u_2 and u_3 which increase $\hat{P}_{\max,1}$ and decrease $\hat{P}_{\max,2}$, c.f.,

$$\hat{P}_{\max,1} = \left(1 + \frac{U2_{\max,4} u_2}{U2C_{50} + u_2}\right)\left(1 + \frac{U3_{\max,2} u_3}{U3C_{50} + u_3}\right) P_{\max,1},$$
$$\hat{P}_{\max,2} = \left(1 - \frac{U2_{\max,5} u_2}{U2C_{50} + u_2}\right)\left(1 - \frac{U3_{\max,3} u_3}{U3C_{50} + u_3}\right) P_{\max,2}.$$

So drug administration shifts the balance towards $P_{\max,1}$ and this creates an asymptotically stable positive equilibrium point (Q_*, P_*, E_*), hopefully with low values for P_* and Q_*.

Figure 4 shows how the positive equilibrium values change as (only) one of the controls is varied. Note that the equilibrium values for Q and E do not change if only the control u_1 is varied. Also for changes in the controls u_2 and u_3 these equilibrium values change little and in the graphs the corresponding curves are almost constant. However, in these cases the equilibrium values for Q and P are well-controlled by the therapies. Contrary to the case when $Q = 0$, the u_2 and u_3 controls have strong effects by cutting down the influx of cells from the Q into the P compartment. All equilibria shown in these graphs satisfy the conditions of Theorem 2 and are locally asymptotically stable.

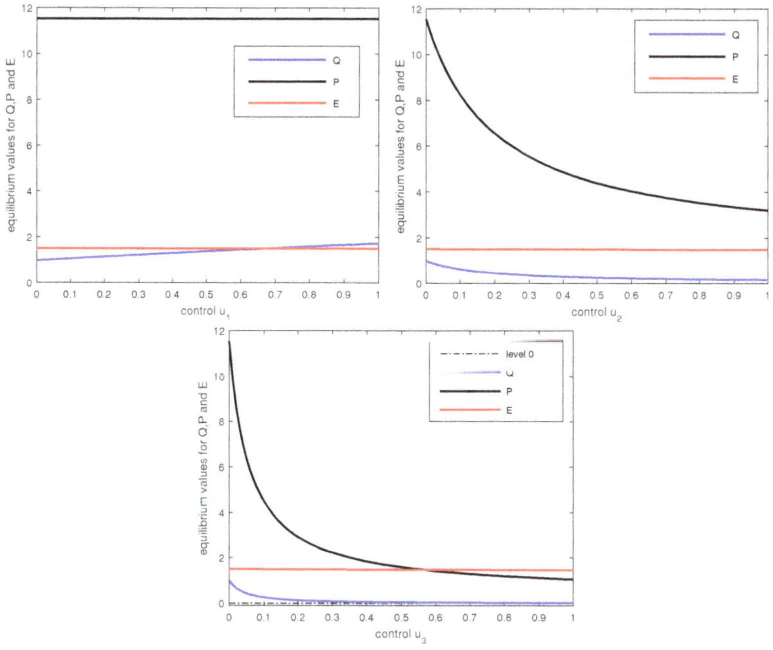

Figure 4. The values of the positive equilibrium point (Q_*, P_*, E_*) as the values for a single control are varied from 0 to 1. The parameter values used in the computations are given in Tables 1 and 2.

4. Discussion and Conclusions

We considered the dynamical behavior of a mathematical model for CML that incorporated three types of therapies defined by targeted effects on proliferating cells and immunomodulatory properties. We analyzed the long-term dynamical behavior of quiescent and proliferating leukemic cells and immune effects (represented by effector T cells). General parameter values were considered to capture a range of possible scenarios. Some thresholds in the parameter space have been determined analytically that separate different types of dynamical behavior that may correspond to the chronic and the accelerated/blast phases of the disease. It has been illustrated how increasing levels of the therapies affect the equilibrium solutions and their stability. As Q becomes small, the analysis of the dynamics in the plane $Q = 0$ indicates that a tyrosine kinase inhibitor can effectively control the disease. However, for larger values of Q, the behavior of the equilibrium solutions shown in Figure 4 suggests that the immunomodulatory properties of the controls u_2 and/or u_3 are essential in controlling the disease, since u_1 alone cannot move the equilibrium value P_* if Q_* slowly increases. Thus this analysis for constant controls already gives some interesting insights into the roles of the various therapies. Indeed, this analysis for constant parameters and controls is a natural first step towards formulating the model as an optimal control problem where treatment constraints and an objective functional incorporating leukemic cell populations and toxicity for the therapeutic agents will be introduced. Although optimal control solutions such as those computed in [11] can provide insight, optimization of the system under clinical dosing constraints (such as only allowing certain dose levels, and only allowing them to change at certain intervals) would be useful [18].

Acknowledgments: The first author received financial support from Bristol-Myers Squibb for this work. The second author is employed by Bristol-Myers Squibb. The authors thank reviewers at Bristol-Myers Squibb for comments related to clinical information included in this paper.

Author Contributions: Helen Moore led the construction of the initial disease and therapy model. Urszula Ledzewicz led the dynamical system analysis. Urszula Ledzewicz and Helen Moore wrote the paper together.

Conflicts of Interest: The funding sponsors had no role in the analysis or conclusions of this work, or the decision to publish the results.

References

1. Faderl, S.; Talpaz, M.; Estrov, Z.; O'Brien, S.; Kurzrock, R.; Kantarjian, H. The biology of chronic myeloid leukemia. *N. Engl. J. Med.* **1999**, *341*, 164–172.
2. Deininger, M.; O'Brien, S.G.; Guilhot, F.; Goldman, J.M.; Hochhaus, A.; Hughes, T.P.; Radich, J.P.; Hatfield, A.K.; Mone, M.; Filian, J.; et al. International randomized study of interferon vs STI571 (IRIS) 8-year follow up: sustained survival and low risk for progression or events in patients with newly diagnosed chronic myeloid leukemia in chronic phase (CML-CP) treated with imatinib. *Blood* **2009**, *114*, 1126.
3. Sawyers, C.L. Chronic myeloid leukemia. *N. Engl. J. Med.* **1999**, *340*, 1330–1340.
4. Weiden, P.L.; Sullivan, K.L.; Flournoy, N.; Storb, R.; Thomas, E.D.; Seattle Marrow Transplant Team. Antileukemic effect of chronic graft-versus-host disease: Contribution to improved survival after allogeneic marrow transplantation. *N. Engl. J. Med.* **1981**, *304*, 1529–1533.
5. Talpaz, M.; Kantarjian, H.; McCredie, K.; Trujillo, J.; Keating, M.; Gutterman, J.U. Therapy of chronic myelogenous leukemia. *Cancer* **1987**, *59*, 664–667.
6. Bristol-Myers Squibb. A Phase 1B Study to Investigate the Safety and Preliminary Efficacy for the Combination of Dasatinib Plus Nivolumab in Patients With Chronic Myeloid Leukemia (CML). Available online: http://clinicaltrials.gov/show/NCT02011945 (accessed on 7 March 2016).
7. Rubinow, S.I. A simple model of steady state differentiating cell system. *J. Cell Biol.* **1969**, *43*, 32–39.
8. Rubinow, S.I.; Lebowitz, J.L. A mathematical model of neutrophil production and control in normal men. *J. Math. Biol.* **1975**, *1*, 187–225.
9. Fokas, A.S.; Keller, J.B.; Clarkson, B.D. Mathematical model of granulocytopoesis and chronic myelogeneous leukemia. *Cancer Res.* **1999**, *51*, 2084–2091.

10. Moore, H.; Li, N.K. A mathematical model for chronic myelogenous leukemia (CML) and T cell interaction. *J. Theor. Biol.* **2004**, *227*, 513–523.
11. Nanda, S.; Moore, H.; Lenhart, S. Optimal control of treatment in a mathematical model of chronic myelogenous leukemia. *Math. Biosci.* **2007**, *210*, 143–156.
12. Moore, H.; Strauss, L.; Ledzewicz, U. Mathematical optimization of combination therapy for Chronic Myeloid Leukemia (CML). In Presented at the 6th American Conference on Pharmacometrics, Crystal City, VA, USA, 4–7 October 2015.
13. Afenya, E.K.; Calderón, C. Diverse ideas on the growth kinetics of disseminated cancer cells. *Bull. Math. Biol.* **2000**, *62*, 527–542.
14. Nakamura-Ishizu, A.; Takizawa, H.; Suda, T. The analysis, roles and regulation of quiescence in hematopoietic stem cells. *Development* **2014**, *141*, 4656–4666.
15. Gabrielsson, J.; Weiner, D. *Pharmacokinetic and Pharmacodynamic Data Analysis: Concepts and Applications*, 5th ed.; Apotekarsocieteten: Stockholm, Sweden, 2016.
16. Shudo, E.; Ribeiro, R.M.; Perelson, A.S. Modelling hepatitis C virus kinetics: the relationship between the infected cell loss rate and the final slope of viral decay. *Antivir. Ther.* **2009**, *14*, 459–464.
17. Branford, S.; Yeung, D.T.; Prime, J.A.; Choi, S.Y.; Bang, J.H.; Park, J.E.; Kim, D.W.; Ross, D.M.; Hughes, T.P. BCR-ABL1 doubling times more reliably assess the dynamics of CML relapse compared with the BCR-ABL1 fold rise: implications for monitoring and management. *Blood* **2012**, *119*, 4264–4271.
18. Schättler, H.; Ledzewicz, U. *Optimal Control for Mathematical Models of Cancer Therapies*; Springer: New York, NY, USA, 2015.

applied sciences

MDPI

Article

Chronic Inflammation in the Epidermis: A Mathematical Model

Shinji Nakaoka [1], Sota Kuwahara [2], Chang Hyeong Lee [3], Hyejin Jeon [4,5], Junho Lee [5], Yasuhiro Takeuchi [2] and Yangjin Kim [5,6,*]

1 Institute of Industrial Science, University of Tokyo, Tokyo 153-8505, Japan; snakaoka@m.u-tokyo.ac.jp
2 College of Science and Engineering, Aoyama Gakuin University, Kanagawa 252-5258, Japan; shinzy.nakaoka@gmail.com (S.K.); takeuchi@gem.aoyama.ac.jp (Y.T.)
3 Department of Mathematical Sciences, Ulsan National Institute of Science and Technology, Ulsan 44919, Korea; leechanghyeong@gmail.com
4 Department of Radiology, Seoul National University Hospital, Seoul 03080, Korea; jhjisthebest@gmail.com
5 Department of Mathematics, Konkuk University, Seoul 05029, Korea; juneho2222@gmail.com
6 Department of Mathematics, Ohio State University, Columbus, OH 43210, USA
* Correspondence: ahyouhappy@konkuk.ac.kr; Tel.: +82-02-450-0450; Fax: +82-02-458-1952

Academic Editor: Yang Kuang
Received: 11 June 2016; Accepted: 31 August 2016; Published: 9 September 2016

Abstract: The epidermal tissue is the outmost component of the skin that plays an important role as a first barrier system in preventing the invasion of various environmental agents, such as bacteria. Recent studies have identified the importance of microbial competition between harmful and beneficial bacteria and the diversity of the skin surface on our health. We develop mathematical models (M1 and M2 models) for the inflammation process using ordinary differential equations and delay differential equations. In this paper, we study microbial community dynamics via transcription factors, protease and extracellular cytokines. We investigate possible mechanisms to induce community composition shift and analyze the vigorous competition dynamics between harmful and beneficial bacteria through immune activities. We found that the activation of proteases from the transcription factor within a cell plays a significant role in the regulation of bacterial persistence in the M1 model. The competition model (M2) predicts that different cytokine clearance levels may lead to a harmful bacteria persisting system, a bad bacteria-free state and the co-existence of harmful and good bacterial populations in Type I dynamics, while a bi-stable system without co-existence is illustrated in the Type II dynamics. This illustrates a possible phenotypic switch among harmful and good bacterial populations in a microenvironment. We also found that large time delays in the activation of immune responses on the dynamics of those bacterial populations lead to the onset of oscillations in harmful bacteria and immune activities. The mathematical model suggests possible annihilation of time-delay-driven oscillations by therapeutic drugs.

Keywords: epidermis; mathematical model; bacterial inflammation; bacterial competition

1. Introduction

The skin is the largest tissue, which is composed of several different layers. The epidermis is located at the outmost part of the skin tissue, which acts as a first barrier for the invasion of physical (water), chemical (proteins) and biological (virus and bacteria) agents. A population of keratinocytes is the major cell type in the epidermis, which constitutes stratum basale, stratum spinosum, stratum granulosum and stratum corneum. Keratinocytes release anti-microbial peptides or pro-inflammatory cytokines to prevent bacterial or viral infection [1]. The second outmost layer, the dermis, is situated between the epidermis and subcutaneous tissues, which are composed of fibroblasts, macrophages and adipocytes [2]. The dermis contains extracellular matrix components, including collagen and elastin, as well as lymph, blood vessels and many skin-resident immune cell types. The homeostasis of the skin tissue is maintained by appropriate elimination of invading agents and tight regulation of cellular activities. On the other hand, the breakdown of the homeostasis of the skin tissue induces numerous diseases, including cancer, complications after an injury and inflammatory symptoms. Atopic dermatitis (AD) is one of the major skin inflammatory diseases, which are characterized by the elevated level of serum IgE and chronic allergic immune responses [3]. Incidence of AD has been increasing in developed countries. Notably, recent genetic studies have revealed that barrier dysfunction of the epidermis due to filaggrin mutation is a major triggering factor of disease progression [4,5]. Filaggrin is synthesized in keratinocytes at the stratum granulosum, which is degraded to become a major component of natural moisturizing factor (NMF). Lack of NMFs is associated with water loss skin dryness, leading to the progression of AD and ichthyosis vulgaris [4]. Not only dry condition, but also excessive proteolytic activities of the epidermis are implicated as causal factors of AD. Patients with Netherton syndrome who exhibit atopic dermatitis-like chronic inflammation indicate a genetic defect causing excessive serine proteases [6].

The inference of bacteria as an environmental factor has been implicated as a possible factor for the progression of skin inflammatory diseases. Two major pathogenic bacteria species for skin diseases are *Staphylococcus aureus* (*S. aureus*) and *Streptococcus pyogenes* (*S. pyogenes*). Importantly, *S. aureus* is a major virulent species, which is implicated to be associated with the progression of atopic dermatitis [7]. *S. aureus* also induces impetigo and another serious symptoms. Methicillin-resistant *S. aureus* (MRSA) is an antibiotic-resistant strain of *S. aureus*, the incidence of which is often reported as a nosocomial infection [8]. On the other hand, some commensal bacteria can exhibit mutualistic behaviors through the suppression of potentially pathogenic bacterial species via direct and indirect interactions, known as probiotic effects. For instance, *Staphylococcus epidermititis*(*S. epidermititis*), a major commensal bacterial species in the skin, can support the host defense by releasing antimicrobial peptides [9,10]. The other beneficial microbial species include species belonging to the *Lactobacillus* genus. *Lactobacillus reuteri* helps keratinocyte survival from *S. aureus*-induced cell death by outcompeting *S. aureus* [11].

The immune system plays a major role in preventing the invasion of numerous agents, including bacteria, fungi, virus and foreign proteins [12]. Not only foreign antigens, but also antigens presented by commensal bacteria can be an antigenic stimulation for the host immune system [13]. In fact, the number of bacteria in abundance is controlled by the host immune system under normal conditions [14]. The pathogenicity of *S. aureus* can be conferred by numerous immune evasion strategies. In fact, several virulent factors of *S. aureus* have been reported in [15–17]. On the contrary, several beneficial roles of *S. epidermititis* have been reported, although *S. epidermititis* can be virulent as a nosocomial pathogen for immunocompromised patients [18]. *S. epidermititis* triggers innate immune responses via toll-like receptor (TLR)-2, which mediate the killing of pathogenic bacteria, such as *S. aureus* [19]. *Lactobacillus plantarum* can utilize the host innate immune system mediated by epithelial cells by modulating the IL 17, IL-23 and TLR-2/4 expressions [20].

Regardless of the fact that many causal and preventive factors for the progression of AD and other skin inflammatory diseases have been identified, each experimental and clinical research only focuses on a specific aspect of the skin biology. The integration of knowledge in each sub-domain is needed in order to achieve a comprehensive understanding of the progression of AD. As described above, the detection of the manifestation of atopic dermatitis requires the integration of weakened barrier function due to a genetic defect or excessive proteolytic activity, the inflammatory response triggered by some of commensal bacteria and abnormal recruitment of immune cells via irregular cellular communication with respect to cytokine signaling. Mathematical modeling and simulation study enable integration and provide some basic insights into the maintenance mechanism underlying the skin homeostasis and disease development as a defect of the homeostatic condition.

We focus on competitive bacterial interactions among *S. epidermatitis* as good bacteria and *S. aureus* as bad bacteria that occur at the skin tissue to specifically reflect experimental and empirical observations, such as [19]. Although our primary focus is to investigate the effects of bacterial competition on the dynamics of inflammatory responses in the epidermis, the mathematical models presented here can be useful as a general scheme to describe the interactions among bacterial species as an environmental factor with host immune responses on the surface of the body, such as the epidermis and gastro-intestinal (GI) tract. See [21,22] for the diverse roles of proteases in the GI tract in the maintenance of intestinal homeostasis. Hence, we focus on constructing mathematical models to represent a less detailed, but the general manner of interactions among inflammation-related molecules, such as protease, transcription factors and extracellular cytokines with bacterial species in epidermis.

In the present paper, we investigate how chronic inflammation can occur at the skin tissue. Simple mathematical models are employed to describe the invasion of bacteria, the proteolytic activity of keratinocytes, the activation of innate immune response and the release of antimicrobial peptide and cytokines. The organization of the present paper is as follows. In Section 2, two mathematical models (M1 and M2) are formulated. The M1 model is formulated to model the dynamics of inflammation in response to bacterial infection via a transcription factor and extracellular cytokines, as well as active proteases. Time delays may play a central role in the regulation of the bacterial-immune system in this study due to possible delays in the regulation of the intracellular transcription factors, protease induction and secretion of extracellular cytokines from bi-directional communication between a cell in the tissue and the microenvironment. Artificial manipulation (inhibition or enhancement of molecular players) of signaling pathways by therapeutic drugs typically induces time delays [23,24] in generating the final production of immune responses, i.e., extracellular cytokines, such as IL-4, IL-12, TNF-α and IFN-γ [25,26]. To explicitly describe the competition between harmful and good bacteria, a formulation via an ordinary differential equation (ODE) and a delay differential equation (DDE) is employed in the M2 model. Mathematical analyses on these models, including the existence and stability of equilibria, are discussed in Section 3. In Section 3.1, numerical simulations are performed to investigate how prominently the chronic inflammatory state is established and maintained. Moreover, we investigate how bacterial competition can lead to a high chronic inflammation state as an imbalanced state (dysbiosis). In Section 3.2, we analyze the competition dynamics of harmful and good bacteria in the absence and presence of time delays and immune boosting drugs. We also perform several in silico experiments, which include the investigation of: (i) the effect of the clearance speed of cytokines on generating three regimes (harmful bacteria persistence, good bacteria persistence and co-existence) in Type I dynamics and a bi-stability system as a possible phenotypic switch in Type II dynamics; (ii) the effect of time delays on generating the oscillatory behaviors of bacterial populations; (iii) the impact of different drug injection regimes on the bacterial populations. In Section 4, we provide a discussion on the fundamental mechanism of the bacterial attacks and immune response, as well as the survival schemes of harmful bacteria in competition with good bacteria and future work in detail. Nondimensionalization and sensitivity analysis of the model are given in the Appendix.

2. Materials and Methods

In this paper, we present two kinds of mathematical models, the M1 basic model in Section 2.1 and the M2 competition model in Section 2.2.

2.1. M1 Model

In this section, we develop a mathematical model based on the schematic diagram in Figure 1. As indicated in Section 1, the key main players of the bacterial infection network in absence of competition with other bacteria are the following variables:

$B =$ density of harmful bacteria,
$P =$ concentration of protease,
$A_I =$ concentration of the intracellular transcription factor,
$A_E =$ concentration of extracellular cytokines,

Figure 1A illustrates the dynamical regulation of immune activities in response to bacterial infection. Bacteria grow in the system with a carrying capacity and induce the secretion of proteases for enhanced bacterial invasion. These proteolytic activities are suppressed by protease inhibitors under normal conditions, but the perturbed balances between a protease and its inhibitors induce the activation of transcriptional factors, causing skin troubles, such as atopy. The upregulated transcriptional factors induce immune activities (extracellular cytokines), which in turn try to kill bacteria. In order to incorporate the biological interactions shown in Figure 1A into our model of bacteria-immune dynamics, we began by simplifying this network. Figure 1B shows a representation of Figure 1A. The kinetic interpretation of the arrows and hammerheads in the given network represents induction (arrow) and inhibition (hammerhead). We merged all complex networks between proteases and their inhibitors into one component (blue dotted box in Figure 1A), while we kept the components of bacteria, transcriptional factor and external allergic immune responses (cytokines) in one module (red dotted boxes in Figure 1A), respectively. The scheme includes bacterial growth, the secretion of proteases from bacterial invasion and stimulated transcriptional factors, the activation of the intracellular transcription factor from upregulated proteases and secreted cytokines, the activation of extracellular cytokines from the transcription factor, protein degradation of those key molecules and eradication of those bacteria by cytokines.

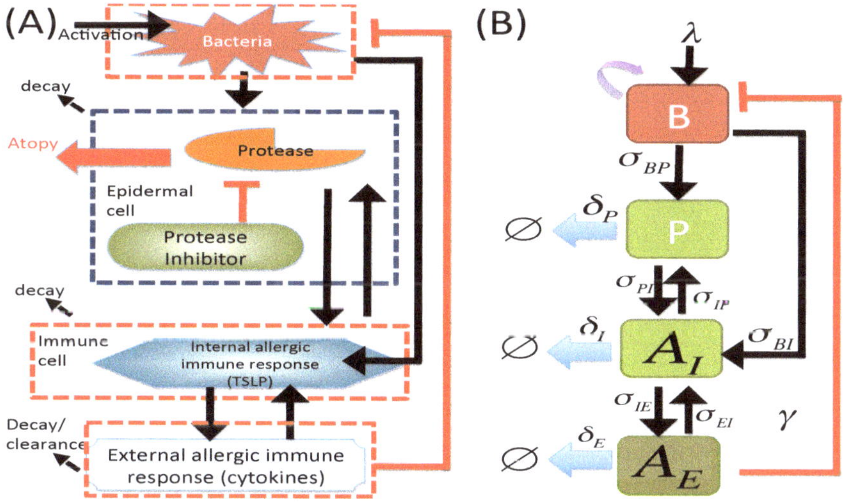

Figure 1. A schematic of the M1 model. (**A**) A schematic of immune responses to bacterial infection; (**B**) the final network model that abstracts the key structure of the interaction network in (A). By merging a multi-species compartment (blue dashed box in (A) including proteases and their inhibitors) into a compartment ('P' in (B)), we get a simpler model in (B). Densities of bacteria, proteases, intracellular transcription factor and extracellular cytokines are represented by 'B', 'P', 'A_I' and 'A_E', respectively.

Activation of transcription factors that are associated with immune responses, such as a member of the interferon regulatory factor (IRF) family, is often mediated by positive feedbacks among these transcription factors [27]. It is also known that activation of protease is mediated by molecular interactions among the members of the kallikrein family [28]. Production of the cytokines is also facilitated by a positive feedback, which is often referred to as a cytokine storm [29]. These activations with positive feedbacks can be modeled by Hill functions.

In this work, we consider the following specific type of functional response known as the Hill function:

$$\sigma_\circ(X) := \frac{m_\circ X^n}{a_\circ^n + (s_\circ X)^n}, \tag{1}$$

where $\circ \in \{\mathrm{IP}, \mathrm{BP}, \mathrm{PI}, \mathrm{BI}, \mathrm{EI}, \mathrm{IE}\}$. Assume that the activation of protease is mediated by transcription factor A_I and bacteria B. Based on these observations, we write the phenomenological equations for the rate change of those key players (B, P, A_I, A_E) as follows:

$$\begin{aligned}
\frac{dB}{dt} &= \lambda + r_B B\left(1 - \frac{B}{K}\right) - \gamma A_E B,\\
\frac{dP}{dt} &= \sigma_{IP}(A_I) + \sigma_{BP}(B) - \delta_P P,\\
\frac{dA_I}{dt} &= \sigma_{PI}(P) + \sigma_{BI}(B) + \sigma_{EI}(A_E) - \delta_I A_I,\\
\frac{dA_E}{dt} &= \sigma_{IE}(A_I) - \delta_E A_E.
\end{aligned} \tag{2}$$

where λ is the source of bacterial populations in the tissue from the air, r_B is the growth rate of bacteria, K is the carrying capacity of the bacterial population, γ is the killing rate of bacteria by immune cytokines and, finally, $\delta_P, \delta_I, \delta_E$ are the decay/clearance rates of proteases, transcriptional factors and cytokines, respectively.

A list of parameters is summarized in Table 1.

Table 1. Dimensionless parameter values in the M1 model. TF = transcription factor.

Parameter	Description	Value
m_{IP}	Maximum activation rate of proteases by TFs	8–100
a_{IP}	Half saturation constant of proteases by TFs	3.0
s_{IP}	Inhibitory strength of proteases activation by TFs	1
n	Hill cooperativity coefficient	2
m_{BP}	Maximum activation rate of proteases by bacteria	8
a_{BP}	Half saturation constant of proteases by bacteria	3.0
s_{BP}	Inhibitory strength of proteases activation by bacteria	1
m_{PI}	Maximum activation rate of TFs by proteases	8
a_{PI}	Half saturation constant of TFs by proteases	3.0
s_{PI}	Inhibitory strength of TF activation by proteases by	1
m_{BI}	Maximum activation rate of TFs by bacteria	8
a_{BI}	Half saturation constant of TFs by bacteria	3.0
s_{BI}	Inhibitory strength of TFs by bacteria	1
m_{EI}	Maximum activation rate of TFs by cytokines	8
a_{EI}	Half saturation constant of TFs by cytokines	3.0
s_{EI}	Inhibitory strength of TFs by cytokines	1
m_{IE}	Maximum activation rate of cytokines by TFs	8
a_{IE}	Half saturation constant of cytokines by TFs	3.0
s_{IE}	Inhibitory strength of cytokines by TFs	1

Table 1. *Cont.*

Parameter	Description	Value
δ_P	Degradation rate of protease	3.0
δ_I	Degradation rate of transcription factor	3.0
δ_E	Degradation rate of extracellular cytokines	3.0
λ	Migration rate of bacteria	0.1
r_B	Population growth rate of bacteria	0.1
K	Carrying capacity of bacteria	10.0
γ	Per capita elimination rate of bacteria	1.0

2.2. M2 Model

In this section, we consider two different bacterial strains. Figure 2A illustrates the dynamical regulation of bacterial infection and immune responses. There exists competition between harmful and good bacteria for bacterial growth. Bacterial infection induces upregulation of the transcriptional factor within the cell for immune activity and the secretion of proteases for enhanced bacterial invasion. Induced extracellular cytokines from the transcription factor then suppress both of those bacteria. In order to incorporate the interaction network shown in Figure 2B into our model of bacterial competition and inflammation, we began by simplifying this network. As indicated in Section 1, five main players of the bacterial infection network are harmful bacteria, good bacteria, protease, intracellular transcription factors and extracellular cytokines. Let the variables B_1, B_2, P, A_I and A_E be the densities or concentrations of harmful bacteria, good bacteria, protease, transcription factor and cytokines, respectively. Figure 2B shows a representation of Figure 2A. The kinetic interpretation of the arrows and hammerheads in the given network represents induction (arrow) and inhibition (hammerhead). The scheme includes bacterial growth of both harmful and good bacteria, mutual inhibition between harmful bacteria and good bacteria, secretion of proteases from those bacterial invasions, activation of the intracellular transcription factor from these bacterial infections, activation of extracellular cytokines via the transcription factor, protein degradation of those key molecules and eradication of those bacteria by cytokines.

It has been known that: (i) the half-life of proteases (P) is short, which indicates the large decay rate of P and that protease reactions occur quickly; and (ii) typical chemical reactions among proteins and genes at a fast time scale lead to the fast internal dynamics. This allows us to use quasi-steady state approximation (QSSA) to simplify the complex models (or interaction networks in Figure 2B). Based on the topological structure and uni-directional activation flows in the immune reaction module (transcription factor activities (A_I), proteolytic activities (P), cytotoxic levels (A_E); gray box in Figure 2B) and the corresponding QSSA, we merged all complex networks between transcriptional factors (A_I), proteases (P) and extracellular cytokines (A_E) into one component (gray box in Figure 2B,C), while we kept the harmful (B_1) and bad (B_2) bacterial components in one module (yellow dotted boxes in Figure 2B), respectively. Based on these observations, we write the phenomenological equations for the rate change of those key players (B_1, B_2, A_E) as follows:

$$\frac{dB_1}{dt} = r_1 B_1 (1 - \alpha_{11} B_1 - \alpha_{12} B_2) - \gamma A_E B_1, \quad (3)$$

$$\frac{dB_2}{dt} = r_2 B_2 (1 - \alpha_{21} B_1 - \alpha_{22} B_2) - \gamma A_E B_2, \quad (4)$$

$$\frac{dA_E}{dt} = \beta_1 B_1 + \beta_2 B_2 - \delta A_E. \quad (5)$$

where r_1 and r_2 are the growth rate of harmful and good bacteria, respectively, α_{ii} ($i = 1, 2$) and α_{ij} ($i \neq j, i, j = 1, 2$) represent intra-specific and inter-specific competition coefficients between harmful and good bacteria, respectively, γ is the killing rate of those bacteria by the extracellular cytokines, β_1 and β_2 are the activation of immune responses (level of cytokines) from harmful and good bacteria, respectively, and δ is the clearance rate of the immune cytokines. In many cases, measured values of these parameter values ($r_1, r_2, \alpha_{ii}, \gamma, \beta_1, \beta_2, \delta$) are not available due to technical reasons. In order to determine appropriate ranges of parameter values for correct dynamical behavior reflecting a real biological system and to investigate the sensitivity of the bacterial populations and immune responses to these parameters, we have performed sensitivity analysis for a mathematical model (3)–(5) in Appendix B. Some of the parameter values ($\alpha_{11}, r_2, \gamma, \beta_1$) are very sensitive, but others (α_{22}) are not sensitive to these changes. See the Appendix for more details.

For computational purposes, we nondimensionalize the variables and parameters of the M1 (Equation (2)) and M2 models (Equations (3)–(5)) in Appendix A.

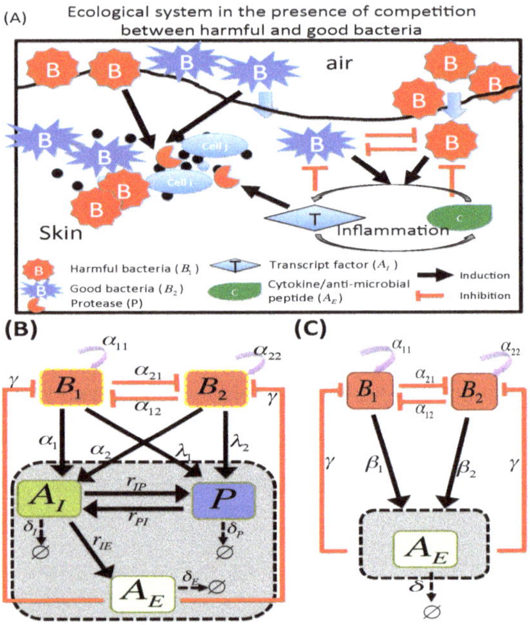

Figure 2. A schematic of the M2 model. (**A**) A schematic of the biological system for the competition between harmful and good bacteria and immune responses. There exists mutual antagonism between harmful and good bacteria. On the other hand, bacterial infection induces upregulation of the transcriptional factor (blue diamond) within the cell for immune activity and secretion of proteases (red quarter pie) for enhanced bacterial invasion. Induced extracellular cytokines (green) from transcription factor (blue) then suppress both harmful (red star) and good (blue star) bacteria. All arrows refer to the induction of gene expression or proteins. The hammerheads from and to bacteria (B_1, B_2) refer to the inhibition or suppression of bacterial growth. (**B**) Topological networks representing the biological observations in (A). Densities of harmful bacteria, good bacteria, proteases, intracellular transcription factor and extracellular cytokines are represented by 'B_1', 'B_2', 'P', 'A_I' and 'A_E', respectively. (**C**) The final network model that abstracts the key structure of the network in (B). By merging a multi-species compartment (gray box in (B) including 'A_I', 'P' and 'A_E') into a compartment 'A_E' (gray box in (C)), we get a simpler model in (C).

3. Results

In this section, we analyze the dynamics of two models (M1 and M2). In the next section, we first investigate the dynamics of M1 model for bacterial infection and its implications on immune responses.

3.1. *Dynamics of the M1 Model*

The M1 model deals with immune responses to bacterial infection via the regulation of proteases (P), internal allergic immune responses (A_I) and the external allergic immune response in terms of cytokines (A_E). We first classify the steady states of the M1 model (2) in the next section.

3.1.1. Classification of Steady States

There exist three types of equilibria for System (2). Let $E_B := (\bar{B}, 0, 0, 0)$ denote a steady state representing no protease and immune responses under bacterial persistence, where $\bar{B} > K$ is a positive root of $\bar{B}^2 - K\bar{B} - K\lambda/r = 0$. Let $E^* := (B^*, P^*, A_I^*, A_E^*)$ denote an inflammatory state that is maintained by additional external stimuli from bacteria. The components of E^* are determined by the solution of the following system of equations:

$$\begin{aligned}
&\lambda + r_B B^* \left(1 - \frac{B^*}{K}\right) - \gamma A_E^* B^* = 0, \\
&\sigma_{IP}(A_I^*) + \sigma_{BP}(B^*) - \delta_P P^* = 0, \\
&\sigma_{BI}(B^*) + \sigma_{PI}(P^*) + \sigma_{EI}(A_E^*) - \delta_I A_I^* = 0, \\
&\sigma_{IE}(A_I^*) - \delta_E A_E^* = 0.
\end{aligned} \tag{6}$$

By substituting the first and third equations into the second equation of (6), we obtain the following equation with respect to A_I^*:

$$\delta_I A_I^* = \sigma_{PI}((\sigma_{IP}(A_I^*) + \sigma_{BP}(B^*))/\delta_P) + \sigma_{EI}(\sigma_{IE}(A_I^*)/\delta_E) + \sigma_{BI}(B^*). \tag{7}$$

It follows from the first and fourth equations of (6) that B^* is explicitly written as a positive root of the following quadratic equation with respect to B:

$$B^2 - K\left\{1 - \frac{\gamma}{r_B \delta_E}\sigma_{IE}(A_I^*)\right\} B - \frac{K\lambda}{r_B} = 0. \tag{8}$$

Note that $B^* > 0$. A_I^* must satisfy:

$$\sigma_{IE}(A_I^*) < \frac{r_B \delta_E}{\gamma}. \tag{9}$$

Then, there exists a unique positive root of (8), denoted by $B = B^* > 0$. Since $\sigma_{IE}(A_I)$ is continuous and monotonically increasing with respect to A_I, (9) is rewritten as:

$$A_I^* < \sigma_{IE}^{-1}\left(\frac{r_B \delta_E}{\gamma}\right). \tag{10}$$

Then, (7) is rewritten as:

$$\begin{aligned}
\delta_I A_I^* =& \sigma_{PI}((\sigma_{IP}(A_I^*) + \sigma_{BP}(K - \gamma K \sigma_{IE}(A_I^*)/r_B\delta_E))/\delta_P) \\
&+ \sigma_{EI}(\sigma_{IE}(A_I^*)/\delta_E) + \sigma_{BI}(K - \gamma K \sigma_{IE}(A_I^*)/r_B\delta_E).
\end{aligned} \tag{11}$$

The existence of positive equilibrium E^* is determined by the root of (11) with Constraint (10).

Let E_L^*, E_U^* and E_H^* denote three equilibria of (2) ordered by the value of A_Is: $A_{I,L}^* < A_{I,U}^* < A_{I,H}^*$. From the biological point of view, $A_{I,\cdot}^*$ represents the strength of inflammation triggered by bacterial antigenic stimuli.

Figure 3 indicates that there are two possible cases: the existence of a unique equilibrium for weak activation of proteases ($m_{IP} = 8$; Figure 3A) or multiple equilibria for enhanced activation of proteolytic activation ($m_{IP} = 50$; Figure 3B). Here, m_{IP} is the activation rate of proteases from the transcription factor in the cell. In the upper panels of Figure 3, the straight solid line (blue) and the dotted curve (green) show the left- and right-hand sides of (11) as a function of (A_I^*), respectively. The intersection of those two curves represents the equilibria (E_L^*, E_U^* and E_H^*). Stability analysis indicates that: (i) when m_{IP} is small ($m_{IP} = 8.0$, the upper panel of Figure 3A), E_L^* (black filled circle) is stable; (ii) when m_{IP} is relatively large ($m_{IP} = 50.0$, the upper panel of Figure 3B), two steady states (E_L^* and E_U^*; empty circles) are unstable, but one steady state (E_H^*; black filled circle) is stable. Figure 3A shows the emergence of the bacterial persistence phenotype in response to the weak activation of proteases ($m_{IP} = 8.0$). The unique positive stable equilibrium E_L^* resides in the region (pink box) where the transcription factor activities are suppressed and bacterial growth is active. On the other hand, when protease activation is enhanced more than six-fold ($m_{IP} = 50$), there exist three positive equilibria (E_L^* (unstable), E_U^* (unstable) and E_H^* (stable)) simultaneously (Figure 3B). The stable steady state E_H^* resides in the region (blue box in Figure 3B) where bacterial activities are inhibited by persistent internal immune responses. These results predict the dynamical changes for various levels of protease activation and illustrate the importance of protease activation in the regulation of bacterial growth under the surveillance of the immune system in the tissue.

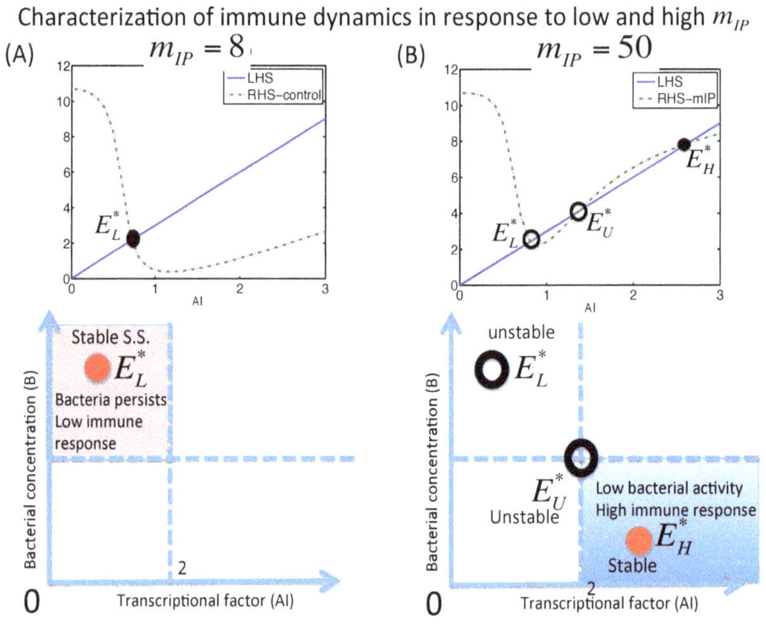

Figure 3. Characterization of the protease activation and immune response in the M1 model. Circles in the lower panels represent the steady state solutions of (11) for high and low values of a control parameter m_{IP}, the activation rate of proteases from the transcription factor in the cell. The intersection (black circles) of the straight line (left-hand side of (11); blue solid line) and curve (right-hand side of (11); green dotted curve) corresponds to the numerical value of A_I^* in the upper panel. * Black filled circle = stable, empty circle = unstable. (**A**) The bacterial persistence phenotype in response to the weak activation of proteases ($m_{IP} = 8.0$). There exists a unique positive stable equilibrium E_L^* in the region (pink box) where transcription factor activities are reduced and bacterial persistence is observed. (**B**) Suppression of bacterial growth by immune activities in response to enhanced activation of proteases ($m_{IP} = 50$). There exist three positive equilibria E_L^*, E_U^* and E_H^* simultaneously. While two equilibria E_L^* and E_U^* are unstable, the equilibrium E_H^* is stable. The equilibrium E_H^* resides in the region (blue box) where bacterial activities are reduced and high internal immune responses persist. All other parameters are fixed as in Table 1.

In the next section, we analyze the dynamics of the M1 model and discuss the implication of the internal and external immune responses on the regulation of the harmful bacteria population.

3.1.2. Dynamics of the M1 Model System

When the M1 model system (2) is in equilibrium, we can solve bacteria density (B) as a function of the activation rate of proteases from the transcription factor (m_{IP}) for any set of parameters. Figure 4A shows the graph $B = B(m_{IP})$ as the S-shaped curves when other parameter values are fixed as in Table 1. While a portion of the upper branch in the lower protease activation range is stable, the remainder of the upper branch corresponding to the intermediate range of the protease activation rate is unstable. On the other hand, the lower branch is stable, and the middle branch is unstable. If m_{IP} is small, then the system (2) is in the upper branch, B is high and the bacterial persistence phenotype emerges. This situation continues to hold as m_{IP} is increased until it reaches criticality. At this point, the system jumps down to the low branch, with a suppressed level of bacterial activities, and the bacterial growth is inhibited (while immune activities are increased). As m_{IP} is decreased from a high level of protease activation, the bacterial growth remains suppressed, until m_{IP} is decreased to the left knee point (red arrow; ∼42), at which time, the bacterial population jumps to the upper branch, and the bacteria return to the growth phase. Figure 4B–D also shows the graphs $A_I = A_I(m_{IP})$, $P = P(m_{IP})$ and $A_E = A_E(m_{IP})$ as the hysteresis loops, as well. One notes that the bifurcation curves for those variables in immune responses (intracellular transcription factors (A_I), protease level (P) and extracellular cytokines (A_E)) show the flipped images of the $B - m_{IP}$ hysteresis loop in Figure 4A, reflecting the bacteria-immune competition system. In other words, the immune activities (levels of A_I, P and A_E) are suppressed compared to bacterial persistence in response to the weak protease activation, while high levels of immune responses are shown compared to inhibited bacterial growth in response to strong protease activation.

Based on the dynamics of the bacterial activities and cytokine levels observed above, we shall define two adaptive types of the bacterial infection system (bacterial persist ($\mathbb{T}_B$) and immune boosting ($\mathbb{T}_I$) systems) as follows:

$$\begin{aligned} \mathbb{T}_B &= \{(B, A_E) \in \mathbb{R}^2 : B > th_B,\ \ 0 \leq A_E < th_{AE}\}, \\ \mathbb{T}_I &= \{(B, A_E) \in \mathbb{R}^2 : 0 \leq B < th_B,\ \ A_E > th_{AE}\} \end{aligned} \tag{12}$$

where th_B and th_{AE} are the threshold values of bacterial activities and cytokine level, respectively. With this definition (12), the unique stable equilibrium E_L^* in Figure 3A belongs to the region $\mathbb{T}_B$ in the case of low m_{IP}, while the stable steady state E_H^* in Figure 3B resides in the region $\mathbb{T}_I$ in the case of high m_{IP}.

In Figure 5, we show how the system adapts to the changes in the parameter m_{IP} as predicted in the analysis above (Figure 4). Figure 5A–C illustrates two distinct patterns of the steady state (SS; red filled circles) of the dynamical system in response to a low (m_{IP} = 30, Figure 5A), intermediate (m_{IP}= 50, Figure 5B) and high (m_{IP}= 80, Figure 5C) activation rate of protease (m_{IP}) in the B-A_E phase diagram. A low level of protease activation from the transcription factor (m_{IP} = 30) induces low cytokine levels and high bacterial infection (Figure 5A), while the intermediate or high activation level (m_{IP} = 50, 80) leads to significant immune response and suppressed bacterial activities (Figure 5B,C, respectively) regardless of initial conditions. Figure 5D illustrates two distinct modes in the B-A_E plane as described in (12): (i) the bacterial persist region ($\mathbb{T}_B$) where bacterial growth is enhanced and cytokine levels are suppressed; (ii) the immune boosting zone ($\mathbb{T}_I$ where the extracellular cytokine levels are increased and bacterial activities are inhibited. In Figure 5E–G, we show the time courses of bacteria density (B; red) and the concentrations of protease (P, pink), transcription factor (A_I; green) and cytokines (A_E, blue) in response to three protease activation rates from the transcription factor (m_{IP} = 30, 50, 80) corresponding to Figure 5A–C, respectively, with the initial condition $B(0) = 1.7, P(0) = 0.1, A_I(0) = 0.1, A_E(0) = 0.1$.

In the next section, we investigate the dynamics of the competition M2 model.

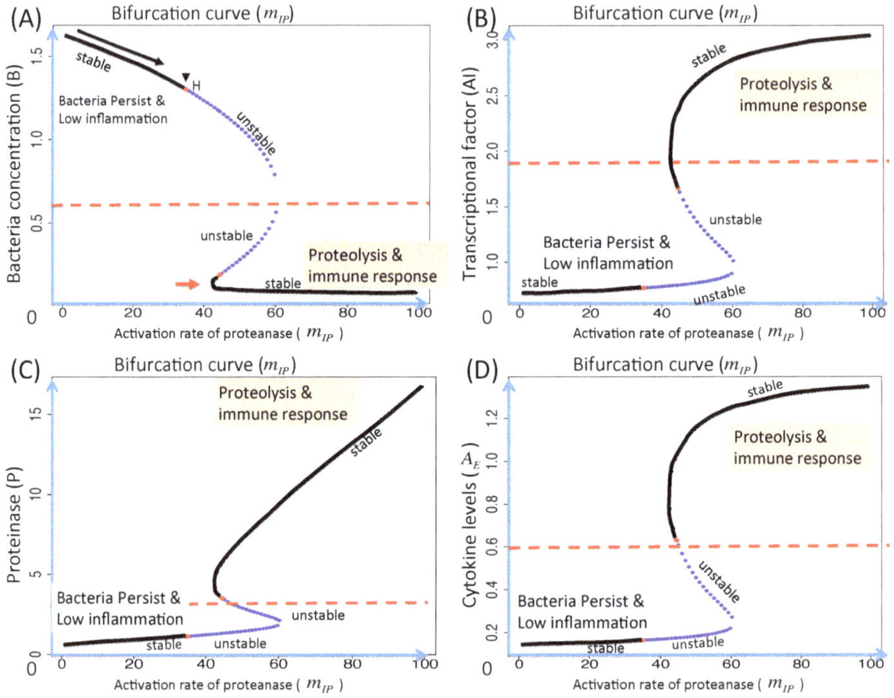

Figure 4. Bifurcation curves on the M1 model (2). (**A**) The B-m_{IP} hysteresis loop: bacterial growth is active when m_{IP} varies in the upper stable branch and suppressed when m_{IP} varies in the lower stable branch. We define the bacterial persistence types by $B > th_B$ and the immune boosting region by $B < th_B$ and take $th_B = 0.7$. As m_{IP} is increased from a low value (black arrow) in the upper branch, the system loses stability at a Hopf bifurcation point (black arrowhead, marked with 'H'). (**B–D**) The corresponding hysteresis loops for intracellular transcription factors (A_I), protease level (P) and extracellular cytokines (A_E). Other parameters are fixed as in Table 1. Black = stable, blue = unstable, red dots = Hopf bifurcation point.

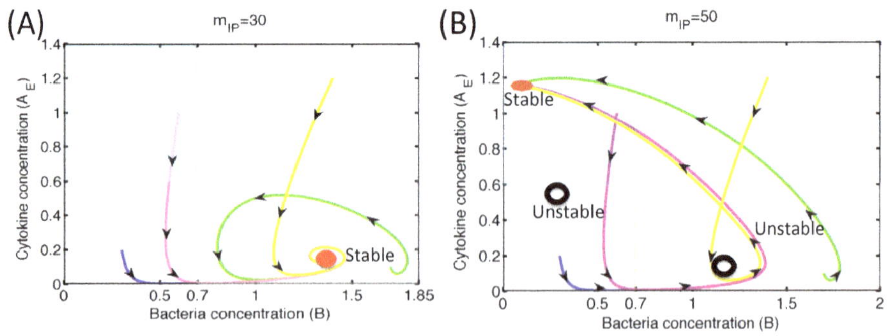

Figure 5. *Cont.*

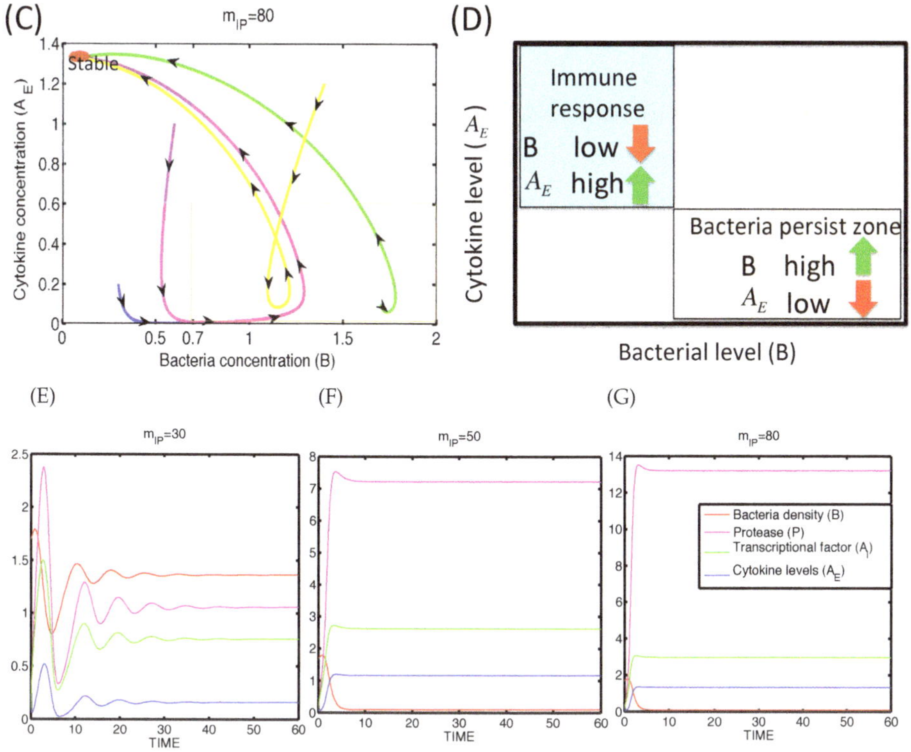

Figure 5. Dynamics of the M1 model (2). (**A–C**) Dynamics of the M1 system in the B-A_E phase plane to a low ($m_{IP} = 30$; (A)), intermediate ($m_{IP} = 50$; (B)) and high ($m_{IP} = 80$; (C)) activation rate (m_{IP}) of protease by the transcription factor. * Filled red circles in (A–C) = stable steady state (S.S.), empty black circle in (B) = unstable S.S. (**D**) A schematic of two adaptive types of bacterial infection systems (bacteria persist ($\mathbb{T}_B$) and immune boosting ($\mathbb{T}_I$) systems): $\mathbb{T}_B = \{(B, A_E) \in \mathbb{R}^2 : B > th_B,\ 0 \le A_E < th_{AE}\}$, $\mathbb{T}_I = \{(B, A_E) \in \mathbb{R}^2 : 0 \le B < th_B,\ A_E > th_{AE}\}$. All other parameters are fixed as in Table 1. (**E–G**) Time courses of the main variables (B, P, A_I, A_E) for various activation rates ($m_{IP} = 30, 50, 80$) with the initial condition: $B(0) = 1.7, P(0) = 0.1, A_I(0) = 0.1, A_E(0) = 0.1$. All other parameters are fixed as in Table 1.

3.2. *Dynamics of Two Bacterial Strains Model (M2 Model)*

We first investigate the existence of the equilibria of the M2 model (3)–(5).

3.2.1. Existence of Equilibria

By replacing the variable A_E with I for notational purposes, the M2 model system is given by:

$$
\begin{aligned}
\frac{dB_1}{dt} &= r_1 B_1 \left(1 - \frac{B_1}{K_1} - \alpha_{12} B_2\right) - \gamma B_1 I, \\
\frac{dB_2}{dt} &= r_2 B_2 \left(1 - \frac{B_2}{K_2} - \alpha_{21} B_1\right) - \gamma B_2 I, \qquad (13) \\
\frac{d}{dt} I(t) &= \beta_1 B_1 + \beta_2 B_2 - \delta I.
\end{aligned}
$$

where $\alpha_{11} = \frac{1}{K_1}$, $\alpha_{22} = \frac{1}{K_2}$. There are four types of equilibria of System (13). $E_0 = (0,0,0)$ is a trivial equilibrium representing that neither bacteria nor the immune response exist. Let $E_1 = (\bar{B}_1, 0, \bar{I}_1)$ and $E_2 = (0, \bar{B}_2, \bar{I}_2)$ denote dominant equilibria in which either B_1 or B_2 exists. The explicit values of each component of E_1 and E_2 are given as follows, respectively.

$$\begin{aligned} E_1 &= \left(\frac{K_1 r_1 \delta}{\delta r_1 + \gamma \beta_1 K_1}, 0, \frac{K_1 r_1 \beta_1}{\delta r_1 + \gamma \beta_1 K_1} \right), \\ E_2 &= \left(0, \frac{K_2 r_2 \delta}{\delta r_2 + \gamma \beta_2 K_2}, \frac{K_2 r_2 \beta_2}{\delta r_2 + \gamma \beta_2 K_2} \right). \end{aligned} \tag{14}$$

Let $E_+ := (B_1^*, B_2^2, I^*)$ denote a positive equilibrium representing the coexistence of two bacterial species under the pressure of the immune response. It follows from the third equation of (13) that I^* is given by:

$$I^* = \frac{\beta_1}{\delta} B_1^* + \frac{\beta_2}{\delta} B_2^*. \tag{15}$$

By substituting the explicit value of I^* into the first and second equations of (13), we obtain the following linear system of equations with respect to B_1^* and B_2^*:

$$\begin{pmatrix} r_1\delta + \gamma\beta_1 K_1 & K_1(r_1\alpha_{12}\delta + \gamma\beta_2) \\ K_2(r_2\alpha_{21}\delta + \gamma\beta_1) & r_2\delta + \gamma\beta_2 K_2 \end{pmatrix} \begin{pmatrix} B_1 \\ B_2 \end{pmatrix} = \begin{pmatrix} r_1 K_1 \delta \\ r_2 K_2 \delta \end{pmatrix} \tag{16}$$

Hence, the explicit values of each component of E_+ are given by:

$$\begin{aligned} B_1^* &= \frac{K_1\delta\{r_1 r_2(1 - K_2\alpha_{12})\delta + \gamma\beta_2 K_2(r_1 - r_2)\}}{D_0}, \\ B_2^* &= \frac{K_2\delta\{r_1 r_2(1 - K_1\alpha_{21})\delta + \gamma\beta_1 K_1(r_2 - r_1)\}}{D_0}, \\ I^* &= \frac{r_1 r_2\delta\{\beta_1 K_1(1 - \alpha_{12}K_2) + \beta_2 K_2(1 - \alpha_{21}K_1)\}}{D_0}, \end{aligned} \tag{17}$$

where D_0 is given by:

$$\begin{aligned} D_0 :=& (r_1\delta + \gamma\beta_1 K_1)(r_2\delta + \gamma\beta_2 K_2) \\ & - K_1 K_2 (r_1\alpha_{12}\delta + \gamma\beta_2)(r_2\alpha_{21}\delta + \gamma\beta_1). \end{aligned} \tag{18}$$

For equilibrium E_+ to be a positive equilibrium requires $B_1^* > 0$ and $B_2^* > 0$. Define matrix $A = \{a_{ij}\}$ and vector $\mathbf{b} = (b_1, b_2)^T$ $(i,j = 1,2)$, such that:

$$A \begin{pmatrix} B_1 \\ B_2 \end{pmatrix} = \mathbf{b}. \tag{19}$$

Then, $a_{11} = r_1\delta + \gamma\beta_1 K_1$, $a_{12} = K_1(r_1\alpha_{12}\delta + \gamma\beta_2)$, $a_{21} = K_2(r_2\alpha_{21}\delta + \gamma\beta_1)$, $a_{22} = r_2\delta + \gamma\beta_2 K_2$, $b_1 = r_1 K_1 \delta$ and $b_2 = r_2 K_2 \delta$. An internal equilibrium exists if and only if:

$$\frac{b_2}{a_{22}} < \frac{b_1}{a_{12}} \text{ and } \frac{b_1}{a_{11}} < \frac{b_2}{a_{21}} \tag{20}$$

or:

$$\frac{b_2}{a_{22}} > \frac{b_1}{a_{12}} \text{ and } \frac{b_1}{a_{11}} > \frac{b_2}{a_{21}} \tag{21}$$

Note that (20) and (21) are equivalent to:

$$\begin{aligned} & r_1 r_2 (K_2\alpha_{12} - 1)\delta - \gamma\beta_2 K_2 (r_1 - r_2) < 0 \text{ and} \\ & r_1 r_2 (K_1\alpha_{21} - 1)\delta - \gamma\beta_1 K_1 (r_2 - r_1) < 0 \end{aligned} \tag{22}$$

and:

$$r_1 r_2 (K_2\alpha_{12} - 1)\delta - \gamma\beta_2 K_2 (r_1 - r_2) > 0 \text{ and}$$
$$r_1 r_2 (K_1\alpha_{21} - 1)\delta - \gamma\beta_1 K_1 (r_2 - r_1) > 0, \tag{23}$$

respectively. Define D_1, D_2, w_1 and w_2 by:

$$D_1 := K_1\alpha_{21} - 1, \;\; D_2 := K_2\alpha_{12} - 1. \tag{24}$$

$$w_1 = \frac{\gamma\beta_1 K_1 (r_2 - r_1)}{r_1 r_2 D_1}, \;\; w_2 := \frac{\gamma\beta_2 K_2 (r_1 - r_2)}{r_1 r_2 D_2}. \tag{25}$$

Finally, we consider the conditions for the existence of E_+. It follows from (22) and (23) that $E_+ \in \mathbb{R}^3_+$ if and only if:

$$r_1 r_2 D_1 \delta + \gamma\beta_1 K_1 (r_1 - r_2) < 0 \text{ and } r_1 r_2 D_2 \delta + \gamma\beta_2 K_2 (r_2 - r_1) < 0 \tag{26}$$

or:

$$r_1 r_2 D_1 \delta + \gamma\beta_1 K_1 (r_1 - r_2) > 0 \text{ and } r_1 r_2 D_2 \delta + \gamma\beta_2 K_2 (r_2 - r_1) > 0. \tag{27}$$

Note that the stability conditions of E_1 and E_2 are given by (34) and (37), respectively (see the next subsection for details). Conditions (34) and (37) are mutually exclusive with (26), but are identical to (27). In other words, coexistent equilibrium E_+ exists only if both E_1 and E_2 are unstable or locally stable. In the later case, the system would be expected to exhibit bistability between E_1 and E_2.

In summary, the existence conditions of internal equilibrium E_+ are classified in Tables 2 and 3 according to the sign of $r_1 - r_2$ and $w_1 - w_2$.

Table 2. Existence condition of E_+ when $r_1 > r_2$.

Case	$w_1 < w_2$	$w_2 < w_1$
$D_1 > 0$ & $D_2 > 0$	$w_2 < \delta$	nonexistence
$D_1 < 0$ & $D_2 > 0$	$w_1 < \delta < w_2$	$w_2 < \delta < w_1$
$D_1 > 0$ & $D_2 < 0$	nonexistence	nonexistence
$D_1 < 0$ & $D_2 < 0$	nonexistence	$w_1 < \delta$

Table 3. Existence condition of E_+ when $r_1 < r_2$.

Case	$w_1 < w_2$	$w_2 < w_1$
$D_1 > 0$ & $D_2 > 0$	nonexistence	$w_1 < \delta$
$D_1 < 0$ & $D_2 > 0$	nonexistence	nonexistence
$D_1 > 0$ & $D_2 < 0$	$w_1 < \delta < w_2$	$w_2 < \delta < w_1$
$D_1 < 0$ & $D_2 < 0$	$w_2 < \delta$	nonexistence

In the next section, we check the stability of the equilibria of the M2 model (13).

3.2.2. Stability of Equilibria

Mathematical conditions for local stability of equilibria are derived based on the linearized equations around any of the equilibria $E_\circ = (B_1^\circ, B_2^\circ, I^\circ)$. The Jacobian matrix for $E_\circ$ is given by:

$$J(E_\circ) = \begin{pmatrix} r_1 - \frac{2r_1 B_1^\circ}{K_1} - r_1\alpha_{12} B_2^\circ - \gamma I^\circ & -r_1\alpha_{12} B_1^\circ & -\gamma B_1^\circ \\ -r_2\alpha_{21} B_2^\circ & r_2 - \frac{2B_2^\circ r_2}{K_2} - r_2\alpha_{21} B_1^\circ - \gamma I^\circ & -\gamma B_2^\circ \\ \beta_1 & \beta_2 & -\delta \end{pmatrix}. \tag{28}$$

The Jacobi matrix for E_0 is given by:

$$J(E_0) = \begin{pmatrix} r_1 & 0 & 0 \\ 0 & r_2 & 0 \\ \beta_1 & \beta_2 & -\delta \end{pmatrix}. \tag{29}$$

Since $r_1 > 0$ and $r_2 > 0$, E_0 is always unstable.
The Jacobi matrix for E_1 is given by:

$$J(E_1) = \begin{pmatrix} -\frac{r_1{}^2\delta}{r_1\delta+\gamma\beta_1K_1} & -\frac{r_1{}^2\alpha_{12}K_1\delta}{r_1\delta+\gamma\beta_1K_1} & -\frac{\gamma K_1 r_1\delta}{r_1\delta+\gamma\beta_1K_1} \\ 0 & \frac{r_1r_2(1-\alpha_{21}K_1)\delta+\gamma\beta_1K_1(r_2-r_1)}{r_1\delta+\gamma\beta_1K_1} & 0 \\ \beta_1 & \beta_2 & -\delta \end{pmatrix}. \tag{30}$$

Characteristic equation $P_1(\lambda) = 0$ defined for $J(E_1)$ is given by:

$$P_1(\lambda) = \left\{\lambda - \frac{r_1r_2(1-\alpha_{21}K_1)\delta+\gamma\beta_1K_1(r_2-r_1)}{r_1\delta+\gamma\beta_1K_1}\right\} \begin{vmatrix} \lambda + \frac{r_1{}^2\delta}{r_1\delta+\gamma\beta_1K_1} & \frac{\gamma K_1r_1\delta}{r_1\delta+\gamma\beta_1K_1} \\ -\beta_1 & \lambda+\delta \end{vmatrix}. \tag{31}$$

Let $A_1(\lambda)$ be defined by:

$$A_1 = \begin{vmatrix} -\frac{r_1{}^2\delta}{r_1\delta+\gamma\beta_1K_1} & -\frac{\gamma K_1r_1\delta}{r_1\delta+\gamma\beta_1K_1} \\ \beta_1 & -\delta \end{vmatrix}. \tag{32}$$

Note that the trace and determinant of A_1 satisfy $\mathrm{tr}(A_1) < 0$ and $\det(A_1) > 0$. Hence, E_1 is locally asymptotically stable if:

$$r_1r_2(1-\alpha_{21}K_1)\delta + \gamma\beta_1K_1(r_2-r_1) < 0. \tag{33}$$

Note that (33) is rewritten as:

$$r_1r_2D_1\delta + \gamma\beta_1K_1(r_1-r_2) > 0. \tag{34}$$

In other words, (34) is equivalent to one of the following two conditions:

(E_1S_1) $r_1 > r_2$ and $D_1 > 0$,
(E_1S_2) $r_2 > r_1$, $D_1 > 0$ and $0 < w_1 < \delta$.

Stability conditions of E_2 are derived from the Jacobi matrix for E_2:

$$J(E_2) = \begin{pmatrix} \frac{r_1r_2(1-\alpha_{12}K_2)\delta+\gamma\beta_2K_2(r_1-r_2)}{r_2\delta+\gamma\beta_2K_2} & 0 & 0 \\ -\frac{r_2{}^2\alpha_{21}K_2\delta}{r_2\delta+\gamma\beta_2K_2} & -\frac{r_2{}^2\delta}{r_2\delta+\gamma\beta_2K_2} & -\frac{\gamma K_2r_2\delta}{r_2\delta+\gamma\beta_2K_2} \\ \beta_1 & \beta_2 & -\delta \end{pmatrix}. \tag{35}$$

In a similar way to E_1, E_2 is locally asymptotically stable if:

$$r_1r_2(1-\alpha_{12}K_2)\delta + \gamma\beta_2K_2(r_1-r_2) < 0. \tag{36}$$

Note that (36) is rewritten as:

$$r_1r_2D_2\delta + \gamma\beta_2K_2(r_2-r_1) > 0. \tag{37}$$

In other words, (37) is equivalent to one of the following two conditions:

(E_2S_1) $r_2 > r_1$ and $D_2 > 0$,
(E_2S_2) $r_1 > r_2$, $D_2 > 0$ and $0 < w_2 < \delta$.

We note that the system of differential equations for bacterial strains without any anti-microbial killing and recruitment and with the same growth rates of bacteria ($r_1 = r_2 = r_B$):

$$\begin{aligned}\frac{dB_1}{dt} =& r_B B_1 (1 - \alpha_{11} B_1 - \alpha_{12} B_2)\\ \frac{dB_2}{dt} =& r_B B_2 (1 - \alpha_{21} B_1 - \alpha_{22} B_2)\end{aligned} \tag{38}$$

reduces to the classical two-dimensional Lotka–Volterra competition model. If we assume that the magnitude of inter-specific competition is stronger than intra-specific competition, i.e.,

$$\alpha_{11} < \alpha_{21} \text{ and } \alpha_{22} < \alpha_{12}, \tag{39}$$

then a unique positive equilibrium of (38) exists, and it is unstable. It can be shown that solutions converge to either $(1/\alpha_{11}, 0)$ or $(0, 1/\alpha_{22})$ depending on the choice of the initial state.

In the next section, we investigate the dynamics of the competition model (M2) and immune system.

3.2.3. Dynamics of the Competition Model M2 in Response to the Immune System

We shall define three adaptive types of the competition system (harmful bacteria-persist ($\mathbb{T}_B$), harmful bacteria-free ($\mathbb{T}_F$) and co-existence ($\mathbb{T}_c$)) corresponding to regions in the B_1-B_2 plane, including equilibria points (E_1, E_2, E_+) discussed in the previous section:

$$\begin{aligned}\mathbb{T}_B :=& \{(B_1, B_2) \in \mathbb{R}^2 : B_1 > 0,\ \ B_2 = 0\}, (\text{corresponding to } E_1)\\ \mathbb{T}_F =& \{(B_1, B_2) \in \mathbb{R}^2 : B_1 = 0,\ \ B_2 > 0\}, (\text{corresponding to } E_2)\\ \mathbb{T}_c =& \{(B_1, B_2) \in \mathbb{R}^2 : B_1 > 0,\ \ B_2 > 0\}, (\text{corresponding to } E_+)\end{aligned} \tag{40}$$

The basic parameter set for the M2 model is given in Table 4.

Table 4. Parameters used in the M2 model.

Parameter	Description	Type I	Type II
	Inter- and intra-competition		
a_{11}	Inter-specific competition coefficient	$1/a_{11} = K_1 = 0.5$	$K_1 = 2.0$
a_{22}	Inter-specific competition coefficient	$1/a_{22} = K_2 = 2.0$	$K_2 = 1.5$
a_{12}	Intra-specific competition coefficient	1.0	1.0
a_{21}	Intra-specific competition coefficient	1.0	1.0
	Activation/production rates		
r_1	Growth rate of harmful bacteria	1.5	1.5
r_2	Growth rate of good bacteria	1.0	1.0
β_1	activation of cytokines by harmful bacteria	0.1	1.0
β_2	activation of cytokines by good bacteria	1.0	0.1
	Inhibition/decay Rates		
γ	Per capita elimination rate of bacteria	1.0	1.0
δ	Decay rate of cytokines	0.35	0.25

In Figure 6, we investigate the dynamics of the two-species model (3)–(5) in the presence of the immune response for a base parameter set (Type I). $r_1 = 1.5$, $r_2 = 1.0$, $K_1 = 0.5$, $K_2 = 2.0$, $\alpha_{12} = 1.0$, $\alpha_{21} = 1.0$, $\beta_1 = 0.1$, $\beta_2 = 1.0$, $\gamma = 1.0$. Analysis in Section 3.2.2 indicates that: (i) E_2 is stable if $\delta > 0.66666$, whereas E_1 is stable if $\delta < 0.03333$; and (ii) E_+ is expected to be stable if $0.03333 < \delta < 0.66666$. Figure 6A–C shows the trajectories $(B_1(t), B_2(t))$ of harmful and good bacteria populations for various decay rates of cytokines (δ = 0.033 (Figure 6A), 0.35 (Figure 6B) and

0.7 (Figure 6C)) with four initial conditions: $B_1(0) = 0.07, B_2(0) = 0.45$ (yellow curve); $B_1(0) = 0.38$, $B_2(0) = 0.58$ (green curve); $B_1(0) = 0.05, B_2(0) = 0.1$ (blue curve); $B_1(0) = 0.35, B_2(0) = 0.02$ (purple curve). The initial condition of the immune response was set to be zero ($A_E(0) = 0$) for all cases. Figure 6E–F shows the trajectories of $(B_1(t), A_E(t))$ corresponding to Figure 6A–C, respectively. When the decay rate of cytokines is small ($\delta = 0.033$), the system converges to E_1 equilibrium where good bacteria are cleared out and harmful bacteria survive in a battle via the immune system (Figure 6A). Initial strong immune responses (black arrow in Figure 6E) due to the weak clearance rate ($\delta \ll 1$) significantly eliminate both harmful and good bacterial populations (black arrowhead in Figure 6A). While the good bacteria are totally eradicated by this strong immune response due to the relatively low growth rate, the harmful bacteria survive due to the higher growth rate and winning the competition battle with the good ones (blue arrow in Figure 6A). The system adapts a harmful bacteria-free equilibrium when δ is large ($\delta = 0.7$ in Figure 6C). The strong clearance of immune activities in the system leads to relatively weak immune responses (black arrow in Figure 6G). This increases the chances of winning the competition for good bacteria and decreases the harmful bacteria population, pushing the system $(B_1(t), B_2(t))$ in the upper left corner (black arrowhead in Figure 6C). Then, the system converges to the E_2 equilibrium, the attractor (red filled circle in Figure 6C), where harmful bacteria are eradicated by the immune system and the helpful bacteria persist. On the other hand, an intermediate immune response ($\delta = 0.3$ in Figure 6B) leads to the co-existence of good and harmful bacteria populations. The immune system initially successfully attacks and decreases the number of both harmful and good bacteria, but this also reduces the activation of the immune system (cf. Equation (5)). The reduced immune activity also increases the chance of the regrowth of both bacterial types (arrowhead in Figure 6B), leading to the coexistence of those two bacterial populations (red filled circle in Figure 6B), corresponding to E_+ equilibrium. In response to small ($\delta = 0.0033$), intermediate ($\delta = 0.35$) and large ($\delta = 0.7$) decay rates of cytokines, the system transits from harmful bacteria-persist region ($\mathbb{T}_B$) to the co-existence zone ($\mathbb{T}_c$) and then to the harmful bacteria-free ($\mathbb{T}_F$) region (yellow curved arrow). Dysbiosis, or bacterial imbalance, represents a state of reduced species diversity with the emergence of a few extraordinary highly abundant species. Dysbiosis is broadly observed for several microbial ecologies, including in aquatic systems or intestinal systems. It has been shown to be associated with illnesses, such as cancer [30,31], bacterial vaginosis [32], inflammatory bowel disease [33,34], chronic fatigue syndrome [35], obesity [36,37] and colitis [38]. Dysbiosis in the gut is known to be associated with major chronic inflammation states [39]. Dysbiosis caused by the imbalance of the skin commensal bacterial species composition has been reported [40,41]. Importantly, dysbiosis characterized by the increase of *S. aureus* has been shown to be associated with atopic dermatitis, one of the major skin inflammatory diseases [7]. It is suggested that dysbiosis and *Staphylococcus aureus* colonization drive inflammation in atopic dermatitis [7]. In our model, different immune reactions from the weak, intermediate and strong clearance strength (δ) of extracellular cytokines, such as IL-4, IL-12 and IFN-γ, result in the imbalance between harmful and good bacteria, co-existence or healthy tissue homeostasis (Figure 6D). In particular, increased levels of harmful bacteria and reduced levels of the beneficial bacteria, i.e., dysbiosis, may be induced when the clearance of the immune reactions is weak ($\delta = 0.033$; Figure 6A). The tipping point of the balance between beneficial and harmful bacteria is the vigorous and subtle competition between those different kinds of bacteria.

Figure 7A shows the steady state of harmful bacteria (B_1^*) at three equilibria (E_1 red; E_2 blue; E_+ black) as a function of δ. Solid and dotted curves illustrate the stable and unstable branches at E_1, E_2, E_+ for the continuous spectrum of δ. Small, intermediate and large values of δ lead to harmful bacteria persisting, co-existence and good bacteria persisting regimes, respectively, as shown in Figure 6A–C. Figure 7B illustrates how the system transits from $\mathbb{T}_B$ to $\mathbb{T}_c$ and then to $\mathbb{T}_F$ in response to an increase in δ by following the stable branches of B_1^* at E_1, E_+ and E_2, respectively, in Figure 7A. This illustrates how phase transitions ($\mathbb{T}_B \to \mathbb{T}_c \to \mathbb{T}_F$) in Figure 6D can be induced by the continuous increase in the clearance rate of the immune system.

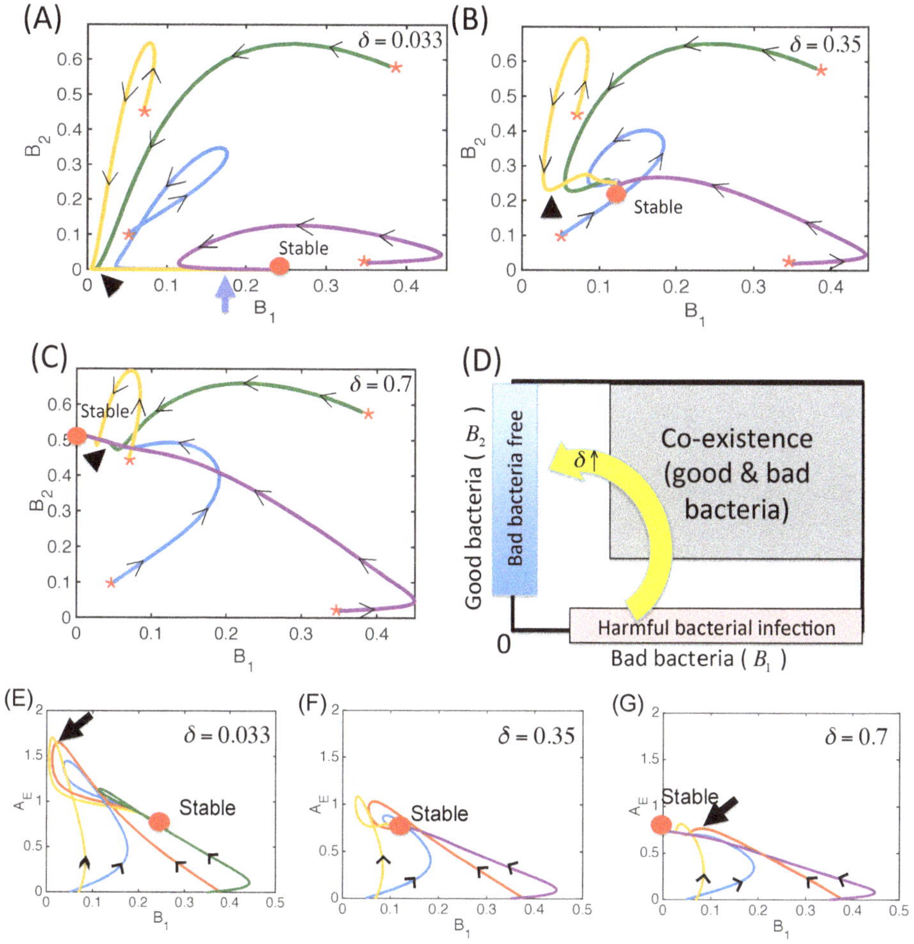

Figure 6. Co-existence and dynamics of harmful and good bacteria in response to various cytokine clearance levels (δ) in the competition M2 model (3)–(5). (**A–C**) Trajectories $(B_1(t), B_2(t))$ of bacterial populations for various decay rates of cytokines (δ = 0.033 (**A**), 0.35 (**B**), 0.7 (**C**)) with four initial conditions; (**D**) characterization of the dynamical system in the B_1-B_2 plane. There exist three phenotypic regions: harmful bacteria-persist ($\mathbb{T}_B$, lower pink box near x-axis), harmful bacteria-free ($\mathbb{T}_F$, left blue box near y-axis) and co-existence ($\mathbb{T}_c$, gray box in the center) regions. As δ is increased (δ = 0.0033 $\rightarrow$ 0.35 $\rightarrow$ 0.7), the system undergoes the transition from $\mathbb{T}_B$ to $\mathbb{T}_c$ and then to $\mathbb{T}_F$. (**E–G**) Trajectories of $(B_1(t), A_E(t))$ corresponding to (A–C), respectively. All other parameters are fixed as in Table 4 (Type I).

In Figure 8, we investigate the dynamics of the bi-stable competition system with the parameter set (Type II): $r_1 = 1.5$, $r_2 = 1.0$, $K_1 = 2.0$, $K_2 = 1.5$, $\alpha_{12} = 1.0$, $\alpha_{21} = 1.0$, $\beta_1 = 1.0, \beta_2 = 0.1$, $\gamma = 1.0$. Analysis in Section 3.2.2 indicates that E_1 is stable if $\delta > 0.1$, whereas E_2 is stable if $\delta > 0$. In comparison to the previous case in Figure 6, the system does not present the co-existence region ($\mathbb{T}_c$). Figure 8A shows the regions of harmful bacteria-persist ($\mathbb{T}_B$) and bacteria-free ($\mathbb{T}_F$) in the $B_1 - B_2$ phase-plane. While initial states (B_1, B_2) in the region R_B (red) converge to the harmful bacteria-persist equilibrium ($\mathbb{T}_B$, red circle), the initial states in the upper-left region (R_F; blue) converge to the harmful bacteria-free equilibrium ($\mathbb{T}_F$; red circle). For example, curves indicate the trajectories

$(B_1(t), B_2(t))$ for two very close initial conditions: $B_1(0) = 0.15$, $B_2(0) = 0.713$ (blue curve) and $B_1(0) = 0.15$, $B_2(0) = 0.71$ (red curve) near the boundary (green dotted curve) between R_B and R_F. The initial condition of cytokines was set to be zero ($A_E(0) = 0$). $E_1 = (0.9375, 0), E_2 = (0, 0.3158)$. Figure 8B,C shows the trajectories in the $B_1 - A_E$ plane (Figure 8B) and time courses (Figure 8C) of bacterial populations (B_1, B_2) and cytokine level (A_E) for two very close initial conditions in Figure 8A: $B_1(0) = 0.15$, $B_2(0) = 0.713$ (dotted curves) and $B_1(0) = 0.15$, $B_2(0) = 0.71$ (solid curves). Figure 8D shows the bi-stable nature of the dynamical system where one kind of the harmful or beneficial bacteria dies out and the other kind survives depending on the initial state $B_1(0)$, $B_2(0)$. Therefore, the initial status of exposure to both harmful and beneficial bacteria determines the dysbiosis or healthy tissue.

In the next section, we investigate the effect of time delays in immune responses on the dynamics of the system.

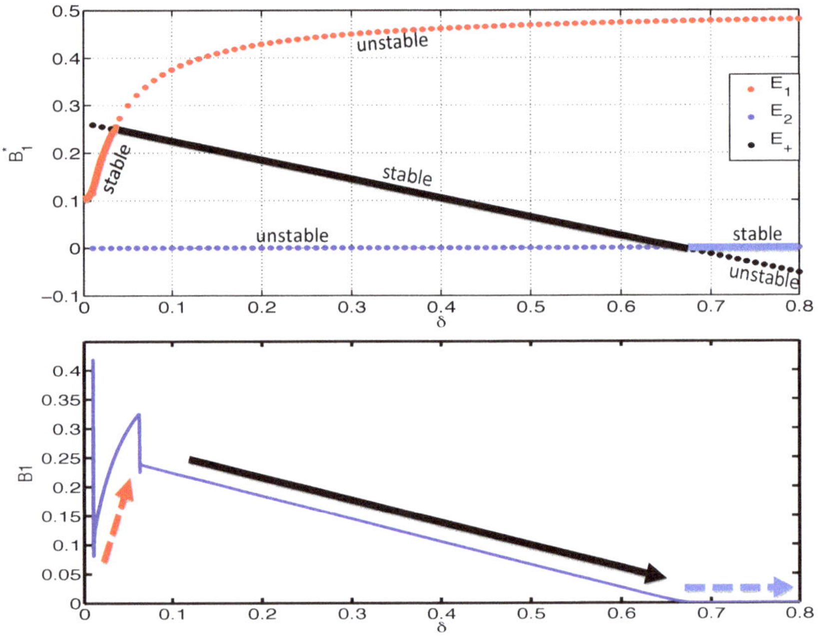

Figure 7. (**A**) Steady state solutions of bad bacteria (B_1^*) as a function of δ corresponding to E_1 (red), E_2 (blue) and E_+ (black) equilibria, respectively. Solid curve = stable, dotted curve = unstable. Green arrows indicate the points $\delta = 0.0033, 0.35, 0.7$ corresponding to Figure 6A–C, respectively. (**B**) Trajectories of $B_1(t)$ in a $B_1 - \delta$ plane when δ is a monotonic increasing function of time, satisfying $\frac{d\delta}{dt} = 0.00002$ with the initial condition $B_1(0) = 0.12, B_2(0) = 0.01, A_E(0) = 0, \delta(0) = 0.01$ near the stable equilibrium point $E_1 = E_1(\delta = 0.01)$. As δ is increased, the system sequentially follows the stable branches in (A), leading to the transition from stable E_1 branch (red solid curve in (A)) to stable E_+ branch (black solid curve in (A)) and then to stable E_2 branch (blue solid curve in (A)). All other parameters are fixed as in Table 4 (Type I).

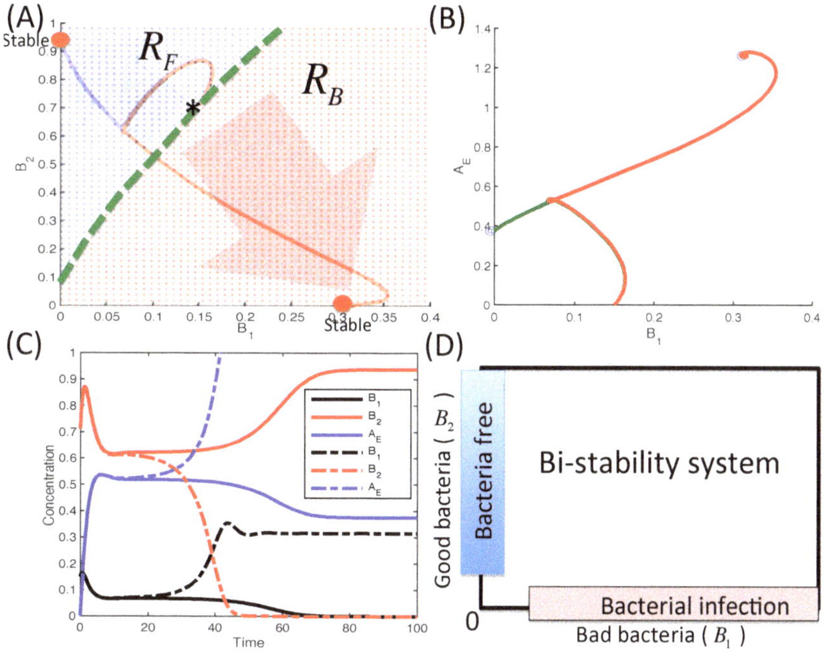

Figure 8. Bi-stability dynamics of harmful and beneficial bacteria in response to the immune system in the two-species M2 model (3)–(5). (**A**) Dynamics of harmful bacteria-persist ($\mathbb{T}_B$) and bacteria-free ($\mathbb{T}_F$) in the B_1-B_2 phase-plane. While initial states (B_1, B_2) in the region R_B (red) converge to the harmful bacteria-persist equilibrium ($\mathbb{T}_B$), the initial states in the upper-left region (R_F; blue) converge to the harmful bacteria-free equilibrium ($\mathbb{T}_F$). Blue and red curves indicate the trajectories $(B_1(t), B_2(t))$ for two very close initial conditions: $B_1(0) = 0.15$, $B_2(0) = 0.713$ (blue curve) and $B_1(0) = 0.15$, $B_2(0) = 0.71$ (red curve) near the asterisk on the boundary (green dotted curve) between R_B and R_F. $A_E(0) = 0$. * Filled red circle in (A,**B**) = stable S.S.: $E_1 = (0.9375, 0), E_2 = (0, 0.3158)$. (B,C) Trajectories $(B_1(t), A_E(t))$ and time courses (C) of bacterial populations (B_1, B_2) and cytokine level (A_E) for two very close initial conditions in (A): $B_1(0) = 0.15$, $B_2(0) = 0.713$ (dotted curves) and $B_1(0) = 0.15$, $B_2(0) = 0.71$ (solid curves). (**D**) Characterization of the system: the dynamics adapts to the bi-stability system where the dynamical system chooses either harmful or good bacteria persisting tissue based on the initial exposure to those bacterial kinds. All other parameters are fixed as in Table 4 (Type II).

3.2.4. Effect of Time Delays in Immune Response on the Competition System

In this section, we introduce time delays in the immune response for the reduced competition system (3)–(5). The governing equations for the simple model with time delays (τ_1, τ_2) are given by:

$$\frac{d}{dt}B_1(t) = r_1 B_1(t)(1 - \alpha_{11}B_1(t) - \alpha_{12}B_2(t)) - \gamma A_E(t)B_1(t), \tag{41}$$

$$\frac{d}{dt}B_2(t) = r_2 B_2(t)(1 - \alpha_{21}B_1(t) - \alpha_{22}B_2(t)) - \gamma A_E(t)B_2(t), \tag{42}$$

$$\frac{d}{dt}A_E(t) = \beta_1 B_1(t - \tau_1) + \beta_2 B_2(t - \tau_2) - \delta A_E(t), \tag{43}$$

where τ_1, τ_2 are time delays in the immune response for harmful and beneficial bacteria attacks, respectively.

In Figure 9, we investigate the effect of small time delays ($\tau_1 = \tau_2 = 1.5$). In the absence of time delays ($\tau_1 = \tau_2 = 0$), the population of bad bacteria (B_1) converges to zero ($B_1^{s,2} = 0$; blue solid line in Figure 9A), while the population of good bacteria (B_2) and the immune system (A_E) converge to positive equilibria $B_2^{s,2}$ (blue solid line in Figure 9B) and $A_E^{s,2}$ (blue solid line in Figure 9C), respectively. On the other hand, in the presence of time delays ($\tau_1 = \tau_2 = 1.5$), the population of good bacteria (B_2) converges to zero ($B_2^{s,1} = 0$; red dashed line in Figure 9C), while the population of bad bacteria (B_1) and the immune system (A_E) persists with positive equilibria $B_1^{s,1}$ (red dashed line in Figure 9B) and $A_E^{s,1}$ (red dashed line in Figure 9C), respectively. Therefore, an introduction of weak time delays in the system induces a switch from the B2-dominant equilibrium $(0, B_2^{s,2}, A_E^{s,2})$ to the B1-dominant equilibrium $(B_1^{s,1}, 0, A_E^{s,1})$. See Figure 9D,E. The initial condition was $B_1(0) = 0.15$, $B_2(0) = 0.72$, $A_E(0) = 0.0$.

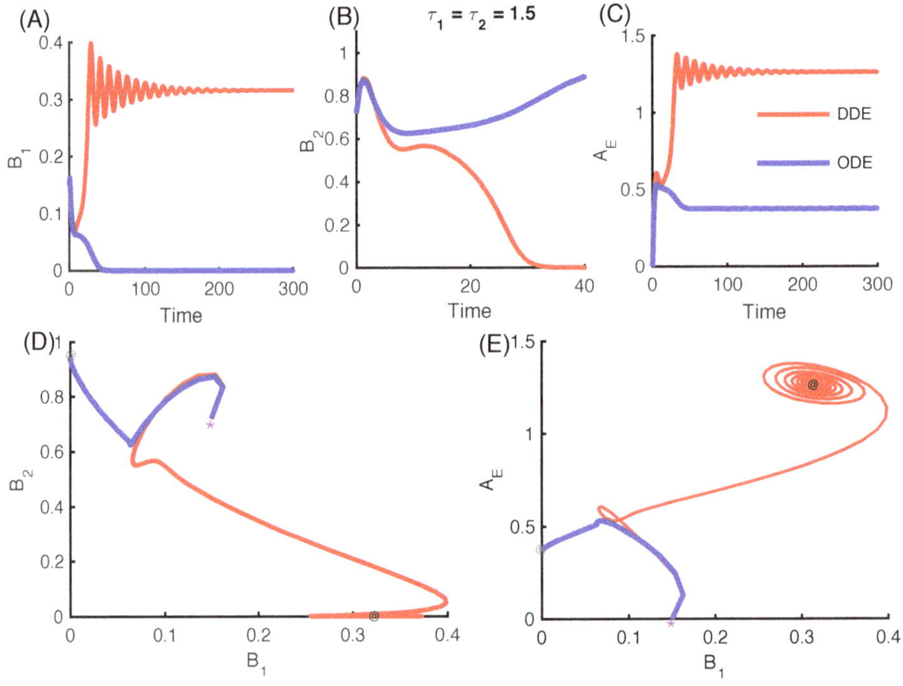

Figure 9. Effect of small time delays ($\tau_1 = \tau_2 = 1.5$) in the competition M2 model (41)–(43). (**A–C**) Time courses of the populations of bad bacteria (B_1 in (A)) and good bacteria (B_2 in (B)) and immune response (A_E in (C)) in the absence (blue solid lines) and presence (red dashed lines) of time delays; (**D**) the corresponding trajectories of B_1 and B_2 in (A,B) in the phase plane; (**E**) the corresponding trajectories of B_1 and A_E in (B,C) in the phase plane. The introduction of weak time delays in the system induces a switch from the B2-dominant equilibrium $(0, B_2^{s,2}, A_E^{s,2})$ to the B1-dominant equilibrium $(B_1^{s,1}, 0, A_E^{s,1})$. Initial condition: $B_1(0) = 0.15$, $B_2(0) = 0.72$, $A_E(0) = 0.0$. All other parameters are fixed as in Table 4 (Type II).

However, for larger time delays, the DDE system completely changes the dynamics. The system induces oscillations in both the population of bad bacteria (B_1; Figure 10A) and the levels of immune cytokines (A_E; Figure 10C) in the presence of larger time delays ($\tau_1 = \tau_2 = 3.5$). The system maintains the extinction of the good bacterial population under this condition (Figure 10B) as in the small time delays in Figure 9. The dynamics of the ODE case is shown in the blue solid curves in Figure 10 and is the same as in Figure 9. This oscillatory behavior of the in vivo pathogens and specific/non-specific immunity was observed in experiments [42,43], and the time delay may have existed in the specific bacterial kinds.

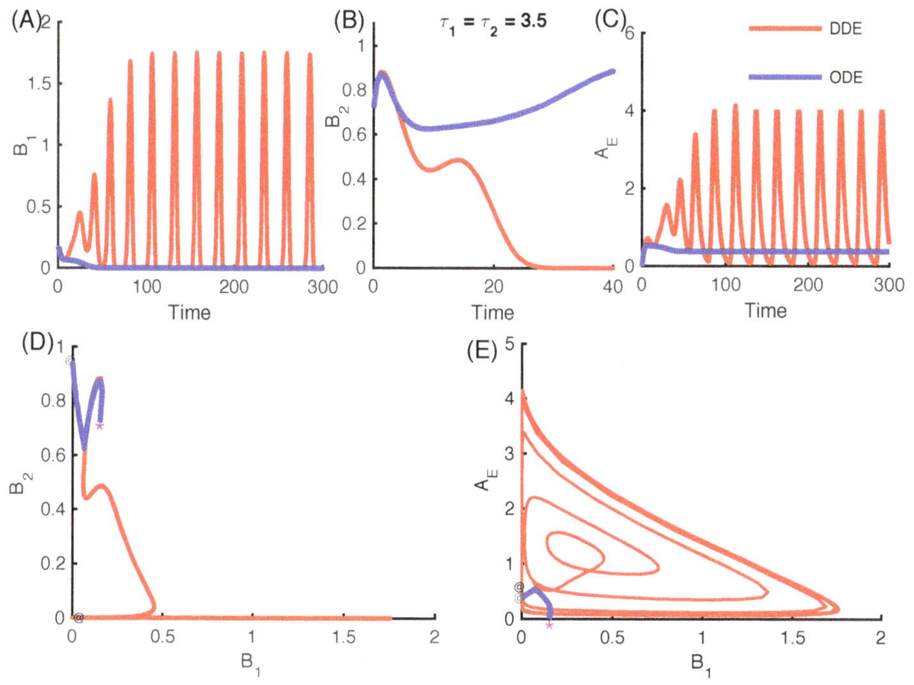

Figure 10. Effect of large time delays ($\tau_1 = \tau_2 = 3.5$) in the competition M2 model (41)–(43). The introduction of time delays in the system induces oscillatory behaviors of both bad bacteria (B_1) and immune cytokines (A_E) and the extinction of good bacteria. (**A–C**) Time courses of the populations of bad bacteria (B_1 in (A)) and good bacteria (B_2 in (B)) and the immune response (A_E in (C)) in the absence (blue solid lines) and presence (red dashed lines) of time delays; (**D**) the corresponding trajectories of B_1 and B_2 in (A,B) in the phase plane; (**E**) the corresponding trajectories of B_1 and A_E in (B,C) in the phase plane; Initial condition: $B_1(0) = 0.15$, $B_2(0) = 0.72$, $A_E(0) = 0.0$. All other parameters are fixed as in Table 4 (Type II).

Our investigation illustrates that the system undergoes dynamical changes as the time delays ($\tau = \tau_1 + \tau_2$) are increased. In the absence ($\tau = 0$) or small values of time delays, the bi-stable system induces either the imbalance state ($B_1^{s,1}, 0, A_E^{s,1}$) or disease-free state ($0, B_2^{s,2}, A_E^{s,2}$). As the time delays are further increased, the system induces the oscillatory behaviors of harmful bacteria and immune activities for some initial conditions ($B_1, B_2, 0$). This indicates that the strength of time delays in either the induction of extracellular cytokines or the manipulation of intracellular signaling pathways may be enough to perturb the bistable pathogen-immune dynamics and leads to the recurrence of the harmful bacteria population.

In the next section, we investigate the local stability of the reduced system of DDEs (41)–(43) and present the onset of Hopf bifurcation.

3.2.5. Local Stability Analysis of Delay Differential Equations

For simplicity, we consider the simplified version of the model for analysis by ignoring the good bacterial dynamics:

$$\begin{cases} \frac{d}{dt}B_1(t) = rB_1(t)(1-\alpha_{11}B_1(t)) - \gamma A_E(t)B_1(t), \\ \frac{d}{dt}A_E(t) = \beta B_1(t-\tau) - \delta A_E(t), \end{cases} \tag{44}$$

Assume that $\delta \gg 1$. Quasi steady state approximation is applied to (44) to obtain two-dimensional system:

$$\frac{d}{dt}B_1(t) = rB_1(t)(1-\alpha_{11}B_1(t) - \beta_{11}B_1(t-\tau)) \tag{45}$$

where β_{11} is explicitly given by:

$$\beta_{11} = \frac{\gamma\beta}{r\delta}. \tag{46}$$

The linearized system of (45), which is defined for $E_1 = (\bar{B}_1, 0)$, is given by:

$$\left\{ \frac{d}{dt}x(t) = -r\bar{B}_1(\alpha_{11}x(t) + \beta_{11}x(t-\tau)) \right. \tag{47}$$

Characteristic equation $P(z) = 0$ defined for (47) is explicitly given by:

$$P(z) = z + r\bar{B}_1\alpha_{11} + r\bar{B}_1\beta_{11}e^{-z\tau} = 0. \tag{48}$$

Let us investigate whether $P(z) = 0$ has a pair of pure imaginary roots $z = \pm i\omega$, where without loss of generality, we can assume that $\omega > 0$. Then, the real and imaginary parts of $P(+i\omega) = 0$ are given by:

$$\begin{aligned} &r\bar{B}_1\alpha_{11} + r\bar{B}_1\beta_{11}\cos\omega\tau = 0, \\ &r\bar{B}_1\beta_{11}\sin\omega\tau = \omega \end{aligned} \tag{49}$$

By adding the square of real and imaginary parts, we obtain that:

$$\omega^2 = (r\bar{B}_1)^2(\beta_{11}^2 - \alpha_{11}^2). \tag{50}$$

The equality in (50) holds if and only if $\beta_{11} > \alpha_{11}$. It follows from the first equation of (49) that:

$$\tau = \frac{1}{\omega}\cos^{-1}\left(-\frac{\alpha_{11}}{\beta_{11}}\right). \tag{51}$$

By numerical computations, critical time delay τ^* at which a system undergoes Hopf bifurcation is determined for the parameter set in Figure 11. More precisely, $\tau^* \simeq 1.418$. Figure 11 illustrates the dynamic changes of stability at a steady state in Equation (45) as the time delay (τ) passes through the critical Hopf bifurcation point $\tau^* \simeq 1.418$. The stable state of the equilibrium for smaller τ's ($\tau = 0.0$ (Figure 11A), 1.0 (Figure 11B) and 1.417 (Figure 11C)) becomes the unstable state around τ^* ($\tau = 1.4181$ in Figure 11C).

3.2.6. Therapeutic Approaches

Results in the previous section indicate that harmful bacteria may not be completely removed by typical immune responses due to recurrence in the presence of time delays, and one has to introduce a drug that enhances the immune system for the eradication of harmful bacteria. We introduce a drug (D) under the following assumptions: (i) drugs enhance the immune activities of extracellular cytokines by inhibiting signaling networks involving the intracellular transcription factors and protease activation; (ii) the drug is administrated with either a constant rate or periodic injection of drug; (iii) drug compounds have a natural decay. The governing equations then are given by:

$$\frac{d}{dt}B_1(t) = r_1 B_1(t)(1 - \alpha_{11}B_1(t) - \alpha_{12}B_2(t)) - \gamma A_E(t)B_1(t), \quad (52)$$

$$\frac{d}{dt}B_2(t) = r_2 B_2(t)(1 - \alpha_{21}B_1(t) - \alpha_{22}B_2(t)) - \gamma A_E(t)B_2(t), \quad (53)$$

$$\frac{d}{dt}A_E(t) = D(t) + \beta_1 B_1(t - \tau_1) + \beta_2 B_2(t - \tau_2) - \delta A_E(t), \quad (54)$$

$$\frac{d}{dt}D(t) = \lambda_D - \delta_D D(t). \quad (55)$$

where τ_1, τ_2 are time delays in the immune response for harmful and beneficial bacteria, respectively, as in the previous section, λ_D is the injection rate of drugs and δ_D is the decay rate of the drug.

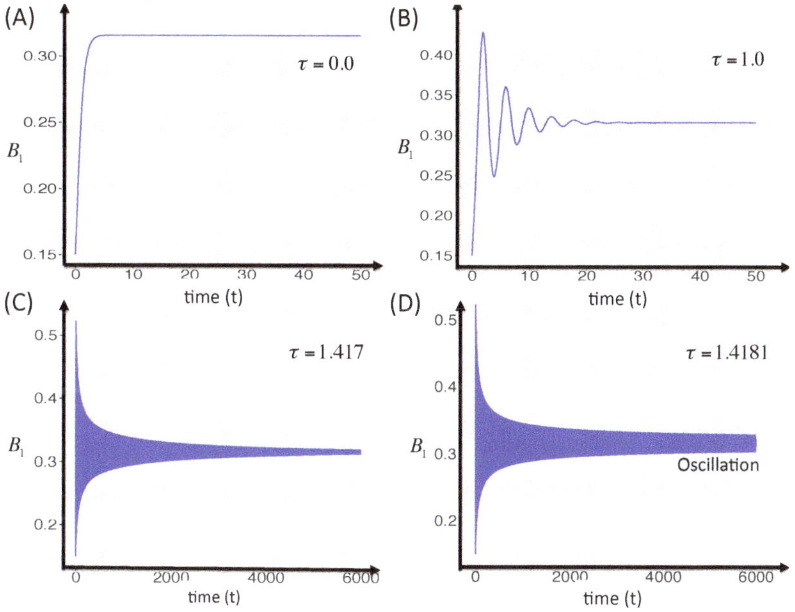

Figure 11. Characterization of the simpler delay differential equation (DDE) immune system (45). **(A–D)** Time courses of harmful bacteria population (B_1) in the absence ($\tau = 0.0$ in (A)) and presence of different time delays ($\tau = 1.0$ in (B); $\tau = 1.417$ in (C), and $\tau = 1.4181$ in (D)). The stable equilibrium becomes unstable as τ passes through the Hopf bifurcation point $\tau^* \simeq 1.418$.

In Figure 12, we investigate the effect of constant drug injection on the dynamics of the DDE system (52)–(55) in the presence of large time delays ($\tau_1 = \tau_2 = 3.5$). Figure 12A–C shows populations of harmful (B_1, red curve) and beneficial (B_2, blue curve) bacteria in the upper panels and the levels of cytokines (A_E, solid black curve) and drugs (D, dotted green curve) in the lower panels for low ($\lambda_D = 0.01$), intermediate ($\lambda_D = 0.03$) and large ($\lambda_D = 0.1$) injection rates, respectively. A small injection ($\lambda_D = 0.01$) of drugs is not so effective to remove the harmful bacteria and still maintain the oscillatory patterns of both harmful bacteria (B_1) and immune cytokines (A_E). This still leads to the extinction of beneficial bacteria ($B_2(t) \rightarrow 0$ as $t \rightarrow \infty$). See Figure 12A. For an intermediate level of injection ($\lambda_D = 0.03$; Figure 12B), drugs annihilate the oscillations of the harmful bacteria population and immune responses, leading to the infection-persist state ($B_2^s = 0,\ B_1^s > 0$). On the other hand, a large amount of drug injection ($\lambda_D = 0.1$; Figure 12C) significantly enhances the immune activity and eliminates both harmful and beneficial pathogens ($B_1^s = 0,\ B_2^s = 0$). Figure 12D–E shows the trajectories in the B_1-B_2 (Figure 12D) and B_1-A_E (Figures 12E) planes, respectively, for the corresponding λ_D's (λ_D = 0.01 (red), 0.03 (blue) and 0.1 (black)).

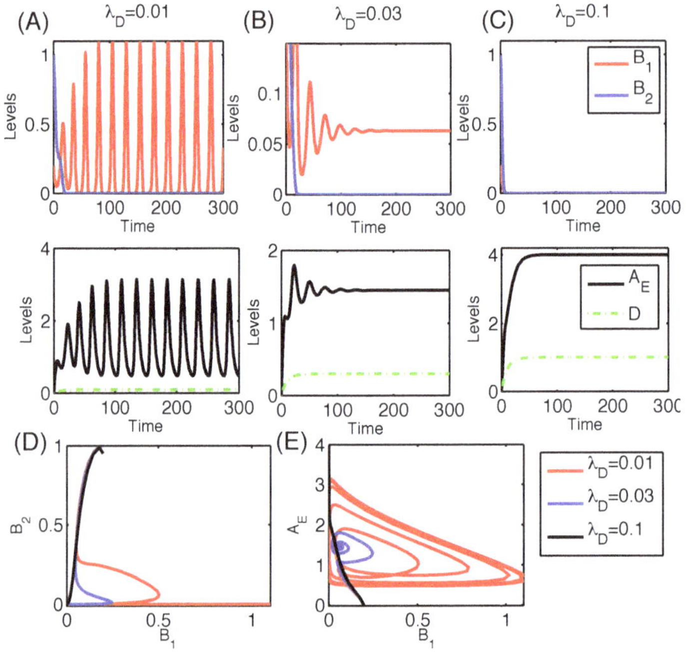

Figure 12. Dynamics of system in response to constant drug injection in the presence of time delays ($\tau_1 = \tau_2 = 3.5$) on immune response in the two-species model (52)–(55). (**A**) Small amount of drug injection ($\lambda_D = 0.01$) still results in the oscillatory behaviors of both harmful bacteria (B_1) and immune cytokines (A_E) and the extinction of good bacteria; (**B**) for an intermediate level of injection ($\lambda_D = 0.03$), this oscillation disappears, and the system leads to the infection-persist state ($B_2^s = 0,\ B_1^s > 0$); (**C**) large amount of injection ($\lambda_D = 0.1$) significantly enhances the immune activity and removes both good and bad pathogens ($B_1^s = 0,\ B_2^s = 0$); (**D**) trajectories of bad (B_1) and good (B_2) bacteria for various drug injection rates ($\lambda_D = 0.01, 0.03, 0.1$); (**E**) trajectories of bad bacteria (B_1) and immune activity (A_E) for various drug injection rates ($\lambda_D = 0.01, 0.03, 0.1$). Initial condition: $B_1(0) = 0.2,\ B_2(0) = 0.95, A_E(0) = 0.0, D(0) = 0$. All other parameters are fixed as in Table 4 (Type II).

In real intravenous injection, the drug is administrated in a periodic infusion; we investigate the effect of drugs in a more realistic situation in the clinical setting. For this, we replace Equation (55) with the following:

$$\frac{d}{dt}D(t) = \sum_{i=1}^{N_D} \lambda_D I_{[t_i, t_i+t_d]}(t) - \delta_D D(t). \tag{56}$$

where λ_D is the injection rate, δ_D is the decay rate of drugs, N_D is the number of drug injections and $I_{[t_i,t_i+t_d]}(\cdot)$ is the characteristic equations that give one over the time interval $[t_j, t_i + t_d]$ with duration t_d and zero otherwise. Here, the injection period is fixed: $\tau_D = t_{j+1} - t_j,\ \forall j = 1, \ldots, N_D$.

In Figure 13, we investigate the dynamics of the system in response to the periodic injection of the high and low doses of drugs that boost patients' immunity. For a low dose of drugs

($\lambda_D = 0.2,\ t_d = 1\ h,\ N_D = 10$), the system still maintains the oscillatory behaviors of bad bacteria, and the immune-boosting effect from drugs is not significant enough to eradicate the bad bacteria (red curve (B_1) in Figure 13A). See the relatively low immune responses during IV injection periods in Figure 13D (blue curve (A_E) for $t < 240$). However, the promoted immune activity in response to a higher dose of drugs ($\lambda_D = 1.0,\ t_d = 1\ h,\ N_D = 10$) results in bacterial extinction (red curve (B_1) in Figure 13C) due to elevated levels of initial immune response (blue curve (A_E) in Figure 13D).

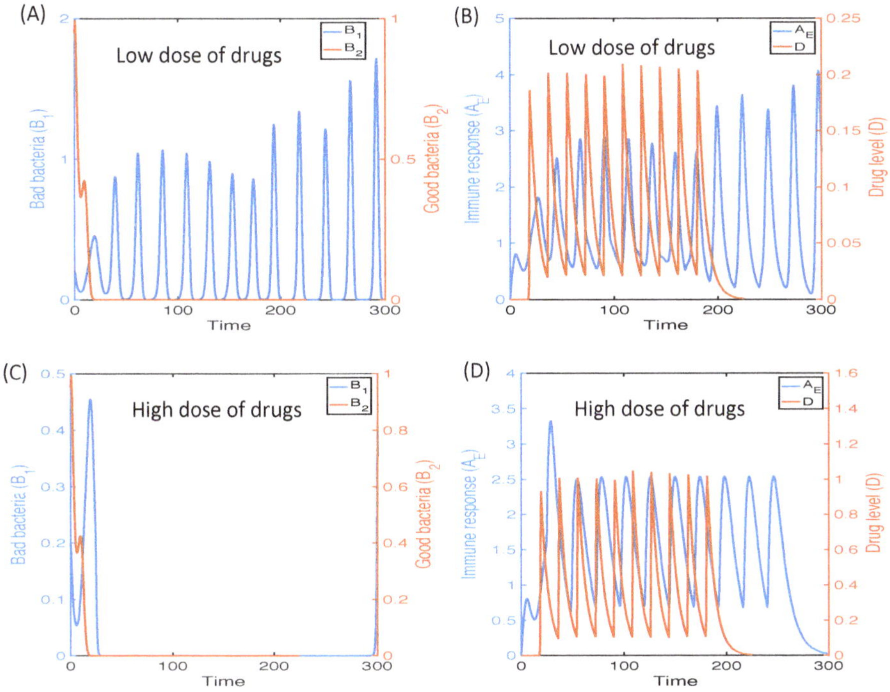

Figure 13. Dynamics of the system in response to low and high doses of drugs in the two-species M2 model (52)–(56). Immune activity was enhanced by the injection of drugs in a periodic fashion with an injection period 24 h ($\tau_D = 24\ h$). (**A**) Time courses of populations of bad (B_1) and good bacteria (B_2) in response to periodic injection of drugs with a lower infusion rate $\lambda_D = 0.2$ (duration $t_d = 1\ h$ fixed); (**B**) time courses of immune response (A_E) and drug levels (D) corresponding to (A); (**C**) time courses of populations of bad (B_1) and good bacteria (B_2) in response to periodic injection of drugs with higher infusion rate $\lambda_D = 1.0$ (duration $t_d = 1\ h$); (**D**) time courses of immune response (A_E) and drug levels (D) corresponding to (C). Initial condition: $B_1(0) = 0.2,\ B_2(0) = 0.95, A_E(0) = 0.0, D(0) = 0$. Parameter values: $N_D = 10,\ \delta_D = 0.1,\ \tau_1 = \tau_2 = 3.5$; all other parameters are fixed as in Table 4 (Type II).

In Figure 14, we investigate the effect of therapeutic drugs on the regulation of the eradication or recurrence of harmful bacteria in response to various combinations of infusion rates (λ_D) and injection periods (τ_D). For a fixed value of injection period (τ_D), the system switches from the recurrence phase to the eradication phase for harmful bacteria as the injection rate (λ_D) is increased. For a fixed λ_D, the larger interval length between drug injections tends to increase the chance of the recurrence of harmful bacteria. The model predicts that the larger infusion rate (λ_D) and shorter injection interval would lead to the elimination of harmful bacteria.

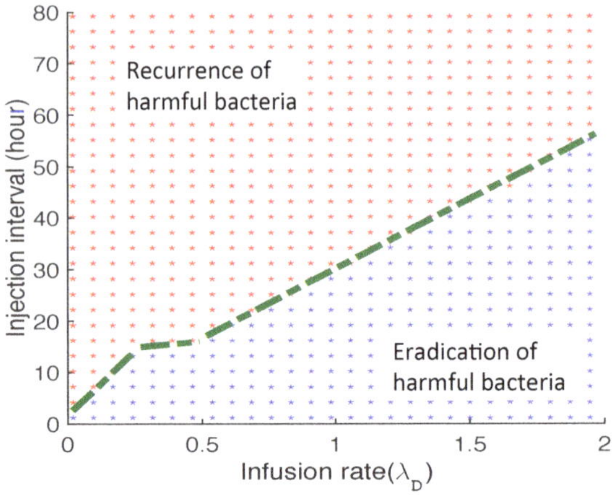

Figure 14. Therapeutic strategies of eradicating harmful bacteria by the injection of immune-boosting drugs for various infusion rates (λ_D) and injection periods (τ_D) in a two-species M2 model (52)–(56). Eradication (blue asterisk) and recurrence (red asterisk) of harmful bacteria in the λ_D-τ_D parameter space ($t_d = 1.0$ fixed) at $t = 300$. Initial condition: $B_1(0) = 0.2$, $B_2(0) = 0.95$, $A_E(0) = 0.0$, $D(0) = 0$. Parameter values: $N_D = 60$, $\delta_D = 0.1$, $\tau_1 = \tau_2 = 3.5$; all other parameters are fixed as in Table 4 (Type II).

4. Conclusions and Discussion

This paper investigates the progression of skin inflammatory disease by mathematical modeling and simulation. Three mathematical models were built to investigate how bacterial antigenic stimuli initiate and maintain the inflammatory response of keratinocytes. First, the effect of positive feedback regulation among protease and the transcription factor was incorporated, where we consider a single bacterial species as the source of antigenic stimuli (model M1). The existence of multiple positive equilibria indicated by equilibrium analysis implies that feedback switch occurs for model M1. To investigate how high inflammatory response is maintained, parameter m_{IP} representing the activation rate of proteases from the transcription factor was varied. Model M1 exhibits qualitatively different types of behaviors: one is the persistence of bacteria under a low inflammatory state; another is a high inflammatory state. Numerical computations indicate that the transition from low to high inflammatory states can occur when parameter m_{IP} varies (Figures 3–5). From the biological point of view, these computational results suggest that excessive protease activity can lead to a high inflammatory response. A general scheme presented in this paper applies to at least two different types of chronic inflammatory diseases. The first one is atopic dermatitis, which is often caused by primarily defection of the barrier function at the epidermis via excessive protease activity. In the previous study, the switch-like behavior was extensively investigated, which qualitatively explains the progression of atopic dermatitis [44]. The second one is an inflammatory bowel disease, which is recognized as a major inflammatory disease in the gut. Several research works imply the association of excessive protease activity with the progression of chronic inflammation [21].

To investigate the effect of species competition among bacterial species, model M1 was extended to include commensal or beneficial bacteria, which compete with a harmful bacterial species. Mathematical model M2 was constructed to investigate the bacterial competition under immune suppression. Model M2 consists of a classical Lotka–Volterra competition model with immune suppression as a negative feedback effect on both species. Numerical simulations were implemented extensively to understand the qualitative behavior of two bacteria under immune suppression (Figures 6 and 8). The outcomes of competition among two species were classified by means of the existence and stability conditions of equilibria (Section 3.2). The mathematical condition for the stability of the dominant equilibrium in which only a single bacteria species exists was derived. Condition (E_1S_1) represents a situation in which harmful bacteria outcompete beneficial bacteria when the growth rate of harmful bacteria is faster than that of beneficial bacteria ($r_2 > r_1$). Moreover, the second condition $D_1 = \alpha_{21}K_1 - 1 > 0$ represents the transversal eigenvalue of equilibrium E_1 in the B_2-direction. $D_1 > 0$ implies that the direction of the transversal eigenvalue is negative. Hence, beneficial bacteria cannot invade and grow when the harmful bacteria exist and have reached their carrying capacity. Hence, the condition (E_1S_1) implies the non-invasibility of beneficial bacteria. On the other hand, the condition (E_1S_2) represents an interesting situation. Despite the fact that harmful bacteria have a slower growth rate than beneficial bacteria, this can prevent the invasion of beneficial bacteria ($D_1 > 0$) and, importantly, suppress the growth rates of both bacteria by utilizing the boosting of immune suppression. In other words, harmful bacteria take advantage of growth inhibition by immune suppression, which leads to the dominance of harmful bacteria. From the biological point of view, there exists a possibility that some of the bacteria favor the inflammatory condition that suppresses potential competitors. In [7], it is shown that *S. aureus* increases in abundance during the process of dysbiosis, which can drive the inflammatory response, leading to atopic dermatitis progression. Hence, the implication derived from the mathematical computation and analysis results can partly explain how *S. aureus* grows in abundance while suppressing potential competitors.

Finally, time delays were introduced to represent the time required to activate an immune response. In the skin tissue, this is generally mediated by innate immune cell types, such as neutrophils, which have to be recruited from peripheral blood to an infection site. Hence, a time delay is inevitable to consider the process of immune activation. By the introduction of a time delay, interestingly, beneficial bacteria can outcompete harmful bacteria even though having a disadvantage in competition (Figures 9–11). These numerical computation results indicate the possibility that the time delay in immune cell recruitment can affect the outcome when two species are competitively bistable. A case study on *Francisella tularensis* infection, which may causes hypercytokinemia, reported at least a one-day delay in neutrophil recruitment post infection [45]. The introduction of a time delay in the recruitment of activated innate immune cells to the infection site in the skin exhibits interesting behaviors. Bacterial antigenic stimuli trigger the immune response via TLR4, antigen recognition receptors, which specifically detect lipopolysaccharide (LPS) expressed on the surface of Gram-negative bacteria. If we assume that good and bad bacteria are both Gram-negative bacteria, then it would be reasonable to assume that the time required to activate immune responses in response to bacterial stimuli would be the same. Hence, the same value was utilized for the two delays in this work. We also investigated the effects of the immune-boosting drugs on the selection of harmful or beneficial bacteria and developed strategies to prevent the cycles or recurrence of harmful bacteria populations (Figures 12–14).

The results of this work can serve as a starting point for better comprehensive modeling and experimentation. Some problems and extensions of the model that can be addressed in the future are as follows.

- In the present model, we concentrated on two major pathogenic and commensal bacterial species to obtain basic insight into how microbial interactions mediated chronic inflammation. However, more than hundreds of bacterial species have been demonstrated to coexist in the skin tissue. Metagenomic analysis targeting the gut- and skin-resident microbiome has revealed that numerous uncultured species exist and potentially affect the maintenance of skin homeostasis, as well as the progression of skin inflammatory disease [46]. The existence of spatial compartmentalization by forming heterogeneous clusters of colonies across the epidermis and dermis has been shown [47,48]. Although a few numbers of dominant species exist in terms of population abundance, bacterial diversity in the skin is highly maintained [49]. Complex interactions among commensal bacteria, the host immune system and different sources of environmental fluctuations should be essential factors for the maintenance of species diversity. Therefore, the incorporation of more than two bacterial species into the model would be more realistic. Colonization of harmful bacteria would be prevented by community-level resistance by a bacterial community. The incorporation of multiple species interactions will provide new insights on how the loss of bacterial diversity would lead to high inflammatory states.

- We considered the same time delays in the M2 model in this work. There exists the possibility of an immune escape mechanism, which might justify the use of different time delays. For instance, certain types of bacteria downregulate antigenicity when they invade tissue in order to escape from immune surveillance [50]. This would lead to a time delay in the activation of the immune system. Major extensions of the current model to include different time delays are warranted.
- The mathematical models presented here do not distinguish immune cell types, which are crucial to determine the difference between the epidermis and the GI tract. For instance, Langerhans cells are the major resident immune cell type that stays below the second layer of the stratum granulosum (below the tight junction) and captures the antigen. After capturing the antigen, Langerhans cells move to a draining lymph node to present the antigen to lymphocytes, known as homing. In the intestine, invading bacteria that attach to the gut epithelial cells trigger inflammatory responses, and finally, these bacteria are eliminated by immune cells recruited from the Payer's patch or gastric mucosal lymphoid follicles.
- Explicit incorporation of spatial structure is essential to represent specific and unique information to the epidermis or the GI tract. In the present paper, however, we focused on the role of bacterial species to induce inflammatory responses rather than spatial structure, which forms specific and unique interactions among invading bacteria and immune cells. The ongoing project aims to incorporate spatial structure and heterogeneity in immune cell subtypes, but it is currently under investigation.
- The major signaling networks that control the intracellular regulation of transcriptional factor, proteases and protease inhibitors need to be addressed.
- The microenvironment also plays an important role in the regulation of epidermis and stem cell dynamics [51]. These include other immune cells, endothelial cells and stromal cells, such as fibroblasts, as well as growth factors secreted by these cells.
- Cell-mechanical regulations, such as actin and serum response factor, were also shown to transduce bio-physical cues from the microenvironment to control epidermal stem cell fate [52].

Our understanding of the complex biochemical interactions between the epidermal cells and the microenvironment is very limited. Hybrid approaches may be used to take into account these intracellular signaling pathways in addition to the mechanical interactions of cells with microenvironment for detailed proteolytic activities, growth and invasion at the cellular level and viscoelastic response of the whole tissue [23,53–57]. Yet a more comprehensive understanding of the role of the microenvironment in epidermal homeostasis may lead to the development of new therapeutic agents. We will discuss these in future work.

Acknowledgments: Shinji Nakaoka was partly supported by (i) the Japan Society for the Promotion of Science (JSPS) through the Grant-in-Aid for Young Scientists (B25871132) and 15H05707. Yangjin Kim is supported by the Basic Science Research Program through the National Research Foundation of Korea funded by the Ministry of Education (NRF- 2015R1D1A1A01058702). Chang Hyeong Lee was supported by the Basic Science Research Program through the National Research Foundation of Korea (NRF) funded by the Ministry of Education (2014R1A1A2054976). This work was supported by the A3Foresight Program of China (NSF), Japan (JSPS) and Korea (NRF 2014K2A2A6000567).

Author Contributions: Design of models: Shinji Nakaoka, Yangjin Kim. Writing the manuscript: Shinji Nakaoka, Yangjin Kim, Chang Hyeong Lee. Stability analysis of the model: Sota Kuwahara, Yasuhiro Takeuchi. Numerical simulations: Hyejin Jeon, Junho Lee, Shinji Nakaoka. Discussion: Shinji Nakaoka, Yangjin Kim, Chang Hyeong Lee.

Conflicts of Interest: The authors declare no conflict of interest.

Appendix A. Nondimensionalization

For the M1 model, we nondimensionalize the variables and parameters in the governing Equation (2) as follows:

$$\bar{t} = \frac{t}{T}, \ \bar{B} = \frac{B}{B^*}, \ \bar{P} = \frac{P}{P^*}, \ \bar{A}_I = \frac{A_I}{A_I^*}, \ \bar{A}_E = \frac{A_E}{A_E^*}, \ \bar{\lambda} = \frac{T\lambda}{B^*}, \ \bar{r}_B = T\, r_B, \ \bar{K} = \frac{K}{B^*},$$
$$\bar{\gamma} = T\gamma A_E^*, \ \bar{\delta}_P = T\delta_P, \ \bar{\delta}_I = T\delta_I, \ \bar{\delta}_E = T\delta_E, \ \bar{m}_\circ = \frac{Tm_\circ}{Y^*}, \ \bar{a}_\circ = \frac{a_\circ}{X^*}, \ \bar{s}_\circ = s_\circ. \tag{A1}$$

where $\circ \in \{\mathrm{IP}, \mathrm{BP}, \mathrm{PI}, \mathrm{BI}, \mathrm{EI}, \mathrm{IE}\}$ and $Y \in \{P, P, A_I, A_I, A_I, A_E\}$ for $X \in \{A_I, B, P, B, A_E, A_I\}$.

For the two-species M2 model, we nondimensionalize the variables and parameters in the governing Equations (3)–(5) as follows:

$$\bar{t} = \frac{t}{T}, \ \bar{B}_1 = \frac{B_1}{B_1^*}, \ \bar{B}_2 = \frac{B_2}{B_2^*}, \ \bar{A}_E = \frac{A_E}{A_E^*}, \ \bar{r}_1 = T\, r_1, \ \bar{r}_2 = T\, r_2, \ \bar{\alpha}_{11} = \alpha_{11} B_1^*,$$
$$\bar{\alpha}_{12} = \alpha_{12} B_2^*, \ \bar{\alpha}_{21} = \alpha_{21} B_1^*, \ \bar{\alpha}_{22} = \alpha_{22} B_2^*, \ \bar{\gamma} = T\gamma A_E^*, \ \bar{\beta}_1 = \frac{T\beta_1 B_1^*}{A_E^*}, \tag{A2}$$
$$\bar{\beta}_2 = \frac{T\beta_2 B_2^*}{A_E^*}, \ \bar{\delta} = T\delta.$$

Table A1 lists the reference values in the model.

Table A1. Reference value. tw = estimated in this work.

Var	Description	Dimensional Value	Reference
T	Time scale	3.5 h	tw
B^*	Bacteria density (=B_1, B_2)	2.4×10^9 CFU/mL	[58]
P^*	Protease concentration	50 mU/mL	[59–62]
A_I^*	Transcriptional factor concentration	10 nM	[63]
A_E^*	Cytokine concentration	2.0×10^1 pg/cm^3	[64,65]

Appendix B. Sensitivity Analysis

In the mathematical model developed in this paper, there are a few parameters for which no experimental data are known due to abstract mathematical terms or that may affect significantly the computational results and predictions. In order to determine the sensitivity of the bacterial populations and the concentration of cytokines for these parameters, we have performed sensitivity analysis for a mathematical model (3)–(5). We have chosen a range for each of these parameters and divided each range into 10,000 intervals of uniform length. The base values and the ranges of the perturbed parameters are as in Table B1. For each of the ten parameters, a partial rank correlation coefficient (PRCC) value is calculated [66]. The calculated PRCC values range between −1 and 1 with the sign indicating whether an increase in the parameter value will decrease (−) or increase (+) the bacterial populations (B_1 and B_2) and cytokine concentration (A_E) at a given time [66].

Figure B1 shows sensitivity analysis of all variables for ten key parameters $(r_1, K_1, \alpha_{12}, r_2, \alpha_{21}, K_2, \gamma, \beta_1, \beta_2, \delta)$ in the mathematical model (3)–(5) at the selected time ($t = 1, 40, 80$). For example, we show the sensitivity of bacterial populations in response to the changes of parameter values in Figure B1. It is natural to predict the positive correlation of the bad bacterial population (B_1) with the growth rate (r_1) and carrying capacity ($K_1 = 1/\alpha_{11}$). Indeed, the bad bacterial population is very sensitive to the changes in the growth rate and the carrying capacity, i.e., strong positive correlations with r_1 and K_1. The initial strong correlations of the bad bacterial population with γ, β_1 at $t = 1$ turn into weak correlations at later times ($t = 40, 80$). On the other hand, the good bacterial population is positively correlated with r_2, but is negatively correlated with K_1, γ. The initial strong correlations of good bacterial population with α_{21}, β_1 at $t = 1$ turn into weak correlations at later times ($t = 40, 80$). On the contrary, the weak initial correlation of populations of bad and good bacteria with the decay rate of cytokines (δ) turns into a strong positive correlation. This indirectly indicates the role of suppressed immune response in boosting the bacterial growth. In our model, the immune response plays a key role in the regulation of bacterial dynamics. The strength of the immune response is positively correlated with K_1, β_1, β_2, but is negatively correlated with the decay rate (δ). In particular, the immune response is strongly negatively correlated with the killing rate (γ) of both bad and good bacteria at later times ($t = 40, 80$) due to decreased bacterial populations in response to increased immune response. The parameters $\alpha_{12}, \beta_2, K_2 (= 1/\alpha_{22})$ have little correlation with all variables (B_1, B_2, A_E). Table B1 summarizes the results of the sensitivity analysis in terms of the populations of bad (B_1) and good (B_2) bacteria and cytokine concentration (A_E), at $t = 1, 40, 80$.

Table B1. Sensitivity analysis for the local ODE system (3)–(5) at time t = 1, 40, 80. Parameters used in sensitivity analysis and PRCC values of populations of bad bacteria (B_1) and good bacteria (B_2) and the concentration of the immune system (A_E) at various time points (t = 1, 40, 80) are shown for 10 perturbed parameters $r_1, K_1, \alpha_{12}, r_2, \alpha_{21}, K_2, \gamma, \beta_1, \beta_2, \delta$. A range (minimum/maximum) of these ten perturbed (non-dimensional) parameters ($[r_1^{min}, r_1^{max}]$, $[K_1^{min}, K_1^{max}]$, $[\alpha_{12}^{min}, \alpha_{12}^{max}]$, $[r_2^{min}, r_2^{max}]$, $[\alpha_{21}^{min}, \alpha_{21}^{max}]$, $[K_2^{min}, K_2^{max}]$, $[\gamma^{min}, \gamma^{max}]$, $[\beta_1^{min}, \beta_1^{max}]$, $[\beta_2^{min}, \beta_2^{max}]$, $[\delta^{min}, \delta^{max}]$) and their baseline ($r_1^{base}$, K_1^{base}, α_{12}^{base}, r_2^{base}, α_{21}^{base}, K_2^{base}, γ^{base}, β_1^{base}, β_2^{base}, δ^{base}) are given in the lower section. Sample size = 10,000. * Significant (p-value < 0.01).

Par PRCC	r_1	K_1	α_{12}	r_2	α_{21}	K_2	γ	β_1	β_2	δ
$B_1(1)$	−0.469 *	0.9802 *	−0.206 *	−0.048 *	0.0397 *	−0.0018	−0.448 *	−0.464 *	−0.050 *	0.0694 *
$B_2(1)$	0.1245 *	−0.567 *	0.0294 *	0.6956	−0.693 *	0.0289 *	−0.399 *	−0.410 *	−0.060 *	0.0541 *
$A_E(1)$	−0.237 *	0.8100 *	−0.033 *	0.1185 *	−0.114 *	0.0091	−0.093 *	0.9596 *	0.6476 *	−0.393 *
$B_1(40)$	0.4098 *	0.5636 *	−0.295 *	−0.500 *	0.1479 *	−0.088 *	−0.248 *	−0.291 *	0.0100	0.3772 *
$B_2(40)$	−0.264 *	−0.574 *	0.0706 *	0.6627 *	−0.292 *	0.0687 *	−0.512 *	−0.384 *	−0.308 *	0.6232 *
$A_E(40)$	0.3034 *	0.3626 *	−0.033 *	0.4340 *	−0.082 *	0.0782 *	−0.739 *	0.4905 *	0.3453 *	−0.743 *
$B_1(80)$	0.4035 *	0.5614 *	−0.292 *	−0.509 *	0.1360 *	−0.085 *	−0.232 *	−0.284 *	0.0212	0.3412 *
$B_2(80)$	−0.204 *	−0.478 *	0.0704 *	0.6061 *	−0.241 *	0.0592 *	−0.406 *	−0.283 *	−0.247 *	0.5178 *
$A_E(80)$	0.3059 *	0.3551 *	−0.031 *	0.4312 *	−0.069 *	0.0715 *	−0.728 *	0.4730 *	0.3195 *	−0.730 *
Min	0.5	0.1	0.1	0.1	0.1	1.0	0.1	0.05	0.05	0.01
Base	1.5	0.5	1.0	1.0	1.0	2.0	1.0	0.1	1.0	0.35
Max	2.5	2.5	2.0	2.0	2.0	2.5	2.0	2.0	2.0	1.0

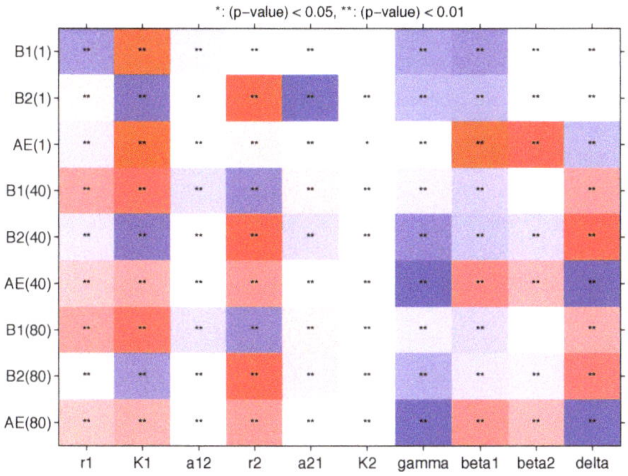

Figure B1. Sensitivity analysis of a mathematical model, Equations (3)–(5). General Latin hypercube sampling (LHS) scheme and partial rank correlation coefficient (PRCC) performed on the model. The reference outputs are the densities of bad bacteria (B_1) and good bacteria (B_2) and the concentration of inflammatory cytokines (A_E) at time t = 1, 40, 80. The colors indicate PRCC values of all variables (B_1, B_2, A_E) in the simple model (Equations (3)–(5)) for ten model parameters ($r_1, K_1, \alpha_{12}, r_2, \alpha_{21}, K_2, \gamma, \beta_1, \beta_2, \delta$). Red color indicates positive correlations, and blue color indicates negative correlations between the main variable and each parameter at the given time. The analysis was carried out using the method of Marino et al. (2008) [66] with a sample size of 10,000.

The sensitivity analysis described above was carried out using the method from [66] and MATLAB files available from the website of Denise Kirschner's Lab [67].

References

1. Chung, W.O.; Dale, B.A. Innate immune response of oral and foreskin keratinocytes: Utilization of different signaling pathways by various bacterial species. *Infect. Immun.* **2004**, *72*, 352–358.
2. Graham-Brown, R.; Burns, T. *Lecture Notes: Dermatology*, 10th ed.; Wiley-Blackwell: Hoboken, NJ, USA, 2011.
3. Bieber, T. Atopic Dermatitis. *N. Engl. J. Med.* **2008**, *358*, 1483–1494.
4. Smith, F.J.D.; Irvine, A.D.; Terron-Kwiatkowski, A.; Sandilands, A.; Campbell, L.E.; Zhao, Y.; Liao, H.; Evans, A.T.; Goudie, D.R.; Lewis-Jones, S.; et al. Loss-of-function mutations in the gene encoding filaggrin cause ichthyosis vulgaris. *Nat. Genet.* **2006**, *38*, 337–342.
5. Kubo, A.; Nagao, K.; Amagai, M. Epidermal barrier dysfunction and cutaneous sensitization in atopic diseases. *J. Clin. Investig.* **2012**, *122*, 440–447.
6. Briot, A.; Deraison, C.; Lacroix, M.; Bonnart, C.; Robin, A.; Besson, C.; Dubus, P.; Hovnanian, A. Kallikrein 5 induces atopic dermatitis-like lesions through PAR2-mediated thymic stromal lymphopoietin expression in Netherton syndrome. *J. Exp. Med.* **2009**, *206*, 1135–1147.
7. Kobayashi, T.; Glatz, M.; Horiuchi, K.; Kawasaki, H.; Akiyama, H.; Kaplan, D.H.; Kong, H.H.; Amagai, M.; Nagao, K. Dysbiosis and Staphylococcus aureus Colonization Drives Inflammation in Atopic Dermatitis. *Immunity* **2015**, *42*, 756–766.
8. Tenover, F.C.; Pearson, M.L. Methicillin-resistant Staphylococcus aureus. *Emerg. Infect. Dis.* **2004**, *10*, 2052–2053.
9. Gallo, R.L.; Nakatsuji, T. Microbial symbiosis with the innate immune defense system of the skin. *J. Investig. Dermatol.* **2011**, *131*, 1974–1980.
10. Christensen, G.J.M.; Brüggemann, H. Bacterial skin commensals and their role as host guardians. *Benef. Microbes* **2014**, *5*, 201–215.

11. Prince, T.; McBain, A.J.; O'Neill, C.A. Lactobacillus reuteri protects epidermal keratinocytes from Staphylococcus aureus-induced cell death by competitive exclusion. *Appl. Environ. Microbiol.* **2012**, *78*, 5119–5126.
12. Murphy, K. *Janeway's Immunobiology*, 8th ed.; Immunobiology: The Immune System (Janeway); Garland Science: New York, NY, USA, 2012.
13. Shen, W.; Li, W.; Hixon, J.A.; Bouladoux, N.; Belkaid, Y.; Dzutzev, A.; Durum, S.K. Adaptive immunity to murine skin commensals. *Proc. Natl. Acad. Sci. USA* **2014**, *111*, E2977–E2986.
14. Nakamizo, S.; Egawa, G.; Honda, T.; Nakajima, S.; Belkaid, Y.; Kabashima, K. Commensal bacteria and cutaneous immunity. *Semin. Immunopathol.* **2015**, *37*, 73–80.
15. Thammavongsa, V.; Kim, H.K.; Missiakas, D.; Schneewind, O. Staphylococcal manipulation of host immune responses. *Nat. Rev. Microbiol.* **2015**, *13*, 529–543.
16. Lo, C.W.; Lai, Y.K.; Liu, Y.T.; Gallo, R.L.; Huang, C.M. Staphylococcus aureus hijacks a skin commensal to intensify its virulence: Immunization targeting β-hemolysin and CAMP factor. *J. Investig. Dermatol.* **2011**, *131*, 401–409.
17. Son, E.D.; Kim, H.J.; Park, T.; Shin, K.; Bae, I.H.; Lim, K.M.; Cho, E.G.; Lee, T.R. Staphylococcus aureus inhibits terminal differentiation of normal human keratinocytes by stimulating interleukin-6 secretion. *J. Dermatol. Sci.* **2014**, *74*, 64–71.
18. Schoenfelder, S.M.K.; Lange, C.; Eckart, M.; Hennig, S.; Kozytska, S.; Ziebuhr, W. Success through diversity—How Staphylococcus epidermidis establishes as a nosocomial pathogen. *Int. J. Med. Microbiol.* **2010**, *300*, 380–386.
19. Lai, Y.; Cogen, A.L.; Radek, K.A.; Park, H.J.; Macleod, D.T.; Leichtle, A.; Ryan, A.F.; Di Nardo, A.; Gallo, R.L. Activation of TLR2 by a small molecule produced by Staphylococcus epidermidis increases antimicrobial defense against bacterial skin infections. *J. Investig. Dermatol.* **2010**, *130*, 2211–2221.
20. Rizzo, A.; Losacco, A.; Carratelli, C.R.; Domenico, M.D.; Bevilacqua, N. Lactobacillus plantarum reduces Streptococcus pyogenes virulence by modulating the IL-17, IL-23 and Toll-like receptor 2/4 expressions in human epithelial cells. *Int. Immunopharmacol.* **2013**, *17*, 453–461.
21. Vergnolle, N. Protease inhibition as new therapeutic strategy for GI diseases. *Gut* **2016**, *65*, 1215–1224.
22. Biancheri, P.; Di Sabatino, A.; Corazza, G.R.; MacDonald, T.T. Proteases and the gut barrier. *Cell Tissue Res.* **2013**, *351*, 269–280.
23. Kim, Y.; Roh, S. A hybrid model for cell proliferation and migration in glioblastoma. *Discret. Contin. Dyn. Syst. B* **2013**, *18*, 969–1015.
24. Kim, Y.; Kang, H.; Powathil, G.; Kim, H.; Trucu, D.; Lee, W.; Lawler, S.; Chaplain, M. MicroRNA regulation of a cancer network in glioblastoma: The role of miR-451-AMPK-mTOR in regulation of cell proliferation and infltration. *J. Roy. Soc. Interface* **2015**, submitted.
25. Lee, S.; Hwang, H.; Kim, Y. Modeling the role of TGF-beta in regulation of the Th17 phenotype in the LPS-driven immune system. *Bull. Math. Biol.* **2014**, *76*, 1045–1080.
26. Lim, J.; Lee, S.; Kim, Y. Hopf bifurcation in a model of TGF-beta in regulation of the Th17 phenotype. *Discret. Contin.s Dyn. Syst. B* **2016**, in press.
27. Sato, M.; Hata, N.; Asagiri, M.; Nakaya, T.; Taniguchi, T.; Tanaka, N. Positive feedback regulation of type I IFN genes by the IFN-inducible transcription factor IRF-7. *FEBS Lett.* **1998**, *441*, 106–110.
28. Prassas, I.; Eissa, A.; Poda, G.; Diamandis, E.P. Unleashing the therapeutic potential of human kallikrein-related serine proteases. *Nat. Rev. Drug Discov.* **2015**, *14*, 183–202.
29. D'Elia, R.V.; Harrison, K.; Oyston, P.C.; Lukaszewski, R.A.; Clark, G.C. Targeting the "cytokine storm" for therapeutic benefit. *Clin. Vaccine Immunol.* **2013**, *20*, 319–327.
30. Castellarin, M.; Warren, R.; Freeman, J.; Dreolini, L.; Krzywinski, M.; Strauss, J.; Barnes, R.; Watson, P.; Allen-Vercoe, E.; Moore, R.; et al. Fusobacterium nucleatum infection is prevalent in human colorectal carcinoma. *Genome Res.* **2012**, *22*, 299–306.
31. Kostic, A.; Gevers, D.; Pedamallu, C.; Michaud, M.; Duke, F.; Earl, A.; Ojesina, A.; Jung, J.; Bass, A.; Tabernero, J.; et al. Genomic analysis identifies association of Fusobacterium with colorectal carcinoma. *Genome Res.* **2012**, *22*, 292–298.
32. Africa, C.; Nel, J.; Stemmet, M. Anaerobes and Bacterial Vaginosis in Pregnancy: Virulence Factors Contributing to Vaginal Colonisation. *Int. J. Environ. Res. Public Health* **2014**, *11*, 6979–7000.
33. Marteau, P. Bacterial Flora in Inflammatory Bowel Disease. *Dig. Dis.* **2009**, *27*, 99–103.

34. Lepage, P.; Leclerc, M.; Joossens, M.; Mondot, S.; Blottiere, H.; Raes, J.; Ehrlich, D.; Dore, J. A metagenomic insight into our gut's microbiome. *Gut* **2012**, *62*, 146–158.
35. Lakhan, S.; Kirchgessner, A. Gut inflammation in chronic fatigue syndrome. *Nutr. Metab.* **2010**, *7*, 79, doi:10.1186/1743-7075-7-79.
36. Turnbaugh, P.; Ley, R.; Mahowald, M.; Magrini, V.; Mardis, E.; Gordon, J. An obesity-associated gut microbiome with increased capacity for energy harvest. *Nature* **2006**, *444*, 1027–1031.
37. Turnbaugh, P.; Hamady, M.; Yatsunenko, T.; Cantarel, B.; Duncan, A.; Ley, R.; Sogin, M.; Jones, W.; Roe, B.; Affourtit, J.; et al. A core gut microbiome in obese and lean twins. *Nature* **2009**, *457*, 480–484.
38. Mazmanian, S. Capsular polysaccharides of symbiotic bacteria modulate immune responses during experimental colitis. *J. Pediatr. Gastroenterol. Nutr.* **2008**, *46*, 11–12.
39. Kamada, N.; Seo, S.U.; Chen, G.Y.; Núñez, G. Role of the gut microbiota in immunity and inflammatory disease. *Nat. Rev. Immunol.* **2013**, *13*, 321–335.
40. Kubica, M.; Hildebrand, F.; Brinkman, B.M.; Goossens, D.; Del Favero, J.; Vercammen, K.; Cornelis, P.; Schröder, J.M.; Vandenabeele, P.; Raes, J.; et al. The skin microbiome of caspase-14-deficient mice shows mild dysbiosis. *Exp. Dermatol.* **2014**, *23*, 561–567.
41. Schommer, N.N.; Gallo, R.L. Structure and function of the human skin microbiome. *Trends Microbiol.* **2013**, *21*, 660–668.
42. Pugliese, A.; Gandolfi, A. A simple model of pathogen-immune dynamics including specific and non-specific immunity. *Math. Biosci.* **2008**, *214*, 73–80.
43. Malka, R.; Shochat, E.; Rom-Kedar, V. Bistability and bacterial infections. *PLoS ONE* **2010**, *5*, e10010.
44. Tanaka, R.J.; Ono, M.; Harrington, H.A. Skin barrier homeostasis in atopic dermatitis: Feedback regulation of kallikrein activity. *PLoS ONE* **2011**, *6*, e19895.
45. Mares, C.A.; Ojeda, S.S.; Morris, E.G.; Li, Q.; Teale, J.M. Initial delay in the immune response to Francisella tularensis is followed by hypercytokinemia characteristic of severe sepsis and correlating with upregulation and release of damage-associated molecular patterns. *Infect. Immun.* **2008**, *76*, 3001–3010.
46. Hannigan, G.D.; Grice, E.A. Microbial ecology of the skin in the era of metagenomics and molecular microbiology. *Cold Spring Harb. Perspect. Med.* **2013**, *3*, a015362.
47. Naik, S.; Bouladoux, N.; Wilhelm, C.; Molloy, M.J.; Salcedo, R.; Kastenmuller, W.; Deming, C.; Quinones, M.; Koo, L.; Conlan, S.; et al. Compartmentalized control of skin immunity by resident commensals. *Science* **2012**, *337*, 1115–1119.
48. Belkaid, Y.; Naik, S. Compartmentalized and systemic control of tissue immunity by commensals. *Nat. Immunol.* **2013**, *14*, 646–653.
49. Grice, E.A.; Segre, J.A. The human microbiome: Our second genome. *Annu. Rev. Genom. Hum. Genet.* **2012**, *13*, 151–170.
50. Van Avondt, K.; van Sorge, N.M.; Sorge, N.M.V.; Meyaard, L. Bacterial immune evasion through manipulation of host inhibitory immune signaling. *PLoS Pathog.* **2015**, *11*, e1004644.
51. Nie, J.; Fu, X.; Han, W. Microenvironment-dependent homeostasis and differentiation of epidermal basal undifferentiated keratinocytes and their clinical applications in skin repair. *J. Eur. Acad. Dermatol. Venereol.* **2013**, *27*, 531–535.
52. Connelly, J.; Gautrot, J.; Trappmann, B.; Tan, D.; Donati, G.; Huck, W.; Watt, F. Actin and serum response factor transduce physical cues from the microenvironment to regulate epidermal stem cell fate decisions. *Nat. Cell Biol.* **2010**, *12*, 711–718.
53. Kim, Y.; Stolarska, M.; Othmer, H. A hybrid model for tumor spheroid growth in vitro I: Theoretical development and early results. *Math. Models Methods Appl. Sci.* **2007**, *17*, 1773–1798.
54. Kim, Y.; Stolarska, M.; Othmer, H. The role of the microenvironment in tumor growth and invasion. *Prog. Biophys. Mol. Biol.* **2011**, *106*, 353–379.
55. Kim, Y. Regulation of cell proliferation and migration in glioblastoma: New therapeutic approach. *Front. Mol. Cell. Oncol.* **2013**, *3*, 53, doi:10.3389/fonc.2013.00053.
56. Kim, Y.; Othmer, H. A hybrid model of tumor-stromal interactions in breast cancer. *Bull. Math. Biol.* **2013**, *75*, 1304–1350.
57. Kim, Y.; Powathil, G.; Kang, H.; Trucu, D.; Kim, H.; Lawler, S.; Chaplain, M. Strategies of eradicating glioma cells: A multi-scale mathematical model with miR-451-AMPK-mTOR control. *PLoS ONE* **2015**, *10*, e0114370.

58. Lee, M.H.; Arrecubieta, C.; Martin, F.J.; Prince, A.; Borczuk, A.C.; Lowy, F.D. A postinfluenza model of Staphylococcus aureus pneumonia. *J. Infect. Dis.* **2010**, *201*, 508–515.
59. Lin, R.; Kwok, J.; Crespo, D.; Fawcett, J. Chondroitinase ABC has a long-lasting effect on chondroitin sulphate glycosaminoglycan content in the injured rat brain. *J. Neurochem.* **2008**, *104*, 400–408.
60. Gu, W.; Fu, S.; Wang, Y.; Li, Y.; Lu, H.; Xu, X.; Lu, P. Chondroitin sulfate proteoglycans regulate the growth, differentiation and migration of multipotent neural precursor cells through the integrin signaling pathway. *BMC Neurosci.* **2009**, *10*, 1–15.
61. Bruckner, G.; Bringmann, A.; Hartig, W.; Koppe, G.; Delpech, B.; Brauer, K. Acute and long-lasting changes in extracellular-matrix chondroitin-sulphate proteoglycans induced by injection of chondroitinase ABC in the adult rat brain. *Exp. Brain Res.* **1998**, *121*, 300–310.
62. Kim, Y.; Lee, H.; Dmitrieva, N.; Kim, J.; Kaur, B.; Friedman, A. Choindroitinase ABC I-mediated enhancement of oncolytic virus spread and anti-tumor efficacy: A mathematical model. *PLoS ONE* **2014**, *9*, e102499.
63. Milo, R.; Phillips, R. *Cell Biology by the Numbers*, 1st ed.; Taylor & Francis: Philadelphia, PA, USA, 2015. ISBN: 9780815345374.
64. Kim, Y.K.; Oh, S.Y.; Jeon, S.G.; Park, H.W.; Lee, S.Y.; Chun, E.Y.; Bang, B.; Lee, H.S.; Oh, M.H.; Kim, Y.S.; et al. Airway exposure levels of lipopolysaccharide determine type 1 versus type 2 experimental asthma. *J. Immunol.* **2007**, *178*, 5375–5382.
65. Kim, Y.; Lee, S.; Kim, Y.; Kim, Y.; Gho, Y.; Hwang, H.; Lawler, S. Regulation of Th1/Th2 cells in asthma development: A mathematical model. *Math. Biosci. Eng.* **2013**, *10*, 1095–1133.
66. Marino, S.; Hogue, I.; Ray, C.; Kirschner, D. A methodology for performing global uncertainty and sensitivity analysis in systems biology. *J. Theor. Biol.* **2008**, *254*, 178–196.
67. Kirschner, D. Uncertainty And Sensitivity Analysis. Available online: http://malthus.micro.med.umich.edu/lab/usadata/ (accessed on 1 March 2016).

applied sciences

MDPI

Article

Global Dynamics of Modeling Flocculation of Microorganism

Wei Wang [1], Wanbiao Ma [1,*] and Hai Yan [2]

1 Department of Applied Mathematics, School of Mathematics and Physics, University of Science and Technology Beijing, Beijing 100083, China; wei_wang@163.com
2 Department of Biological Science and Technology, School of Chemical and Biological Engineering, University of Science and Technology Beijing, Beijing 100083, China; haiyan@ustb.edu.cn
* Correspondence: wanbiao_ma@ustb.edu.cn; Tel.: +86-136-2129-8550

Academic Editor: Yang Kuang
Received: 31 May 2016; Accepted: 26 July 2016; Published: 5 August 2016

Abstract: From a biological perspective, a dynamic model describing the cultivation and flocculation of a microorganism that uses two different kinds of nutrients (carbon source and nitrogen source) is proposed. For the proposed model, there always exists a boundary equilibrium, i.e., *Rhodopseudomonas palustris* -free equilibrium. Furthermore, under additional conditions, the model also has five positive equilibria at most, i.e., the equilibria for which carbon source, nitrogen source, *Rhodopseudomonas palustris* and flocculants are coexistent. The phenomena of backward and forward bifurcations are extensively discussed by using center manifold theory. The global stability of the boundary equilibrium of the proposed model is deeply investigated. Moreover, the local stability of the positive equilibrium and the uniform persistence of the proposed model are discussed. Under additional conditions, the global stability of the positive equilibrium is studied. Some control strategies are given by the theoretical analysis. Finally, some numerical simulations are performed to confirm the correctness of the theoretical results.

Keywords: dynamic model; flocculation; global stability; uniform persistence

1. Introduction

Photosynthetic bacteria, which are common microorganisms in the natural environment, have been applied in the field of environmental protection, such as in the treatment of sewage, domestic wastewater and the bioremediation of sediment mud polluted with organic matter (see, for example, [1–4]). On the other hand, photosynthetic bacteria can produce relatively large amounts of physiologically-active substances, such as vitamin B_{12} , ubiquinone (coenzyme $Q10$), 5-aminolevulinic acid (ALA) and RNA (see, for example [5]). In particular, vitamin B_{12} has been used in treating anemia and as an eye lotion. Recently, applications as health food supplements have received considerable attention. Coenzyme $Q10$ has been used in treating heart diseases for many years. Further, coenzyme $Q10$ has been used not only as a medicine, but also as some food supplements, because of its physiological activities. One of the developments of ALA applications is in the area of photodynamic diagnosis. RNA is an attractive source of 5′-ribonucleotides for use as a flavor enhancer in the food industry. In recent years, the production of RNA has been used as a dietary source of pyrimidine for human immune functions (see, for example [6,7]).

Some photosynthetic bacteria, such as *Rhodopseudomonas palustris*, are extensively used in the production of lycopene, aquaculture, and so on [8]. It can use sunlight, inorganic and organic compounds for energy. Further, *Rhodopseudomonas palustris* can have practical value for removing microcystin from the water body during algal blooms [9]. It can also degrade 2,4,6-trinitrotoluene (TNT), which has negative effects on the human body and aquatic life, resulting in a major threat to drinking and irrigation water supplies, as well as the recreational use of surface waters worldwide. Moreover, *Rhodopseudomonas palustris* is regarded as the most promising microbial system for the biological production of hydrogen, which has been extensively developed because of its high-energy content and clean product after combustion [10]. However, the concentration of *Rhodopseudomonas palustris* is very low under anaerobic light culture conditions (see, for example, [11,12]). Therefore, cost-efficient harvesting of *Rhodopseudomonas palustris* is a new challenge. In order to harvest *Rhodopseudomonas palustris* from the liquid, it is necessary to flocculate the single cells into large cell aggregates. Flocculation is a chemically-based separation process that requires less energy than centrifugation and ultrafiltration and, thus, is regarded as the most promising means for degrading microorganisms. Since algal toxins of blooms have happened occasionally in recent years, the problems of degrading microorganisms have received wide attentions (see, for example, [13–16]).

Flocculants are a kind of important water treatment reagent, which can be divided into organic flocculants and inorganic flocculants according to the chemical compositions [17]. Although organic flocculants, such as polyacrylamide, are frequently used in wastewater treatment and industrial downstream processes because of their high efficiency, some of them are not easily degraded in nature [18,19], and some of the monomers derived from synthetic polymers are harmful to the human body (see, for example, [20,21]). To solve these environmental problems, inorganic flocculants are increasingly being seen as an alternative in the settlement of microorganisms, more specifically in wastewater treatment owing to their inexpensive and nontoxic characteristics. Thus, inorganic flocculants may be used as nontoxic, cost-effective and widely-available flocculants for harvesting *Rhodopseudomonas palustris* (see, for example, [22–27] and the references therein).

Mathematical models have played an important role in better understanding microbiology and population biology (see, for example, [28,29]). In recent years, the dynamics of the chemostat models has received considerable attentions (see, for example, [29]). The article of Smith and Waltman has played an important role in the development of the chemostat models [30]. From then on, much research on the chemostat models has been extensively studied by many authors. A model describing two populations of microorganisms competing for one single limiting nutrients was proposed in [31]. Later, the model was extended to an arbitrary number of populations in [32,33]. These models, which were studied in articles [31–33], have proven that they all include a competitive exclusion effect. In the articles [34–39], some further developments have been performed on the chemostat models to place the relevant models in a naturally more sensible manner.

Let $S(t)$ and $X(t)$ denote the concentration of the nutrients and *Rhodopseudomonas palustris*, respectively, in the culture vessel at time t. $P(t)$ denotes the concentration of inorganic flocculants, which are used for harvesting *Rhodopseudomonas palustris* (see Figure 1). The constants $S^0 > 0$ and $P^0 > 0$ denote the input concentration of the nutrients and flocculants, respectively. For simplicity, it is assumed that the input of the nutrients and flocculants is continuous. The constant $D > 0$ is the dilution rate of the chemostat.

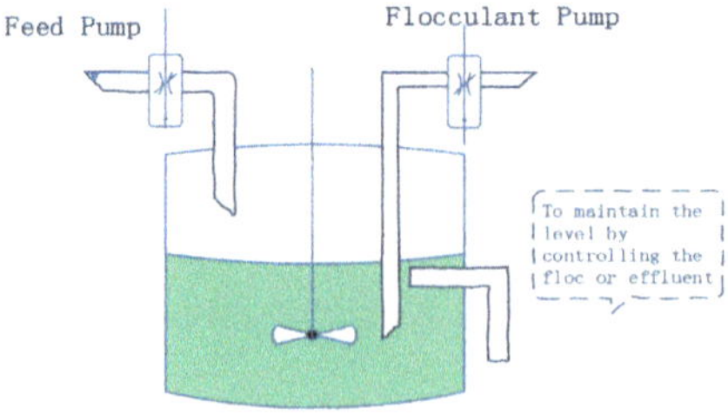

Figure 1. The device for collecting *Rhodopseudomonas palustris* in the chemostat by inputting inorganic flocculants.

The constant $\tau \geq 0$ denotes the time delay involved in the conversion of nutrients to *Rhodopseudomonas palustris*. The flocculation rate of microorganisms is assumed to be a bilinear mass-action function response $m_1X(t)P(t)$, where $m_1 \geq 0$ is the per capita contact rate. At the same time, flocculants produce loss or consumption [17], and the loss rate of flocculants is also assumed to be a bilinear mass-action function response $m_2X(t)P(t)$, where $m_2 \geq 0$ is constant. Thus, in [40], the following dynamic model has been proposed:

$$\begin{cases} \frac{dS(t)}{dt} = (S^0 - S(t))D - r_1S(t)X(t), \\ \frac{dX(t)}{dt} = rS(t-\tau)X(t-\tau) - DX(t) - m_1X(t)P(t), \\ \frac{dP(t)}{dt} = (P^0 - P(t))D - m_2X(t)P(t). \end{cases} \tag{1}$$

In Model (1), $r \geq 0$ and $r_1 \geq 0$ are constants, and the bilinear mass-action uptake function $S(t)X(t)$ has been used.

It should be mentioned here that, the analysis reveals that Model (1) proposed in [40] exhibits the phenomenon of backward bifurcation for the existence of positive equilibria. Moreover, the local stability properties of the equilibria have been dealt with in detail.

People found that the influence of different nutrients has played an important role in the culture of microorganisms. In order to take this into consideration, appropriate combinations of nutrients are considered in chemostat models. Models with two competitors and two perfectly-complemented growth-limiting nutrients are studied in [41,42]. Local asymptotic conditions for the equilibria are derived.

When there is a microorganism to compete for two or more resources, it may become necessary to consider how the resources, once consumed, interact to promote growth. In [41], the authors employ consumer needs to provide a criterion to classify resources and classify resources as perfectly complementary, perfectly substitutable or imperfectly substitutable. Perfectly-complementary resources are different essential substances that must be taken together. In this case, each substance fulfills different functions with respect to the growth of microorganisms. For example, carbon source and nitrogen source may be complementary for the growth of bacterium.

Motivated by the papers mentioned above, in this paper, we further consider a dynamic model describing the cultivation and flocculation of *Rhodopseudomonas palustris*, and the nutrients presented in [40] will be divided into carbon source and nitrogen source, which are perfectly complementary in the culture of *Rhodopseudomonas palustris* (see, for example, [5,41–44]). We assume that the growth of *Rhodopseudomonas palustris* is always co-limited by carbon and nitrogen for all possible nutrient conditions. We do not consider the case where only one of these nutrients limits growth, for example, in environmental scenarios of high carbon, but very low nitrogen loads, the growth of bacteria may be purely nitrogen limited.

Let $C(t)$ and $N(t)$ denote the concentration of carbon source and nitrogen source, respectively, in the culture vessel at time t. The constants $C^0 > 0$ and $N^0 > 0$ denote the input concentration of carbon source and nitrogen source, respectively. We assume that the conversion of nutrients to microorganism biomass occurs instantly. That is, the time delay τ in Model (1) equals zero. Hence, we have the following dynamic model describing the cultivation and flocculation of a microorganism:

$$\begin{cases} \frac{dC(t)}{dt} = (C^0 - C(t))D - \frac{X(t)}{\delta_1}r_1\mu_1(C(t))\mu_2(N(t)), \\ \frac{dN(t)}{dt} = (N^0 - N(t))D - \frac{X(t)}{\delta_2}r_2\mu_1(C(t))\mu_2(N(t)), \\ \frac{dX(t)}{dt} = r\mu_1(C(t))\mu_2(N(t))X(t) - DX(t) - m_1X(t)P(t), \\ \frac{dP(t)}{dt} = (P^0 - P(t))D - m_2X(t)P(t). \end{cases} \tag{2}$$

In Model (2), the parameters D, r, m_1, m_2 and P^0 are the same as Model (1). The term $r\mu_1(C(t))\mu_2(N(t))$ is the growth rate of *Rhodopseudomonas palustris*, and the terms $r_1\mu_1(C(t))\mu_2(N(t))$ and $r_2\mu_1(C(t))\mu_2(N(t))$ represent the quantity of the decreasing of the carbon source and nitrogen source, respectively, where r_1 and r_2 are non-negative constants; the functions $\mu_1(C(t))$ and $\mu_2(N(t))$ are nonnegative and continuous for $C(t) \geq 0$, $N(t) \geq 0$. For the simplicity of the theoretical analysis, in this paper, the functions $\mu_1(C(t))$ and $\mu_2(N(t))$ are chosen as Monod-type functions, i.e.,

$$\mu_1(C(t)) = \frac{C(t)}{K_1 + C(t)}, \ \mu_2(N(t)) = \frac{N(t)}{K_2 + N(t)},$$

where $K_1 > 0$ and $K_2 > 0$ are the half-saturation constants with respect to the carbon source and nitrogen source, respectively. δ_i $(i = 1, 2)$ are yield coefficients (see, for example, [29]), which are defined as:

$$\delta_i = \frac{mass\ of\ organism\ formed}{mass\ of\ substrate\ consumed}, \ (i = 1, 2).$$

Therefore, the dynamic Model (2) can be rewritten in the following form:

$$\begin{cases} \frac{dC(t)}{dt} = (C^0 - C(t))D - \frac{r_1 C(t)N(t)X(t)}{\delta_1(K_1+C(t))(K_2+N(t))}, \\ \frac{dN(t)}{dt} = (N^0 - N(t))D - \frac{r_2 C(t)N(t)X(t)}{\delta_2(K_1+C(t))(K_2+N(t))}, \\ \frac{dX(t)}{dt} = \frac{rC(t)N(t)X(t)}{(K_1+C(t))(K_2+N(t))} - DX(t) - m_1X(t)P(t), \\ \frac{dP(t)}{dt} = (P^0 - P(t))D - m_2X(t)P(t). \end{cases} \tag{3}$$

It is convenient to introduce dimensionless variables. In particular, we define:

$$\overline{C} = \frac{C}{C^0}, \ \overline{N} = \frac{N}{N^0}, \ \overline{X} = X, \ \overline{P} = \frac{P}{P^0}, \ \overline{K}_1 = \frac{K_1}{C^0}, \ \overline{K}_2 = \frac{K_2}{N^0}, \ \overline{t} = tD,$$

$$\overline{r}_1 = \frac{r_1}{\delta_1 DC^0}, \ \overline{r}_2 = \frac{r_2}{\delta_2 DN^0}, \ \overline{r} = \frac{r}{D}, \ \overline{m}_1 = \frac{m_1 P^0}{D}, \ \overline{m}_2 = \frac{m_2}{D},$$

and still denote $\overline{C}, \ \overline{N}, \ \overline{X}, \ \overline{P}, \overline{K}_1, \overline{K}_2, \overline{t}, \overline{r}_1, \overline{r}_2, \overline{r}, \overline{m}_1$ and $\overline{m}_2$ with $C, \ N, \ X, \ P, K_1, K_2, t, r_1, r_2, r, m_1$ and m_2, then Model (3) becomes:

$$\begin{cases} \frac{dC(t)}{dt} = 1 - C(t) - \frac{r_1 C(t)N(t)X(t)}{(K_1+C(t))(K_2+N(t))}, \\ \frac{dN(t)}{dt} = 1 - N(t) - \frac{r_2 C(t)N(t)X(t)}{(K_1+C(t))(K_2+N(t))}, \\ \frac{dX(t)}{dt} = \frac{rC(t)N(t)X(t)}{(K_1+C(t))(K_2+N(t))} - X(t) - m_1X(t)P(t), \\ \frac{dP(t)}{dt} = 1 - P(t) - m_2X(t)P(t). \end{cases} \tag{4}$$

According to the biological considerations, the initial condition of Model (4) is given as:

$$C(0) = C_0 \geq 0, \ N(0) = N_0 \geq 0, \ X(0) = X_0 \geq 0, \ P(0) = P_0 \geq 0, \tag{5}$$

where the constants C_0, N_0, X_0 and P_0 represent the initial concentration of the carbon source, nitrogen source, *Rhodopseudomonas palustris* and flocculants respectively.

The purpose of this paper is to tackle the existence of backward and forward bifurcations by using center manifold theory and to investigate the global stability properties of the two classes of equilibria by constructing the suitable Lyapunov functions.

The organization of the paper is as follows. The global existence, nonnegativity and boundedness of the solutions of Model (4) are investigated in Section 2. In Section 3, the existence of the equilibria and the phenomena of backward and forward bifurcations are extensively discussed. In Section 4, the global stability of the boundary equilibrium of Model (4) is discussed by the stability theory of ordinary differential equations. Furthermore, we consider the local stability of positive equilibrium, the uniform persistence of Model (4) and the global asymptotic stability of the positive equilibrium in Section 5. In Section 6, some control strategies are given by the theoretical analysis. Some discussions are given in Section 7.

2. The Global Existence, Nonnegativity and Boundedness of Solutions

From the biological considerations, it is necessary to show that all of the solutions of Model (4) are nonnegative and bounded for all $t \geq 0$. By using the basic theory of ordinary differential equations [45] and some simple calculations, it is not difficult to show the following result.

Theorem 1. *The solution $(C(t), N(t), X(t), P(t))$ of Model (4) with the initial Condition (5) is existent, unique and nonnegative for all $t \geq 0$ and satisfies:*

$$\limsup_{t\to\infty} C(t) \leq 1,\ \limsup_{t\to\infty} N(t) \leq 1,\ \limsup_{t\to\infty} M(t) \leq \alpha,$$

$$\limsup_{t\to\infty} P(t) \leq 1,\ \liminf_{t\to\infty} C(t) \geq \underline{C},\ \liminf_{t\to\infty} N(t) \geq \underline{N},\ \liminf_{t\to\infty} P(t) \geq \underline{P},$$

where $M(t) = \frac{r}{2r_1}C(t) + \frac{r}{2r_2}N(t) + X(t)$, $\alpha = \frac{r}{2r_1} + \frac{r}{2r_2}$, $\underline{C} = \frac{K_1(K_2+1)}{K_1(K_2+1)+r_1\alpha}$, $\underline{N} = \frac{K_2(K_1+1)}{K_2(K_1+1)+r_2\alpha}$, $\underline{P} = \frac{1}{1+m_2\alpha}$.

Proof. From the theory of the local existence of solutions for ordinary differential equations, it can be obtained that the solution $(C(t), N(t), X(t), P(t))$ of Model (4) is existent and unique for $t \in [0, \delta)$. Here, δ is some positive constant [29,45,46]. Furthermore, we also have that the solution $(C(t), N(t), X(t), P(t))$ is nonnegative for $t \in [0, \delta)$.

We can easily show that $C(t)$, $N(t)$ and $P(t)$ are bounded on $t \in [0, \delta)$. Let us further show that $X(t)$ is also bounded on $t \in [0, \delta)$. For $t \geq 0$, define:

$$M(t) = \tfrac{r}{2r_1}C(t) + \tfrac{r}{2r_2}N(t) + X(t).$$

From Model (4), we obtain that, for $t \geq 0$,

$$M(t) \leq \tfrac{r}{2r_1} + \tfrac{r}{2r_2} - M(t). \tag{6}$$

From (6) and the well-known comparison principle, we have that $M(t)$ is also bounded for $t \in [0, \delta)$. Hence, by employing the continuation theorems of the solutions [29,45,46], the solution $(C(t), N(t), X(t), P(t))$ is existent and unique for any $t \geq 0$. Similarly, the solution is nonnegative for any $t \geq 0$.

Thus, from the comparison principle, we have that:

$$\limsup_{t\to\infty} C(t) \leq 1,\ \limsup_{t\to\infty} N(t) \leq 1,\ \limsup_{t\to\infty} M(t) \leq \tfrac{r}{2r_1} + \tfrac{r}{2r_2} = \alpha,\ \limsup_{t\to\infty} P(t) \leq 1.$$

By the first equation of Model (4), we obtain that, for $t \geq 0$,

$$C(t) = 1 - C(t) - \frac{r_1 C(t)N(t)X(t)}{(K_1 + C(t))(K_2 + N(t))} \geq 1 - \left(1 + \frac{r_1\alpha}{K_1(K_2+1)}\right)C(t).$$

Again, we can conclude that $\liminf_{t\to\infty} C(t) \geq \underline{C}$. By using the technique similar above, we can show that $\liminf_{t\to\infty} N(t) \geq \underline{N}$, $\liminf_{t\to\infty} P(t) \geq \underline{P}$. We complete the proof of Theorem 1. □

Theorem 2. *The compact set:*

$$\Omega = \left\{(C, N, X, P) \in R_+^4 \,\middle|\, \underline{C} \leq C \leq 1,\ \underline{N} \leq N \leq 1,\ 0 \leq M \leq \alpha,\ \underline{P} \leq P \leq 1\right\}$$

attracts all of the solutions of Model (4) and is positively invariant with respect to Model (4).

Proof. According to Theorem 1, it only needs to be proven that Ω is positively invariant with respect to Model (4). That is, it needs to be shown that $\underline{C} \leq C(t) \leq 1$, $\underline{N} \leq N(t) \leq 1$, $M(t) \leq \alpha$, $\underline{P} \leq P(t) \leq 1$ for any $t \geq 0$ if $(C(0), N(0), X(0), P(0)) \in \Omega$. Let us show $M(t) \leq \alpha$ for any $t \geq 0$.

In fact, if there exists some $t_1 > 0$, such that $M(t_1) > \alpha$, then $t_1^* = \sup\{t|M(t) = \alpha, t \in [0, t_1]\}$ is existent and $t_1^* \geq 0$. Hence, we obtain that $M(t_1^*) = \alpha$, $M(t_1) > \alpha$ and $M(t) > \alpha$ for $t \in (t_1^*, t_1)$. By the Lagrange mean-value theorem, there exists some $t_2 \in (t_1^*, t_1)$, such that $M(t_2) = \frac{M(t_1)-M(t_1^*)}{t_1-t_1^*} > 0$.

On the other hand, from (6), we have that $M(t_2) \leq \alpha - M(t_2) < \alpha - \alpha = 0$, which is a contradiction. Thus, $M(t) \leq \alpha$ for any $t \geq 0$. Therefore, from Model (4), we have that for any $t \geq 0$, $C(t) \leq 1 - C(t)$, $N(t) \leq 1 - N(t)$, $P(t) \leq 1 - P(t)$, from which we easily have that $C(t) \leq 1$, $N(t) \leq 1$, $P(t) \leq 1$ for any $t \geq 0$.

Next, we prove that $C \geq \underline{C}$ for any $t \geq 0$ if $(C(0), N(0), X(0), P(0)) \in \Omega$.

In fact, if there exists some $t_3 > 0$, such that $C(t_3) < \underline{C}$, then $t_2^* = \sup\{t|C(t) = \underline{C},\ t \in [0, t_3]\}$ is existent and $t_2^* \geq 0$. Hence, we have that $C(t_2^*) = \underline{C}$, $C(t_3) < \underline{C}$ and $C(t) < \underline{C}$ for $t \in (t_2^*, t_3)$. By the Lagrange mean-value theorem, there exists some $t_4 \in (t_2^*, t_3)$, such that $C(t_4) = \frac{C(t_3)-C(t_2^*)}{t_3-t_2^*} < 0$.

On the other hand, from Model (4), we obtain that:

$$C(t_4) \geq 1 - C(t_4) - \frac{r_1 C(t_4)\alpha}{K_1(K_2+1)} > 1 - \left(1 + \frac{r_1\alpha}{K_1(K_2+1)}\right)\underline{C} = 0,$$

which is a contradiction. Thus, $C(t) \geq \underline{C}$ for any $t \geq 0$. Therefore, from Model (4), we obtain that, for any $t \geq 0$,

$$N(t) \geq 1 - \left(1 + \frac{r_2\alpha}{K_2(K_1+1)}\right)N(t),\ \ P(t) \geq 1 - (1 + m_2\alpha)P(t),$$

from which we easily obtain that $\underline{N} \leq N(t)$, $\underline{P} \leq P(t)$, for any $t \geq 0$. This completes the proof of Theorem 2. □

3. The Existence of the Equilibria and Its Classification

Let (C, N, X, P) be any equilibrium of Model (4). Then, (C, N, X, P) satisfies the following nonlinear algebraic equations,

$$\begin{cases} 1 - C - \frac{r_1 CNX}{(K_1+C)(K_2+N)} = 0, \\ 1 - N - \frac{r_2 CNX}{(K_1+C)(K_2+N)} = 0, \\ \frac{rCNX}{(K_1+C)(K_2+N)} - X - m_1 XP = 0, \\ 1 - P - m_2 XP = 0. \end{cases} \tag{7}$$

Model (4) always has the boundary equilibrium $E_0(1, 1, 0, 1)$. The existence of E_0 indicates that, if there is no *Rhodopseudomonas palustris* to be added into the culture vessel at the beginning of the culture, the concentrations of the carbon source, nitrogen source and flocculants always maintain the constant values 1, 1 and 1, respectively. The equilibrium $E_0(1, 1, 0, 1)$ is also called *Rhodopseudomonas palustris*-free equilibrium.

Define the basic bifurcation parameter as:

$$R_0 = \frac{r}{(K_1+1)(K_2+1)(m_1+1)}.$$

Let (C^*, N^*, X^*, P^*) be any positive equilibrium of Model (4). From (7), we have that:

$$\begin{cases} (K_1 + C^*)(K_2 + N^*)\left(1 + \frac{m_1}{1+m_2X^*}\right) - rC^*N^* = 0, \\ P^* = \frac{1}{1+m_2X^*}, \\ C^* = \frac{-r_1 m_2 (X^*)^2 + (m_2 r - r_1(m_1+1))X^* + r}{r(1+m_2X^*)}, \\ N^* = \frac{-r_2 m_2 (X^*)^2 + (m_2 r - r_2(m_1+1))X^* + r}{r(1+m_2X^*)}. \end{cases} \tag{8}$$

Clearly, X^* should satisfy the following conditions:

$$\begin{aligned} -r_1 m_2 (X^*)^2 + (m_2 r - r_1(m_1+1))X^* + r > 0, \\ -r_2 m_2 (X^*)^2 + (m_2 r - r_2(m_1+1))X^* + r > 0. \end{aligned} \tag{9}$$

Substituting the second, third and forth equations of (8) into the first equation gives a fifth order algebraic equation,

$$f(X^*) = a(X^*)^5 + b(X^*)^4 + c(X^*)^3 + d(X^*)^2 + eX^* + f = 0,$$

where:

$$\begin{aligned}
a =& \; r_1 r_2 m_2^3 (1-r), \\
b =& \; r_1 m_2^3 r \left(r - (K_2+1) - \tfrac{r_2(m_1+2)}{m_2} \right) + 3 r_1 r_2 m_2^2 (m_1+1) \\
& + r_2 m_2^3 r \left(r - (K_1+1) - \tfrac{r_1(m_1+1)}{m_2} \right), \\
c =& \; m_2^3 r^2 \left((K_1+1)(K_2+1) + \tfrac{(r_1+r_2)(m_1+3)}{m_2} - r \right) + q, \\
d =& \; m_2^2 r R_0 (K_1+1)^2 (K_2+1)^2 (m_1+1)^2 (1-R_0) + d_1, \\
e =& \; r_1 r (m_1+1)(r - (K_2+1)(m_1+1)) + r_2 r (m_1+1)(r - (K_1+1)(m_1+1)) \\
& + m_2 r^2 ((K_1+1)(K_2+1)(2m_1+3) - 3r), \\
f =& \; r R_0 (K_1+1)^2 (K_2+1)^2 (m_1+1)^2 (1-R_0), \\
q =& \; r_1 r_2 m_2 (3-r)(1+m_1)^2 - 2 q_1 r (1+m_1) - r m_2^2 (K_1 r_2 + K_2 r_1), \\
q_1 =& \; r_1 m_2^2 + r_1 r_2 m_2 + r_2 m_2^2 + K_1 r_2 m_2^2 + K_2 r_1 m_2^2, \\
d_1 =& \; m_2^2 r^2 \left(2(K_1+1)(K_2+1) + \tfrac{(r_1+r_2)(2m_1+3)}{m_2} - 2r \right) + h, \\
h =& \; -m_2^2 r (2(m_1+1)+1)(r_2 K_1 + r_1 K_2) + 2 r_1 r_2 m_2 (1+m_1)^2 \\
& - r_1 m_2^2 r - r_2 m_2^2 r - h_1 r (1+m_1), \\
h_1 =& \; 2 r_1 m_2^2 + 3 r_1 r_2 m_2 + r_2 m_2^2 + r_2 m_2^2 + K_2 m_2^2.
\end{aligned}$$

Let us consider the necessary condition for the existence of the positive equilibria of Model (4).

From the first equation in (8), we obtain the following function:

$$F(X^*) = \frac{m_1(K_1+C^*)(K_2+N^*)}{1+m_2X^*} + K_1K_2 + K_1N^* + K_2C^* + (1-r)C^*N^*,$$

which implies that $r > 1$ is a necessary condition for a positive equilibrium to exist.

Using the methods similar to [47], we can give the sufficient conditions of the existence of the positive equilibria of Model (4). The following results (Theorem 3) follow from the various possibilities enumerated in Table A1 (see Appendix A):

Theorem 3.

(*i*) *Model (4) has a unique positive equilibrium if $R_0 < 1$ and Conditions (9) hold and whenever Cases 1, 9, 13, 15 and 16 in Table A1 are satisfied;*

(*ii*) *Model (4) could have more than one positive equilibrium if $R_0 < 1$ and Conditions (9) hold and whenever Cases 2–8, 10–12 and 14 in Table A1 are satisfied;*

(*iii*) *Model (4) could have five positive equilibria at most if $R_0 > 1$ and Conditions (9) hold and whenever Cases 1–16 in Table A1 are satisfied.*

Hence, under suitable conditions, there may be at most five different positive roots for the fifth order algebraic equation. Let $X = X^*$ be any such positive root, which also satisfies Conditions (9). Thus, from (8), $C = C^* > 0$, $N = N^* > 0$ and $P = P^* > 0$ can be obtained. Therefore, Model (4) at most has five positive equilibria of the type of $E^*(C^*, N^*, X^*, P^*)$. The equilibrium $E^*(C^*, N^*, X^*, P^*)$

indicates that the carbon source, nitrogen source, *Rhodopseudomonas palustris* and flocculants may be coexistent for any time $t \geq 0$.

Remark 1. *The existence of multiple positive equilibria of Model (4) when $R_0 < 1$ (shown in Table A1; see Appendix A) indicates the possibility of the existence of backward bifurcation (see, for example, [40,47,48]), where the stable boundary equilibrium co-exists with a stable positive equilibrium. This can be explored below via numerical simulations (see Figure 2a,b). A rigorous result can be obtained using center manifold theory [49]. The detailed proof is given in Appendix B.*

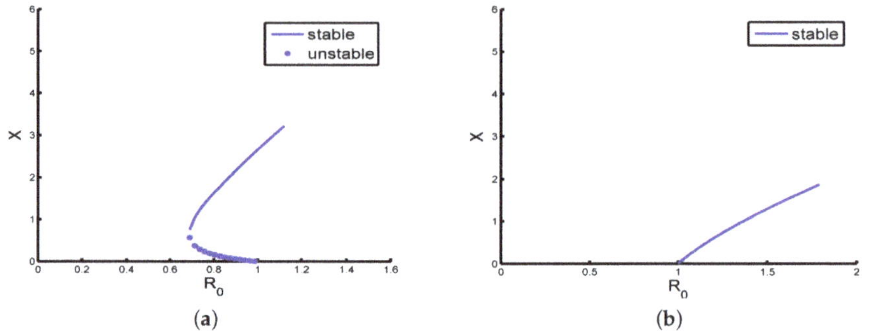

Figure 2. (**a**) Simulations of backward bifurcation for Model (4) with $K_1 = 0.36$, $K_2 = 0.3$, $m_1 = 0.0001$, $m_2 = 4$, $r_1 = 0.96$, $r_2 = 1.0001$; (**b**) simulations of forward bifurcation for Model (4) with $K_1 = 0.36$, $K_2 = 0.3$, $m_1 = 0.0001$, $m_2 = 4$, $r_1 = 0.96$, $r_2 = 1.0001$. Solid curves represent stable equilibrium and dashed curves represent unstable equilibrium.

From the analysis given in Appendix B, we have established the following result.

Theorem 4.

(*i*) *If:*

$$m_1 m_2 > \frac{rr_1 K_1}{(K_1+1)^3 (K_2+1)^2} + \frac{rr_2 K_2}{(K_1+1)^2 (K_2+1)^3},$$

then Model (4) undergoes a backward bifurcation at $R_0 = 1$.

(*ii*) *If:*

$$m_1 m_2 < \frac{rr_1 K_1}{(K_1+1)^3 (K_2+1)^2} + \frac{rr_2 K_2}{(K_1+1)^2 (K_2+1)^3},$$

then Model (4) undergoes a forward bifurcation at $R_0 = 1$.

The existence of backward bifurcation implies that stable boundary equilibrium and stable positive equilibrium may be coexistent. In biology, this means that the basic bifurcation parameter R_0 is not the threshold value, which is used to determine whether *Rhodopseudomonas palustris* can be harvested successfully or not. In this case, more complicated dynamic properties may occur. There may exist a new threshold value, which is less than R_0 and used to determine whether *Rhodopseudomonas palustris* can be harvested successfully or not.

4. The Global Stability of the Boundary Equilibrium

Global stability properties of the equilibria E_0 or E^* imply that the asymptotic properties of the carbon source, nitrogen source, *Rhodopseudomonas palustris* and flocculants in the culture vessel are not dependent on the initial values C_0, N_0, X_0 and P_0. For the global stability property of the boundary equilibrium E_0 of Model (4), we have the following result.

Theorem 5. *If $R_0 < 1$, then the boundary equilibrium $E_0(1,1,0,1)$ of Model (4) is locally asymptotically stable. Further, if:*

$$R_0 \leq \frac{m_1 \underline{P}+1}{m_1+1} < 1, \tag{10}$$

then the boundary equilibrium E_0 of Model (4) is globally asymptotically stable.

Proof. It is easy to show that the boundary equilibrium $E_0(1,1,0,1)$ of Model (4) is locally asymptotically stable by the characteristic equation of the linearization of Model (4). Next, we prove the global asymptotic stability of the boundary equilibrium E_0 of Model (4).

Since Ω is attractive and positively forward invariant for Model (4), hence it just considers Model (4) in Ω. Define:

$$V_1 = X.$$

Apparently, $V_1(1,1,0,1) = 0$ and V_1 is continuous on Ω. If $R_0 \leq \frac{m_1 \underline{P}+1}{m_1+1} < 1$, then the derivative of V_1 along the solutions of Model (4) is:

$$\begin{aligned} V_1 =& \frac{rC(t)N(t)X(t)}{(K_1+C(t))(K_2+N(t))} - X(t) - m_1 X(t)P(t) \\ \leq& \left(\frac{rC(t)N(t)}{(K_1+C(t))(K_2+N(t))} - 1 - m_1 \underline{P}\right)X(t) \\ \leq& (m_1+1)\left(\frac{rC(t)N(t)}{(K_1+C(t))(K_2+N(t))}\frac{1}{m_1+1} - R_0\right)X(t) \\ \leq& (m_1+1)\left(\frac{r}{(K_1+1)(K_2+1)}\frac{1}{m_1+1} - R_0\right)X(t) = 0 \end{aligned}$$

for any $t \geq 0$. Hence, V_1 is a Lyapunov function of Model (4) on Ω.

Define $E = \left\{(C,N,X,P)\middle|(C,N,X,P) \in \Omega,\ V_1 = 0\right\}$. We have that:

$$E \subset \{(C,N,X,P)|(C,N,X,P) \in \Omega,\ X = 0,\ \text{or } C = 1 \text{ and } N = 1\}.$$

Let $\hat{M}$ be the largest set in E, which is invariant with respect to Model (4). Clearly, $\hat{M}$ is not empty, since $(1,1,0,1) \in \hat{M}$. For any $(C_0, N_0, X_0, P_0) \in \hat{M}$, let $(C(t), N(t), X(t), P(t))$ be the solution of Model (4) with the initial Condition (5). From the invariance of $\hat{M}$, we get $(C(t), N(t), X(t), P(t)) \in \hat{M} \subseteq E$ for any $t \in R$. Thus, we get, for each t, $X(t) = 0$, or $C(t) = 1$, and $N(t) = 1$.

If for some $\tilde{t}$, $C(\tilde{t}) = 1$ and $N(\tilde{t}) = 1$, then we have that $C(\tilde{t}) = N(\tilde{t}) = 0$. Hence, from the first or second equation of Model (4), we obtain that $X(\tilde{t}) = 0$. Thus, for any $t \in R$, we have that $X(t) \equiv 0$.

Subsequently, from the first, second and forth equations of Model (4), we have that, for any $t \in R$,

$$C(t) = 1 - C(t), \quad N(t) = 1 - N(t), \quad P(t) = 1 - P(t).$$

Furthermore, for any $t \in R$, we have that:

$$C(t) = 1 - (1 - C(0))e^{-t}, \quad N(t) = 1 - (1 - N(0))e^{-t}, \quad P(t) = 1 - (1 - P(0))e^{-t}.$$

If $C(0) < 1$ or $N(0) < 1$ or $P(0) < 1$, then $C(t)$, $N(t)$ and $P(t)$ become negative values $(t \to -\infty)$. Then, we obtain that $C(0) = 1$, $N(0) = 1$ and $P(0) = 1$. Hence, we obtain that, for any $t \in R$, $C(t) = N(t) = P(t) \equiv 1$. Therefore, we obtain $\hat{M} = \{(1,1,0,1)\}$. The classical Lyapunov–LaSalle invariance principle shows that E_0 is globally attractive. Since it has been shown that, if $R_0 \leq \frac{m_1 \underline{P}+1}{m_1+1} < 1$, the boundary equilibrium E_0 of Model (4) is locally asymptotically stable. Hence, the boundary equilibrium E_0 of Model (4) is globally asymptotically stable. □

Remark 2. *Let us use some numerical simulations to check the correctness of the theoretical analyses. We set*

$$r = 1.5,\ K_1 = 0.36,\ K_2 = 0.3,\ m_1 = 1.5,\ m_2 = 4,\ r_1 = 0.96,\ r_2 = 1.0001.$$

By the analysis of Theorem 5, we obtain that the boundary equilibrium E_0 of Model (4) is globally asymptotically stable (see Figure 3a).

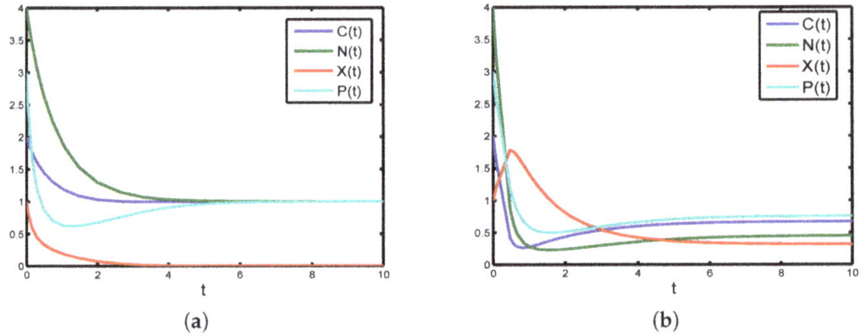

Figure 3. (**a**) Simulations of the global stability of the boundary equilibrium with $r = 1.5,\ K_1 = 0.36,\ K_2 = 0.3,\ m_1 = 1.5,\ m_2 = 4,\ r_1 = 0.96,\ r_2 = 1.0001$; (**b**) simulations of the global stability of the positive equilibrium $E^* \approx (0.6709, 0.4515, 0.3181, 0.7587)$ with $r = 5.8,\ K_1 = 1,\ K_2 = 0.6,\ m_1 = 0.0001,\ m_2 = 1,\ r_1 = 6,\ r_2 = 10$. The initial conditions are $C_0 = 2, N_0 = 4, X_0 = 1, P_0 = 3$.

Furthermore, if the parameters are chosen as follows, $r = 3.5,\ K_1 = 0.36,\ K_2 = 0.3,\ m_1 = 1.5,\ m_2 = 4,\ r_1 = 0.96,\ r_2 = 1.0001$, it is easy to check that the condition $R_0 < 1$ holds, but there are two positive equilibria, $E_1^*(C^*, N^*, X^*, P^*) \approx (0.4783, 0.4565, 1.5782,\ 0.1376)$ and $E_2^*(C^*, N^*, X^*, P^*)$ $\approx (0.9100, 0.9063, 0.1741,\ 0.5895)$. Then, the backward bifurcation phenomenon is illustrated. Therefore, sufficient Condition (10) is reasonable.

5. Stability and Uniform Persistence

In this section, we give the local asymptotic stability of the positive equilibrium and uniform persistence of Model (4). Further, under additional conditions, the global asymptotic stability of the positive equilibrium is obtained by using some techniques of constructing Lyapunov functions.

5.1. The Local Asymptotic Stability of the Positive Equilibrium

For the local asymptotic stability of the positive equilibrium of Model (4), we have the following result.

Theorem 6. *If the positive equilibrium $E^*(C^*, N^*, X^*, P^*)$ of Model (4) exists and the condition:*

$$m_1 < \frac{K_1(1-C^*)}{P^*C^*(K_1+C^*)} + \frac{K_2(1-N^*)}{P^*N^*(K_2+N^*)} \tag{11}$$

holds, then the positive equilibrium $E^*(C^*, N^*, X^*, P^*)$ of Model (4) is locally asymptotically stable.

Proof. For convenience, we assume that:

$$\Delta = \frac{K_1 r_1 N^* X^*}{(K_1 + C^*)^2 (K_2 + N^*)} + \frac{K_2 r_2 C^* X^*}{(K_1 + C^*)(K_2 + N^*)^2}.$$

For the linearized system, the corresponding characteristic equation of Model (4) can be expressed as follows:

$$\lambda^4 + A_1\lambda^3 + A_2\lambda^2 + A_3\lambda + A_4 = 0,$$

where:

$$\begin{aligned}
A_1 =& 3 + m_2X^* + \Delta,\\
A_2 =& \Delta(m_1P^* + 2 + m_2X^*) + 1 + \Delta + 2(1 + m_2X^*) - m_1m_2P^*X^*,\\
A_3 =& \Delta(2m_1P^* + 3 + 2m_2X^*) - 2m_1m_2P^*X^* + 1 + m_2X^*,\\
A_4 =& m_2X^*(\Delta - m_1P^*) + (1 + m_1P^*)\Delta.
\end{aligned}$$

According to the Routh-Hurwitz criterion, we need to show that:

$$A_1A_2 - A_3 > 0, \quad A_3(A_1A_2 - A_3) - A_1^2A_4 > 0, \quad A_4 > 0.$$

For the sake of simplicity, we define:

$$\Delta_1 = A_1, \ \Delta_2 = A_1A_2 - A_3, \ \Delta_3 = A_3(A_1A_2 - A_3) - A_1^2A_4, \ \Delta_4 = A_4\Delta_3.$$

By computation, it can be obtained that:

$$\begin{aligned}
\Delta_1 &= A_1 > 0,\\
\Delta_2 &= m_2X^*(1 + \Delta + m_2X^*)(\Delta - m_1P^*) + (2 + 2m_2X^* + \Delta)(3 + m_2X^*)\\
&\quad + \Delta(6 + m_1P^*) + \Delta(m_2X^* + \Delta)(3 + m_1P^*),\\
\Delta_3 &= P_1(\Delta - m_1P^*) + Q,\\
P_1 &= 2m_2X^*\Delta(1 + \Delta)(3 + m_1P^* + m_2X^*) + 4m_2X^*(1 + \Delta(1 + m_2X^*))\\
&\quad + 4m_2X^*(1 + m_2X^*)(m_2X^* + 3) + 2(m_2X^*)^2\Delta(1 + m_2X^*)\\
&\quad + m_2X^*\Delta(1 + m_2X^*)(2m_1P^* + 3 + 2m_2X^*),\\
Q &= a_1(1 + m_1P^*) + a_2(3 + m_1P^* + m_2X^*) + a_3(1 + m_2X^*) + a_4m_1P^*m_2X^*\Delta\\
&\quad + 2a_5m_1P^*m_2X^* + a_6m_2X^*\Delta + \Delta^2(2 + 3m_2X^*)(3 + 2m_2X^*) + 14\Delta + 4\Delta^2,\\
a_1 &= 2\Delta^3 + 8\Delta^2 + 7\Delta + m_2X^*\Delta(m_2X^* + 7) + m_1P^*\Delta^2(\Delta + 2),\\
a_2 &= \Delta(1 + \Delta)(1 + m_2X^*) + \Delta^3(2 + m_1P^*) + \Delta^2,\\
a_3 &= 2(1 + (1 + m_2X^*)(3 + m_2X^*)),\\
a_4 &= 2(m_2X^*)^2 + (1 + \Delta)^2,\\
a_5 &= 4 + (m_2X^*)^2 + m_1P^*m_2X^* + 2m_2X^*,\\
a_6 &= 4m_2X^* + 4\Delta + 13,\\
\Delta_4 &= A_4\Delta_3.
\end{aligned}$$

If:

$$\Delta = \frac{K_1r_1N^*X^*}{(K_1 + C^*)^2(K_2 + N^*)} + \frac{K_2r_2C^*X^*}{(K_1 + C^*)(K_2 + N^*)^2} > m_1P^*,$$

that is,

$$m_1 < \frac{K_1(1-C^*)}{P^*C^*(K_1+C^*)} + \frac{K_2(1-N^*)}{P^*N^*(K_2+N^*)},$$

we obtain that $\Delta_1 > 0$, $\Delta_2 > 0$, $\Delta_3 > 0$, and $\Delta_4 > 0$. Hence, from the Routh-Hurwitz criterion, we obtain that the positive equilibrium $E^*(C^*, N^*, X^*, P^*)$ of Model (4) is locally asymptotically stable. This completes the proof of Theorem 6. □

5.2. Uniform Persistence

As pointed out in [29,45,50], uniform persistence is an important concept in the cultivation of microorganisms. From a biological perspective, one basic question about biological models involves the long-time survival of the species. To this end, we may want to know whether the constructed model is uniformly persistent with respect to one or more species. That is, the number of the species will keep positive and bounded away from zero for any positive time. In recent years, this kind of research has been particularly common in microorganisms (see, for example, [45]), especially in the simulation of harmful algae growth, where we usually want to predict if the harmful algae will go towards extinction or if blooms of the harmful algae will remain, which directly impacts human health and food webs in aquatic ecosystems. From a mathematical point of view, uniform persistence can give sufficient criteria for the existence of a positive equilibrium for the dissipative system [51].

Model (4) is said to be uniformly persistent if there are positive constants n_1, n_2, n_3, n_4, N_1, N_2, N_3, N_4, such that each positive solution $(C(t), N(t), X(t), P(t))$ of Model (4) satisfies:

$$\begin{aligned} n_1 \le \liminf_{t\to\infty} C(t) \le \limsup_{t\to\infty} C(t) \le N_1, \\ n_2 \le \liminf_{t\to\infty} N(t) \le \limsup_{t\to\infty} N(t) \le N_2, \\ n_3 \le \liminf_{t\to\infty} X(t) \le \limsup_{t\to\infty} X(t) \le N_3, \\ n_4 \le \liminf_{t\to\infty} P(t) \le \limsup_{t\to\infty} P(t) \le N_4. \end{aligned}$$

Now, we give a result on the uniform persistence of Model (4). To proceed, we introduce the following notation and terminology. Denote by $R(t)$ $(t \ge 0)$ the family of solution operators corresponding to Model (4). The ω-limit set $\omega(x)$ of x consists of y , such that there exists a sequence $t_n \to \infty$ as $n \to \infty$ with $R(t_n)x \to y$ as $n \to \infty$. Define:

$$G = \left\{(C, N, X, P) \in R_+^4 \mid \underline{C} \le C \le 1,\ \underline{N} \le N \le 1,\ X \ge 0,\ \underline{P} \le P \le 1\right\},$$

$$G_0 = \{(C, N, X, P) \in G \mid \underline{C} \le C \le 1,\ \underline{N} \le N \le 1,\ X > 0,\ \underline{P} \le P \le 1\},$$

$$\partial G_0 = G \backslash G_0,$$

$M_\partial = \{(C, N, X, P) \in \partial G_0 \mid R(t)(C, N, X, P)$ satisfies Model (4) and $R(t)(C, N, X, P) \in \partial G_0,\ \forall t \ge 0\}$,

$$\Omega(M_\partial) = \cup_{x \in M_\partial} \omega(x).$$

From Theorem 2, we obtain that G is positively invariant corresponding to $R(t)$. It is easy to see that G_0 is also positively invariant.

Clearly, ∂G_0 is relatively compact in G. Now, we show that $\Omega(M_\partial) = \{(1, 1, 0, 1)\}$.

In fact, $\{(1, 1, 0, 1)\} \subseteq \Omega(M_\partial)$. For any $(C(0),\ N(0),\ X(0),\ P(0)) \in M_\partial$, it has that, for all $t \ge 0$, $X(t) \equiv 0$ and $\lim_{t\to+\infty} C(t) = \lim_{t\to+\infty} N(t) = \lim_{t\to+\infty} P(t) = 1$. Thus, $\Omega(M_\partial) = \{(1, 1, 0, 1)\}$.

Theorem 7. *If $R_0 > 1$, then Model (4) is uniformly persistent.*

Proof. By Theorem 1, it can be obtained that nonnegative solutions of Model (4) are point dissipative. According to the definitions above, it suffices to show that ∂G_0 repels uniformly nonnegative solutions of Model (4). It is obvious that there is only one equilibrium E_0 in M_∂.

We now show that $W^s(E_0) \cap G_0 = \varnothing$. Assume $W^s(E_0) \cap G_0 \ne \varnothing$, then there exists a positive solution of Model (4), such that $\lim_{t\to\infty} (C(t), N(t), X(t), P(t)) = (1, 1, 0, 1)$. Assume:

$$U(C(t), N(t), P(t)) = \frac{rC(t)N(t)}{(K_1 + C(t))(K_2 + N(t))} - 1 - m_1 P(t),$$

then:

$$U(1, 1, 1) = \frac{r}{(K_1 + 1)(K_2 + 1)} - 1 - m_1.$$

Then, $r > (K_1 + 1)(K_2 + 1)(1 + m_1)$ is equivalent to $U(1, 1, 1) > 0$.

Since $\lim_{t\to\infty} U(C(t), N(t), P(t)) = U(1, 1, 1) > 0$, there exists $T > 0$, such that, for any $t \ge T$,

$$U(C(t), N(t), P(t)) > \tfrac{1}{2} U(1, 1, 1) > 0.$$

By the third equation of Model (4), we obtain that, for any $t \geq T$,

$$\begin{aligned} X(t) &= \tfrac{rC(t)N(t)X(t)}{(K_1+C(t))(K_2+N(t))} - X(t) - m_1 P(t)X(t) \\ &= U(C(t), N(t), P(t))X(t) \\ &> \tfrac{1}{2}U(1,1,1)X(t). \end{aligned}$$

This implies that $X(t) \to +\infty$ as $t \to +\infty$, which leads to a contradiction. Thus, $W^s(E_0) \cap G_0 = \varnothing$.

In the following, let us show that E_0 is factually isolated. That is, there exists some neighborhood U of E_0, such that E_0 is the largest invariant set in U. In fact, for sufficiently small positive constant ε, let us choose:

$$U = U(E_0) = \{(C, N, X, P) \in G | 1 - C < \varepsilon,\ 1 - N < \varepsilon,\ X < \varepsilon,\ 1 - P < \varepsilon\}.$$

We show that E_0 is the largest invariant set of U for some ε.

If not, for any sufficiently small ε, there exists some invariant set $W(W \subset U)$, such that $W \backslash E_0$ is not empty. Let $(C_0, N_0, X_0, P_0) \in W \backslash E_0$ and $(C(t), N(t), X(t), P(t))$ be the solution of Model (4) with the initial function (5). Then, we have that $(C(t), N(t), X(t), P(t)) \in U, t \in (-\infty,\ +\infty)$. From the third equation of Model (4), we obtain that, for $t \in (-\infty, +\infty)$,

$$X(t) \geq \frac{r(1-\varepsilon)(1-\varepsilon)X(t)}{(K_1+1-\varepsilon)(K_2+1-\varepsilon)} - X(t) - m_1 X(t).$$

Since $r > (K_1+1)(K_2+1)(m_1+1)$, we can choose sufficiently small ε, such that:

$$r(1-\varepsilon)(1-\varepsilon) > (K_1+1-\varepsilon)(K_2+1-\varepsilon)(1+m_1).$$

If $X_0 > 0$, then we have $X(t) \to +\infty\ (t \to +\infty)$, which contracts the boundedness of $X(t)$. Hence, we get $X_0 = 0$. From the third equation of Model (4), we obtain that $X(t) \equiv 0$. From the first, second and forth equations of Model (4), for $t \in (-\infty, +\infty)$, we obtain that $C(t) = 1 - C(t)$, $N(t) = 1 - N(t)$ and $P(t) = 1 - P(t)$. Hence, we must have that $C_0 = N_0 = P_0 = 1$. Therefore, $(C_0, N_0, X_0, P_0) = (1, 1, 0, 1)$, which is a contradiction. Thus, we obtain that E_0 is factually isolated.

Clearly, E_0 is acyclic in M_∂. By paper [52], ∂G_0 repels uniformly nonnegative solutions of Model (4). It then follows that Model (4) is persistent.

Define $p : X \to R_+$ by $p(C, N, X, P) = X,\ (C, N, X, P) \in G$. Obviously, we obtain that $G_0 = p^{-1}$ $(0, +\infty)$ and $\partial G_0 = p^{-1}(0)$. Thus, by ([53], Theorem 3), we have $\liminf_{t\to\infty}(C(t), N(t), X(t), P(t)) \geq (\eta, \eta, \eta, \eta)$. It then follows that Model (4) is uniformly persistent. The proof of Theorem 7 is completed. □

According to [51,53], we easily obtain the following result.

Corollary 8. *If $R_0 > 1$, then Model (4) has one global attractor A_0 and also has at least one positive equilibrium $E^*(C^*, N^*, X^*, P^*) \in A_0$.*

5.3. The Global Asymptotic Stability of the Positive Equilibrium

In this subsection, the global asymptotic stability of the positive equilibrium $E^*(C^*, N^*, X^*, P^*)$ of Model (4) is studied.

Theorem 9. *Assume that $R_0 > 1$ and Model (4) has a unique positive equilibrium $E^*(C^*, N^*, X^*, P^*)$. Let:*

$$\begin{aligned}
D_1 &= r_1(2 + m_1 P^*)(K_1 + C^*)(K_2 + N^*) - r^2 N^*, \\
D_2 &= r_2(2 + m_1 P^*)(K_1 + C^*)(K_2 + N^*) - r^2, \\
D_3 &= b_{13}^2 + b_{23}^2 + 4b_{33}, \\
D_4 &= 4b_{33}b_{44} + \left(b_{23}^2 + b_{13}^2\right)b_{44} + \left(b_{14}^2 + b_{24}^2\right)b_{33} - b_{34}^2 + b_{23}b_{24}b_{34} + b_{13}b_{14}b_{34}.
\end{aligned}$$

If $D_1 \geq 0$, $D_2 \geq 0$, $D_3 < 0$ and $D_4 > 0$ hold, then the positive equilibrium $E^*(C^*, N^*, X^*, P^*)$ of Model (4) is globally asymptotically stable. Here:

$$\begin{aligned}
b_{13} &= \frac{2r_1 + r_1 m_1 P^*}{r} + \frac{r}{(K_1+1)(K_2+1)(K_1+C^*)} - \frac{rN^*}{(K_1+C^*)(K_2+N^*)}, \\
b_{14} &= \frac{r_1 m_1 \alpha}{r}, \ b_{23} = \frac{2r_2 + r_2 m_1 P^*}{r}, \ b_{24} = \frac{r_2 m_1 \alpha}{r}, \ b_{33} = -\frac{(r_1^2 + r_2^2)(1 + m_1 P^*)}{r^2}, \\
b_{34} &= \frac{(r_1^2 + r_2^2) m_1 \alpha}{r^2} + m_1 + m_2, \ b_{44} = -(1 + m_2 X^*).
\end{aligned}$$

Proof. Let us consider the following Lyapunov function on A_0,

$$\begin{aligned}
V_2 = \ & \tfrac{1}{2}\left(C - C^* + \tfrac{r_1}{r}(X - X^*)\right)^2 + \tfrac{1}{2}\left(N - N^* + \tfrac{r_2}{r}(X - X^*)\right)^2 \\
& + X - X^* - X^* \ln \tfrac{X}{X^*} + P - P^* - P^* \ln \tfrac{P}{P^*}.
\end{aligned}$$

Clearly, $V_2(C^*, N^*, X^*, P^*) = 0$, and V_2 is positive definite with respect to $E^*(C^*, N^*, X^*, P^*)$. The derivative of V_2 along the solutions of Model (4) is:

$$\begin{aligned}
V_2 = \ & \left(C(t) - C^* + \tfrac{r_1}{r}(X(t) - X^*)\right)\left(C(t) + \tfrac{r_1}{r}X(t)\right) \\
& + \left(N(t) - N^* + \tfrac{r_2}{r}(X(t) - X^*)\right)\left(N(t) + \tfrac{r_2}{r}X(t)\right) \\
& + \tfrac{X(t) - X^*}{X(t)} X(t) + \tfrac{P(t) - P^*}{P(t)} P(t) \\
= \ & \left(C(t) - C^* + \tfrac{r_1}{r}(X(t) - X^*)\right)\left(1 - C(t) - \tfrac{r_1}{r}X(t) - \tfrac{r_1 m_1 X(t) P(t)}{r}\right) \\
& + \left(N(t) - N^* + \tfrac{r_2}{r}(X(t) - X^*)\right)\left(1 - N(t) - \tfrac{r_2}{r}X(t) - \tfrac{r_2 m_1 X(t) P(t)}{r}\right) \\
& + (X(t) - X^*)\left(\tfrac{rC(t)N(t)}{(K_1 + C(t))(K_2 + N(t))} - 1 - m_1 P(t)\right) \\
& + \tfrac{P(t) - P^*}{P(t)}\left(1 - P(t) - m_2 X(t) P(t)\right)
\end{aligned}$$

for any $t \geq 0$.

Noting that:

$$\begin{cases}
1 - C^* - \frac{r_1 C^* N^* X^*}{(K_1 + C^*)(K_2 + N^*)} = 0, \\
1 - N^* - \frac{r_2 C^* N^* X^*}{(K_1 + C^*)(K_2 + N^*)} = 0, \\
\frac{rC^* N^* X^*}{(K_1 + C^*)(K_2 + N^*)} - X^* - m_1 X^* P^* = 0, \\
1 - P^* - m_2 X^* P^* = 0.
\end{cases}$$

Thus, we have that:

$$\begin{aligned}
V_2 = \ & \left(C(t) - C^* + \tfrac{r_1}{r}(X(t) - X^*)\right)\left(C^* - C(t) + \tfrac{r_1}{r}(X^* - X(t)) + \tfrac{r_1 m_1}{r}(X^* P^* - X(t)P(t))\right) \\
& + \left(N(t) - N^* + \tfrac{r_2}{r}(X(t) - X^*)\right)\left(N^* - N(t) + \tfrac{r_2}{r}(X^* - X(t)) + \tfrac{r_2 m_1}{r}(X^* P^* - X(t)P(t))\right) \\
& + (X(t) - X^*)\left(\tfrac{rC(t)N(t)}{(K_1 + C(t))(K_2 + N(t))} - \tfrac{rC^* N^*}{(K_1 + C^*)(K_2 + N^*)}\right) \\
& + m_1(X(t) - X^*)(P^* - P(t)) + \tfrac{P(t) - P^*}{P(t)}\left(P^* - P(t) + m_2(X^* P^* - X(t)P(t))\right)
\end{aligned}$$

for any $t \geq 0$.

For any $t \geq 0$, we have that:

$$\begin{aligned}
&\frac{rC(t)N(t)}{(K_1+C(t))(K_2+N(t))} - \frac{rC^*N^*}{(K_1+C^*)(K_2+N^*)} \\
&= -\frac{rC^*N^*}{(K_1+C^*)(K_2+N^*)} + \frac{rC(t)N^*}{(K_1+C^*)(K_2+N^*)} - \frac{rC(t)N^*}{(K_1+C^*)(K_2+N^*)} \\
&+\frac{rC(t)N(t)}{(K_1+C^*)(K_2+N^*)} - \frac{rC(t)N(t)}{(K_1+C^*)(K_2+N^*)} + \frac{rC(t)N(t)}{(K_1+C^*)(K_2+N(t))} \\
&-\frac{rC(t)N(t)}{(K_1+C^*)(K_2+N(t))} + \frac{rC(t)N(t)}{(K_1+C(t))(K_2+N(t))}.
\end{aligned}$$

Hence, it can be obtained that, for any $t \geq 0$,

$$\begin{aligned}
V_2 = \; & (C(t) - C^* + \tfrac{r_1}{r}(X(t) - X^*))\big(C^* - C(t) + \tfrac{r_1}{r}(X^* - X(t)) \\
&+\tfrac{r_1 m_1 P^*}{r}(X^* - X(t)) + \tfrac{r_1 m_1 X(t)}{r}(P^* - P(t))\big) \\
&+(N(t) - N^* + \tfrac{r_2}{r}(X(t) - X^*))\big(N^* - N(t) + \tfrac{r_2}{r}(X^* - X(t)) \\
&+\tfrac{r_2 m_1 P^*}{r}(X^* - X(t)) + \tfrac{r_2 m_1 X(t)}{r}(P^* - P(t))\big) \\
&+(X(t) - X^*)\Big(-\tfrac{rC^*N^*}{(K_1+C^*)(K_2+N^*)} + \tfrac{rC(t)N^*}{(K_1+C^*)(K_2+N^*)} \\
&-\tfrac{rC(t)N^*}{(K_1+C^*)(K_2+N^*)} + \tfrac{rC(t)N(t)}{(K_1+C^*)(K_2+N^*)} - \tfrac{rC(t)N(t)}{(K_1+C^*)(K_2+N^*)} \\
&+\tfrac{rC(t)N(t)}{(K_1+C^*)(K_2+N(t))} - \tfrac{rC(t)N(t)}{(K_1+C^*)(K_2+N(t))} + \tfrac{rC(t)N(t)}{(K_1+C(t))(K_2+N(t))}\Big) \\
&-\tfrac{(P(t)-P^*)^2}{P(t)} + \tfrac{m_2(P(t)-P^*)}{P(t)}(X^*(P^* - P(t)) + P(t)(X^* - X(t))) \\
&+m_1(X(t) - X^*)(P^* - P(t)).
\end{aligned}$$

Motivated by the papers mentioned in [50,54], the derivative of V_2 along the solutions of Model (4) is represented in quadratic form. Then, we obtain that:

$$\begin{aligned}
V_2 = \; & -(C(t) - C^*)^2 - (N(t) - N^*)^2 + \Big(\tfrac{2r_1 + r_1 m_1 P^*}{r} + \tfrac{rC(t)N(t)}{(K_1+C(t))(K_2+N(t))(K_1+C^*)} \\
&- \tfrac{rN^*}{(K_1+C^*)(K_2+N^*)}\Big)(C(t) - C^*)(X^* - X(t)) + \Big(\tfrac{2r_2 + r_2 m_1 P^*}{r} \\
&+ \tfrac{rC(t)N(t)}{(K_2+N(t))(K_2+N^*)(K_1+C^*)} - \tfrac{rC(t)}{(K_1+C^*)(K_2+N^*)}\Big)(N(t) - N^*)(X^* - X(t)) \\
&- \tfrac{1+m_2 X^*}{P(t)}(P(t) - P^*)^2 + \tfrac{r_1 m_1 X}{r}(C(t) - C^*)(P^* - P(t)) \\
&- \tfrac{(r_1^2 + r_2^2)(1+m_1 P^*)}{r^2}(X(t) - X^*)^2 + \left(\tfrac{(r_1^2 + r_2^2) m_1 X(t)}{r^2} + m_1 + m_2\right)(X^* - X(t))(P^* - P(t)) \\
&+ \tfrac{r_2 m_1 X(t)}{r}(N(t) - N^*)(P^* - P(t))
\end{aligned}$$

for any $t \geq 0$.

If the condition $D_1 \geq 0$ holds, we obtain that, for any $t \geq 0$,

$$\frac{2r_1 + r_1 m_1 P^*}{r} + \frac{rC(t)N(t)}{(K_1+C(t))(K_2+N(t))(K_1+C^*)} - \frac{rN^*}{(K_1+C^*)(K_2+N^*)} \geq 0.$$

Furthermore, from the condition $D_2 \geq 0$ and $C(t) \leq 1$ for any $t \geq 0$, we obtain that:

$$\frac{2r_2 + r_2 m_1 P^*}{r} + \frac{rC(t)N(t)}{(K_2 + N(t))(K_2 + N^*)(K_1 + C^*)} - \frac{rC(t)}{(K_1 + C^*)(K_2 + N^*)} \geq 0.$$

These conditions ensure that the coefficients of $(C(t)-C^*)(X(t)-X^*)$ and $(N(t)-N^*)(X(t)-X^*)$ are nonnegative. Thus, we obtain that, for any $t \geq 0$,

$$\begin{aligned} V_2 \leq \; & -(C(t)-C^*)^2-(N(t)-N^*)^2+\Big(\tfrac{2r_1+r_1m_1P^*}{r}+\tfrac{r}{(K_1+1)(K_2+1)(K_1+C^*)} \\ & -\tfrac{rN^*}{(K_1+C^*)(K_2+N^*)}\Big)|C(t)-C^*||X^*-X(t)| \\ & +\Big(\tfrac{2r_2+r_2m_1P^*}{r}\Big)|N(t)-N^*||X^*-X(t)|-(1+m_2X^*)(P(t)-P^*)^2 \\ & +\tfrac{r_1m_1\alpha}{r}|C(t)-C^*||P^*-P(t)|-\tfrac{(r_1^2+r_2^2)(1+m_1P^*)}{r^2}(X(t)-X^*)^2 \\ & +\left(\tfrac{(r_1^2+r_2^2)m_1\alpha}{r^2}+m_1+m_2\right)|X^*-X(t)||P^*-P(t)|+\tfrac{r_2m_1\alpha}{r}|N(t)-N^*||P^*-P(t)| \\ = \; & Y^T(t)WY(t), \end{aligned}$$

where $Y(t)=(|C(t)-C^*|,|N(t)-N^*|,|X(t)-X^*|,|P(t)-P^*|)$. W is a symmetric 4×4 matrix and $W=\{b_{ij}\}_{1\leq i,j\leq 4}$ with:

$$W=\begin{pmatrix} b_{11} & \frac{1}{2}b_{12} & \frac{1}{2}b_{13} & \frac{1}{2}b_{14} \\ \frac{1}{2}b_{12} & b_{22} & \frac{1}{2}b_{23} & \frac{1}{2}b_{24} \\ \frac{1}{2}b_{13} & \frac{1}{2}b_{23} & b_{33} & \frac{1}{2}b_{34} \\ \frac{1}{2}b_{14} & \frac{1}{2}b_{24} & \frac{1}{2}b_{34} & b_{44} \end{pmatrix}.$$

Here, $b_{11}=-1,\ b_{12}=0,\ b_{22}=-1$. The other parameters have the same definitions as in Theorem 9. Then, W is negative definite. By Lemma 6.2 provided in [50], we can obtain that the real quadratic form $Y^T(t)WY(t)$ is negative definite. Then, by using some classical analysis techniques of differential equations, the positive equilibrium $E^*(C^*,N^*,X^*,P^*)$ of Model (4) is globally asymptotically stable. □

Remark 3. *Obviously, it is easy to obtain that* $b_{11}=-1<0$, *and:*

$$\begin{vmatrix} b_{11} & \frac{1}{2}b_{12} \\ \frac{1}{2}b_{12} & b_{22} \end{vmatrix} = 1 > 0.$$

If the conditions $D_3<0$ *and* $D_4>0$ *hold, we obtain that:*

$$\begin{vmatrix} b_{11} & \frac{1}{2}b_{12} & \frac{1}{2}b_{13} \\ \frac{1}{2}b_{12} & b_{22} & \frac{1}{2}b_{23} \\ \frac{1}{2}b_{13} & \frac{1}{2}b_{23} & b_{33} \end{vmatrix} < 0,$$

and:

$$\begin{vmatrix} b_{11} & \frac{1}{2}b_{12} & \frac{1}{2}b_{13} & \frac{1}{2}b_{14} \\ \frac{1}{2}b_{12} & b_{22} & \frac{1}{2}b_{23} & \frac{1}{2}b_{24} \\ \frac{1}{2}b_{13} & \frac{1}{2}b_{23} & b_{33} & \frac{1}{2}b_{34} \\ \frac{1}{2}b_{14} & \frac{1}{2}b_{24} & \frac{1}{2}b_{34} & b_{44} \end{vmatrix} > 0.$$

Next, we illustrate that the conditions given in Theorem 9 are reasonable. We set $r=5.8,\ K_1=1,\ K_2=0.6,\ m_1=0.0001,\ m_2=1,\ r_1=6,\ r_2=10$. *It is easy to see that there exists a unique positive equilibrium* $E^*(C^*,N^*,X^*,P^*)\approx(0.6709,0.4515,0.3181,0.7587)$, *and the conditions* D_1*–*D_4 *hold. By Theorem 9, the positive equilibrium of Model (4) is globally asymptotically stable (see Figure 3b).*

In general, it is very difficult to obtain the global stability properties of the positive equilibrium of Model (4). In this paper, we have obtained the sufficient conditions to ensure the global stability properties of the positive equilibrium by constructing a suitable Lyapunov function. Because of the complexity of the conditions, it is difficult to grasp biological intuition. However, from the numerical simulations, we have found some interesting biological phenomena. The conditions D_1*–*D_4 *can be satisfied if the flocculation rate of microorganisms* (m_1) *is small enough (see, Figure 3b). From a biological point of view, these conditions are reasonable. For more detailed biological considerations, we will leave it for further investigation.*

6. Control Strategies

In this section, some control strategies are provided by suitable theoretical analysis. If $R_0 < 1$ holds, then, we have from Theorem 5 that the boundary equilibrium E_0 of Model (4) is locally asymptotically stable. We note that the condition $R_0 < 1$ is equivalent to the following inequality,

$$R_0 = \frac{r}{(K_1+1)(K_2+1)(m_1+1)} < 1. \tag{12}$$

All of the parameters r, K_1, K_2 and m_1 in (12) are defined as in Model (4). (12) can be further written as the following form,

$$\frac{r}{\left(\frac{K_1}{C^0}+1\right)\left(\frac{K_2}{N^0}+1\right)(m_1 P^0+D)} < 1. \tag{13}$$

Here, all of the parameters r, K_1, K_2, m_1, C^0, N^0, P^0 and D in (13) are defined as in Model (3).

In view of the biological meanings of the parameters in Model (3) and Condition (13), Theorem 5 indicates that the concentration of *Rhodopseudomonas palustris* in the chemostat tends to zero, and the concentration of the carbon source, nitrogen source and flocculants may tend to the constant values C^0, N^0 and P^0, respectively, as time t increases, if one of the following two cases occurs: (a) reducing the absorption of *Rhodopseudomonas palustris* or the carbon input concentration, or the nitrogen input concentration; (b) improving the velocity or the flocculation effect or flocculant input concentration. These cases are reasonable, since they imply the insufficient sources for *Rhodopseudomonas palustris* to grow. Hence, in the environmental science field, it can be used to remove algae and heavy metals.

From Theorem 7, we have that Model (3) is uniformly persistent if $R_0 > 1$. This means that the concentrations of the carbon source, nitrogen source, *Rhodopseudomonas palustris* and flocculants in the chemostat may be ultimately maintained at some positive constant values, as time t increases, if one of the following two cases occurs: (a) improving the absorption of *Rhodopseudomonas palustris*, or the carbon input concentration, or the nitrogen input concentration; (b) reducing the velocity or the flocculation effect or flocculant input concentration.

These control strategies can be performed by numerical simulations.

In the following, for convenience, we simulate the extinction or persistence of microorganism ($X(t)$) numerically by using (12) and Model (4).

If the parameters are chosen as in Table A2 (see Appendix A), *Rhodopseudomonas palustris* in the chemostat will tend to extinction (see Figure 4a).

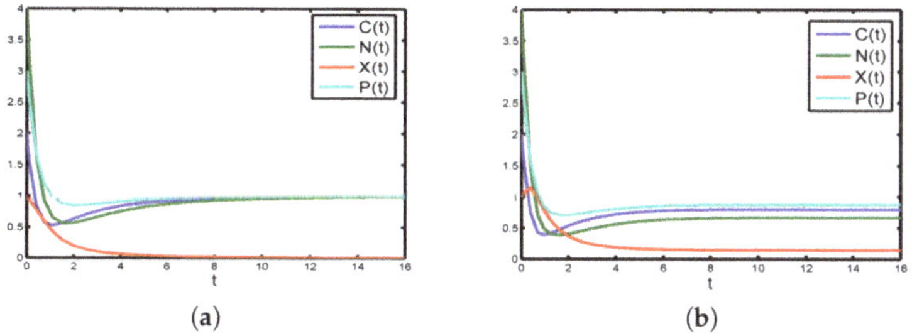

Figure 4. (**a**) *Rhodopseudomonas palustris* in the chemostat will tend to extinction with the parameters in Table A1. (**b**) *Rhodopseudomonas palustris* in the chemostat will tend to be constant if the absorption of *Rhodopseudomonas palustris* (r) is improved. The initial conditions are $C_0 = 2$, $N_0 = 4$, $X_0 = 1$, $P_0 = 3$.

If the absorption of *Rhodopseudomonas palustris* (r) is improved from $r = 5.8$ to $r = 8$ and the other parameters are the same as Table A2, *Rhodopseudomonas palustris* in the chemostat will tend to be constant (see Figure 4b).

If the flocculation effect (m_1) is reduced from $m_1 = 1$ to $m_1 = 0.1$, *Rhodopseudomonas palustris* in the chemostat will tend to be constant (see Figure 5a).

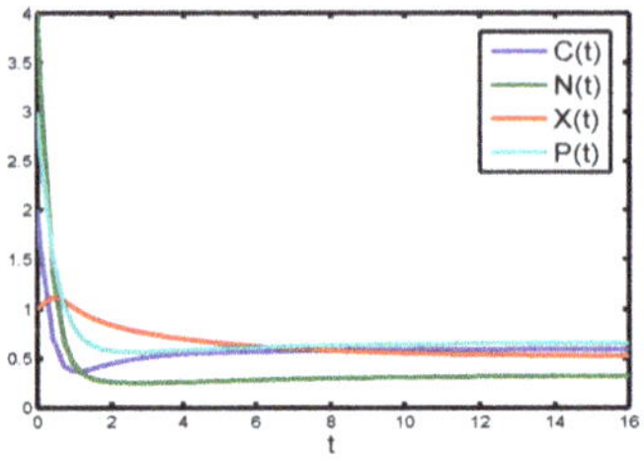

(a)

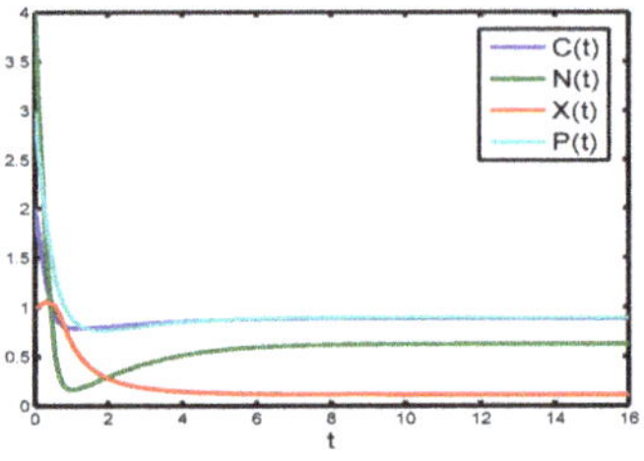

(b)

Figure 5. (**a**) *Rhodopseudomonas palustris* in the chemostat will tend to be constant if the flocculation effect (m_1) is reduced. (**b**) *Rhodopseudomonas palustris* in the chemostat will tend to be constant if the Michaelis–Menten constant of carbon (K_1) is reduced. The initial conditions are $C_0 = 2$, $N_0 = 4$, $X_0 = 1$, $P_0 = 3$.

If Michaelis–Menten constant of carbon (K_1) is reduced from $K_1 = 1$ to $K_1 = 0.5$, *Rhodopseudomonas palustris* in the chemostat will tend to be constant (see Figure 5b).

If Michaelis–Menten constant of nitrogen (K_2) is reduced from $K_2 = 1$ to $K_2 = 0.25$, *Rhodopseudomonas palustris* in the chemostat will tend to be constant (see Figure 6).

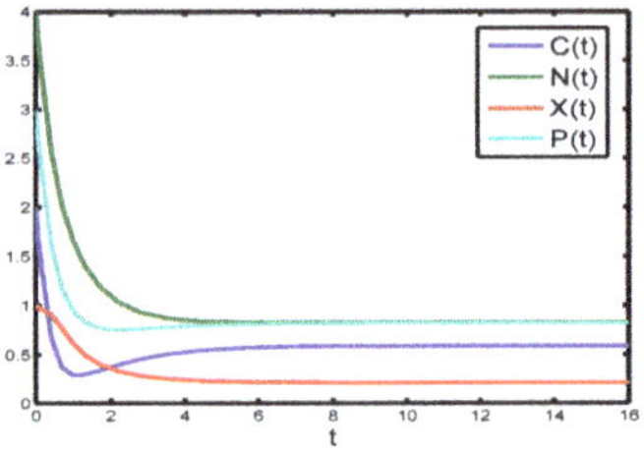

Figure 6. *Rhodopseudomonas palustris* in the chemostat will tend to be constant if the Michaelis–Menten constant of nitrogen (K_2) is reduced. The initial conditions are $C_0 = 2$, $N_0 = 4$, $X_0 = 1$, $P_0 = 3$.

7. Discussion and Conclusions

In the paper, based on some biological considerations and chemostat models, a dynamic model governed by ordinary differential equations with four variables (carbon source, nitrogen source, *Rhodopseudomonas palustris* and flocculants) is presented. There is a boundary equilibrium and at most five positive equilibria for the proposed model. To give a theoretical analysis for the existence of all of the positive equilibria of Model (4), the method of the Descartes rule of signs is applied to the classifications of the positive roots of a fifth order algebraic equation.

The local and global stability properties of the boundary equilibrium of Model (4) have been studied in detail. An interesting phenomenon of backward and forward bifurcations is observed. That is, there may exist two positive equilibria even if the condition $R_0 < 1$ holds. Hence, sufficient Condition (10) to ensure the global stability of the boundary equilibrium is reasonable in mathematics.

The local stability of the positive equilibrium of Model (4) is also carried out. From Condition (11), we have that the positive equilibrium is locally asymptotically stable when the flocculation coefficient m_1 is small enough. Hence, Condition (11) is also reasonable in biology.

Uniform persistence of Model (4) has also been completely studied under the condition $R_0 > 1$. Uniform persistence has very important significance both in mathematics and biology, and it characterizes the long-term survival of some microorganisms [45].

Finally, some control strategies are provided by simple theoretical analysis. From Theorem 5, we have that *Rhodopseudomonas palustris* in the chemostat will tend to extinction if $R_0 < 1$. In this case, these control strategies can be applied to remove *Cyanobacteria*, which are well known to produce a variety of toxins and have serious harm on human health. From Theorem 7, we have that *Rhodopseudomonas palustris* in the chemostat will tend to be positive constant if $R_0 > 1$. In this case, these control strategies can be widely used for the collection of useful microorganisms.

It is well-known that the existence of time delays is inevitable in biology. For example, in the cultivation of microorganisms, there are always time delays in the process of transferring nutrients and the uptake of nutrients. Hence, chemostat models with time delays that account for the time lapsing between the uptake of nutrients by cells and the incorporation of these nutrients as biomass have been given much attention [40,55–57]. Based on Model (3), it may have the following more general form with time delays,

$$\begin{cases} \frac{dC(t)}{dt} = \left(C^0 - C(t)\right)D - \frac{r_1 C(t)N(t)X(t)}{\delta_1(K_1+C(t))(K_2+N(t))} + \rho_1 X(t-\sigma), \\ \frac{dN(t)}{dt} = \left(N^0 - N(t)\right)D - \frac{r_2 C(t)N(t)X(t)}{\delta_2(K_1+C(t))(K_2+N(t))} + \rho_2 X(t-\sigma), \\ \frac{dX(t)}{dt} = \frac{re^{-d_1\tau}C(t-\tau)N(t-\tau)X(t-\tau)}{(K_1+C(t-\tau))(K_2+N(t-\tau))} - DX(t) - d_1X(t) - m_1X(t)P(t), \\ \frac{dP(t)}{dt} = \left(P^0 - P(t)\right)D - m_2X(t)P(t). \end{cases} \tag{14}$$

In Model (14), the constants $\rho_1 \geq 0$ and $\rho_2 \geq 0$ are the rate constants at which the carbon source and nitrogen source are recycled because of the death of *Rhodopseudomonas palustris*. The constant $\sigma \geq 0$ is a fixed time during which the carbon source and nitrogen source are released completely from dead *Rhodopseudomonas palustris*. The constant $\tau \geq 0$ denotes the time delay involved in the conversion of nutrients to *Rhodopseudomonas palustris*. The factor $e^{-d_1\tau}$ is the probability constant at which *Rhodopseudomonas palustris* remains in the culture vessel during the conversion process. The theoretical analysis of Model (14) will be studied separately.

Acknowledgments: The authors thank the editors and anonymous reviewers for their valuable comments, especially for the biological considerations, numerical simulations and the proofs of the main results. The second author is supported by the National Natural Science Foundation of China (11471034).

Author Contributions: Wei Wang and Wanbiao Ma conceived of the study, drafted the manuscript and finished the mathematical analysis and numerical simulations. Hai Yan gave some suggestions for constructing the model.

Conflicts of Interest: The authors declare no conflict of interest.

Appendix

In this section, let us apply the Descartes rule of signs to the classifications of the positive roots of $f(X^*)$ [47]. Let m represent the number of sign changes of the coefficients a, b, c, d, e, f of $f(X^*)$ and n represent the number of the positive roots.

Table A1. The number of the positive roots of $f(X^*)$ for $R_0 < 1$ and $R_0 > 1$.

Cases	a	b	c	d	e	f	R_0	m	n
1	−	+	+	+	+	+	$R_0 < 1$	1	1
	−	+	+	+	+	−	$R_0 > 1$	2	0, 2
2	−	+	+	+	−	+	$R_0 < 1$	3	1, 3
	−	+	+	+	−	−	$R_0 > 1$	2	0, 2
3	−	+	+	−	+	+	$R_0 < 1$	3	1, 3
	−	+	+	−	+	−	$R_0 > 1$	4	0, 2, 4
4	−	+	+	−	−	+	$R_0 < 1$	3	1, 3
	−	+	+	−	−	−	$R_0 > 1$	2	0, 2
5	−	+	−	+	+	+	$R_0 < 1$	3	1, 3
	−	+	−	+	+	−	$R_0 > 1$	4	0, 2, 4
6	−	+	−	+	−	+	$R_0 < 1$	5	1, 3, 5
	−	+	−	+	−	−	$R_0 > 1$	4	0, 2, 4
7	−	+	−	−	+	+	$R_0 < 1$	3	1, 3
	−	+	−	−	+	−	$R_0 > 1$	4	0, 2, 4
8	−	+	−	−	−	+	$R_0 < 1$	3	1, 3
	−	+	−	−	−	−	$R_0 > 1$	2	0, 2
9	−	−	+	+	+	+	$R_0 < 1$	1	1
	−	−	+	+	+	−	$R_0 > 1$	2	0, 2
10	−	−	+	+	−	+	$R_0 < 1$	3	1, 3
	−	−	+	+	−	−	$R_0 > 1$	2	0, 2
11	−	−	+	−	+	+	$R_0 < 1$	3	1, 3
	−	−	+	−	+	−	$R_0 > 1$	4	0, 2, 4
12	−	−	+	−	−	+	$R_0 < 1$	3	1, 3
	−	−	+	−	−	−	$R_0 > 1$	2	0, 2
13	−	−	−	+	+	+	$R_0 < 1$	1	1
	−	−	−	+	+	−	$R_0 > 1$	2	0, 2
14	−	−	−	+	−	+	$R_0 < 1$	3	1, 3
	−	−	−	+	−	−	$R_0 > 1$	2	0, 2
15	−	−	−	−	+	+	$R_0 < 1$	1	1
	−	−	−	−	+	−	$R_0 > 1$	2	0, 2
16	−	−	−	−	−	+	$R_0 < 1$	1	1
	−	−	−	−	−	−	$R_0 > 1$	0	0

Table A2. Parameter values used in the simulations of control strategies shown in Figures 3–6.

Description	Parameter	Value
the growth rate of *Rhodopseudomonas palustris*	r	5.8
the quantity of decreasing of carbon source	r_1	6
the quantity of decreasing of nitrogen source	r_2	10
the flocculation effect	m_1	1
the flocculation ratio	m_2	1
Michaelis–Menten constant of carbon	K_1	1
Michaelis–Menten constant of nitrogen	K_2	0.6

Appendix

In Section 3, we have discussed the phenomena of backward and forward bifurcations by numerical simulations. In this section, the center manifold theory is used on Model (4) to obtain the rigorous result (see, for example, [40,47,48]).

Let $C = x_1$, $N = x_2$, $X = x_3$, $P = x_4$, so that Model (4) can be re-written in the following form:

$$\begin{cases} \frac{dx_1(t)}{dt} = 1 - x_1 - \frac{r_1 x_1 x_2 x_3}{(K_1+x_1)(K_2+x_2)} = g_1, \\ \frac{dx_2(t)}{dt} = 1 - x_2 - \frac{r_2 x_1 x_2 x_3}{(K_1+x_1)(K_2+x_2)} = g_2, \\ \frac{dx_3(t)}{dt} = \frac{r x_1 x_2 x_3}{(K_1+x_1)(K_2+x_2)} - x_3 - m_1 x_3 x_4 = g_3, \\ \frac{dx_4(t)}{dt} = 1 - x_4 - m_2 x_3 x_4 = g_4. \end{cases} \tag{B1}$$

The Jacobian matrix of Model (15) at $E_0(1,1,0,1)$ is given by:

$$J(E_0) = \begin{pmatrix} -1 & 0 & \frac{-r_1}{(K_1+1)(K_2+1)} & 0 \\ 0 & -1 & \frac{-r_2}{(K_1+1)(K_2+1)} & 0 \\ 0 & 0 & \frac{r}{(K_1+1)(K_2+1)} - 1 - m_1 & 0 \\ 0 & 0 & -m_2 & -1 \end{pmatrix}.$$

Suppose r is chosen as a bifurcation parameter. Solving $R_0 = 1$ gives:

$$r = r^* = (K_1+1)(K_2+1)(m_1+1).$$

Eigenvectors of $J(E_0)|_{r=r^*}$

It can be shown that the Jacobian matrix of Model (15) at $r = r^*$ has a right eigenvector (corresponding to the zero eigenvalue) given by $\omega = (\omega_1, \omega_2, \omega_3, \omega_4)^T$, where:

$$\omega_1 = \frac{-r_1}{(K_1+1)(K_2+1)}\omega_3, \quad \omega_2 = \frac{-r_2}{(K_1+1)(K_2+1)}\omega_3, \quad \omega_3 = \omega_3 > 0, \quad \omega_4 = -m_2\omega_3.$$

Further, the Jacobian matrix of Model (15) at $r = r^*$ has a left eigenvector (associated with the zero eigenvalue) given by $v = (v_1, v_2, v_3, v_4)^T$, where:

$$v_1 = 0, \quad v_2 = 0, \quad v_3 = v_3 > 0, \quad v_4 = 0.$$

Computations of $\hat{a}$ and $\hat{b}$

For Model (15), the associated non-zero partial derivatives of $g = (g_1, g_2, g_3, g_4)^T$ (at E_0) are given by:

$$\frac{\partial^2 g_3}{\partial x_1 \partial x_3} = \frac{rK_1}{(K_1+1)^2(K_2+1)}, \quad \frac{\partial^2 g_3}{\partial x_2 \partial x_3} = \frac{rK_2}{(K_1+1)(K_2+1)^2},$$
$$\frac{\partial^2 g_3}{\partial x_3 \partial x_4} = -m_1, \quad \frac{\partial^2 g_3}{\partial x_3 \partial r^*} = \frac{1}{(K_1+1)(K_2+1)}.$$

It follows from the above expressions that:

$$\begin{aligned} \hat{a} &= v_3 \sum_{i=1,j=1}^{4} \frac{\omega_i \omega_j \partial^2 g_3}{\partial x_i \partial x_j} \\ &= 2v_3\left(\omega_1\omega_3 \frac{rK_1}{(K_1+1)^2(K_2+1)} + \omega_2\omega_3 \frac{rK_2}{(K_1+1)(K_2+1)^2} - \omega_3\omega_4 m_1\right) \\ &= 2v_3\omega_3^2\left(m_1 m_2 - \frac{rr_1K_1}{(K_1+1)^3(K_2+1)^2} - \frac{rr_2K_2}{(K_1+1)^2(K_2+1)^3}\right), \end{aligned}$$

from which it can be shown that $\hat{a} > 0$ if:

$$m_1 m_2 > \frac{rr_1K_1}{(K_1+1)^3(K_2+1)^2} + \frac{rr_2K_2}{(K_1+1)^2(K_2+1)^3}.$$

For the sign of $\hat{b}$, it can be shown that the associated non-vanishing partial derivatives of g are:

$$\hat{b} = v_3 \sum_{i=1}^{4} \frac{\omega_i \partial^2 g_3}{\partial x_i \partial r^*} = \frac{2 v_3 \omega_3}{(K_1+1)(K_2+1)} > 0.$$

Thus, we have established Theorem 4 in view of [48]. The proof is completed.

References

1. Wang, L.; Liao, L. Separation and identification of photosynthetic bacteria (PSB) and purifying effection aquiculture water. *J. Microbiol.* **2004**, *2*, 7–10.
2. Takeno, K.; Sasaki, K.; Nishio, N. Removal of phosphorus from oyster farm mud sediment using a photosynthetic bacterium, Rhodobacter shaeroides IL106. *J. Biosci. Bioeng.* **1999**, *88*, 410–415. [CrossRef]
3. Nagadomi, H.; Kitamura, T.; Watanabe, M.; Sasaki, K. Simultaneous removal of chemical oxygen demand (COD), phosphate, nitrate and hydrogen sulphide in the synthetic sewage wastewater using porous ceramic immobilized photosynthetic bacteria. *Biotechnol. Lett.* **2000**, *22*, 1369–1374. [CrossRef]
4. Chen, F.; Jiang, Y. *Microalgal Biotechnology*; Chinese Light Industry Press: Beijing, China, 1999.
5. Xu, F.; Zhang, S.; Zhang, P. Study of extracting CoQ10 from photosynthetic bacteria by ultrasonic assisted with zymolysis or freezing-thawing. *Chem. Eng.* **2008**, *8*, 43–45.
6. Sasaki, K.; Watanabe, M.; Tanaka, T. Biosynthesis, biotechnological production and applications of 5-aminolevulinic acid. *Appl. Microbiol. Biotechnol.* **2012**, *31*, 119–123.
7. Sasaki, K.; Watanabe, K.; Tanaka, T.; Tanaka, T. Effects of light sources on growth and carotenoid content of photosynthetic bacteria *Rhodopseudomonas palustris*. *Bioresour. Technol.* **2002**, *58*, 23–29.
8. Carlozzi, P.; Sacchi, A. Biomass production and studies on *Rhodopseudomonas palustris* grown in an outdoor, temperature controlled, underwater tubular photobioreactor. *J. Biotechnol.* **2001**, *3*, 239–249. [CrossRef]
9. Li, B.; Feng, J.; Xie, S. Degradation of microcystin by *Rhodopseudomonas palustris*. *Chin. J. Ecol.* **2012**, *31*, 119–123.
10. Wu, S.C.; Liou, S.Z.; Lee, C.M. Correlation between bio-hydrogen production and polyhydroxybutyrate (PHB) synthesis by *Rhodopseudomonas palustris* WP3-5. *Bioresour. Technol.* **2012**, *113*, 44–50. [CrossRef] [PubMed]
11. Kuo, F.S.; Chien, Y.H.; Chen, C. Effects of light sources on growth and carotenoid content of photosynthetic bacteria *Rhodopseudomonas palustris*. *Bioresour. Technol.* **2012**, *113*, 315–318. [CrossRef] [PubMed]
12. Zhou, Q.; Zhang, P.; Zhang, G. Biomass and carotenoid production in photosynthetic bacteria wastewater treatment: Effects of light intensity. *Bioresour. Technol.* **2014**, *171*, 330–335. [CrossRef] [PubMed]
13. Kong, F.; Song, L. *Algal Blooms Process and Its Environmental Characteristics*; Science Press: Beijing, China, 2011.
14. Pan, G.; Zhang, M.; Chen, H.; Zou, H.; Yan, H. Removal of cyanobacterial blooms in Taihu Lake using local soils I. Equilibrium and kinetic screening on the flocculation of Microcystis aeruginosa using commercially available clays and minerals. *Environ. Pollut.* **2006**, *141*, 195–200. [CrossRef] [PubMed]
15. Ghernaout, B.; Ghernaout, D.; Saiba, A. Algae and cyanotoxins removal by coagulation/flocculation: A review. Desalin. *Water Treat.* **2010**, *20*, 133–143. [CrossRef]
16. Li, L.; Pan, G. A universal method for flocculating harmful algal blooms in marine and fresh waters using modified sand. *Environ. Sci. Technol.* **2013**, *47*, 4555–4562. [CrossRef] [PubMed]
17. O'Melia, C.R. *The Scientific Basis of Flocculation*; Sijthoff and Noordhoff: Amsterdam, The Netherlands, 1978; pp. 219–269.
18. Bolto, B.; Gregory, J. Organic polyelectrolytes in water treatment. *Water Res.* **2007**, *41*, 2301–2324. [CrossRef] [PubMed]
19. Brostow, W.; Lobland, H.E.H.; Sagar Pal Singh, R.P. Polymeric flocculants for wastewater and industrial effluent treatment. *J. Mater. Educ.* **2009**, *31*, 157–166.
20. Sharma, B.R.; Dhuldhoya, N.C.; Merchant, U.C. Flocculants-an ecofriendly approach. *J. Polym. Environ.* **2006**, *14*, 195–202. [CrossRef]

21. Singh, R.P.; Karmakar, G.P.; Rath, S.K.; Karmakar, N.C.; Pandey, S.R.; Tripathy, T.; Panda, J.; Kanan, K.; Jain, S.K.; Lan, N.T. Biodegradable drag reducing agents and flocculants based on polysaccharides: Materials and applications. *Polym. Eng. Sci.* **2000**, *40*, 46–60. [CrossRef]
22. Martin, M.A.; Gonzalez, I.; Berrios, M.; Siles, J.A.; Martin, A. Optimization of coagulation-flocculation process for waste water derived from sauce manufacturing using factorial design of experiments. *J. Chem. Eng.* **2011**, *172*, 771–782.
23. Yan, H.; Ma, C.; Sun, X.; Chen, J.; Wang, D. Study on flocculation of *Rhodopseudomonas palustris* by aluminum flocculants. *Chem. Eng.* **2008**, *6*, 53–55.
24. Tang, H.; Luan, Z. Features and mechanism for coagulation-flocculation processes of polyaluminum chloride. *J. Environ. Sci.* **1995**, *7*, 204–211.
25. Salehizadeh, H.; Shojaosadati, S.A. Isolation and characterisation of a bioflocculant produced by *Bacillus firmus*. *Biotechnol. Lett.* **2002**, *24*, 35–40. [CrossRef]
26. Xu, Y.; Purton, S.; Baganz, F. Chitosan flocculation to aid the harvesting of the microalga *Chlorella* sorokiniana. *Bioresour. Technol.* **2013**, *129*, 296–301. [CrossRef] [PubMed]
27. Vandamme, D.; Foubert, I.; Meesschaer, B.; Muylaert, K. Flocculation of microalgae using cationic starch. *J. Appl. Phycol.* **2010**, *22*, 525–530. [CrossRef]
28. Packer, A.; Li, Y.; Andersen, T.; Hu, Q.; Kuang, Y.; Sommerfeld, M. Growth and neutral lipid synthesis in green microalgae: A mathematical model. *Bioresour. Technol.* **2011**, *102*, 111–117. [CrossRef] [PubMed]
29. Smith, H.L.; Waltman, P. *The Theory of the Chemostat. Dynamics of Microbial Competition, Cambridge Studies in Mathematical Biology 13*; Cambridge University Press: Cambridge, UK, 1995.
30. Smith, H.L.; Waltman, P. Competition for a single limiting resource in continuous culture: the variable-yield model. *SIAM J. Appl. Math.* **1994**, *54*, 1113–1131. [CrossRef]
31. Herbert, D.; Elsworth, R.; Telling, R.C. The continuous culture of bacteria, a theoretical and experimental study. *J. Gen. Microbiol.* **1956**, *14*, 601–622. [CrossRef] [PubMed]
32. Hsu, S.B.; Hubbel, S.P.; Waltman, P. A mathematical theory for single nutrient competition in continuous cultures of microorganisms. *SIAM J. Appl. Math.* **1977**, *32*, 366–383. [CrossRef]
33. Hsu, S.B.; Hubbel, S.P.; Waltman, P. Limiting behavior for competing species. *SIAM J. Appl. Math.* **1978**, *34*, 760–763. [CrossRef]
34. Wolkowicz, G.S.K.; Lu, Z. Global dynamics of a mathematical model of competition in the chemostat: General response functions and differential death rates. *SIAM J. Appl. Math.* **1992**, *52*, 222–233. [CrossRef]
35. Li, B.; Smith, H.L. Global dynamics of microbial competition for two resources with internal storage. *J. Math. Biol.* **2007**, *55*, 481–515. [CrossRef] [PubMed]
36. Hsu, S.B.; Waltman, P.; Ellermeyer, S.F. A remark on the global asymptotic stability of a dynamical system modeling two species competition. *Hiroshima Math. J.* **1994**, *24*, 435–445.
37. Ruan, S.; He, X. Global stability in chemostat-type competition models with nutrient recycling. *SIAM J. Appl. Math.* **1998**, *1*, 170–192. [CrossRef]
38. Butler, G.J.; Hsu, S.B.; Waltman, P. Coexistence of competing predators in a chemostat. *J. Math. Biol.* **1983**, *17*, 133–151. [CrossRef]
39. Pilyugin, S.S.; Waltman, P. Multiple limit cycles in the chemostat with variable yield. *Math. Biosci.* **2003**, *182*, 151–166. [CrossRef]
40. Tai, X.; Ma, W.; Guo, S.; Yan, H.; Yin, C. Dynamic model describing flocculation of micoorganism and its theoretical analysis. *Math. Pract. Theory* **2015**, *45*, 198–209.
41. Leon, J.A.; Tumpson, D.B. Competition between two species for two complementary or substitutable resources. *J. Theor. Biol.* **1975**, *50*, 185–201. [CrossRef]
42. Harder, W.; Dijkhuizen, L. Strategies of mixed substrate utilization in microorganisms. *Philos. Trans. R. Soc. Lond. B* **1982**, *297*, 459–480. [CrossRef]
43. Zhang, Y.; Ma, W.; Yan, H.; Takeuchi, Y. A dynamic model describing heterotrophic cultures of *Chlorella* and its stability analysis. *Math. Biosci. Eng.* **2011**, *8*, 1117–1133. [PubMed]
44. Li, B.; Wolkowicz, G.S.K.; Kuang, Y. Global asymptotic behavior of a chemostat model with two perfectly complementary resources and distributed delay. *SIAM J. Appl. Math.* **2000**, *60*, 2058–2086. [CrossRef]
45. Hale, J.K. *Ordinary Differential Equations*, 2nd ed.; Robert E. Krieger Publishing Company, Inc.: Huntington, NY, USA, 1980.

46. Kuang, Y. *Delay Differential Equations with Applications in Population Dynamics*; Academic Press, Inc.: Boston, MA, USA, 1993.
47. Sharomi, O.; Gumel, A.B. Re-infection-induced backward bifurcation in the transmission dynamics of Chlamydia trachomatis. *J. Math. Anal. Appl.* **2009**, *356*, 96–118. [CrossRef]
48. Sharomi, O.; Podder, C.N.; Gumel, A.B.; Song, B. Mathematical analysis of the transmission dynamics of HIV/TB co-infection in the presence of treatment. *Math. Biosci. Eng.* **2008**, *5*, 145–174. [PubMed]
49. Carr, J. *Applications Centre Manifold Theory*; Springer-Verlag: New York, NY, USA, 1981.
50. Nani, F.; Freedman, H.I. A mathematical model of cancer treatment by immunotherapy. *Math. Biosci.* **2000**, *163*, 159–199. [CrossRef]
51. Zhao, X. *Dynamical Systems in Population Biology*; Springer: New York, NY, USA, 2003.
52. Thieme, H.R. Persistence under relaxed point-dissipativity (with application to an endemic model). *SIAM J. Math. Anal.* **1993**, *24*, 407–435. [CrossRef]
53. Smith, H.L.; Zhao, X. Robust persistence for semidynamical systems. *Nonl. Anal.: Theory Methods Appl.* **2001**, *47*, 6169–6179. [CrossRef]
54. Liu, M.; Bai, C. Global asymptotic stability of a stochastic delayed predator-prey model with Beddington-DeAngelis functional response. *Appl. Math. Comput.* **2014**, *226*, 581–588. [CrossRef]
55. Cunningham, A.; Nisbet, R.M. Time lag and co-operativity in the transient growth dynamics of microalgae. *J. Theor. Biol.* **1980**, *84*, 189–203. [CrossRef]
56. Beretta, E.; Takeuchi, Y. Qualitative properties of chemostat equation with time delays: Boundedness, local and global asymptotic stability. *Differ. Equ. Dyn. Syst.* **1994**, *2*, 263–288. [CrossRef]
57. Xia, H.; Wolkowicz, G.S.K.; Wang, L. Transient oscillation induced by delayed growth response in the chemostat. *J. Math. Biol.* **2005**, *50*, 489–530. [CrossRef] [PubMed]

applied sciences

MDPI

Article

Optimal Control of Drug Therapy in a Hepatitis B Model

Jonathan E. Forde [1,*], Stanca M. Ciupe [2,*], Ariel Cintron-Arias [3] and Suzanne Lenhart [4]

1 Department of Mathematics and Computer Science, Hobart and William Smith Colleges, 300 Pulteney St., Geneva, NY 14456, USA
2 Department of Mathematics, Virginia Tech, 416 McBryde Hall, Blacksburg, VA 24061, USA
3 Department of Mathematics and Statistics, East Tennessee State University, Johnson City, TN 37614, USA; CINTRONARIAS@mail.etsu.edu
4 Department of Mathematics, University of Tennessee, Knoxville, TN 37996, USA; lenhart@math.utk.edu
* Correspondence: forde@hws.edu (J.E.F.); stanca@vt.edu (S.M.C.); Tel.: +1-315-781-3814 (J.E.F.); +1-540-231-3190 (S.M.C.)

Academic Editor: Yang Kuang
Received: 23 May 2016; Accepted: 26 July 2016; Published: 3 August 2016

Abstract: Combination antiviral drug therapy improves the survival rates of patients chronically infected with hepatitis B virus by controlling viral replication and enhancing immune responses. Some of these drugs have side effects that make them unsuitable for long-term administration. To address the trade-off between the positive and negative effects of the combination therapy, we investigated an optimal control problem for a delay differential equation model of immune responses to hepatitis virus B infection. Our optimal control problem investigates the interplay between virological and immunomodulatory effects of therapy, the control of viremia and the administration of the minimal dosage over a short period of time. Our numerical results show that the high drug levels that induce immune modulation rather than suppression of virological factors are essential for the clearance of hepatitis B virus.

Keywords: optimal control; hepatitis B; delay differential equations (DDE); immune response; drug therapy

---

1. Introduction

Hepatitis B virus (HBV) is the leading viral cause of liver disease, affecting 250–350 million people worldwide. Chronic HBV leads to the development of liver cirrhosis and liver cancer [1]. Despite the availability of effective HBV vaccination [2], the prevalence of chronic HBV has only marginally declined [3,4]. The natural course of chronic HBV includes an immune-tolerant phase, hepatitis B e-antigen positive immuno-active and -inactive phases and a hepatitis B e-antigen negative immuno-active phase [5]. Understanding the virological and immunological characteristics of each of these stages can provide a useful framework for the management of chronic HBV [6,7].

Currently, seven drugs have been approved for treating chronic HBV disease: standard and PEGylated interferon (IFN-α) and five nucleo(t)side analogs (NAs) [8]. These medications suppress HBV replication and liver inflammation, but do not lead to cure. The interferon-α treatments modulate immune responses that may lower viral levels. It is given for a finite time (usually 12 months) due to its toxic side effects [8,9]. The nucleo(t)side analogues are administrated for many years and sometimes for life and are responsible for viral suppression. However, life-long therapy is difficult due to costs, side effects, compliance and, most importantly, the development of antiviral drug resistance [10]. Combination therapy has not shown an increased effect on treatment response (but has reduced the rate of drug resistance) [11]. This has made it difficult to establish a universal guideline for treatment start, duration, which type or combinations of drugs to use [7] and how to define the success of therapy (virologically, serologically and/or immunologically) [7,12].

To provide insight into the optimal combination therapy and, in particular, into the immune-mediated effects of interferon-α, we design an optimal control study for a mathematical model of hepatitis B infection. Mathematical models have been used to address transition from acute to chronic HBV infections [13–16] and to study the effects of drug therapy [17–19]. Optimal control theory was developed by Pontryagin et al. for obtaining necessary conditions to characterize optimal controls for systems of ordinary differential equations (ODEs) [20]. Optimal control models have been used previously to design treatment strategies for disease models described by systems of ordinary differential equations with no delays [21–25] and systems of delay differential equations (DDEs) [26]. Kharatishvili [27] developed the extension of the Pontryagin's maximum principle for systems of DDEs with constant time delays. For approximation and numerical methods for such problems, see [28,29].

In this paper, we modify an in-host DDE model of immune responses to hepatitis B infection introduced in [14] by adding the effects of combination drug therapy. In particular, we consider the drug effects in blocking viral production, reducing viral infection, enhancing the killing of infected cells by immune responses and removing immune cell exhaustion. We will use this as a starting model for designing an optimal control problem that advises what is the best combination therapy to ensure viral clearance, immune activation and the least amount of liver damage. Time-varying rates for therapies will be the controls in the system. Optimal control has been applied to the study of hepatitis B therapy using ODE [30,31] and DDE [32] models that did not consider an immune system component and using an ODE model that considers the enhancement of immune responses following administration of traditional Chinese medicine [33].

Here, we introduce the DDE system with constant time delay from [14], derive its corresponding control formulation and use it to determine the temporal, quantitative and qualitative effects of the drugs that lead to hepatitis B virus control, balancing the goals of reducing viral load and minimizing the negative side effects of therapy. Our results indicate that early drug therapy that mainly modulates and restores the immune responses against the virus is mandatory for the success of the therapy.

2. Mathematical Model and Analysis

2.1. Model of Hepatitis B Virus (HBV) Drug Therapy

To understand the various modes of action of antiviral therapy, we modify the model of acute infection published in [14] by considering the combined effects of NAs and interferon-α drugs. As in [14], we consider five state variables, corresponding to uninfected hepatocytes (T), productively-infected hepatocytes (I), free virus (V), immune effector cells (E) and a population of refractory hepatocytes (R).

All hepatocyte populations, uninfected, infected and refractory to reinfection, are maintained by homeostasis described by a logistic equation, with carrying capacity K and maximal per capita growth rate r. Virus infects target cells at rate β. Infected cells are killed by the immune responses at rate μ. As in [14], we assume that infected cells can be lost due to the non-cytolytic response at rate ρ, dependent on the effector cell population E [34], and move into a refractory class R. Refractory cells will still be assayed as infected, since surface antigens persist for some time [35]. We assume that they have lost their viral replicates and do not produce virus, which also makes them poor targets for cytotoxic T lymphocyte (CTL) responses. Therefore, the refractory population will be killed at a smaller rate, $\nu < \mu$. The refractory state is not permanent, and the R population may eventually become susceptible to reinfection. We model this by allowing R cells to move into the uninfected population at rate q. Free virus is produced at rate π and cleared at rate c.

In the absence of infection, we assume the immune effector cells E are at equilibrium value s/d, where s corresponds to a source of effector cells specific for HBV and $1/d$ is their average life-span.

Upon encountering antigen (on the surface of infected liver cells), these cells expand at rate α, with a constant time delay τ accounting for the lag between antigen encounter and effector cell expansion.

We model drug therapy as interference with virus production and infection and as immune modulation, as follows. The contributions of interferon-α are complex, with both direct antiviral and immunomodulatory effects. Here, we consider three effects of IFN-α: enhancement of effector cell killing rates, reduction of viral production rate and the delay of effector cell programmed death [36,37]. We model this by changing rates μ, π and d to $\mu_1 = \mu(1 + a_1\epsilon)$, $\pi_1^0 = \pi(1 - a_2\epsilon)$ and $d_1 = d(1 - a_3\epsilon)$, respectively. In each case, $0 \le \epsilon \le 1$ represents the efficacy of IFN-α, and $a_i \ge 0$ are scalar parameters representing the strength of the corresponding effect of interferon. a_1 can be any nonnegative number, while a_2 and a_3 are between 0 and 1.

The nucleos(t)ide analogues, on the other hand, interfere with both the ability of virus to infect a cell and with the generation of HBV DNA by an infected cell [38]. We model this by assuming that the infectivity rate in the presence of NAs becomes $\beta_1 = \beta(1 - b_1\eta)$, and the viral production rate becomes $\pi_1 = \pi(1 - fa_2\epsilon - (1 - f)b_2\eta)$. Here, $0 \le \eta \le 1$ represents NAs' efficacy; $b_i \ge 0$ are scalar parameters representing the relative strength of the corresponding effect of NAs; and f and $1 - f$ represent the relative contribution of interferon-α and NAs, respectively, to reducing viral production, for $0 \le f \le 1$.

For $t > t_d$, where t_d represents the time of therapy onset, the dynamics of these populations is governed by the following differential equations

$$
\begin{aligned}
\frac{dT}{dt} &= rT\left(1 - \frac{T+I+R}{K}\right) - (1 - b_1\eta)\beta TV + qR \\
\frac{dI}{dt} &= rI\left(1 - \frac{T+I+R}{K}\right) + (1 - b_1\eta)\beta TV - ((1 + a_1\epsilon)\mu + \rho)IE \\
\frac{dV}{dt} &= \pi(1 - fa_2\epsilon - (1 - f)b_2\eta)I - cV \\
\frac{dE}{dt} &= s + \alpha I(t - \tau)E(t - \tau) - d(1 - a_3\epsilon)E \\
\frac{dR}{dt} &= \rho IE + rR\left(1 - \frac{T+I+R}{K}\right) - qR - \nu RE
\end{aligned}
\tag{1}
$$

As this is a system of delay differential equations with fixed delay τ, the initial conditions for the system are defined on the interval $[t_d - \tau, t_d]$ by functions:

$$T(t) > 0, \quad V(t) \ge 0, \quad R(t) \ge 0, \quad I(t) \ge 0 \text{ and } \quad E(t) \ge 0 \tag{2}$$

In many cases, these will be assumed to be constant and equal to the value of the state variables at $t = t_d$.

2.2. Model Analysis

We will consider that therapy leads to viral removal and only investigate the long-term behavior of the equilibrium where $V = 0$. Model (1) has such a disease free steady state, $S_0 = (K, 0, 0, s/d_1, 0)$. In the absence of delay ($\tau = 0$), the Jacobian matrix associated with (1) is:

$$
\begin{bmatrix}
r(1 - \frac{2T+I+R}{K}) - \beta_1 V & -\frac{rT}{K} & -\beta_1 T & 0 & q - \frac{rT}{K} \\
-\frac{rI}{K} + \beta_1 V & r(1 - \frac{T+2I+R}{K}) - (\rho + \mu_1)E & \beta_1 T & -\rho I - \mu_1 I & -\frac{rI}{K} \\
0 & \pi_1 & -c & 0 & 0 \\
0 & \alpha E & 0 & \alpha I - d_1 & 0 \\
-\frac{rR}{K} & -\frac{rR}{K} + \rho E & 0 & \rho I - \nu R & r(1 - \frac{T+I+2R}{K}) - q - \nu E
\end{bmatrix}
$$

We evaluate $J(S_0)$ and obtain the Jacobian matrix:

$$J(S_0) = \begin{bmatrix} -r & -r & -\beta_1 K & 0 & q-r \\ 0 & -(\rho+\mu_1)s/d_1 & \beta_1 K & 0 & 0 \\ 0 & \pi_1 & -c & 0 & 0 \\ 0 & \alpha s/d_1 & 0 & -d_1 & 0 \\ 0 & \rho s/d_1 & 0 & 0 & -q-\nu s/d_1 \end{bmatrix}$$

and the characteristic equation:

$$(r+\lambda)(\lambda^4 + A_1\lambda^3 + A_2\lambda^2 + A_3\lambda + A_4) = 0$$

where

$$\begin{aligned} A_1 &= c + d_1 + q + (\rho + \mu_1 + \eta)\frac{s}{d_1} \\ A_2 &= \frac{s}{d_1}\left(\nu(c+d_1) + (\mu_1+\rho)(d_1+q+\eta\frac{s}{d_1})\right) + qd_1 + cd_1 + qc + \left(c(\mu_1+\rho)\frac{s}{d_1} - \pi_1\beta_1 K\right) \\ A_3 &= \frac{s}{d_1}(bd_1q + cd_1\nu + \rho d_1 q + \nu s(\rho+\mu_1)) + cd_1q + \left(d_1 + q + \nu\frac{s}{d_1}\right)\left(c(\mu_1+\rho)\frac{s}{d_1} - \pi_1\beta_1 K\right) \\ A_4 &= d_1(q + \nu\frac{s}{d_1})\left(c(\mu_1+\rho)\frac{s}{d_1} - \pi_1\beta_1 K\right) \end{aligned} \tag{3}$$

By the Routh–Hurwitz criteria, the characteristic equation has roots with negative real parts if and only if $A_i > 0$, $A_1A_2 - A_3 > 0$ and $A_1A_2A_3 - A_3^2 - A_1^2A_4 > 0$. We can show that these conditions are satisfied when $A_4 > 0$.

Therefore, when $\tau = 0$, S_0 is locally asymptotically stable when:

$$\frac{\pi_1\beta_1 K d_1}{c(\mu_1+\rho)s} = \frac{(1-b_1\eta)(1-fa_2\epsilon-(1-f)b_2\eta)(1-a_3\epsilon)\pi\beta K d}{(a_1\epsilon\mu+\mu+\rho)cs} < 1 \tag{4}$$

and is unstable otherwise. This condition gives us a minimal drug efficacy for viral clearance in the absence of delay in immune activation.

When $\tau > 0$, the characteristic equation at state S_0 is:

$$\det(B + e^{-\lambda\tau}C - \lambda I_5) = 0 \tag{5}$$

where:

$$B = \begin{bmatrix} -r & -r & -\beta_1 K & 0 & q-r \\ 0 & -(\rho+\mu_1)s/d_1 & \beta_1 K & 0 & 0 \\ 0 & \pi_1 & -c & 0 & 0 \\ 0 & 0 & 0 & -d_1 & 0 \\ 0 & \rho s/d_1 & 0 & 0 & -q-\nu s/d_1 \end{bmatrix}$$

and:

$$C = \begin{bmatrix} 0 & 0 & 0 & 0 & 0 \\ 0 & 0 & 0 & 0 & 0 \\ 0 & 0 & 0 & 0 & 0 \\ 0 & \alpha s/d_1 & 0 & 0 & 0 \\ 0 & 0 & 0 & 0 & 0 \end{bmatrix}$$

We see that the transcendental Equation (5) reduces to polynomial:

$$(r+\lambda)(\lambda^4 + A_1\lambda^3 + A_2\lambda^2 + A_3\lambda + A_4) = 0$$

as in the $\tau = 0$ case. Therefore, the infection-free equilibrium S_0 is locally asymptotically stable for all τ when (4) holds.

2.3. Numerical Results

We assume that in the absence of drug therapy, the virus will persist, and the patient will experience chronic infection. We therefore set all parameters values and initial conditions to the values for the chronically-infected Patient 7 in [14] (see Table 1). The dynamics of the five variables in the absence of drug therapy, $\epsilon = \eta = 0$, throughout acute infection and transition to chronic disease is presented in Figure 1.

Table 1. Variables, parameters and values used in simulations with mililiter (mL) and day (d).

Variables		
T	Target cells	$T_0 = 13.6 \times 10^6$ per mL
I	Infected cells	$I_0 = 0$ per mL
V	Free virus	$V_0 = 0.33$ per mL
E	Effector cells	$E_0 = 60$ per mL
R	Refractory cells	$R_0 = 0$ per mL
Parameters		
r	Hepatocyte maximum proliferation rate	1 day^{-1}
β	Infectivity rate constant	1.22×10^{-10} mL (virion $\times$ day)$^{-1}$
K	Hepatocyte carrying capacity	13.6×10^6 cells per mL
μ	Infected cell killing rate	1.2×10^{-4} mL (cell $\times$ day)$^{-1}$
ν	Refractory cell killing rate	1.27×10^{-5} mL (cell $\times$ day)$^{-1}$
ρ	Cure rate	3.38×10^{-4} mL (cell $\times$ day)$^{-1}$
α	Effector cell expansion rate	2×10^{-7} mL (cell $\times$ day)$^{-1}$
τ	Delay	33.4 days
π	Virus production rate	164 virion (cell $\times$ day)$^{-1}$
c	Virus clearance rate	0.67 day^{-1}
s	Effector cell production	10 day^{-1}
d	Effector cell clearance rate	0.5 day^{-1}
q	Waning of refractory cell immunity	2×10^{-5} day^{-1}

We next investigate the change in the model dynamics in the presence of treatment. Initially, we set time-constant drug efficacy, $\eta = 0.5$ and $\epsilon = 0.9$ as in [17]; assume scalars a_i and b_i to be either 1 or 0, representing an effect or lack of effect in therapy; and set $f = 0.5$. We define HBV DNA clearance as the presence of less than one HBV DNA in the host serum. Since we assume that HBV DNA can distribute throughout the 3 liters of serum in an average 70-kg person, the viral extinction becomes $V \leq V_{ext} = 3 \times 10^{-4}$ copies per mL. We notice that the speed of viral extinction is dependent on both the timing of therapy initiation and on the type of effects considered.

Indeed, when the therapy is started at $t_d = T_{peak} = 96$ days post-infection (corresponding to the peak HBV DNA), our model predicts HBV DNA clearance 145 days after the start of therapy when $a_1 = a_2 = a_3 = b_1 = b_2 = 1$ and $f = 0.5$ (see Figure 2a, dashed line) and 2226 days (6.1 years) after the start of therapy when $a_1 = a_2 = b_1 = b_2 = 1$, $a_3 = 0$ and $f = 0.5$ (see Figure 2a, dotted line). This suggests that the immune modulation effect of interferon is important in clearing the HBV DNA in a short amount of time.

Similar immune-mediated effects of interferon are observed when drug therapy is started at the chronic state of Model (1), $t_d = T_{ss} = 20$ years post-infection. Indeed, viremia clearance occurs faster when $a_1 = a_2 = a_3 = b_1 = b_2 = 1$ and $f = 0.5$ (see Figure 3a, dashed line) than when $a_1 = a_2 = b_1 = b_2 = 1$, $a_3 = 0$ and $f = 0.5$ (see Figure 3a, dotted lines). However, viral clearance is delayed by 12 days compared to the case in which therapy is initiated at peak HBV DNA. The rapidity of clearance during peak therapy is due to the transient effects of immune cells, which are activated and expanding at the peak HBV DNA (see Figure 2b, solid lines) and tolerant at the chronic HBV infection (see Figure 3b, solid lines).

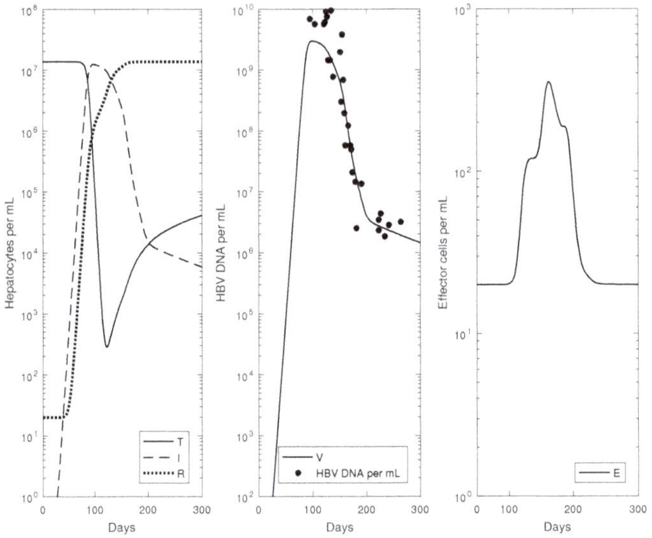

Figure 1. Temporal evolution for the variables in Model (1) without drug therapy, i.e., $\epsilon = \eta = 0$, and the parameters in Table 1. The circles represent patient data.

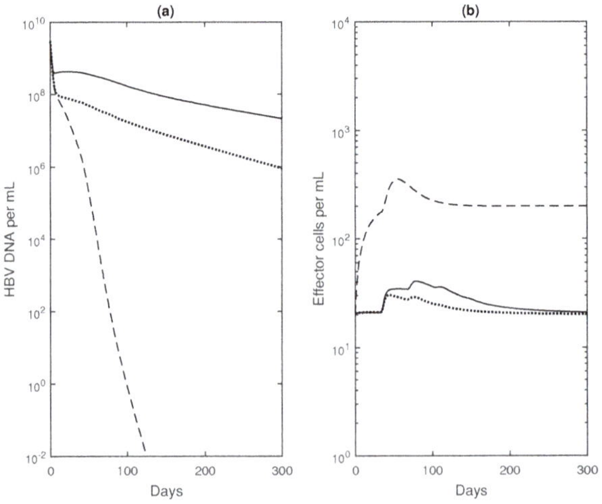

Figure 2. (**a**) Virus V per mL and (**b**) effector cells E per mL given by Model (1) and the parameters in Table 1 for: $\eta = \epsilon = 0$ (solid lines); $\eta = 0.5$, $\epsilon = 0.9$, $a_1 = a_2 = a_3 = b_1 = b_2 = 1$, $f = 0.5$ (dashed lines); and $\eta = 0.5$, $\epsilon = 0.9$, $a_1 = a_2 = b_1 = b_2 = 1$, $a_3 = 0$, $f = 0.5$ (dotted lines). Here, time $t = 0$ corresponds to both the start of therapy and the peak viral load, i.e., the initial conditions are $T(0) = 1.32 \times 10^5$ cells per mL, $I(0) = 1.23 \times 10^5$ cells per mL, $V(0) = 3 \times 10^9$ HBV DNA per mL, $E(0) = 20.23$ cells per mL and $R(0) = 1.23 \times 10^6$ cells per mL.

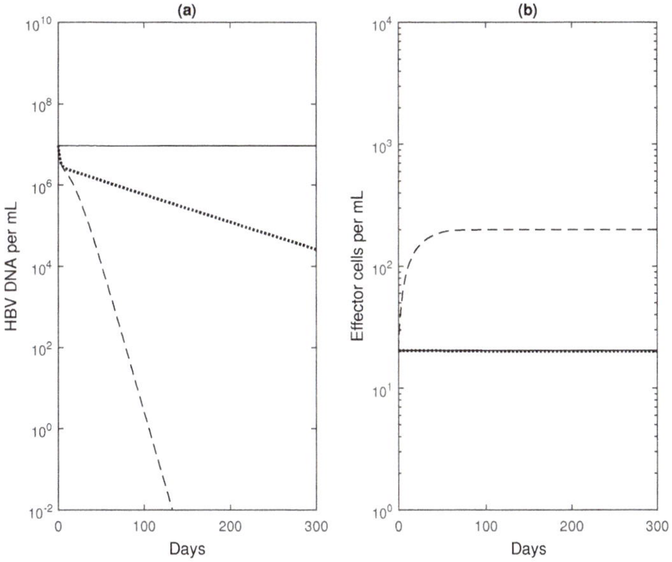

Figure 3. (**a**) Virus V per mL and (**b**) Effector cells E per mL given by model (1) and parameters in Table 1 for $\eta = \epsilon = 0$ (solid lines); $\eta = 0.5$, $\epsilon = 0.9$, $a_1 = a_2 = a_3 = b_1 = b_2 = 1$, $f = 0.5$ (dashed lines); and $\eta = 0.5$, $\epsilon = 0.9$, $a_1 = a_2 = b_1 = b_2 = 1$, $a_3 = 0$, $f = 0.5$ (dotted lines). Here, $t = 0$ corresponds to both the start of therapy and the steady state of the viral load, i.e., the initial conditions are $T(0) = 3 \times 10^5$ cells per mL, $I(0) = 3.8 \times 10^4$ cells per mL, $V(0) = 9.33 \times 10^6$ HBV DNA per mL, $E(0) = 20.3$ cells per mL and $R(0) = 1.33 \times 10^7$ cells per mL.

These results are supported by relative sensitivity curves. Briefly, sensitivity functions are numerical solutions of the following system:

$$\frac{dx}{dt} = g(x(t,\xi),\ z(t,\xi),\ \xi) \tag{6}$$

$$\frac{d}{dt}\frac{\partial x}{\partial \xi} = \frac{\partial g}{\partial x}\frac{\partial x}{\partial \xi} + \frac{\partial g}{\partial z}\frac{\partial z}{\partial \xi} + \frac{\partial g}{\partial \xi} \tag{7}$$

where $x(t,\xi) = (T(t,\xi), I(t,\xi), V(t,\xi), E(t,\xi), R(t,\xi))$, $z(t,\xi) = x(t-\tau,\xi) \in \mathbb{R}^5$, the function g represents the right-hand side of Model (1) and:

$$\xi = (r,\ K,\ \beta,\ q,\ \rho,\ \pi,\ c,\ s,\ \alpha,\ d,\ \eta,\ \nu,\ \mu,\ \epsilon,\ a_1,\ a_2,\ a_3,\ b_1,\ b_2,\ f)$$

The partial derivatives $\partial x_i/\partial \xi_j$, for $i = 1,\ldots,5$ and $j = 1,\ldots,20$, are time functions denoting the rate of change in a state variable with respect to variations in model parameters (see [39] and the references therein). The functions $\xi_j/x_i \times \partial x_i/\partial \xi_j$ are the relative sensitivity curves (similar to relative error), which allow for the comparison between the sensitivity of two variables x_i^1 and x_i^2 with respect to the same parameter ξ_j. Here, we are interested in the sensitivity of $V(t,\xi)$ and $E(t,\xi)$ with respect to model parameters $\xi = \{a_2, a_3, b_2\}$. As the numerical solutions displayed in Figure 4 show, the a_3 effect is the most important drug effect in both reducing virus load and increasing immune response.

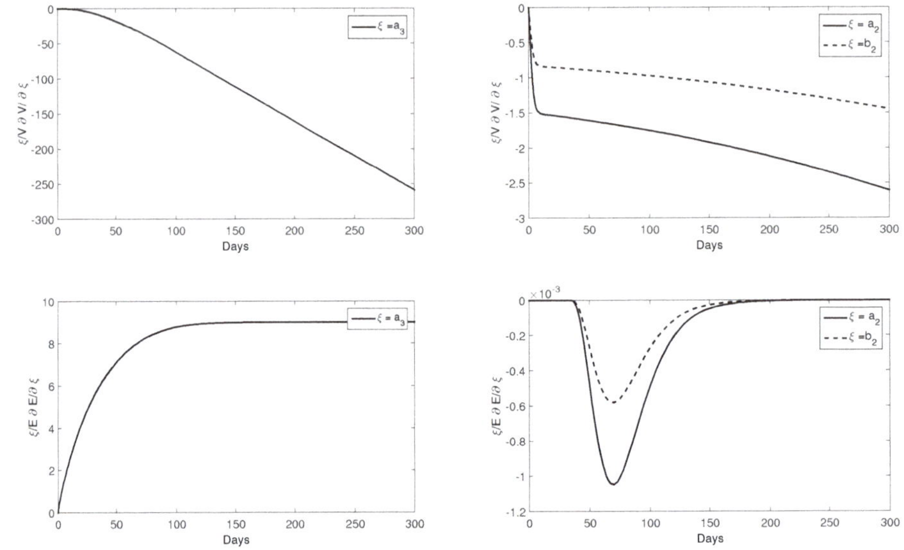

Figure 4. Relative sensitivity curves $\frac{\xi}{V}\frac{\partial V}{\partial \xi}$ (top row) and $\frac{\xi}{E}\frac{\partial E}{\partial \xi}$ (bottom row) for $\xi = a_3$ (left) and $\xi = a_2$ (solid line), $\xi = b_2$ (dashed line) (right). The parameters and initial conditions are as in Figure 3.

Our model suggests that the timing and the type of interferon effects are important in the success of the treatment. In the next sections, we will consider a time-dependent effect of both interferon and nucleo(s)tide analogues, and we will formulate an optimal control problem for our model to determine the connection between successful therapy and the optimal temporal efficacy of the effects considered.

3. Optimal Control Problem

To better understand the interactions between drug therapy and the host immune reaction, we allow the effects of the two drug types (NAs and IFN-α) to vary with time. We replace the NAs' efficacy parameter η with $u_1(t)$ to represent the time-varying effective drug dosage needed for optimal therapy. Similarly, $u_2(t)$ will replace the parameter ϵ to represent the time-varying dosage of IFN-α. As before, a_1, a_2 and a_3 indicate the relative strength of the three effects of IFN-α under consideration; b_1 and b_2 represent the relative strengths of the two effects of the NA; and f and $1 - f$ are the relative contribution of interferon and NAs, respectively, in reducing viral production.

For computational convenience and consistency of notation, we will relabel the state variables as in Table 2. With these adjustments, our model becomes as follows.

Table 2. Correspondence between labeling of the state variables in Models (1) and (8).

$x_1(t) = T(t)$	
$x_2(t) = I(t)$	$z_2(t) = I(t - \tau)$
$x_3(t) = V(t)$	
$x_4(t) = E(t)$	$z_4(t) = E(t - \tau)$
$x_5(t) = R(t)$	

$$\begin{aligned}
\frac{dx_1}{dt} &= rx_1\left(1 - \frac{x_1 + x_2 + x_5}{K}\right) - \beta(1 - b_1u_1)x_1x_3 + qx_5 \\
\frac{dx_2}{dt} &= rx_2\left(1 - \frac{x_1 + x_2 + x_5}{K}\right) + \beta(1 - b_1u_1)x_1x_3 - ((1 + a_1u_2)\mu + \rho)x_2x_4 \\
\frac{dx_3}{dt} &= (1 - fa_2u_2 - (1 - f)b_2u_1)\pi x_2 - cx_3 \\
\frac{dx_4}{dt} &= s + \alpha z_2z_4 - d(1 - a_3u_2)x_4 \\
\frac{dx_5}{dt} &= \rho x_2x_4 + rx_5\left(1 - \frac{x_1 + x_2 + x_5}{K}\right) - qx_5 - \nu x_4x_5
\end{aligned} \tag{8}$$

Our goal is to find the regime of drug therapy that minimizes the objective functional:

$$J(u_1, u_2) = \int_0^T \left(c_1x_3 + c_2x_2 + c_3u_1 + c_4u_2 + \varepsilon_1u_1^2 + \varepsilon_2u_2^2\right) dt + c_5x_3(T) \tag{9}$$

subject to the system of delay differential Equation (8) with initial conditions (2). Here, T represents the duration of therapy. We will study two different scenarios for drug treatment by considering two different sets of initial conditions. For the case of chronic HBV infection, the initial values of the state variables x_i will be constant and equal to their chronic infection equilibrium values for the entire interval $[t_d - \tau, t_d]$. We will also study the case of drug therapy initiation at the peak of viral load, during acute infection. In this case, the initial functions for the state variables x_i will be set by their trajectories during acute infection in the absence of treatment. This is described in more detail in Section 3.3.

Given upper bounds M_i on u_i (for $i = 1, 2$), we seek to find an optimal pair in the control set:

$$U = \{(u_1, u_2) \in L^\infty(0, T) \mid 0 \leq u_i \leq M_i, \; i = 1, 2\}$$

such that

$$J(u_1^*, u_2^*) = \inf_{(u_1, u_2) \in U} J(u_1, u_2)$$

The integral portion of the objective functional encapsulates the goal of minimizing total virus concentration (x_3), infected cell concentration (x_2) and the amount of drug used (u_1 and u_2 terms) over the entire treatment period. The final term $c_5x_3(T)$ represents the goal of minimizing the final viral concentrations at time T, which ideally would be below clearance levels. The parameters $c_i > 0$ and $\varepsilon_j > 0$ give the relative weight of each of these factors.

3.1. Analysis of the Optimal Control Problem

Pontryagin's key idea was to use adjoint functions to attach the state dynamics to the objective functional. This converts the problem into minimizing the Hamiltonian and generates the adjoint differential equations and final-time transversality conditions. In this paper, we use a generalization of Pontryagin's maximum principle to systems of delay differential equations (developed by [27]).

Letting f denote the integrand of the objective functional (9) and $g_1, \ldots, g_5$ be the right-hand sides of System (8), the Hamiltonian for our optimal control problem is:

$$H = f + \lambda_1g_1 + \lambda_2g_2 + \lambda_3g_3 + \lambda_4g_4 + \lambda_5g_5 \tag{10}$$

where the adjoint functions λ_i correspond to the states x_i, for $i \in \{1, \ldots, 5\}$.

From the Hamiltonian, we derive the adjoint equations. For $i = 1, 3$ and 5, we have:

$$\frac{d\lambda_i}{dt} = -\frac{\partial H}{\partial x_i} \tag{11}$$

on $[0, T]$. For $i = 2$ and 4, on the interval $[0, T - \tau]$ we have

$$\begin{aligned} \frac{d\lambda_2}{dt} &= -\frac{\partial H}{\partial x_2} - \left.\frac{\partial H}{\partial z_2}\right|_{t+\tau} \\ \frac{d\lambda_4}{dt} &= -\frac{\partial H}{\partial x_4} - \left.\frac{\partial H}{\partial z_4}\right|_{t+\tau} \end{aligned} \tag{12}$$

where $z_i(t) = x_i(t - \tau)$ represent the delayed state variables, while on the interval $[T - \tau, T]$, we have:

$$\begin{aligned} \frac{d\lambda_2}{dt} &= -\frac{\partial H}{\partial x_2} \\ \frac{d\lambda_4}{dt} &= -\frac{\partial H}{\partial x_4} \end{aligned} \tag{13}$$

just as in [11].

The equations containing delayed state variables produce adjoint equations with a "forward" delay, due to the opposite time orientation of the adjoint differential equation. The adjoint equations are subject to the final condition that $\lambda_3(T) = c_5$, $\lambda_i(T) = 0$ and $i \in \{1, 2, 4, 5\}$. On the interior of the control set, minimizing the Hamiltonian gives:

$$\frac{\partial H}{\partial u_i} = 0$$

at (u_1^*, u_2^*), and then, the optimal control pair becomes:

$$\begin{aligned} u_1^* &= \frac{-c_3 + \beta x_1 x_3(-\lambda_1 + \lambda_2) + \lambda_3(1 - f) b_2 \pi x_2}{2\varepsilon_1} \\ u_2^* &= \frac{-c_4 + \lambda_2 a_1 \mu x_2 x_4 + \lambda_3 f a_2 \pi x_2 - \lambda_4 d a_3 x_4}{2\varepsilon_2} \end{aligned} \tag{14}$$

Using the bounds on the controls, we obtain the following characterization of the optimal control:

$$\begin{aligned} u_1^* &= \min\left\{M_1, \max\left\{0, \frac{-c_3 + \beta x_1 x_3(-\lambda_1 + \lambda_2) + \lambda_3(1 - f) b_2 \pi x_2}{2\varepsilon_1}\right\}\right\} \\ u_2^* &= \min\left\{M_2, \max\left\{\frac{-c_4 + \lambda_2 a_1 \mu x_2 x_4 + \lambda_3 f a_2 \pi x_2 - \lambda_4 d a_3 x_4}{2\varepsilon_2}\right\}\right\} \end{aligned} \tag{15}$$

3.2. Implementation of the Optimal Control Problem

We wish to find optimal controls numerically by applying a forward-backward iterative method [25]. Initially, a constant control is assumed, and the state equations are solved in the forward-time direction from our standard set of initial conditions. Given this solution to the state equations, the adjoint equations are then solved in the backwards-time direction, beginning with the final time T and final condition $\lambda_3(T) = c_5$ and $\lambda_i(T) = 0$ for $i \in \{1, 2, 4, 5\}$. The controls to be used for the next forward run are then updated using the characterization of the optimal control given in (15), and the forward-backward solution process is repeated with the updated control functions. This process is iterated until the controls and the solutions to all of the differential equations converge to within acceptable numerical tolerances. See [28,29,40] for the background on this procedure and the approximation of delay equations.

To use the forward-backward sweep, we rewrite the adjoint equations for our system as:

$$
\begin{aligned}
\frac{d\lambda_1}{dt} &= -\lambda_1\left(r\left(1-\frac{2x_1+x_2+x_5}{K}\right)-(1-b_1u_1)\beta x_3\right)-\lambda_2\left(-\frac{r}{K}x_2+(1-b_1u_1)\beta x_3\right)+\lambda_5\frac{r}{K}x_5,\\
\frac{d\lambda_2}{dt} &= -c_2+\lambda_1\frac{r}{K}x_1-\lambda_2\left(r\left(1-\frac{x_1+2x_2+x_5}{K}\right)-(1+a_1u_2)\mu x_4-\rho x_4\right)\\
&\quad -\lambda_3(1-(1-f)b_2u_1-fa_2u_2)\pi-\lambda_5\left(\rho x_4-\frac{r}{K}x_5\right)-\lambda_4(t+\tau)\alpha x_4\chi_{[0,T-\tau]}\\
\frac{d\lambda_3}{dt} &= -c_1+\lambda_1(1-b_1u_1)\beta x_1-\lambda_2(1-b_1u_1)\beta x_1+\lambda_3 c\\
\frac{d\lambda_4}{dt} &= -\lambda_2\left(-(1+a_1u_2)\mu x_2-\rho x_2\right)+\lambda_4 d(1-a_3u_2)-\lambda_5(\rho x_2-\nu x_5)-\lambda_4(t+\tau)\alpha x_2\chi_{[0,T-\tau]}\\
\frac{d\lambda_5}{dt} &= -\lambda_1\left(-\frac{r}{K}x_1+q\right)+\lambda_2\frac{r}{K}x_2-\lambda_5\left(r\left(1-\frac{x_1+x_2+2x_5}{K}\right)-q-\nu x_4\right)
\end{aligned}
\tag{16}
$$

Notice that in the time interval $[T-\tau, T]$, the advance terms (those with argument $t+\tau$, found in the second and fourth equations) drop out, so we have five ordinary differential equations. On the interval $[0, T-\tau]$, we once again have advance equations, but the solutions to the ODEs on $[T-\tau, T]$ provide the required initial data to solve these equations. Thus, the adjoint equations are advance differential equations on $[0, T-\tau]$ and ordinary differential equations on $[T-\tau, T]$, subject to the final condition $\lambda_3(T) = c_5$ and $\lambda_i(T) = 0$ for $i \in \{1,2,4,5\}$.

For our numerical simulations, we used the built-in MATLAB numerical delay differential equation solver, `dde23`. This tool is not capable of solving advance equations directly, so we made a change of variables to convert the advance differential equations to a system of delay differential equations. Specifically, we define a new reversed-time variable $\sigma = T-t$ and new adjoint variables $L_i(\sigma) = \lambda_i(T-\sigma) = \lambda_i(t)$. In terms of these new variables, adjoint equations for $L_i(\sigma)$ are ordinary differential equations on $[0,\tau]$ and delay differential equations on $[\tau, T]$, subject to the initial condition $L_3(0) = c_5$ and $L_i(0) = 0$ for $i \in \{1,2,4,5\}$. The new system was solved numerically using `dde23` as part of the implementation of the forward-backward method to find the optimal control.

3.3. Numerical Results with Optimal Controls

For optimal therapy during chronic HBV infection, we assume that at the time of treatment initiation, the patient has reached chronic steady state values $\bar{x}_1 = 3\times10^5$ cells per mL, $\bar{x}_2 = 3.8\times10^4$ cells per mL, $\bar{x}_3 = 9.33\times10^6$ HBV DNA per mL, $\bar{x}_4 = 20.3$ cells per mL and $\bar{x}_5 = 1.33\times10^7$ cells per mL. The terms in the objective functional (9) have different scales. In order to weigh each term equally, we choose parameters c_i, such that c_ix_i (for $i = 1,2$) and c_iu_i (for $i = 3,4$) are one at the start of the optimal control. When the optimal control is started at the viral steady state, we normalize variables x_1 and x_2 by $c_1 = \frac{1}{\bar{x}_3}$ and $c_2 = \frac{1}{\bar{x}_2} = \frac{\pi}{c\bar{x}_3}$. As a result, the factors c_1x_1 and c_2x_2 are one at the beginning of treatment and between zero and one from there on. Since we assume $0 \le u_1(t), u_2(t) \le 0.9$ for all t, their corresponding weights are $c_3 = c_4 = 1$. Lastly, we choose the effects for the convex terms to be $\epsilon_1 = \epsilon_2 = \mathbf{0.1}$. The treatment will be given for $T_1 = 400$ days, which is sufficient for virus clearance and is still within the timeline of the HBV therapy guidelines [7].

The ideal clinical outcome of HBV drug therapy is to achieve viral suppression, such that $x_3 \le V_{ext} = 3\times10^{-4}$ HBV DNA per mL, corresponding to one HBV DNA in the body (as in Section 2.3). To account for this final condition, we set $c_5 = 3\times10^3$ so that the final time condition is normalized to one. Under these assumptions, we run the optimal control problem over $T_1 = 400$ days with the aim of finding the best temporal drug usage at each time point that ensures viral clearance while minimizing the drug levels.

We will be considering several scenarios by varying the levels of control effects, described by constants a_i and b_i. Our simulations show that for high virus levels (above extinction), the optimal control will always be the maximal drug dosage for the scenario considered. For example, for a case with any $a_i > 0$ and $b_1 = b_2 = 0$, the optimal control is $u_1(t) = 0$ and $u_2(t) = 0.9$ for all time points.

Similarly, if $a_i > 0$ and $b_j > 0$, then the optimal control is $u_1(t) = u_2(t) = 0.9$ for all time points. Therefore, for the first 325 days, we assume the maximal drug dosage for a given scenario and only run the optimal control with variable $u_1(t)$ and $u_2(t)$ for the final 75 of the 400 days of treatment.

Initially, we include all effects of nucleos(t)ide analogues u_1 and interferon u_2, i.e., $a_1 = a_2 = a_3 = b_1 = b_2 = 1$ and $f = 0.5$. For this choice of parameters, optimal control requires: (i) maximal NA drug dosage $u_1 = 0.9$ for $0 \leq t \leq 347$ days and zero drug dosage for $t \geq 347$ days; and (ii) maximal $u_2 = 0.9$ interferon dosage for $0 \leq t \leq 366.5$ days, zero dosage for $366.5 \leq t \leq 399$ days and maximal dosage after that (see Figure 5a, top row). Under these drug regimes, HBV DNA is cleared 331 days after the start of treatment (see Figure 5b, top row).

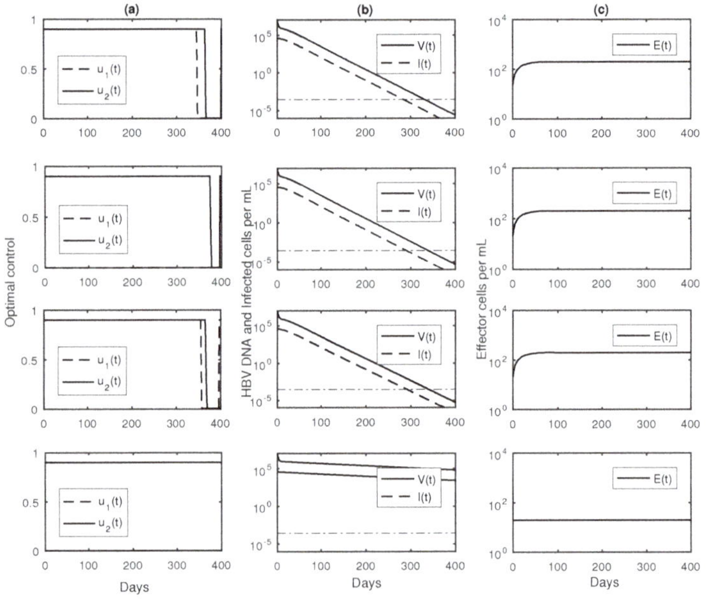

Figure 5. (**a**) Optimal evolution for controls u_1 (dashed lines) and u_2 (solid lines) as given by (15); (**b**) virus V per mL (solid lines) and infected cells I per mL (dashed lines) over time; (**c**) effector cells E per mL over time when drugs are introduced at equilibrium virus concentration and: $a_1 = a_2 = a_3 = b_1 = b_2 = 1, f = 0.5$ (top row); $a_1 = a_2 = a_3 = 1, b_1 = b_2 = 0, f = 1$ (second row); $a_1 = a_2 = 0, a_3 = b_1 = b_2 = 1, f = 0$ (third row); $a_1 = a_2 = a_3 = 0, b_1 = b_2 = 1, f = 0$ (bottom row).

We next assume monotherapy with interferon-α. We set $b_1 = b_2 = 0$, $a_1 = a_2 = a_3 = 1$ and change the anti-viral production effect to $f = 1$. The optimal control problem predicts that the virus is cleared 338 days after the start of therapy, seven days later than under combination therapy (see Figure 5b, second row). Another downside of this monotherapy is the need for interferon dosage to be maximal ($u_2 = 0.9$) for $0 \leq t \leq 377$ and for $t \geq 396.5$ in order to compensate for the lack of NAs (see Figure 5a, second row).

In combination therapy, when u_2 neither increases CTLkilling abilities nor reduces virus production, i.e., $a_1 = a_2 = 0$, and all other effects are maximal $a_3 = b_1 = b_2 = 1$, $f = 0$ (so that NAs have maximal effect in blocking viral production), the HBV DNA will clear 341 days after the start of treatment (see Figure 5b, third row). The controls are: (i) $u_1 = 0.9$ for $0 \leq t \leq 358$ and $t \geq 396$, and zero otherwise; and (ii) $u_2 = 0.9$ for $0 \leq t \leq 369.5$, and zero otherwise (see Figure 5a, third row).

Lastly, we consider monotherapy with NAs alone. We set $a_1 = a_2 = a_3 = 0$, $b_1 = b_2 = 1$ and $f = 0$. The optimal control model does not result in viral clearance even when $u_1 = 0.9$ throughout the duration of the treatment (see Figure 5a,b, bottom row).

For the first three cases, we find that an increase in the effector cells' lifespan is needed for viral suppression. Indeed, when $a_3 = 1$ and $u_2 > 0$, the increase of CTL lifespan to $1/d(1 - a_3u_2)$ leads to elevated CTL concentrations (see Figure 5c, top three rows) and subsequent HBV DNA removal (see Figure 5b, top three rows). By contrast, when $a_3 = 0$, HBV DNA does not reach extinction during the $T_1 = 400$ days of therapy (see Figure 5b, bottom row) due to low HBV-tolerant CTL concentrations, $E = s/d_1 = s/d$, which do not expand in the presence of HBV (see Figure 5c, bottom row).

For optimal therapy during acute HBV infection, to further determine the relationship between CTL activation and the possible success of short-term anti-HBV therapy, we run the optimal control problem during acute HBV disease, where CTL activation has been reported [41]. We start by running the DDE system mainly to the peak virus concentration occurring at time T_{peak}, record the values of all variables for the times $-\tau + T_{peak} \leq t \leq T_{peak}$ and start the optimal control problem at $t_d = T_{peak}$. The treatment will be given for $T_2 = 95$ days and, as before, we include the following weights $c_1 = \frac{1}{x_3(T_{peak})}$, $c_2 = \frac{\pi}{cx_3(T_{peak})}$, $c_3 = c_4 = 1$, $\epsilon_1 = \epsilon_2 = 0.1$ and $c_5 = 3 \times 10^3$, so that all factors are normalized to one.

As in the chronic HBV case, we consider four scenarios. If we include all effects of nucleos(t)ide analogues u_1 and all effects of interferon u_2, i.e., $a_1 = a_2 = b_1 = b_2 = 1$, $f = 0.5$ and $a_3 = 0.9$, we predict that HBV DNA is cleared 90 days after the start of treatment when interferon dosage is maximal at each time step and NAs are zero for $0 \leq t \leq 90$ days and maximal from $t \geq 90$ (see Figure 6a,b, top row). This result implies that NAs do not have a role in viral clearance. This is corroborated by the optimal control solution for interferon monotherapy. Indeed, when we set $a_1 = a_2 = 1$, $a_3 = 0.9$, $b_1 = b_2 = 0$ and $f = 1$, virus is cleared even faster, 87 days after the start of therapy (see Figure 6b, second row) when u_2 dosage is maximal $u_2 = 0.9$ at each time step (see Figure 6a, second row). This is due to the higher efficacy of interferon-α at blocking viral production assumed in this scenario. The results in both of these cases are due to fast expansion and transient maintenance of CTLs to high levels of 2500 cells per mL (see Figure 6c, top two rows).

To further determine which of the interferon effects are the most influential, we remove the first two interferon effects $a_1 = a_2 = 0$, keep $a_3 = 0.9$, $b_1 = b_2 = 1$ and set $f = 0$. Under these conditions, virus is cleared 91 days after the start of therapy, one and four days later than the previous two cases (see Figure 6b, third row) when: (i) NA dosage (u_1) is zero for $0 \leq t \leq 87$ and maximal ($u_1 = 0.9$) afterwards; and (ii) interferon dosage is maximal ($u_2 = 0.9$) for $0 \leq t \leq 91$ and zero afterwards (see Figure 6a, third row). This result implies, again, that interferon is needed for virus clearance.

Lastly, when we consider monotherapy with NAs, i.e., $b_1 = b_2 = 1$, $a_1 = a_2 = a_3 = 0$ and $f = 0$, the HBV DNA will not clear (see Figure 6b, bottom row) in spite of CTL levels being higher than the base of $E = s/d_1 = s/d$ (see Figure 6c, bottom row). Interestingly, our numerical results show that the optimal NA dosage for the first 42 days is zero (see Figure 6a, bottom row).

Our study has used a delay of $\tau = 33.4$ days, since that represented the delay in the CTL activation in the only patient that developed chronic disease in [14]. To check whether the size of the delay affects the results, we have investigated the optimal control problem for a shorter delay of $\tau = 15.2$ days, which is the smallest delay among the patients in [14]. We found that the length of the delay does not influence the results when the therapy is started at the viral steady state (not shown). When the therapy is started at the peak viral load, we find, not surprisingly, that a shorter time lag speeds viral decay and CTL expansion (see Figure 7b,c). Moreover, the optimal control problem for the shorter delay of $\tau = 15.2$ days suggests no therapy ($u_1 = u_2 = 0$) for the first three days post-peak, maximal interferon therapy for $3 \leq t \leq 75$ and $t \geq 90$ and no interferon $u_2 = 0$ otherwise. The role of NAs is transient, with maximal dosages $u_1 = 0.9$ for $20 \leq t \leq 27$, $38 \leq t \leq 51$ and $t > 90$ days and $u_1 = 0$ elsewhere (see Figure 7a, bottom row).

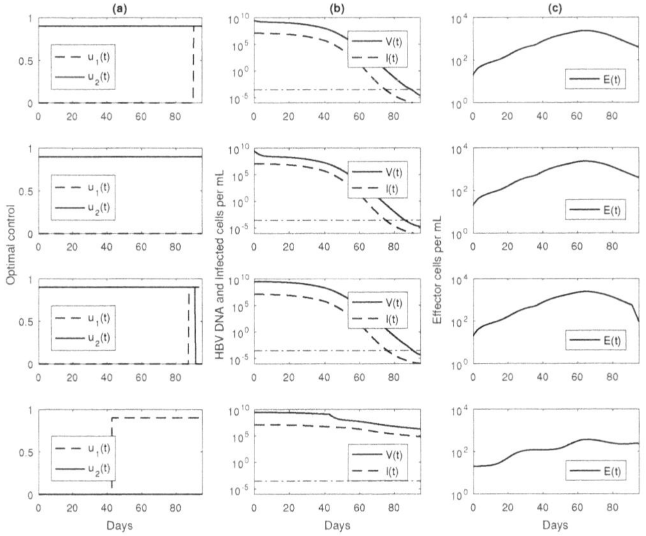

Figure 6. (**a**) Optimal evolution for controls u_1 (dashed lines) and u_2 (solid lines) as given by (15); (**b**) virus V per mL (solid lines) and infected cells I per mL (dashed lines) over time; (**c**) effector cells E per mL over time when drugs are introduced at peak virus concentration and: $a_1 = a_2 = b_1 = b_2 = 1$, $a_3 = 0.9$, $f = 0.5$ (top row); $a_1 = a_2 = 1$, $a_3 = 0.9$, $b_1 = b_2 = 0$, $f = 1$ (second row); $a_1 = a_2 = 0$, $a_3 = 0.9$, $b_1 = b_2 = 1$, $f = 0$ (third row); $a_1 = a_2 = a_3 = 0$, $b_1 = b_2 = 1$, $f = 0$ (bottom row).

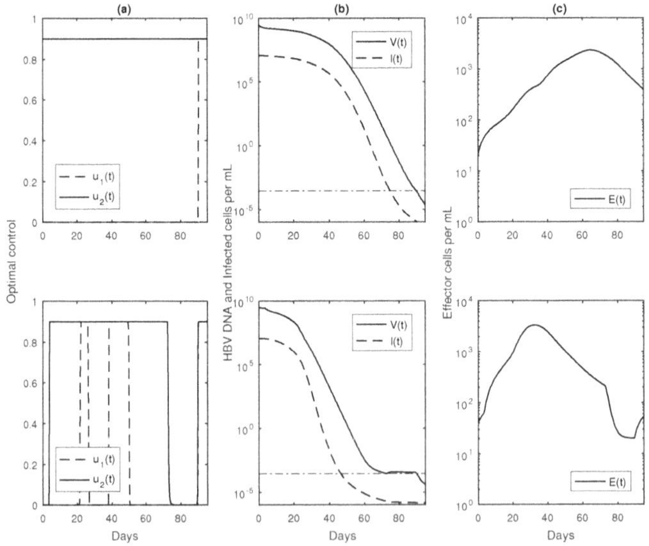

Figure 7. (**a**) Optimal evolution for controls u_1 (dashed lines) and u_2 (solid lines) as given by (15); (**b**) virus V per mL (solid lines) and infected cells I per mL (dashed lines) over time; (**c**) effector cells E per mL over time when drugs are introduced at peak virus concentration for: $a_1 = a_2 = a_3 = b_1 = b_2 = 1$, $f = 0.5$ and $\tau = 33.4$ (top row); $\tau = 15.2$ (bottom row).

Together, these results suggest that the immunomodulatory effects of interferon are needed for HBV DNA control and that immune modulation rather than suppression of virological factors is essential for inducing HBV clearance.

4. Discussion

The management of chronic hepatitis B is based on guidelines regarding screening, diagnosis, duration of treatment, adherence and monitoring of immune and virological markers [12]. Combined antiviral therapy improves the survival rates of patients chronically infected with hepatitis B virus [42]. While the first goal of therapy is to reduce HBV replication, the ultimate goal is the removal of hepatitis B surface and e-antigens and of covalently-closed circular DNA (cccDNA). The removal of the hepatitis B e-antigen is associated with immunological factors, such as removal of the tolerant status of the hepatitis B-specific T cells [36]. Such effects have been demonstrated by interferon therapy, which leads to improved immune function, sustained response rates with combination therapy and improved overall prognosis [37]. These drugs have limitations, such as side effects, the use of injection and poor response in patients with compromised liver function. That makes them unsuitable for long-term administration [43].

To address the positive effects of interferon therapy and to account for its limitations due to adverse effects and limited time usage, we developed a control problem that accounts for: (i) three interferon functions: increased CTL life-span, reduced viremia and increased CTL killing of infected hepatocytes; (ii) two nucleos(t)ide analogue effects: blocking of viral infection and of viral production; (iii) hepatitis B DNA decay below the limits of detection; and (iv) minimal dosage administration over a short time period.

The presence of significant side effects and the need for lengthy treatment make the question of the optimal therapy strategy relevant for the case of chronic hepatitis B. Little work has been previously done on the optimal control of HBV treatment, and the existing models do not include immune system involvement [30]. Ciupe et al. demonstrated that consideration of delayed cytotoxic and non-cytotoxic immune reactions and the presence of cells refractory to infection was necessary to properly understand the dynamics of HBV acute infection and progression to chronic disease [14]. In this study, we derived an optimality system associated with this model, and we constructed the corresponding adjoint equations, which differed from the construction of adjoint equations for systems of ODE [44]. In particular, we found that for a delay τ and a control period $[0, T]$, the adjoint equations are ODEs on $[T-\tau, T]$ and advance differential equations on $[0, T-\tau]$, meaning that there are terms with time argument $t+\tau$. After deriving the proper form of the adjoint equations, we have investigated the optimal control system numerically using a forward-backward sweep method, as in [25].

The control problem indicates the need for high immunomodulatory effects of IFN-α until HBV DNA is cleared. Most importantly, the immunomodulatory effects that increase the survival of effector cells are essential for timely reduction in viremia, which is needed to limit the IFN-α-induced side effects [43]. We predict that starting interferon therapy at the peak viral load, rather than at viral equilibrium, shortens the time to HBV DNA removal. This is due to enhancement of an already activated T cell response. Most interestingly, monotherapy with interferon-α is sufficient for virus control, while the effects of nucleos(t)ide analogues emerge only at the end stages of combination therapy. This result suggests that sequential single therapy (interferon followed by nucleos(t)ide analogues) may be the optimal course of action for both viral suppression and the limitation of drug effects.

Our results imply that an increase in T cell response (in acute infections) and reversal of T cell inactivation (in chronic infections) are essential for fast control of viremia. However, experimental studies have shown that only a small percentage of patients on INF-α therapy experience loss of surface-antigen, e-antigen loss and reduction in virus replication [45]. Moreover, interferon-α is responsible for only partial reversal of T cell inactivation in chronic hepatitis B infections [37] and is not induced naturally during acute hepatitis B infections [46]. However, interferon-α therapy induces cccDNA degradation in cell culture [47] and enhances the innate immune response mediated by natural killer (NK) cells in e-antigen negative patients [45]. Our optimal control study predicts that enhanced cccDNA degradation, which can be incorporated in our model as an effect on the term ρ, has limited effect on the timing of HBV DNA removal (not shown). The NK activation would increase infected cell removal in a complementary fashion to the CTL effect we consider now. Similar to our current results, increased survival of NK cells would be needed for fast HBV DNA removal.

Our study has considered a linear combination effect of interferon and nucleos(t)ide analogues on the blocking of viral production $fa_2u_2 + (1-f)b_2u_1$ with $f = 0.5$. To determine the effect of f on the results, we investigated two control therapies, (i) strong interferon influence on blocking viral production $f = 0.9$ and (ii) strong NA influence on blocking viral production $f = 0.1$, and found that the size of f has little influence one the timing of viral clearance for both peak and equilibrium therapy (not shown).

One limitation in our research comes from ignoring the consequences that prolonged treatment exerts on the evolution of the viral population. Studies have shown that life-long therapy and lack of compliance leads to the development of antiviral drug resistance [10], even though combination therapy helps reduce drug resistance [11]. Our model has not considered the emergence of HBV variants or mutation in the presence of combination therapy. Further work is needed to address the optimal therapy in the presence of these events.

In conclusion, we have designed an optimal control study that shows that a successful short-term anti-HBV therapy requires the modulation of strong innate and/or adaptive immune responses, rather than induction of anti-virological effects. Such therapy needs to be active and elevated for the entire period of viremia.

Acknowledgments: We acknowledge partial support from the National Institute for Mathematical and Biological Synthesis, an Institute sponsored by the National Science Foundation through Award DBI-1300426, with additional support from the University of Tennessee, Knoxville, TN, USA. Lenhart acknowledges partial support from the Boyd Center for Business and Economic Research at the University of Tennessee. We would like to thank the anonymous reviewers for their valuable comments and suggestions.

Author Contributions: Jonathan E. Forde, Stanca M. Ciupe, Ariel Cintron-Arias and Suzanne Lenhart contributed to the formulation and analysis of the control problem and to the preparation of the manuscript. Stanca M. Ciupe and Jonathan E. Forde completed numerical simulations and their interpretations.

Conflicts of Interest: The authors declare no conflict of interest.

References

1. Manzoor, S.; Saalim, M.; Imran, M.; Resham, S.; Ashraf, J. Hepatitis B virus therapy: What's the future holding for us? *World J. Gastroenterol.* **2015**, *21*, 12558–12575.
2. Chen, D.S. Hepatitis B vaccination: The key towards elimination and eradication of hepatitis B. *J. Hepatol.* **2009**, *50*, 805–816.
3. Goldstein, S.T.; Zhou, F.; Hadler, S.C.; Bell, B.P.; Mast, E.E.; Margolis, H.S. A mathematical model to estimate global hepatitis B disease burden and vaccination impact. *Int. J. Epidemiol.* **2005**, *34*, 1329–1339.
4. Ott, J.J.; Stevens, G.A.; Groeger, J.; Wiersma, S.T. Global epidemiology of hepatitis B virus infection: New estimates of age-specific HBsAg seroprevalence and endemicity. *Vaccine* **2012**, *30*, 2212–2219.
5. Lok, A.S.; McMahon, B.J. Chronic hepatitis B. *Hepatology* **2007**, *45*, 507–539.
6. Bhat, M.; Ghali, P.; Deschenes, M.; Wong, P. Prevention and management of chronic hepatitis B. *Int. J. Prev. Med.* **2014**, *5*, S200–S207.
7. Lok, A.S.F.; McMahon, B.J.; Brown, R.S.; Wong, J.B.; Ahmed, A.T.; Farah, W.; Almasri, J.; Alahdab, F.; Benkhadra, K.; Mouchli, M.A.; et al. Antiviral therapy for chronic hepatitis B viral infection in adults: A systematic review and meta-analysis. *Hepatology* **2016**, *63*, 284–306.
8. Lampertico, P.; Aghemo, A.; Viganò, M.; Colombo, M. HBV and HCV therapy. *Viruses* **2009**, *1*, 484–509.
9. Sonneveld, M.J.; Wong, V.; Woltman, A.M.; Wong, G.L.H.; Cakaloglu, Y.; Zeuzem, S.; Buster, E.; Uitterlinden, A.G.; Hansen, B.E.; Chan, H.; et al. Polymorphisms near IL28B and serologic response to peginterferon in HBeAg-positive patients with chronic hepatitis B. *Gastroenterology* **2012**, *142*, 513–520.

10. Hadziyannis, S.J.; Tassopoulos, N.C.; Heathcote, E.J.; Chang, T.-T.; Kitis, G.; Rizzetto, M.; Marcellin, P.; Lim, S.G.; Goodman, Z.; Wulfsohn, M.S.; et al. Adefovir dipivoxil for the treatment of hepatitis B e antigen—Negative chronic hepatitis B. *N. Eng. J. Med.* **2003**, *348*, 800–807.
11. Chan, H.L.; Wang, H.; Niu, J.; Chim, A.M.; Sung, J.J. Two-year lamivudine treatment for hepatitis B e antigen-negative chronic hepatitis B: A double-blind, placebo-controlled trial. *Antivir. Ther.* **2007**, *12*, 345–353.
12. Locarnini, S.; Hatzakis, A.; Chen, D.S.; Lok, A. Strategies to control hepatitis B: Public policy, epidemiology, vaccine and drugs. *J. Hepatol.* **2015**, *62*, S76–S86.
13. Ciupe, S.M.; Catlla, A.; Forde, J.; Schaeffer, D.G. Dynamics of hepatitis B virus infection: What causes viral clearance? *Math. Popul. Stud.* **2011**, *18*, 87–105.
14. Ciupe, S.M.; Ribeiro, R.M.; Nelson, P.W.; Dusheiko, G.; Perelson, A.S. The role of cells refractory to productive infection in acute hepatitis B viral dynamics. *Proc. Natl. Acad. Sci. USA* **2007**, *104*, 5050–5055.
15. Ciupe, S.M.; Ribeiro, R.M.; Nelson, P.W.; Perelson, A.S. Modeling the mechanisms of acute hepatitis B virus infection. *J. Theor. Biol.* **2007**, *247*, 23–35.
16. Ciupe, S.M.; Ribeiro, R.M.; Perelson, A.S. Antibody responses during Hepatitis B viral infection. *PLoS Comput. Biol.* **2014**, *10*, e1003730.
17. Dahari, H.; Shudo, R.M.; Ribeiro, E.; Perelson, A.S. Modeling complex decay profiles of Hepatitis B virus during antiviral therapy. *Hepatology* **2009**, *49*, 32–38.
18. Lewin, S.R.; Ribeiro, R.M.; Walters, T.; Lau, G.K.; Bowden, S.; Locarnini, S.; Perelson, A.S. Analysis of hepatitis B viral load decline under potent therapy: Complex decay profiles observed. *Hepatology* **2001**, *34*, 1012–1020.
19. Tsiang, M.; Rooney, J.F.; Toole, J.J.; Gibbs, C.S. Biphasic clearance kinetics of hepatitis B virus from patients during adefovir dipivoxil therapy. *Hepatology* **1999**, *29*, 1863–1869.
20. Pontryagin, L.S.; Boltyanskii, V.G.; Gamkrelize, R.V.; Mishchenko, E.F. *The Mathematical Theory of Optimal Processes*; Wiley: New York, NY, USA, 1962.
21. Fister, K.R.; Lenhart, S.; McNally, S. Optimizing chemotherapy in an HIV model. *Electron. J. Differ. Eq.* **1998**, *32*, 1–12.
22. Fister, K.R.; Panetta, J.C. Optimal control applied to competing chemotherapeutic cell-kill strategies. *SIAM J. Appl. Math.* **2003**, *63*, 1954–1971.
23. Joshi, H.R. Optimal control of an HIV immunology model. *Opt. Control Appl. Methods* **2003**, *23*, 199–213.
24. Ledzewicz, U.; Brown, T.; Schattler, H. A comparison of optimal controls for a model in cancer chemotherapy with L_1- and L_2-type objectives. *Opt. Methods Softw.* **2004**, *19*, 351–359.
25. Lenhart, S.L.; Workman, J.T. *Optimal Control Applied to Biological Models*; Chapman Hall/CRC: Boca Raton, FL, USA, 2007.
26. Gollman, L.; Mauer, H. Theory and applications of optimal control problems with multiple time-delays. *J. Ind. Manag. Opt.* **2014**, *10*, 413–441.
27. Kharatishvili, G.L. Maximum principle in the theory of optimal time-delay processes. *Dokl. Akad. Nauk USSR* **1961**, *136*, 39–42.
28. Banks, H.T. Approximation of nonlinear functional differential equations control systems. *J. Opt. Theory Appl.* **1979**, *29*, 383–408.
29. Banks, H.T.; Burns, J.A. Hereditary control problems: Numerical methods based on averaging approximations. *SIAM J. Control Opt.* **1978**, *16*, 169–208.
30. Eliaw, A.M.; Alghamdi, M.A.; Aly, S. Hepatitis B virus dynamics: Modeling, analysis and optimal treatment scheduling. *Discret. Dyn. Nat. Soc.* **2013**, *2013*, 1–9.
31. Hattaf, K.; Rachik, M.; Saadi, S.; Yousfi, N. Optimal control in a basic virus infection model. *Appl. Math. Sci.* **2009**, *3*, 949–958.
32. Mouofo, P.T.; Tewa, J.J.; Mewoli, B.; Bowong, S. Optimal control of a delayed system subject to mixed control-state constraints with application to a within-host model of hepatitis virus B. *Ann. Rev. Control* **2013**, *37*, 246–259.
33. Su, Y.; Sun, D. Optimal control of anti-HBV treatment based on combination of Traditional Chinese Medicine and Western Medicine. *Biomed. Signal Proc. Control* **2015**, *15*, 41–48.
34. Guo, J.-T.; Zhou, H.; Liu, C.; Aldrich, C.; Saputelli, J.; Whitaker, T.; Barrasa, M.I.; Mason, M.S.; Seeger, C. Apoptosis and regeneration of hepatocytes during recovery from transient hepadnavirus infections. *J. Virol.* **2000**, *74*, 1495–1505.

35. Guidotti, L.G.; Rochford, R.; Chung, J.; Shapiro, M.; Purcell, R.; Chisari, F.V. Viral clearance without destruction of infected cells during acute HBV infection. *Science* **1999**, *284*, 825–829.
36. Marrack, P.; Kappler, J.; Mitchell, T. Type I interferons keep activated T cells alive. *J. Exp. Med.* **1999**, *189*, 521–529.
37. Rehermann, B.; Bertoletti, A. Immunological aspects of antiviral therapy of chronic hepatitis B virus and hepatitis C virus infections. *Hepatology* **2015**, *61*, 712–721.
38. Perelson, A.S.; Ribeiro, R.M. Hepatitis B virus kinetics and mathematical modeling. *Sem. Liv. Dis.* **2004**, *24*, 11–16.
39. Banks, H.T.; Dediu, S.; Ernstberger, S.E. Sensitivity functions and their uses in inverse problems. *J. Inverse Ill-posed Probl.* **2007**, *15*, 683–708.
40. Hackbush, W. A numerical method for solving parabolic equations with opposite orientations. *Computing* **1978**, *20*, 229–240.
41. Chisari, F.V.; Isogawa, M.; Wieland, S.F. Pathogenesis of hepatitis B virus infection. *Pathol. Biol.* **2010**, *58*, 258–266.
42. Li, G.T.; Yu, Y.-Q.; Chen, S.L.; Fan, P.; Shao, L.-Y.; Chen, J.-Z.; Li, C.-S.; Yi, B.; Chen, W.-C.; Xie, S.-Y.; et al. Sequential combination therapy with pegylated interferon leads to loss of hepatitis B surface antigen and hepatitis B e antigen (HBeAg) seroconversion in HBeAg-positive chronic hepatitis B patients receiving long-term entecavir treatment. *Antimicrob. Agents Chemother.* **2015**, *59*, 4121–4128.
43. Alaluf, M.B.; Shlomai, A. New therapies for chronic hepatitis B. *Liver Int.* **2016**, *36*, 775–782.
44. Gollman, L.; Kern, D.; Maurer, H. Optimal control problems with delays in state and control variables subject to mixed control-state constraints. *Optim. Control Appl. Meth.* **2009**, *30*, 341–365.
45. Micco, L.; Peppa, D.; Loggi, E.; Schurich, A.; Jefferson, L.; Cursaro, C.; Panno, A.M.; Bernardi, M.; Brander, C.; Bihl, F.; et al. Differential boosting of innate and adaptive antiviral responses during pegylated-interferon-alpha therapy of chronic hepatitis B. *J. Hepatol.* **2013**, *58*, 225–233.
46. Dunn, C.; Peppa, D.; Khanna, P.; Nebbia, G.; Jones, M.; Brendish, N.; Lascar, R.M.; Brown, D.; Gilson, R.J.; Tedder, R.J.; et al. Temporal analysis of early immune responses in patients with acute hepatitis B virus infection. *Gastroenterology* **2009**, *137*, 1289–1300.
47. Lucifora, J.; Xia, Y.; Reisinger, F.; Zhang, K.; Stadler, D.; Cheng, X.; Sprinzl, M.F.; Koppensteiner, H.; Makowska, Z.; Volz, T.; et al. Specific and nonhepatotoxic degradation of nuclear hepatitis B virus cccDNA. *Science* **2014**, *343*, 12221–12228.

applied sciences

MDPI

Article

Altered Mechano-Electrochemical Behavior of Articular Cartilage in Populations with Obesity

Sara Manzano [1,2,3], Manuel Doblaré [1,2,3] and Mohamed Hamdy Doweidar [1,2,3,*]

1 Mechanical Engineering Department, School of Engineering and Architecture (EINA), University of Zaragoza, María de Luna s/n, Betancourt Building, 50018 Zaragoza, Spain; manzano@unizar.es (S.M.); mdoblare@unizar.es (M.D.)
2 Aragón Institute of Engineering Research (I3A), University of Zaragoza, Zaragoza 50018, Spain
3 Biomedical Research Networking Center in Bioengineering, Biomaterials and Nanomedicine (CIBER-BBN), Zaragoza 50018, Spain
* Correspondence: mohamed@unizar.es; Tel.: +34-876-555-210

Academic Editor: Yang Kuang
Received: 11 May 2016; Accepted: 21 June 2016; Published: 24 June 2016

Abstract: Obesity, one of the major problems in modern society, adversely affects people's health and increases the risk of suffering degeneration in supportive tissues such as cartilage, which loses its ability to support and distribute loads. However, no specific research regarding obesity-associated alterations in the mechano-electrochemical cartilage environment has been developed. Such studies could help us to understand the first signs of cartilage degeneration when body weight increases and to establish preventive treatments to avoid cartilage deterioration. In this work, a previous mechano-electrochemical computational model has been further developed and employed to analyze and quantify the effects of obesity on the articular cartilage of the femoral hip. A comparison between the obtained results of the healthy and osteoarthritic cartilage has been made. It shows that behavioral patterns of cartilage, such as ion fluxes and cation distribution, have considerable similarities with those obtained for the early stages of osteoarthritis. Thus, an increment in the outgoing ion fluxes is produced, resulting in lower cation concentrations in all the cartilage layers. These results suggest that people with obesity, i.e. a body mass index greater than 30 kg/m^2, should undergo preventive treatments for osteoarthritis to avoid homeostatic alterations and, subsequent, tissue deterioration.

Keywords: obesity; mechano-electrochemical model; articular cartilage; cartilage degeneration; cartilage loading

1. Introduction

In the last decade, obesity has become one of the most serious socioeconomic concerns in developed countries [1,2]. This medical problem is produced by a combination of several factors such as excessive food energy intake, lack of physical activity and/or genetic susceptibility [3–5]. As a result, an increase in the probability of suffering diseases such as osteoarthritis and bone degeneration has been reported [6–8]. In particular, overload on joints generates imbalance in tissue homeostasis [9,10]. This is mainly due to the increment of the force applied on articulations. In the case of normal weight, the applied force is three times the body weight. This force can reach six to ten times the body weight when performing activities such as climbing or running [11]. However, the transmitted force to joints in the case of people with obesity is doubled, generating excessive wear and leading to osteoarthritis [12].

Since articular cartilage is an avascular tissue, the transport of nutrients from synovial fluid to chondrocytes occurs by diffusion and convection when loading or when chemical conditions are changing [13–15]. In this sense, cartilage counts on biomolecules also called proteoglycans. They are responsible for the turgid nature of the tissue and provide the osmotic properties needed to resist compressive loads [16,17]. This occurs mainly as a consequence of the negative charges attached to them, fixed charge density. These negative charges provide a repulsive force that enhances the compressive stiffness of the cartilage [18].

Compression cycle due to body weight generates incoming and outgoing fluxes that helps in nutrients and wastes products exchange [11]. In the case of people with obesity, significant alterations are produced in water and ion fluxes as well as in tissue deformation. These fluxes are considered as essential biomarkers that indicate the degree of cartilage degeneration [19,20]. Therefore, the study of ion fluxes and tissue deformation may reveal alterations in the cartilage tissue.

Despite the great incidence of obesity, there are no exhaustive studies analyzing the correlation between alterations in the biological processes of cartilage and an increase in body weight. Only Travascio et al. [11] have addressed this issue. In their recent work, they describe several changes in protein synthesis and the subsequent reduction in extracellular matrix (ECM) production in hip articular cartilage. They studied these phenomena by means of a self-developed 1D model. However, important questions remain. How does obesity affect water and ion fluxes? Are these alterations a catalyst for osteoarthritis? Is the ion concentration in the tissue similar to that observed in people of normal weight? All these questions need to be addressed.

To analyze and elucidate tissue behavior under these conditions, a previously described 3D computational model of cartilage behavior [19–21] is here extended and employed to study the effects of overweight on the articular joints (Figure 1). The present model considers the influence of ions on electrochemical events as well as proteoglycan repulsion in a loaded 3D environment. The main goals of this study are: (i) to analyze the role of proteoglycan repulsion in the cartilage loading problem; (ii) to study the effect of metabolic alterations (fluxes of water and ions) due to obesity on the homeostasis of femoral hip cartilage; and (iii) to quantify and compare cation and water flux in cartilage of both healthy people and those suffering from obesity when a maximal load is applied on the joint.

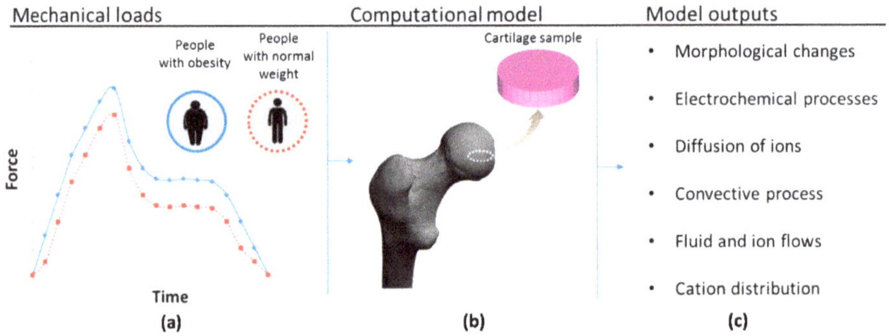

Figure 1. Schematic representation of the simulation process: (**a**) the mechanical load during human gait for people with obesity and with normal weight based on the work of Travascio et al. [11]; (**b**) incorporation of a hip cartilage sample into the computational model; and (**c**) list of the main model outputs.

2. Material and Methods

The main mathematical formulation of the present model is based on the triphasic theory for charged and hydrated soft tissues. This theory has been applied to simulate the behavior of articular cartilage [22,23]). As in our previous works and other related studies in the literature [19,20,24,25], the tissue is considered as a mixture in which four phases are distinguished: A negatively charged poroelastic solid phase including proteoglycans (s), an interstitial fluid phase which includes water (w), cations (+) and anions (−). The tissue behavior depends on the mechano-electrochemical interaction of these phases for cartilage maintenance (for more details see [10,20,26]. The four basic unknowns in this physico-mathematical model correspond to the displacement of the solid matrix, $\mathbf{u}^s$, the chemical potential of the water, ε^w, and the electrochemical potential of the cations and anions, ε^+ and ε^- respectively. The main model equations are summarized below.

2.1. Flow Equations

Outgoing and/or incoming water and ion fluxes are considered in the mathematical formulation through the mass balance equation for the total mixture and the charge balance for each ion. The main fluxes present in the model are water flux ($\mathbf{J}^w$), cation flux ($\mathbf{J}^+$) and anion flux ($\mathbf{J}^-$). The flux equations are detailed below.

Mass balance of the mixture:

$$\nabla \cdot \mathbf{v}^s + \nabla \cdot \mathbf{J}^w = 0 \tag{1}$$

Charge balance of ions:

$$\frac{\partial \left(\Phi^w c^+\right)}{\partial t} + \nabla \cdot \underbrace{\mathbf{J}^+}_{\text{diffusion}} + \nabla \cdot \underbrace{\left(\Phi^w c^+ \mathbf{v}^s\right)}_{\text{convection}} = 0, \tag{2}$$

$$\frac{\partial \left(\Phi^w c^-\right)}{\partial t} + \nabla \cdot \underbrace{\mathbf{J}^-}_{\text{diffusion}} + \nabla \cdot \underbrace{\left(\Phi^w c^- \mathbf{v}^s\right)}_{\text{convection}} = 0, \tag{3}$$

where $\mathbf{v}^s = \frac{\partial \mathbf{u}^s}{\partial t}$ is the velocity of each point of the solid matrix; c^+ and c^- are cation and anion concentrations, respectively; and Φ^w is the porosity of the tissue [26]. Hereafter, the fluxes can be mathematically expressed as a function of the electrochemical potentials:

$$\mathbf{J}^w = -\frac{R\,T\Phi^w}{\alpha}\left(\nabla\varepsilon^w + \frac{c^+}{\varepsilon^+}\nabla\varepsilon^+ + \frac{c^-}{\varepsilon^-}\nabla\varepsilon^-\right) \tag{4}$$

$$\mathbf{J}^+ = -\frac{R\,T\Phi^w c^+}{\alpha}\nabla\varepsilon^w - \left[\frac{\Phi^w c^+ D^+}{\varepsilon^+} + \frac{R\,T\Phi^w \left(c^+\right)^2}{\alpha\varepsilon^+}\right]\nabla\varepsilon^+ - \frac{R\,T\Phi^w c^+ c^-}{\alpha\varepsilon^+}\nabla\varepsilon^-, \tag{5}$$

$$\mathbf{J}^- = -\frac{R\,T\Phi^w c^-}{\alpha}\nabla\varepsilon^w - \left[\frac{\Phi^w c^- D^-}{\varepsilon^-} + \frac{R\,T\Phi^w \left(c^-\right)^2}{\alpha\varepsilon^-}\right]\nabla\varepsilon^- - \frac{R\,T\Phi^w c^+ c^-}{\alpha\varepsilon^+}\nabla\varepsilon^+, \tag{6}$$

where α is the drag coefficient between the solid and the water phases; R is the universal gas constant; T is the absolute temperature; and D^+ and D^- are the cation and anion diffusivities, respectively [27]. The electrochemical potentials are defined as follows:

$$\varepsilon^w = \frac{P}{R\,T} - \Phi\left(c^+ + c^-\right) + \frac{B_w}{R\,T}\theta, \tag{7}$$

$$\varepsilon^+ = \gamma^+ c^+ exp\left(\frac{F_c\psi}{R\,T}\right), \tag{8}$$

$$\varepsilon^- = \gamma^- c^- exp\left(-\frac{F_c\psi}{R\,T}\right) \tag{9}$$

where Φ represents the osmotic coefficient, B_w is the fluid-solid coupling coefficient, P is the fluid pressure, $\theta = div\ \mathbf{u}^s$ is the expansion of the solid matrix related to the infinitesimal strain tensor of the solid matrix, F_c is the Faraday constant and ψ the electrical potential. γ^+ and γ^- refer to the cation and anion activity coefficients, respectively.

2.2. Momentum Balance Equation of the Mixture

In contrast to our previous studies [19,20], the external force applied to the cartilage sample, $\mathbf{f}_{\text{ext}}$, is incorporated into the momentum balance equation to analyze the effect of maximal load on the joint as follows:

$$\nabla \cdot \underbrace{\sigma}_{\sigma^{\text{f}}+\sigma^{\text{c}}+\sigma^{\text{s}}} = \mathbf{f}_{\text{ext}} \,, \tag{10}$$

where σ^{f} corresponds to the stress exerted by the fluid, σ^{c} is the stress due to chemical factors such as proteoglycan repulsion and σ^{s} is the stress of the solid matrix.

In cartilage tissue, the total stress, $\boldsymbol{\sigma}$, can be mathematically formulated as a combination of the osmotic pressure and the elastic stress of the matrix:

$$\sigma = \underbrace{\underbrace{-P\mathbf{I}}_{\sigma^{\text{f}}}}_{\text{osmotic pressure}} \quad \underbrace{\underbrace{-\lambda_{\text{s}}\theta\mathbf{I} + 2\mu_{\text{s}}\epsilon}_{\sigma^{\text{s}}}}_{\text{elastic stress}} \quad \underbrace{\underbrace{-T_{\text{c}}\mathbf{I}}_{\sigma^{\text{c}}}}_{\textbf{chemical expansion}} \,, \tag{11}$$

where $\mathbf{I}$ is the identity tensor. T_{c} is the chemical expansion due to the proteoglycan repulsion phenomenon [20]. λ_{s} and u_{s} are the Lame constants and ϵ is the solid matrix deformation. Note that to accurately simulate obesity-associated alterations in cartilage behavior, the volumetric expansion due to proteoglycan-attached negative charges were introduced into the model formulation. Thus, T_{c} can be expressed as a combination of: the proteoglycan repulsion coefficients, a_0 and k; the ion activity coefficients during the process and at the reference estate, $\gamma^{\mp}$ and $\gamma^{\mp *}$, respectively; and the neutral salt concentration, c.

$$T_{\text{C}} = a_0 c^{\text{F}} exp\left(-k\frac{\gamma^{\mp}}{\gamma^{\mp *}}\right)\sqrt{c\left(c + c^{\text{F}}\right)} \tag{12}$$

For more details, see [20,21].

This phenomenon is demonstrated to be essential for the swelling process when ion concentration variations are generated within the tissue [20,21].

2.3. Simulation of the Articular Cartilage Behavior for People with Obesity

To study the effect of the variation of mechano-electrochemical parameters on cartilage behavior, the experimental design described by Lai et al. [21] is computationally reproduced. Thus, a cartilage specimen of 1.5 mm diameter and 0.5 mm depth is placed inside a circular impermeable confining ring and a loading permeable pattern is located at the top of the sample. Tissue samples are hydrated in NaCl solution of 0.15 M similar to the physiological state. Under these conditions, maximal compression loads relating to healthy people and people with obesity are applied on top of the sample. The applied pressure is that corresponding to the maximal pressure observed after 0.44 seconds of the human gait cycle. For the sake of simplicity, it is considered that, in later cycles, the tissue will exhibit similar behavior to that taken as a reference. The experimental data are extracted from [11]. Note that, for this study, they developed specific measurements in a total of 20 male subjects (10 normal and 10 obese) aged between eighteen and thirty-five volunteered for this study. Prior to experiments, anthropometric and body measurements were collected and recorded.

Clinical guidelines are established to evaluate the grade of obesity in adults by considering the Body Mass Index (BMI). This measure correlates the weight with the height of an adult to assess obesity. Thus, values of BMI between 18.5 and 24.9 kg/m^2 represent a suitable weight for adults, whereas a BMI higher than 30 kg/m^2 indicates obesity. In the model calculations, the applied force and subsequent pressure are extracted from [11] who consider two groups of study: people of normal weight, BMI = 22.262 $\mp$ 1.172 kg/m^2, and people with obesity, BMI = 33.978 $\mp$ 3.629 kg/m^2. The other cartilage properties introduced into the computational model are listed in Table 1.

Table 1. Model parameters used in the computational model.

Description	Symbol	Range or Studied Value	Reference
Young's modulus	E	0.6 MPa	[19]
Poisson coefficient	ν	0.28	[19]
Drag coefficient between the solid and the water phase	α	7×10^{14} N·s·m^{-4}	[22]
Diffusivity of the cations	D^+	5×10^{-1} m·s^{-1}	[22]
Diffusivity of the anions	D^-	8×10^{-1} m·s^{-1}	[22]
Initial FCD	-	0.2 mEq·mL^{-1}	[22]
Activity coefficient of cations	γ^+	0.86	[21]
Activity coefficient of anions	γ^-	0.85	[21]
Gas constant	R	8.314 J·mol^{-1}·K^{-1}	[22]
Absolute temperature	T	298 K	[22]
Osmotic coefficient	Φ	0.8	[21]
Initial amount of water in the tissue	Φ_0^w	0.75	[22]

Accurate quantifications of cation fluxes and distributions within the samples have been performed. Besides, monitoring of tissue changes during the loading processes was also carried out. Note that this model can be employed to study the cartilage behavior in function of the degree of obesity, how long an individual has been obese, co-morbidities, and sex of subjects. Besides, it can be a good tool to study specific cases such as normal and obese rats or rabbits. For all of these studies, experimental data are required to be carried out.

Three linear 8-node hexahedral elements with 2 × 2 × 2 Gaussian integration points are used. The selected average mesh has a total number of 1680 elements. Small-deformation finite element formulation, similar to previous articular cartilage models [11,22,28], has been implemented in a user defined element subroutine (UEL subroutine following Abaqus standard names) of the commercial software package Abaqus 6.11 (Dassault Systemes, Paris, France, 2016). The implementation scheme of the 3D mechano-electrochemical model used for the loading problem is shown in Figure 2.

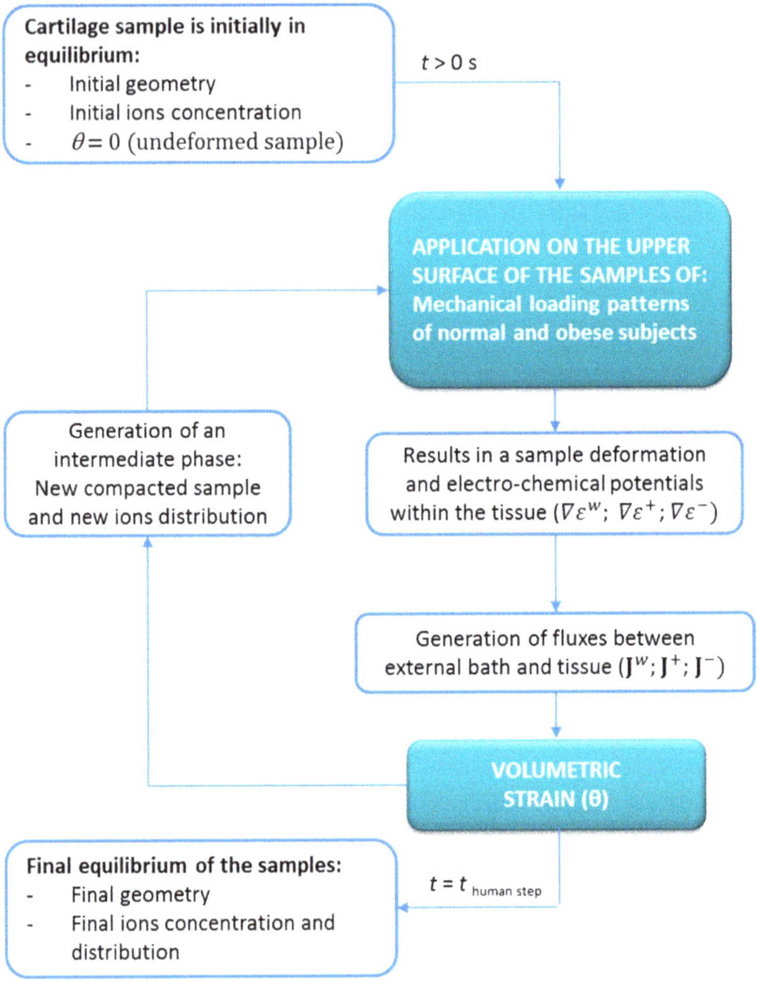

Figure 2. Schematic diagram describing the process of the numerical simulation and the steps of its implementation.

2.3.1. Initial Conditions

Initially, the cartilage sample is equilibrated within a single salt (NaCl) solution with a concentration c^*. The initial conditions for the computational model are,

$$t = 0: \ \mathbf{u} = 0; \ \varepsilon^w = \varepsilon^{w^*}; \ \varepsilon^+ = \varepsilon^{+^*}; \ \varepsilon^- = \varepsilon^{-^*}. \tag{13}$$

The initial equilibrium state of the tissue corresponds to the unloaded undeformed tissue. This has been selected as a reference configuration for strain (time zero seconds, undeformed configuration).

2.3.2. Boundary Conditions

The boundary conditions of the sample in confined configuration (Figure 3) are the following.

Free upper surface:

$$\sigma_z = P_{z,\,ext};\ \varepsilon^{w} = \varepsilon^{w*};\ \varepsilon^{+} = \varepsilon^{+*};\ \varepsilon^{-} = \varepsilon^{-*}. \tag{14}$$

Lateral surface:

$$u_x = u_y = 0;\ J^{w}_{x,y} = J^{+}_{x,y} = J^{-}_{x,y} = 0. \tag{15}$$

Lower surface:

$$\mathbf{u} = \mathbf{0};\ J^{w}_{z} = J^{+}_{z} = J^{-}_{z} = 0. \tag{16}$$

Note that, at $t = 0$ seconds, the sample is unloaded and the concentration of the external solution, c^*, is equal to 0.15 M. When external pressure ($P_{z,\mathrm{ext}}$) is applied, the transient response of the solid displacement is solved using the extended 3D model. A comparison between simulations of cartilage of people with obesity and cartilage affected by osteoarthritic has been made. For simplicity, and due to the lack of experimental data, the cartilage is considered as an isotropic material.

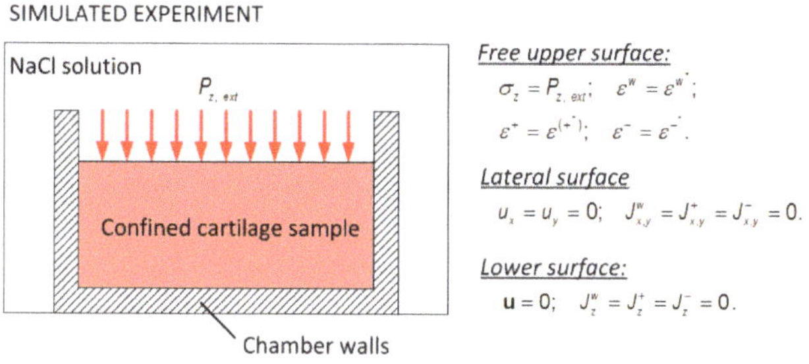

Figure 3. Schematic representation of the experiment simulated by the computational model and the boundary conditions.

3. Results and Discussion

First, the repulsion of negative charges attached to proteoglycans has been studied. This phenomenon, which was neglected in previous models, has been demonstrated to be essential in cartilage free-swelling [20].

Second, water and ion fluxes together with morphological changes in the tissue have been simulated for a critical situation of the human gait (when higher values of forces are experienced by the tissue) in healthy and obesity conditions. The corresponding simulation results of each component (water, cation and ions) are presented after 0.44 s of simulated time, corresponding to the above-mentioned critical phase where maximal cartilage deformation, outgoing water and ion fluxes are observed.

Finally, z-displacement, water and ion flux patterns obtained for people with obesity are compared not only to those relating to people of normal weight but also to degenerated tissue [20].

3.1. Proteoglycan Repulsion

To verify the role of the repulsion phenomenon of the negative charges attached to proteoglycans, the experimental test developed by Chen et al. [29] is here reproduced. They applied 8% of deformation to study the distribution of the Fixed Charge Density (FCD) within the sample. Note that the same confined conditions detailed in the previous section are here considered and the 8% of deformation is applied in the computational model.

Under these conditions, it is observed that when taking into account the proteoglycan repulsion phenomenon, the model gives values closer to those obtained experimentally for the FCD distribution

(Figure 4). In contrast, when this phenomenon is neglected, the results are higher than those obtained experimentally.

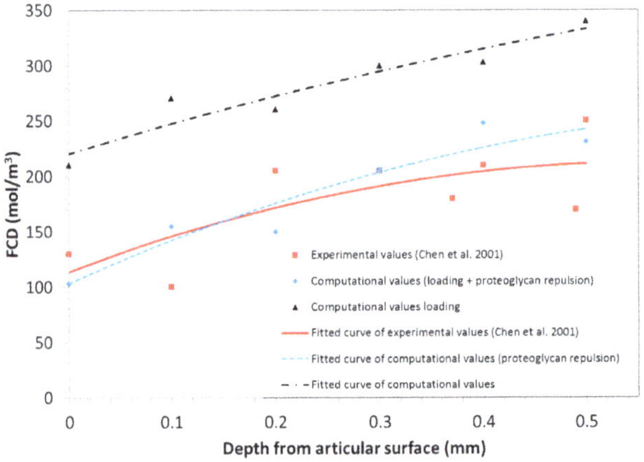

Figure 4. Fixed Charge Density (FCD) distribution along the thickness of a cartilage sample with 8% sample deformation due to compressive loads.

Biologically, this can be explained by the balance of forces that takes place in the sample. When the repulsion phenomenon is not taken into account, the compressive force exerted by loads is balanced with the water contained in the sample. Thus, a higher amount of FCD is observed in all cartilage layers and, subsequently, a higher water content in the sample. However, when the repulsion of proteoglycans is considered, both the water in the sample and the repulsive forces of the attached proteoglycans resist the applied compression load.

3.2. Alterations of Cartilage Tissue in People with Obesity

3.2.1. Displacement and Water Flux

Under normal weight (BMI = 22.262 kg/m^2), the model displays a maximum surface displacement of -1.38×10^{-4} m (Figure 5a.1) after 0.44 s of simulated time, significantly lower than that observed in cases of obesity (BMI = 33.978 kg/m^2), $u_z = -2.4 \times 10^{-4}$ m (Figure 5b.1). This reduction in tissue deformation results in a lower outgoing water flux from the cartilage sample. Thus, simulated cartilage of people with a normal BMI generates a maximum value of $\mathbf{J}^w = 1.31 \times 10^{-8}$ m^3/s (Figure 5a.2,) whereas in people with obesity, the water flux is increased to 1.58×10^{-8} m^3/s (Figure 5b.2).

Biologically, this increase in the outgoing water flux produces pathological dehydration, commonly associated with the joints of obese people and with the majority of cartilage pathologies such as osteoarthritis. These results are consistent with the study of Travascio et al. [11] who found a reduction in cartilage water content in people with obesity. Similarly, they suggested this aspect as an essential promoting agent of cartilage osteoarthritis.

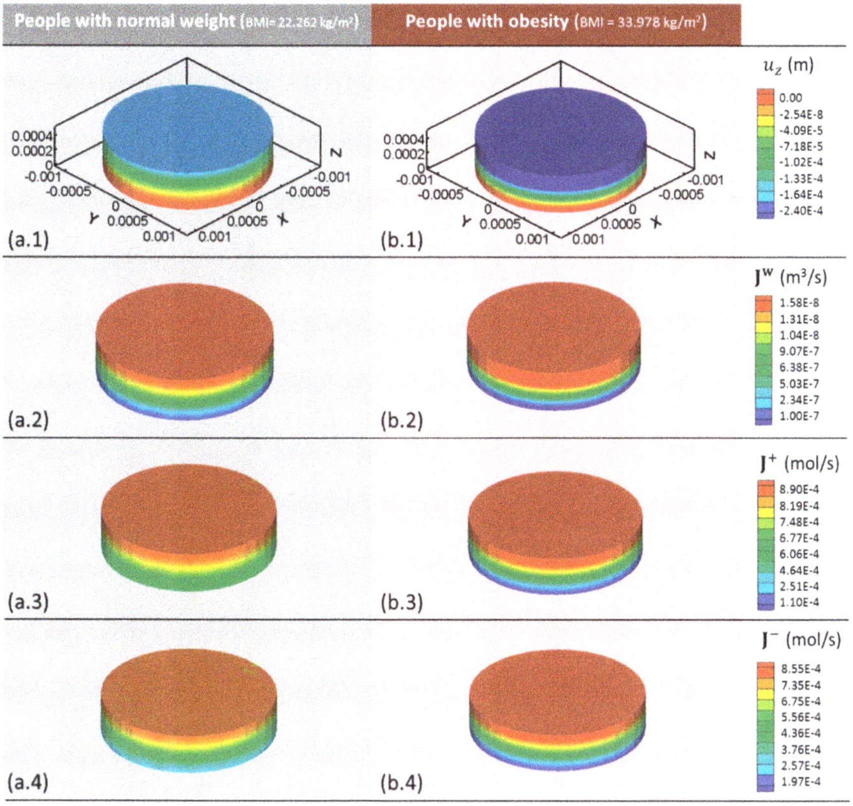

Figure 5. (**1**) *z*-displacement (u_z); (**2**) water ($\mathbf{J}^w$); (**3**) cation ($\mathbf{J}^+$); and (**4**) anion ($\mathbf{J}^-$) fluxes obtained with the 3D computational model after 0.44 s of simulated human gait cycle for (**a**) healthy and (**b**) people with obesity. Note that BMI = 22.262 kg/m^2 corresponds to normal weight and BMI = 33.978 kg/m^2 to obesity. Positive fluxes refer to the emergence of the component from the cartilage sample to the external solution.

3.2.2. Cation Fluxes

To fully understand the alteration in the mechano electrochemical events occurring in overloaded cartilage, ion fluxes have also been studied. A similar trend to that of water outflow is observed for cation fluxes. Cartilage samples with a weight corresponding to BMI = 22.262 kg/m^2 show a maximum cation flux of 8.19 × 10^{-4} mol/s in the upper surface. This flux is reduced to 6.06 × 10^{-4} mol/s in the bottom surface (Figure 5a.3). In obese cartilage simulations, these values are increased reaching a maximum value of 8.9 × 10^{-4} mol/s (Figure 5b.3).

Cartilage degeneration due to osteoarthritis has been widely studied [6–8]. A maximum outgoing cation flux of $\mathbf{J}^+ = 2 \times 10^{-4}$ mol/s was obtained for the early stage of osteoarthritis [20]. Interestingly, the cartilage of the population with obesity exhibits similar values to those obtained for early stages of osteoarthritis.

These findings suggest that people with obesity have an increased risk of suffering cartilage degenerative diseases such as osteoarthritis. The use of preventive treatments to avoid cartilage degradation is thus recommended.

3.2.3. Anion Fluxes

The results obtained for anion fluxes show high similarities with cation fluxes. Anions show the maximum value at the upper surface, $\mathbf{J}^{-} = 7.35 \times 10^{-4}$ mol/s, while the minimum is located at the bottom surface, $\mathbf{J}^{-} = 2.57 \times 10^{-4}$ mol/s after 0.44 s of simulated time for normal weight (BMI = 22.262 kg/m^2) (Figure 5a.4). Under obesity conditions, the maximum outgoing anion flux increases significantly to $\mathbf{J}^{-} = 8.55 \times 10^{-4}$ mol/s (Figure 5b.4). Both cation and anion fluxes directly depend on the applied pressure, thus their increase is directly related to excess in body weight.

3.2.4. Cation Distribution

The gradient of cations was also monitored after 0.44 s of obese loading and compared with results obtained for normal weight and those previously obtained for osteoarthritic conditions [20].

At this phase, the model showed a significant reduction in cation concentration from the lower to the upper surface within the cartilage sample for both obesity and normal weight conditions. In the normal weight case, it ranged from 372 mol/m3 at z = 0 mm to 302 mol/m^3 at 0 mm depth to 372 mol/m^3 at z = 0.5 mm. However, in the obesity case, an important reduction in the cation concentration is observed. The cation concentration at the lower surface (z = 0 mm) had a value of 273 mol/m^3 while a value of 190 mol/m^3 was reached at the upper surface (z = 0.5 mm).

Figure 6 shows how the concentration of cations within the cartilage under obese-loading conditions undergoes alterations, being significantly reduced within the sample.

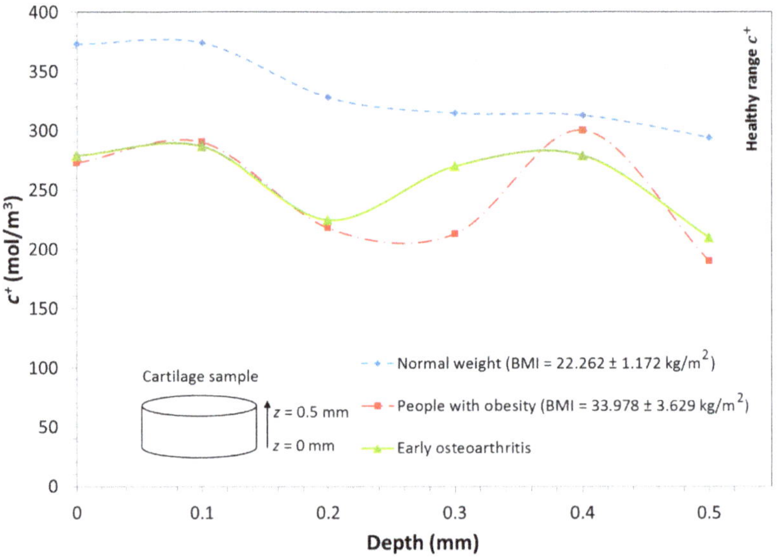

Figure 6. Distribution of cation concentration for people with obesity and of normal weight during human gait. The shaded area represents the range of cation distribution in cartilage of a person of normal weight.

In addition, the obesity results were compared to those obtained for osteoarthritic cartilage in its primary stages of degeneration. Similar alterations were observed in both obese and osteoarthritic tissue. In both cases, there was an increase in the cation concentration in the medium layer and an irregular cation distribution within the sample. These observations open the door to future research to establish and quantify a specific relationship between obesity and an increase in the risk of suffering cartilage diseases such as osteoarthritis. This suggests the necessity of adopting preventive treatments to avoid the progression of cartilage degeneration.

4. Conclusions

In this work, a previously presented three-dimensional mechano-electrochemical model [19,20] has been extended and employed to analyze and quantify the effect of obesity on cartilage behavior. This model can be considered the first 3D computational model for use in the study of the effects of obesity on articular cartilage. This study shows, for the first time, the relation between obesity and cartilage degeneration. It presents the first step when performing a more accurate and sophisticated model to provide patterns that indicate clinician when obesity is starting to affect tissues of the joints. This will let them prevent and treat, at a very early stage, tissue degradation.

The model incorporates important biological and physical aspects of articular cartilage such as proteoglycan repulsion due to attached negative charges, diffusive-convective phenomena and the combined mechano-electrochemical events that occur in healthy as well as pathological tissue. Despite the promising use of the model, it does have some several limitations. First, anisotropic properties should be included in the model in order to analyze the influence of heterogeneity in tissue behavior under loading. Second, in the case of obesity, there is a lack of experimental parameters in the literature. Thus, for future advances, accurate experiments to measure specific cartilage properties are required.

Several interesting insights into cartilage behavior in people with obesity have been evidenced. The results demonstrate that water and ion fluxes within the considered samples present significant alterations, showing a general increase in their values. The simulations also show how the cation concentration is reduced within the sample. Interestingly, these results closely resemble those previously obtained for the early stages of osteoarthritis.

In light of these simulations, it is suggested that people with obesity, i.e., with a BMI greater than 30 kg/m^2, should undergo preventive osteoarthritis treatments to avoid homeostatic cartilage alteration and subsequent tissue deterioration. Besides, as expected, the deformation of the cartilage sample from obese patients is higher than that obtained from people of normal weight. These findings support the hypothesis of Travascio et al. [11] that there is a strong correlation between obesity and an increase in the risk of suffering osteoarthritis.

The present model can be considered as a pioneer 3D computational model for simulating cartilage behavior in people with obesity. It could be a valuable tool for analyzing the effects of several cartilage pathologies taking into consideration mechano-electrochemical tissue behavior. Due to the complexity of in vivo experimental measurements of cartilage behavior in specific loading conditions, such as obesity, this model is presented as a predictive instrument to study the subsequent physiological and pathological processes. This, together with the capacity of the model to display the results in clinically interpretable three-dimensional images, make it an interesting novel tool for the diagnosis, monitoring and efficacy evaluation of potential cartilage therapies.

Acknowledgments: The authors gratefully acknowledge financial support from the Spanish Ministry of Economy and Competitiveness (MINECO MAT2013-46467-C4-3-R and FPU graduate research program AP2010/2557), the Government of Aragon (DGA) and the Biomedical Research Networking Center in Bioengineering, Biomaterials and Nanomedicine (CIBER-BBN). CIBER-BBN is financed by the Instituto de Salud Carlos III with assistance from the European Regional Development Fund.

Author Contributions: The conception and design of the study, or acquisition of data, or analysis and interpretation of data: Sara Manzano and Mohamed Hamdy Doweidar. Drafting the article or revising it critically for important intellectual content: Sara Manzano, Manuel Doblaré and Mohamed Hamdy Doweidar. Final approval of the version to be submitted: Sara Manzano, Manuel Doblaré and Mohamed Hamdy Doweidar.

Conflicts of Interest: The authors declare no conflict of interest.

References

1. Finkelstein, E.A.; Fiebelkorn, I.C.; Wang, G. Nationalmedical spending attributable to overweight and obesity: How much, and who's paying? *Health Aff.* **2003**, *22*, 219–226.
2. Ogden, C.L.; Yanovski, S.Z.; Carroll, M.D.; Flegal, K.M. The epidemiology of obesity. *Gastroenterology* **2007**, *132*, 2087–2102. [CrossRef] [PubMed]
3. Pi-Sunyer, F.X. The practical guide identification, evaluation, and treatment of overweight and obesity in adults. Available online: http://www.nhlbi.nih.gov/files/docs/guidelines/prctgd_c.pdf (accessed on 23 June 2016).
4. Kopelman, P.G. Obesity as a medical problem. *Nature* **2000**, *404*, 635–643. [PubMed]
5. Puhl, R.M.; Heuer, C.A. The stigma of obesity: A review and update. *Obesity* **2009**, *17*, 941–964. [CrossRef] [PubMed]
6. Felson, D.T.; Anderson, J.J.; Naimark, A.; Walker, A.M.; Meenan, R.F. Obesity and knee osteoarthritis. *Ann. Intern. Med.* **1988**, *109*, 18–24. [CrossRef] [PubMed]
7. Felson, D.T.; Chaisson, C.E. Understanding the relationship between body weight and osteoarthritis. *Bailliere Clin. Rheum.* **1997**, *11*, 671–681. [CrossRef]
8. Oliveria, S.A.; Felson, D.T.; Cirillo, P.A.; Reed, J.I.; Walker, A.M. Body weight, body mass index, and incident symptomatic osteoarthritis of the hand, hip, and knee. *Epidemiology* **1999**, *10*, 161–166. [CrossRef] [PubMed]
9. Karvonen, R.L.; Negendank, W.G.; Teitge, R.A.; Reed, A.H.; Miller, P.R.; Fernandez-Madrid, F. Factors affecting articular cartilage thickness in osteoarthritis and aging. *J. Rheum.* **1994**, *21*, 1310–1318. [PubMed]
10. Pottie, P.; Presle, N.; Terlain, B.; Netter, P.; Mainard, D.; Berenbaum, F. Obesity and osteoarthritis: More complex than predicted! *Ann. Rheum. Dis.* **2006**, *65*, 1403–1405. [CrossRef] [PubMed]
11. Travascio, F.; Eltoukhy, M.; Cami, S.; Asfour, S. Altered mechano-chemical environment in hip articular cartilage: Effect of obesity. *Biomech. Mod. Mechanobio.* **2014**, *13*, 945–959. [CrossRef] [PubMed]
12. Felson, D.T. Weight and osteoarthritis. *Am. J. Clin. Nutr.* **1996**, *63*, 430S–432S. [PubMed]
13. O'Hara, B.P.; Urban, J.P.; Maroudas, A. Influence of cyclic loading on the nutrition of articular cartilage. *Ann. Rheum. Dis.* **1990**, *49*, 536–539. [CrossRef] [PubMed]
14. Garcia, A.M.; Frank, E.H.; Grimshaw, P.E.; Grodzinsky, A.J. Contributions of fluid convection and electrical migration to transport in cartilage: Relevance to loading. *Arch. Biochem. Biophys.* **1996**, *333*, 317–325. [CrossRef] [PubMed]
15. Ulrich-Vinther, M.; Maloney, M.D.; Schwarz, E.M.; Rosier, R.; O'Keefe, R.J. Articular cartilage biology. *J. Am. Acad. Orthop. Surg.* **2003**, *11*, 421–430. [CrossRef] [PubMed]
16. Ateshian, G.A.; Maa, S.; Weiss, J.A. Multiphasic Finite Element Framework for Modeling Hydrated Mixtures With Multiple Neutral and Charged Solutes. *J. Biomech. Eng.* **2013**, *135*. [CrossRef] [PubMed]
17. Arbabi, V.; Pouran, B.; Weinans, H.; Zadpoor, A.A. Multiphasic modeling of charged solute transport across articular cartilage: Application of multi-zone finite-bath model. *J. Biomech.* **2016**, *49*, 1510–1517. [CrossRef] [PubMed]
18. Huttunen, J.M.J.; Kokkonen, H.T.; Jurvelin, J.S.; Töyräs, J.; Kaipio, J.P. Estimation of fixed charge density and diffusivity profiles in cartilage using contrast enhanced computer tomography. *Int. J. Numer. Methods Eng.* **2014**, *98*, 371–390. [CrossRef]
19. Manzano, S.; Gaffney, E.A.; Doblare, M.; Doweidar, M.H. Cartilage dysfunction in ALS patients as side effect of motion loss: 3D mechano-electrochemical computational model. *Biomed. Res. Int.* **2014**, *2014*. [CrossRef] [PubMed]
20. Manzano, S.; Manzano, R.; Doblaré, M.; Doweidar, M.H. Altered swelling and ion fluxes in articular cartilage as a biomarker in osteoarthritis and joint immobilization: A computational analysis. *J. Roy. Soc. Int.* **2014**, *12*. [CrossRef] [PubMed]
21. Lai, W.M.; Hou, J.S.; Mow, V.C. A triphasic theory for the swelling and deformation behaviors of articular-cartilage. *J. Biomech. Eng.* **1991**, *113*, 245–258. [CrossRef] [PubMed]
22. Sun, D.N.; Gu, W.Y.; Guo, X.E.; Lai, W.M.; Mow, V.C. A mixed finite element formulation of triphasic mechano-electrochemical theory for charged, hydrated biological soft tissues. *Int. J. Num. Meth. Eng.* **1999**, *45*, 1375–1402. [CrossRef]
23. Mow, V.C.; Guo, X.E. Mechano-electrochemical properties of articular cartilage: Their inhomogeneities and anisotropies. *Ann. Rev. Biomed. Eng.* **2002**, *4*, 175–209. [CrossRef] [PubMed]

24. Huyghe, J.M.; Janssen, J.D. Quadriphasic mechanics of swelling incompressible porous media. *Intern. J. Eng. Sci.* **1997**, *35*, 793–802. [CrossRef]
25. Gu, W.Y.; Lai, W.M.; Mow, V. A mixture theory for charged-hydrated soft tissues containing multi-electrolytes: Passive transport and swelling behaviors. *J. biomechan. eng.* **1998**, *120*, 169–180. [CrossRef]
26. Sun, D.D.; Guo, X.E.; Likhitpanichkul, M.; Lai, W.M.; Mow, V.C. The influence of the fixed negative charges on mechanical and electrical behaviors of articular cartilage under unconfined compression. *J. Biomech. Eng.* **2004**, *126*, 6–16. [CrossRef] [PubMed]
27. Garcia, J.J.; Cortes, D.H. A nonlinear biphasic viscohyperelastic model for articular cartilage. *J. Biomech.* **2006**, *39*, 2991–2998. [CrossRef] [PubMed]
28. Kaasschieter, E.F.; Frijns, A.J.H.; Huyghe, J.M. Mixed finite element modelling of cartilaginous tissues. *Math. Comput. Simul.* **2003**, *61*, 549–560. [CrossRef]
29. Chen, S.S.; Falcovitz, Y.H.; Schneiderman, R.; Maroudas, A.; Sah, R.L. Depth-dependent compressive properties of normal aged human femoral head articular cartilage: Relationship to fixed charge density. *Osteoarthr. Cartil.* **2011**, *9*, 561–569. [CrossRef] [PubMed]

applied sciences

MDPI

Article

The Spotting Distribution of Wildfires

Jonathan Martin and Thomas Hillen*

Centre for Mathematical Biology, Department of Mathematical and Statistical Sciences, University of Alberta, Edmonton, AB T6G2G1, Canada; jmartin2@ualberta.ca

* Correspondence: thillen@ualberta.ca; Tel.: +1-780-492-3395

Academic Editor: Yang Kuang

Received: 12 February 2016; Accepted: 23 May 2016; Published: 17 June 2016

Abstract: In wildfire science, spotting refers to non-local creation of new fires, due to downwind ignition of brands launched from a primary fire. Spotting is often mentioned as being one of the most difficult problems for wildfire management, because of its unpredictable nature. Since spotting is a stochastic process, it makes sense to talk about a probability distribution for spotting, which we call the *spotting distribution*. Given a location ahead of the fire front, we would like to know how likely is it to observe a spot fire at that location in the next few minutes. The aim of this paper is to introduce a detailed procedure to find the spotting distribution. Most prior modelling has focused on the maximum spotting distance, or on physical subprocesses. We will use mathematical modelling, which is based on detailed physical processes, to derive a spotting distribution. We discuss the use and measurement of this spotting distribution in fire spread, fire management and fire breaching. The appendix of this paper contains a comprehensive review of the relevant underlying physical sub-processes of fire plumes, launching fire brands, wind transport, falling and terminal velocity, combustion during transport, and ignition upon landing.

Keywords: spotting; wildfire; transport equations; spotting distribution

1. Introduction

Wildfires are capable of creating powerful updrafts, called convection columns, which launch burning plant materials—referred to as *firebrands*—into the atmosphere [1]. Generally one speaks of coupled fire-atmosphere interactions [2], in which heat and moisture exchanges occur between the fire, the convection column and the atmosphere, resulting in the birth of new fire-driven wind and convective updrafts. Firebrands are then transported by the ambient windflow, simultaneously combusting and decreasing in mass, until they reach the ground. Upon landing, depending on the local fuel and weather conditions at the landing site, a firebrand may ignite the local fuel and start a new fire. Such a fire is called a *spot fire*, and the process is called *spotting* (see Figure 1a for a sketch of the spotting process). Spotting occurs in many ecosystems spanning the Earth, from the Americas to Europe, Africa and Australia.

Spotting can play an important role in wildfire spread, and while many of the subprocesses outlined in Figure 1 have been studied in detail, many remain poorly understood. One thing is certain: where spotting is important, it is a very diffficult spread mechanism to understand and therefore control.

The model framework we present here has the potential to produce realistic spotting *distributions*: spatial maps describing the probability of spot fires downwind of an existing fire. We will show how it is possible to incorporate the vast, but disparate literature on spotting, with some new ideas, to create a very robust modelling framework. We hope to draw the interest of researchers, in particular those working in fire management and experimental or statistical modelling of spotting subprocesses, by highlighting the lesser known subprocesses and demonstrating some potential uses of our approach.

While fully realistic distributions are still some time away, a major advantage of our approach is that as subprocesses become better understood, our spotting distributions will become more accurate.

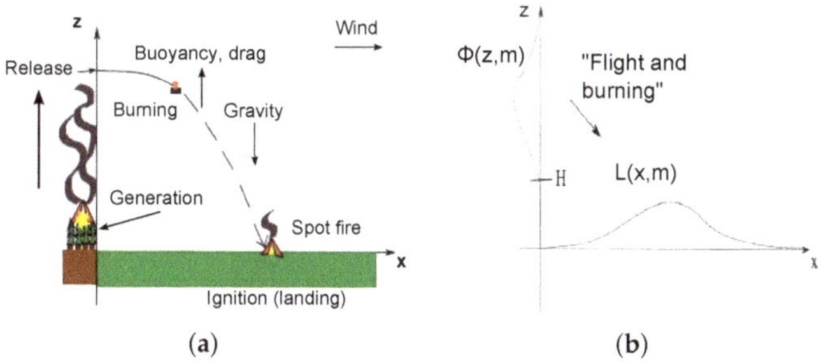

Figure 1. (**a**) Caricature of the spotting process. A cohort of firebrands are launched to various heights z above a spreading ambient forest fire. The x-direction denotes the mean ambient wind direction. The firebrands are then released and transported in the ambient windfield: combusting, falling downward due to gravity, experiencing drag and possibly buoyancy, until they finally come to rest downwind of the main fire. Depending on the local conditions for fuel and weather at the landing site, a new fire may be ignited if the firebrand is still combusting. Such a fire is called a *spot fire*, and the process is called *spotting*; (**b**) Determining the "flight and burning" distribution L from the vertical launch distribution ϕ. The term "flight and burning" refers to the physical details of firebrand flight and combustion, which we will discuss in detail in the Appendix. Shown in the vertical is a cross-section of the launching distribution $\phi(z, m)$ for some fixed $m > 0$. A possible asymptotic landing distribution $\mathbb{L}(x, m)$ is plotted here in the horizontal.

1.1. The Havoc Caused by Spotting

Spotting is not important in all fire contexts, but in a variety of ecosystems worldwide its importance varies from a minor concern, to the primary front spread mechanism or the primary breaching mechanism from wildland to urban structures. The type of fire, classified according to the structure of the fire's convection column, is also very important, and we will discuss this in detail in Section 1.4.

In Boreal forests of North America, for example, there are vast continuous stretches of coniferous forests, which can create incredibly intense fires, called *crown fires* [3,4]. These crown fires are capable of prolific spotting, and highly variable rates of spread. In the most extreme tinder-dry burning conditions, spotting may occur with sufficient frequency beyond a minimum distance, such that it influences the rate of spread. In addition, the prolific release of flaming needles [5] presents another mechanism, which might speed up a fire's local progression. Either of these mechanisms may describe the high variability in observed rates of spread for Boreal crown fires as outlined in [3]. While writing this article, a huge crown fire of high intensity is unfolding in Fort McMurray in Alberta, Canada. Spotting has allowed the fire to enter the city and more than 1600 homes were destroyed.

As another example, in the chapparal brush of California, spotting from brushfires annually threatens property in the wildland-urban interface [6,7]. The warm, dry *foehn* winds which pass through the great basin in Southern California [8], lead to extreme fire spreading scenarios in which spotting can play an important role. In fact, examples of foehn winds are found in many other regions, such as the Chinook winds East of the Rocky Mountains, or the Viento del Sur winds in Spain.

In Australia, spotting is a major issue for fires in Eucalyptus-grassland forests [9–11]. In certain instances, severe burning conditions have led to the observation that the fire's boundary "appears to be moving as a continual coalescence of spot fires" [9]. In these cases, spotting seems to drive the fire front; such situations will be globally described as *spotting dominated cases*. An example of a spotting dominated case are conflagrations, where fire spreads rapidly in an urban setting, as with post-earthquake fires which occurred in San Francisco (1906) or Tokyo (1923) [12].

Spotting is often the cause of a fire *breaching* across an extended obstacle to local spread, such as roads, rivers, or man-made fuel breaks [13]. Naturally, then, spotting is also the most frequent mechanism for the escape of prescribed fires, the latter used by fire ecologists and management, to promote ecological diversity and mitigate potential large fires through fuel management [14]. Spotting can also cause an increased rate of spread across a region which might otherwise slow the fire's progress, as might happen if a crown fire front encounters an extended slash region, across which local spread would be much slower. On the other hand, slash is notorious for spotting, which in this case could allow the fire to reach the slash boundary faster relative to local spread alone.

When conditions aloft are favorable for strong convection column development [15], or large-scale fire-induced vortices (fire whirls) exist [16], spotting may contribute significantly to a fire's spread. A spot fire 29 km downwind of an existing fire in Victoria, Australia appears to be the longest recorded spot fire event [11]. As reported by Ellis [9], the Ash Wednesday fires in Australia, which occurred on the 16th of February 1983, produced spotting distances of between 5 and 12 km, with the most extreme incident measured being 25 km from the primary fire.

While these extreme long-distance dispersal events cannot be seen to drive a front's *rate of spread*, they dramatically increase fire danger. It may take fire suppression crews a long time to reach such fires, increasing their likelihood of growing to a full blaze. Accurate wind-transport models for combusting firebrands, operational in real time and with access to accurate wind information, are in demand to help improve spotting forecasts in such situations. As mentioned above, coupled fire-atmosphere models [2,17,18] may provide an improvement to static fluid dynamic computations currently employed in operational front prediction software (for North American examples see [19,20]). Indeed, Large Eddy Simulations (LES) coupled with firebrand dispersal have shown that the fire plume may be quite different from the plume used in standard models like the Baum and McCaffery plume [21], leading to different firebrand trajectories [22–24]. In addition, there is an extensive literature covering empirical measurement of plume characteristics; extensive measurement of plume heights and characteristics in [25] compared a variety of plume models (different from Baum and McCaffery) against measurements for approximately 2000 wildfire plumes. The results showed an unexpected importance of the lower limit of the atmospheric boundary layer, as well as a perplexing independence of plume injection heights on wind, but also a non-neglibible fraction of plumes whose structure could not be determined by the models employed. Comparison of model output versus experimental observation, coupled with the development of increasingly sophisticated plume models, provides a promising paradigm involving experiment and modelling which must be further employed in addressing the challenging processes which drive wildfire spread.

As a final, very important note, there are examples in the literature, which indicate that spotting can at times lead to the *acceleration* of a fire's rate of advance. During the Beerburrum Fire No. 48, which occurred in Queensland Australia in 1994, the firefighters observed "Spotting ... accelerated its rate of spread and its advance was halted only by Pumecestone Passage" [9]. The issue of acceleration in local spread caused by the addition of a non-local spread mechanism is not new in for invasion ecology since the seminal paper by Kot *et al.*, [26], but the possibility of acceleration is an important open, unaddressed question in the context of wildfire spreading with spotting.

1.2. The Primary Questions of This Article

Motivated by all the problems which spotting causes, the central questions which we address here or in future work are:

1. What is the *spotting distribution*, or the probability of spot fire ignition, at each location downwind of an existing fire front?

In this paper we will focus primarily on a mechanistic approach to the derivation of the spotting distribution. Of equal importance is the development of experimental approaches to *measure* the spotting distribution; we discuss this problem in Section 4.2. As soon as the spotting distribution is found, we can ask important follow-up questions, such as:

2. What is the probability that a fire will *breach* an obstacle?
3. What role does spotting play on the *rate of spread* of a fire front? Can spotting cause a wildfire to quickly traverse a region across which it would spread slower with purely local spread?
4. Can spotting *accelerate* a fire's advance?

In this paper we will address Question 1 in Sections 1.4 and 2. Here we will borrow heavily from methodologies proposed in plant, insect and animal dispersal [27,28]. Questions 2, 3 and 4 have been discussed in detail in the PhD thesis of J. Martin [29], while Problem 3 has been further discussed in the more general mathematical framework of *birth-jump processes* in [30]. The problem of connecting theory to experimental measurement is taken up in the Discussion.

Our approach is based on existing physical principles and experimental results, summarized in Table A1 from the Appendix, to derive the spotting distribution. In Section 2 we present a new transport model for firebrand transport and combustion. We model the time-mean behaviour of trajectories and combustion, since turbulence creates high variability in the atmospheric paths of individual firebrands. The authors have not been able to find published quantitative spotting distributions from more complicated physics-based models (e.g., fire-atmosphere models, or the extensive physics-based model by Sardoy [31]). In fact, the only known article presenting a spotting distribution is that of Wang [32], though our methodology is much more general. Our ability to provide spotting distributions, incorporating realistic sub-processes as discussed in the Appendix, is the greatest strength of the present article.

We also introduce, in Section 2, a transport and combustion model in the form of a hyperbolic partial differential equation (PDE), based on the *launching distribution*, the *horizontal wind profile*, the *terminal falling velocity*, the *burning rate* of the flying brands, ending with a discussion of the *ignition probability* for a landed firebrand to generate a spot fire.

In Section 3 we discuss how the models of Section 2 can be used to determine the spotting distribution. We provide analytical solutions of the transport PDE from Section 2, and employ these to generate examples of the spotting distribution. The analytical solutions are used to illustrate how the characteristic components of the model, such as horizontal wind, terminal falling velocity or combustion model may influence the spotting distribution.

In our discussion of measuring the spotting distribution in Section 4 and in the Appendix, we describe a number of successes which have been achieved in describing some of the above subprocesses. Most of these are fairly recent, as the wildfire research community has become increasingly interested in understanding spotting. We also draw attention to components of the spotting process which are relatively poorly understood, taking the lead from experimentalists in Dispersal Ecology [27]. We suggest potential methodologies which may be employed in experimentally measuring spotting distributions.

1.3. Prior and Concurrent Models Coupling Spotting with Local Spread

A recent cellular automata model [33] was the first of its kind to incorporate spotting. The authors assumed a simple two-dimensional normal distribution for the spotting distribution, where the spotting distribution represents the probability of ignition occurring in a non-burning cell in a given time step. This model was considered first separate from local spread, in which non-burning cells can be ignited by their burning neighbours, with some probability related to the rate of spread indicated by the empirically-based Fire Behaviour Prediction system [34]. In addition, burning cells become burnt-out in each time step with some probability. The authors of the paper then used distributions to describe the spotfire distribution in the plane, neglecting topographical variation. The room for improvement of the normal distributions from [33] provided some of the initial motivation for examining more realistic spotting kernels, as in this paper.

The cellular automata model was unique in that it provided a coherent mathematical model in which spotting and local spread were simultaneously present. Since then other stochastic models have been developed, such as the augmentation of Discrete Event System Specification models (which employ novel Lagrangian point-advancement techniques), to include spotting [35]. This particular reference improved on the spotting distribution of [33], employing the more realistic log-normal distribution.

It must be mentioned that computer implementations of spotting have been employed in computer-based local spread simulators (like PROMETHEUS[20], or FARSITE [19]). Spotting models, such as those developed by Albini [1], have been incorporated in an ad-hoc way [36], focussing on the maximum spotting distance, rather than the downwind spotting *distribution*. Most of the Albini models deal with *torching*, where ladder fuels allow at most a few isolated trees to crown and spot at the same time, and with *line thermals* [13], which are well-mixed rising columns of air which lift firebrands—this framework is more appropriate for a large, very intense forest fire, as we consider here.

The above computational models provide fire front prediction through explicit curve tracking. The level-set method is an alternative and very powerful method for front tracking. Consequently, several authors use level set methods for the propagation of wild fires [37–39]. Our approach for the spotting distribution studied here differs from the level-set approach, as we are not only interested in the location of the fire front, but we also like to understand the distribution of spot fires in a continuous region ahead of the fire. While here we focus on the spotting process, a detailed probabilistic model for wildfire spread including spotting and local effects has been developed in the thesis of J. Martin [29]. We briefly discuss these models in the Discussion Section 4.1, though we emphasize that the present work is devoted to the spotting model.

1.4. The Types of Spotting Considered in This Paper

It should be stressed that the model we develop corresponds to a line fire spreading in a flat, homogeneous medium. This means the fire front represents a division between burned and unburned regions. We consider spotting only in the direction perpendicular to the fire front. This is an idealization, since the vortex structure of the convection column releases fire brands in any direction.

With respect to wind, we do not account for the fire-atmosphere interaction. Describing this analytically is beyond the scope of our current discussion, but poses an interesting challenge for future research. It is perhaps best left to coupled fire-atmosphere models, many of which are outlined in the Appendix.

In general, the convection column from a wildfire will be bent in the wind's direction. Many formulae have been considered in the literature to characterize the bending angle as a function of fire characteristics. For illustrative purposes we will consider the model of Van Wagner (1973), as discussed in the doctoral thesis by Alexander [40], which built on earlier work by Taylor (1961) and Thomas (1962, 1964). This model proposes a burning angle θ, measured relative to the horizontal axis, given by

$$\tan(\theta) = \left(\frac{bI}{w^3}\right)^{\frac{1}{2}}, \tag{1}$$

where I is the fire intensity (in kiloWatts per second) and w is the ambient windspeed (in metres per second). Here b is a 'buoyancy term', $b = \frac{g}{\rho c_p T}$, where g is acceleration due to gravity, ρ is the fuel density, c_p is the specific heat at constant pressure, and T is temperature. Hence the term $(\frac{bI}{w^3})^{1/2}$ is dimensionless, since for linefires we measure intensity I in terms of (J/(m·s)).

The more intense the fire and the weaker the windspeed respectively, the more upright the convection column. Hence our approximation, that our convection column is vertical, is more accurate the more intense the fire. In fact, it is mentioned in the thesis by Alexander [40], that with ambient windspeeds not exceeding one metre per second, the convection column is essentially vertical.

Depending on the convection column, and the fire's surroundings [8], one can consider seven types of spotting, as suggested by S. Ryu [41]. We want to be clear about the types of spotting, which our impulse-release model might describe, as well as, how our models might be modified to account for other spotting scenarios.

A type one fire consists of a very powerful convection column, with light surface winds, which rises vertically into the atmosphere.

A type two fire is similar, but the distinction is the presence of strong surface winds, which can lead to spotting. We would expect the spotting distribution to be highly concentrated along the front, and here the influence of local spread mechanisms may be comparable to the influence of spotting on rates of spread. The Canadian Fire Behavior Prediction System (FBP) [34] accounts for spotting up to about thirty metres downwind in its rate of spread computation, so spotting from a fire insufficient to sustain ignition beyond thirty metres has already been accounted for in terms of the rate of spread computation. On the other hand, there may be a "blocking effect", where the convection column allows a relatively fast and vertically uniform atmospheric flow in the near-field; since we assume firebrands are carried passively by the atmosphere, as the windspeed is comparatively large. Since we model a vertical release, type two fires are of specific interest in this paper.

Fires of type three (as well as type seven) consist of spotting over mountainous topography, so our models as they stand would be inadequate to model this type of situation. Discussing mountainous topography is the subject of future research.

Another extreme spotting situation is **a type four fire**, where strong winds aloft cause the shearing off of the top of the convection column. The result is a nearly horizontal column aloft, which rains firebrands down well ahead of the main fire. These conditions can be incorporated into our model framework by suitable choice of lofting distributions and wind profiles.

For spotting type five, where the convection column leans towards the strong winds but does not break off, both short and long-range spotting is possible. The more intense the fire, the straighter the convection column and the more intense the wind, the more horizontal the convection column. Empirical relationships have been developed to quantify the angle which the column makes with the vertical [34]; in particular, in the case of an extremely intense fire, the column is nearly vertical, corresponding to the idealized launching distributions considered in this work. In future work we could consider initial conditions along a slanted line, as in the work of Wang [32].

Spotting type six situations occur where there are very strong winds above the ground, so that no convection column forms. In this case, spotting could play an important role, and diffusion and non-local dispersal might occur over similar spatial scales. For example, in coniferous forests, chapparal brush, slash, Eucalyptus-grassland forests, or even conflagration fires in cities, enormous amounts of firebrands are generated. In all these cases, the firebrands are literally swept along by the wind. Our launching model L3 from the Appendix covers this situation.

1.5. Ignition of Fuel Beds by Firebrands

One of the most challenging problems in modelling the spotting process, is to determine the ignition probability E, which is the probability for ignition to occur once a firebrand has made contact with a given fuel bed. Since the fire landscape is very heterogeneous, the ignition probability E may vary spatially, and can depend on a variety of factors, like:

- The species of plants emitting firebrands.
- Landed firebrand characteristics like diameter, length, and mass.
- The travel time t^* from launch to landing.
- The moisture content of the fuel bed and local weather.
- The surface area, and thermal conductivity between firebrand and fuel.
- Whether the firebrand is in a "glowing" or "flaming" state upon landing.

- Variability of firebrand type within the launching stand (e.g., a coniferous tree might emit both small brands or cones).
- Whether there is a "re-settling" after landing due to slope or wind.
- Whether there is a shading effect from the sun due to the presence of the convection column.

An ideal model for ignition would account explicitly for all factors just mentioned. Several physics-based models have attempted to do just that (e.g., [31]), but necessarily include many equations whose analysis is mostly limited to computer simulation. Since we are interested in fire fronts, which occur at the macroscopic scale, we can ignore some of the smaller-scale details in our development of ignition models.

Prior to 2006 the experimental investigation of this topic was limited to qualitative descriptions of ignition. The experiments show that fuel bed moisture, firebrand mass and geometry are the most important characteristics determining ignition probability [5,9,42]. The Fine Fuel Moisture Code (FFMC) used by the FBP system [34], which provides a numerical measure for the moisture contained in forest litter, has been a useful standard for determining ignition probabilities. It is determined in turn by the Fire Weather Index, another component of the FBP system [34]. The lower the FFMC, the higher the fine fuel moisture content. Experiments by the Aerospace Corporation [16,43] found that for high fine moisture contents only large flaming brands cause ignition, while for low fine fuel moisture content glowing embers may easily ignite a fuel bed. Albini [13] reports that spotting can be significant when fine fuel moisture content is below ten percent, and confirmed the earlier results by [43] that glowing embers can be sources of ignition.

Following these experiments, in about 2006, Manzello [44] began qualitative experiments on firebrand ignition, for brands emitted during controlled laboratory burning of pine and fir trees. This work is a collaboration between the National Institute of Standards and Technology in the USA, and the Building Research Institute in Japan [45], where ongoing experiments may help further quantify the ignition process.

One important result following from the combustion experiments of Manzello, is that firebrands are not produced if the dead fuel moisture content exceeds thirty percent. One can further postulate that there will always be a maximum moisture above which spotting does not occur. Together these results suggest, as is common knowledge, that lower fine fuel moisture content (high FFMC) can correspond to greater spotfire risk.

A still more recent paper [46], is the first to describe ignition probabilities using regression analysis on systematic experimental data. The authors performed a number of controlled lab experiments to determine the time of ignition, rate of spread, rate of combustion, maximum and mean flame heights, and ignition frequency of fuel beds for a variety of fuel beds, representative of the Mediterranean. Examples of fuel beds include a variety of pine, eucalyptus, and grass beds, which could also be representative of fine fuels from forests in North America and Australia. These fuels were ignited under varying values for fuel moisture, ambient windflow, bulk density, and fuel arrangement (or loading). In terms of firebrands, the authors examined pine cones, Eucalyptus bark, acorns and twigs, and assessed their likelihood to cause ignition in terms of the fuel bed properties just mentioned. The general results of [46] are that grasses present higher flammability risk compared to tree and bush litters, pine litter is more ignitable than hardwood litter, and an increase of the fine fuel moisture and bulk density decreases the time, but not necessarily the likelihood of ignition. Finally, firebrand type and state (*i.e.*, glowing or flaming) are the most important determinants of ignition.

The glowing or flaming state of a firebrand had already been qualitatively discussed in a number of investigations (e.g., [5,7,9]). A rigorous analysis of Eucalyptus firebrands by Ellis [9], and pine or fir firebrands by Manzello [44], confirms the results of [46] that flaming firebrands present greater risk of ignition. The time-to-burnout of flaming was investigated from a theoretical perspective in [31]. Flaming ignition most likely plays a more important role in short-range spotting, where "re"-flaming is possible upon landing, and is important to consider [9].

The results of [47] suggested that there are three firebrand groups which are important for spotting. These include: heavy firebrands with the ability to sustain flames, which are efficient for long-distance spotting (e.g., pine cones, cylindrical brands); light firebrands with high surface area-volume ratios are effective for short-range spotting (e.g., Eucalyptus or pine bark plates). Light firebrands with low surface-volume ratios fall somewhere in the middle of the other two classes.

Finally, experiments using the Commonwealth Scientific and Industrial Research Organization's (CSIRO's) Pyrotron combustion wind tunnel have further quantified the role of fuel moisture on ignition probability [48]. Further quantitative work, as discussed in these final two paragraphs, is underway and will soon be employed by fire management in assessing fire danger.

2. A Transport Model for Firebrand Transport and Combustion

The spotting mechanism can be divided into various physical processes that act in concert. Our mathematical formulation allows us to consider each process separately, and then join them together to get the spotting distribution. The main model ingredients, which we encourage the reader to visualize in Figure 1a, are listed below. We discuss detailed physical models for each of these processes in the Appendix, see also Table A1.

1. **Launching:** The launching distribution $\phi(z, m)$ describes how many fire brands of mass m are launched into the convection column to the height z. We assume a maximal loftable mass of $\bar{m}$, such that $0 \leq m \leq \bar{m}$. We use H to denote the canopy height (in metres) such that lofting is only considered for heights $z \geq H$. The launching distribution is a true probability density on $[H, \infty] \times [0, \bar{m}]$, normalized and dimensionless. We measure heights z in metres, and masses m in kilograms (though it will be noted that typical firebrand masses are on the order of grams). Notice that one may be interested in many more characteristics of the firebrands launched: the firebrand type, for example, could be important [46,47].

2. **Horizontal wind profile:** We describe the horizontal windspeed (in metres per second), parallel to the downwind direction (or perpendicular to the front), by $w(z) > 0$, which, depending on the physical model, might depend on the height z (in metres).

3. **Terminal falling velocity:** We assume that flying fire brands quickly reach their terminal velocity $v(t, m)$ (measured in metres per second), where falling through gravity and frictional drag are in equilibrium. We make the strong assumption that $v < 0$ as soon as the ember leaves the convection column; in reality, we would expect turbulent up-drafts in a neighbourhood of the convection column. It is an interesting challenge to properly describe the vertical and horizontal variation in the strength of such updrafts in a neighbourhood of the convection column, though it is beyond the scope of this paper to do so. However, as discussed in the Appendix, outside the region of significant updrafts, the assumption that the brand will rapidly assume its terminal speed and falling orientation is well-justified, established through wind tunnel experiments [5,9,17].

4. **Burning rate:** With $f(t, z, m)$ we denote the combustion rate of a brand of mass m at height z in a well oxygenated environment. The combustion rate f has units of kilograms per second. While the burning rate depends on the relative firebrand velocity, in most models we will assume this dependence is negligible.

5. **Ignition probability:** The ignition probability $E(m)$ describes the probability that a landed burning mass m starts a spot fire. As a probability density on the space $[0, \bar{m}]$ (with masses in kilograms), it is normalized to take on values between zero and one, and is dimensionless. Of course, ignition generally depends on the local fuel conditions, moisture content and temperature amongst other variables, so we are making a simplifying assumption that ignition is homogeneous in space. Notice further that we are implicitly assuming that thermal energy transfer, proportional to firebrand mass, depends only on mass and not for example on firebrand geometry (the latter being known to influence energy transfer).

The asymptotic landing distribution $\mathbb{L}(x, m)$, determined in Equation (21) below, roughly describes how much mass lands where (here x is downwind distance in metres and m is mass in kilograms). The determination of the asymptotic landing distribution is illustrated in Figure 1b and is captured in the following flow diagram:

$$\underbrace{\phi(z, m)}_{\text{Launching distribution}} \quad \underbrace{\rightarrow}_{\textbf{Transport and combustion}} \quad \underbrace{\mathbb{L}(x, m)}_{\text{Landing distribution}} \tag{2}$$

To determine the spotting distribution, we must be careful. If one examines too far downwind, one will not find spotfires from an impulse release, since there is a finite combustion lifetime for each firebrand.

It is useful at this point to introduce the *combustion operator* and its inverse, where the combustion operator $C(m, t)$ tells us how much mass (in kilograms) remains after t time units, starting from mass $= m$. The inverse combustion operator $C^{-1}(m, t)$ is then defined as usual by

$$C^{-1}(C(m, t), t) = m. \tag{3}$$

Another important concept is the *landing time*, which we denote by $t^*(x)$. The landing time has units of seconds, and quantifies how long it took a firebrand released at $x = 0$ to reach the ground at location x. Provided windspeeds vary monotonically with height, each downwind location will have a unique landing time, hence this quantity is well defined. For example, if we have constant windspeed w and falling speed v, then $t^*(x) = \frac{x}{w}$, as can easily be computed.

In order to apply the ignition operator, it is necessary to determine the total landed mass at given location. This involves integrating the asymptotic landing distribution against m, incorporating the inverse combustion operator evaluated at the landing time into the distribution for the integration. We delay the presentation of this complicated term until Section 2.3.

Ignition, if it occurs, is extremely rapid in tinder-dry burning conditions (observed experimentally for example in the Porter Lake experiments [42]), which corresponds to scenarios we are interested in exploring. In this context we assume that masses ignite instantly upon landing. Hence to determine the spotting distribution, we can apply the ignition operator to the total landed mass, where, as in the previous paragraph, the landing distribution's mass component is evaluated in terms of the inverse combustion operator at the landing time. The exact formula is given in Equation (23).

One can then augment our earlier flow chart to include spotting,

$$\underbrace{\phi(z, m)}_{\text{Launching distribution}} \quad \underbrace{\rightarrow}_{\textbf{Transport and combustion}} \quad \underbrace{\mathbb{L}(x, m)}_{\text{Landing distribution}} \quad \underbrace{\rightarrow}_{\textbf{Ignition}} \quad \underbrace{\mathbb{S}(x)}_{\text{Spotting distribution}} \tag{4}$$

In turn, this flowchart defines a *map* $\mathbb{T} : \phi \rightarrow \mathbb{S}$. This is a map between distributions, which allows for abstract mathematical explorations of the spotting process.

2.1. The Impulse Release IBVP

The distribution of firebrands at time $t \geq 0$, location x, height z and mass m may be described by $p(t, x, z, m) \in \mathbb{R}^+$. Since flying and burning are deterministic processes, we model the whole process with a transport equation

$$\frac{\partial p}{\partial t} + \frac{\partial (wp)}{\partial x} + \frac{\partial (vp)}{\partial z} + \frac{\partial (fp)}{\partial m} = 0 \tag{5}$$

on the domain

$$(t, x, z, m) \in \Omega := [0, \infty) \times [0, \infty) \times [H, \infty) \times [0, \bar{m}]. \tag{6}$$

The functions $w(z), v(t, m), f(t, z, m)$ describe the physical processes of wind transport, falling, and burning, respectively. We will use the appendix to give detailed, physically based, description of these terms.

To obtain a well defined model, we need to specify boundary and initial conditions. The above Model in Equation (5) is hyperbolic and the spatial characteristics are pointing downwards (falling) and to the right (wind). Hence we only need boundary conditions on the left boundary at $x = 0$. Like in a confetti problem, where you throw confetti into the wind and see where the pieces land, we assume a one-time release of fire brands and we are interested to see where they land. When the fire releases fire brands over a period of time, then a simple integration of our model will account for all the landed brands. Hence we set an initial condition of

$$p(0,x,z,m) = \begin{cases} N\phi(z,m) & \text{for} \quad x=0 \\ 0 & \text{else.} \end{cases} \tag{7}$$

where N is the total number of launched brands, and a boundary condition as

$$p(t,0,z,m) = 0, \quad \text{for} \quad t > 0. \tag{8}$$

2.2. Solution of the Transport Model

The above transport model can be explicitly solved using the method of characteristics. We reformulate the above model by using the product rule for the partial derivatives:

$$\frac{\partial p(t,x,z,m)}{\partial t} + w(z)\frac{\partial p(t,x,z,m)}{\partial x} + v(t,m)\frac{\partial p(t,x,z,m)}{\partial z} + f(t,z,m)\frac{\partial p(t,x,z,m)}{\partial m} \tag{9}$$

$$= -\frac{\partial f(t,z,m)}{\partial m} p(t,x,z,m). \tag{10}$$

The characteristic equations are

$$\frac{dx}{dt} = w(z), \tag{11}$$

$$\frac{dz}{dt} = v(t,m), \tag{12}$$

$$\frac{dm}{dt} = f(t,z,m), \tag{13}$$

where Equation (11) describes horizontal transport due to wind, Equation (12) describes vertical movement due to gravity, buoyancy *etc.*, and Equation (13) describes the combustion process of a burning ember. Detailed physical models are given in the Appendix. Each of the above equations needs to be equipped with an initial condition

$$x(0) = x_0, \quad z(0) = z_0, \quad m(0) = m_0, \tag{14}$$

which describes the initial location x_0, the initial height z_0 and the initial mass m_0 of a fire brand. We introduce $Y := (x,z,m)$ and we assume that the functions on the right hand side are globally Lipschitz continuous such that the IVP for the System (11)–(13), together with the initial conditions in Equation (14) has a unique solution

$$Y(t,Y_0) = (x(t;x_0), v(t;v_0), m(t;m_0)), \quad \text{with} \quad Y_0 = (x_0, v_0, m_0), \tag{15}$$

such that $x(0;x_0) = x_0$ etc., which we call the characteristics of the system. Along the characteristics, p satisfies an ordinary differential equation

$$\frac{d}{dt}p(t,Y(t,Y_0)) = \frac{\partial p(t,Y(t,Y_0))}{\partial t} + \nabla_Y p(t,Y(t,Y_0))\frac{\partial Y(t,Y_0)}{\partial t} \tag{16}$$

$$= -\frac{\partial f(t,Y(t,Y_0))}{\partial m} p(t,Y(t,Y_0)), \tag{17}$$

where we used Equation (10) in the last step. Equation (17) is solved by the exponential

$$p(t, Y(t, Y_0)) = p(0, Y_0) \exp\left(-\int_0^t \frac{\partial f(s, Y(s, Y_0))}{\partial m} ds\right). \tag{18}$$

This solution "lives" on the characteristics $Y(t)$. To get a full solution to our problem, we consider a given point $Y(x(t; x_0), z(t; z_0), m(t; m_0))$ and we follow the characteristics backwards to its origin. We denote the backwards solution of the characteristics equation as $Y^{-1}(t, Y)$, such that $Y^{-1}(t, Y(t, Y_0)) = Y_0$. Using this we can write the solution as

$$p(t, Y) = p\left(0, Y^{-1}(t, Y)\right) \exp\left(\int_0^t \frac{\partial f(\sigma, Y^{-1}(\sigma, Y))}{\partial m} d\sigma\right). \tag{19}$$

A schematic for the use of the method of characteristics, in case of constant wind and constant terminal falling velocity, is shown in Figure 2a.

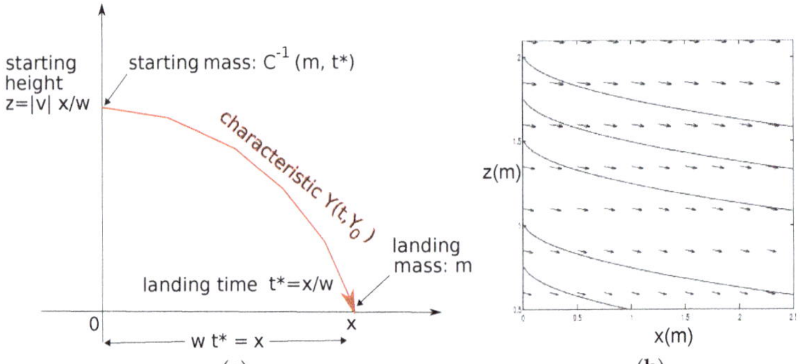

Figure 2. (**a**) Schematic of the method of characteristics. A typical characteristic $Y(t, Y_0)$ is shown as a curved red line; (**b**) Spatial characteristics (solid lines) for the power-law wind profile, with constant vertical velocity, described in Equation (34). The vertical-axis represents height in metres, while the x-axis represents downwind distance in metres. Here $w_H = 5$, $\beta = 0.5$, $H = 0.5$ and $v = -1$.

2.3. From Landed Firebrands to the Spotting Distribution

The "landed brands" are those that reach the canopy at height $z = H$ (where we assume that they fall straight to the ground—meaning we neglect within-stand winds and resettling, although both effects can be considered [49]). We can track these explicitly according to the solution Equation (19) of the impulse-release IBVP for the transport Model (5) described earlier in this Section. We recall from the introduction to Section 2 the distribution keeping track of all brands that have landed before time t, namely the *landing distribution*

$$L(t, x, m) := \int_0^t p(s, x, H, m) ds \tag{20}$$

and recall that the limit as $t \to \infty$ the *asymptotic landing distribution*

$$\mathbb{L}(x, m) = \lim_{t \to \infty} L(t, x, m). \tag{21}$$

To determine the spotting distribution, we must first integrate the asymptotic landing distribution appropriately, to obtain the landed masses in kilograms. Employing the inverse combustion operator C^{-1} and the landing time $t^*(x)$, the mass in kilograms at location x is given by

$$\int_0^{\bar{m}} m \, \mathbb{L}\left(x, C^{-1}(m, t^*(x)\right) \, dm. \tag{22}$$

Finally we can determine the spotting distribution in terms of the latter, by employing the ignition operator E, which we recall maps mass in kilograms to values in $[0,1]$, as follows:

$$\mathbb{S}(x) = E\left(\int_0^{\bar{m}} m \, \mathbb{L}\left(x, C^{-1}(m, t^*(x))\right) \, dm\right). \tag{23}$$

Our use of the spotting distribution is similar to the Green's-function approach that is used in the study of linear partial differential equations [50]. The Green's function describes the evolution of an impulse release, and the full solution to a PDE can be found by integrating the Green's function against initial and boundary conditions.

In the examples in the next Section, we show some illustrative but simple spotting distributions. We have seen that the spotting distribution depends on various physical submodels for firebrand launching, flight, burning, landing and ignition. In the Appendix, we provide a detailed review of the most commonly used submodels, which we summarize in Table A1. Each combination of these submodels leads to a reasonable spotting distribution. This gives more than 500 cases, which we cannot cover in this paper. Hence, in the next Section, we focus on some easy to understand examples, which we identify by their corresponding model number as used in Table A1. We hope that the present work will generate interest, especially in the experimental community, in order to determine the most physically plausible spotting distributions for a variety of important spread scenarios.

3. Examples of the Spotting Distribution

3.1. Case (W1,V1): Constant Wind and Terminal Velocity

(Case (W1,V1) in Table A1). As explained in the Appendix, wind-tunnel experiments show that firebrands quickly take on their terminal speed and configuration [9,51]. Of course, the firebrands are combusting, so the speed and geometric properties will evolve with time. As a first approximation, and in the interest of analytical tractability, we will assume firebrands fall with constant velocity $v < 0$ and constant wind $w > 0$. All combustion models studied here (F0–F6) have a combustion rate that is independent of the mass. Hence here and throughout we assume $\partial f / \partial m = 0$.

The landing time t^* will be important in what follows. To present a concrete example, if $w = 1$, $v = -1$, then the landing time for firebrands launched initially at $x = 0$ to arrive at $x = 2$ is $t^* = 2$.

As a starting point, fix a point (x, z) in the quarter plane $\{(x, z) \mid x > 0, z > H\}$, and assign a time $t > 0$. If $x > wt$, then the firebrands which started at 0 have not yet reached location x, so we can determine the firebrand density $p(t, x, z, m)$ from the initial condition for p. However, if $x < wt$, then we must determine the firebrand density by integrating backwards along characteristics to the boundary at $x = 0$. The latter requires solving Equations (11) and (12) with negative signs in front of the terms on the RHS. It is clear, then, that the firebrand density, provided N firebrands are released above a canopy of height H, is given by

$$p(t, x(t; x_0), z(t; z_0), m(t; m_0)) \tag{24}$$

$$= \begin{cases} N\,\delta(t - t^*)\phi\left(z(t; z_0) - vt, C^{-1}(m(t; m_0), t^*(x(t; x_0)))\right), & x \leq wt; \\ p(0, x(t; x_0) - wt, z(t; z_0) - vt, m_0), & x > wt, \end{cases} \tag{25}$$

where ϕ is the vertical launch distribution, and C^{-1} is the inverse combustion operator. A schematic of this solution is shown in Figure 2a. We remind the reader that at $t = 0$ we have $p = 0$ for $x > 0$, from our initial condition for an impulse release, so we find $p(t, x(t; x_0), z(t; z_0), m(t; m_0)) = 0$ for $x > wt$. Recalling that the distribution of landed brands $L(t, x, m)$ is given by $\int_0^t p(s, x, 0, m)\, ds$ for $t > 0$, so we find:

$$L(t, x(t; x_0), m(t; m_0)) = \begin{cases} N \int_0^t \delta(s - \frac{x(s;x_0)}{w})\phi(H - vs, C^{-1}(m(t; m_0), s))\, ds, & x \leq wt; \\ 0, & x > wt. \end{cases} \quad (26)$$

$$= N\mathbb{H}\left(t - \frac{x(t; x_0)}{w}\right) \phi\left(H + |v|\frac{x(t; x_0)}{w}, C^{-1}\left(m, \frac{x(t; x_0)}{w}\right)\right), \quad (27)$$

where $\mathbb{H}$ represents the Heaviside, or unit-step function, defined by $\mathbb{H}(x) = 0$ for $x < 0$, and $\mathbb{H}(x) = 1$ for $x \geq 0$.

Taking the limit as $t \to \infty$ in Equation (26), we obtain the asymptotic landing distribution $\mathbb{L}(x, m)$,

$$\mathbb{L}(x, m) = N\, \phi\left(|v|\frac{x}{w}, C^{-1}\left(m, \frac{x}{w}\right)\right), \quad (28)$$

This equation maps the launching distribution $\phi(z, m)$ to the asymptotic landing distribution $\mathbb{L}(x, m)$, as depicted on the right of Figure 1. We will compute some explicit examples in Section 3.6.

We impose the assumption, that if:

$$C^{-1}\left(m(t; m_0), \frac{x(t; x_0)}{w}\right) < 0, \quad \text{we set} \quad \mathbb{L}(x, m) = 0. \quad (29)$$

This is to assure that we do not obtain any negative mass density.

We see that the asymptotic landing distribution in Equation (28) is simply the number of firebrands released N, multiplied by the vertical launch distribution $\phi(z, m)$ with its arguments shifted. We employ this remarkably simple result in the next subsections, for important vertically-varying horizontal wind profiles.

3.2. Case (W3,V1): Power-Law for Wind, Constant Terminal Velocity

In the previous subsection we explored the case where $w > 0$ and $v < 0$ are constants, corresponding to constant wind and constant falling terminal velocity. We also made no discussion of the canopy; the approximation that the canopy height $H \approx 0$, can be useful in describing long-distance spotting events, as on this spatial scale the canopy height is negligible. Increasing slightly our model complexity, in this subsection we consider a 'power-law' wind profile, first employed in the context of spotting by Albini [1]. We model the horizontal windspeed as a function of height z above the canopy by

$$w(z) = w_H \left(\frac{z}{H}\right)^\beta, \quad (30)$$

where H is the canopy height, w_H is the windspeed at the canopy's base, and $\beta \in (0, 1)$. We are interested solely in values of $z \geq H$. Further, we will not consider resettling or within-canopy winds, so we assume once a firebrand has reached $z = H$, it drops straight downward, and is immediately capable of igniting a new fire. We will continue to assume $v < 0$.

With all this said, the spatial characteristics read:

$$\frac{dx}{dt} = w_H \left(\frac{z}{H}\right)^\beta, \quad (31)$$

$$\frac{dz}{dt} = v. \quad (32)$$

The solution to the first equation, given an initial condition z_0, is $z(t;z_0) = z_0 + vt$, which we then use to solve the equation for x. Notice that since $v < 0$, the heights z are decreasing with time. The Equation (31) for x becomes:

$$\frac{dx}{dt} = \frac{w_H}{H^\beta}(z_0 + vt)^\beta. \tag{33}$$

Integrating both sides with respect to t, we obtain:

$$x(t) = x_0 + \frac{w_H}{v(\beta+1)H^\beta}\left((z_0 + vt)^{\beta+1} - z_0^{\beta+1}\right). \tag{34}$$

Trajectories for Equation (34) are illustrated on the right of Figure 2.

Now consider a firebrand which reaches the canopy at (x, H). We wish to determine the landing time $t^*(x)$ which it took for the firebrand to travel from $x = 0$ at time $t = 0$ to the top of the canopy at $z = H$. As in the preceding subsection, we will run the characteristics in reverse. To do this, we choose $x_0 = x$, $z_0 = H$, $t = t^*$ and $x(t^*) = 0$, and insert these values into:

$$x(t) = x_0 + \frac{w_H}{|v|(\beta+1)H^\beta}\left((z_0 + |v|t)^{\beta+1} - z_0^{\beta+1}\right). \tag{35}$$

We obtain:

$$0 = x + \frac{w_H}{|v|(\beta+1)H^\beta}\left((H + |v|t^*)^{\beta+1} - H^{\beta+1}\right). \tag{36}$$

Solving for t^* in terms of x in Equation (36), we obtain:

$$t^*(x) = \frac{1}{|v|}\left[\left(H^{\beta+1} - \frac{H^\beta(\beta+1)|v|}{w_H}x\right)^{\frac{1}{\beta+1}} - H\right]. \tag{37}$$

The reader will notice that in the case where $\beta = 0$, Equation (37) reduces to $t^*(x) = \frac{x}{w}$, which is consistent with the constant w, constant v case in Section 3.1.

We find the exact solution p of our impulse IBVP for the transport and combustion process:

$$p(t, x(t;x_0), z(t;z_0), m(t;m_0)) \tag{38}$$

$$= \begin{cases} N\delta(t-t^*)\phi\Big(z(t;z_0) - vt^*(x(t;x_0)), C^{-1}(m,t^*)\Big), & x \le \int_0^t \mathbb{F}(s)\,ds; \\ 0, & x > \int_0^t \mathbb{F}(s)\,ds, \end{cases} \tag{39}$$

where

$$\mathbb{F}(s) \quad := \quad w(z(s;z_0)), \tag{40}$$

so that the bounds on x appearing in Equation (38) can be written in explicit terms, through the following expression:

$$\int_0^t \mathbb{F}(s)\,ds \quad := \quad \frac{w_H}{|v|(\beta+1)H^\beta}\left((z_0 + |v|t)^{\beta+1} - z_0^{\beta+1}\right). \tag{41}$$

We can interpret the latter integral in Equation (41) as the location of the leading edge of the expanding firebrand distribution $p(t)$, since $p(t) = 0$ for $x > \int_0^t \mathbb{F}(s)\,ds$, but $p(t) \ge 0$ when $x \le \int_0^t \mathbb{F}(s)\,ds$.

In particular, from Equation (41) we see that $\lim_{t\to\infty} \int_0^t \mathbb{F}(s)\,ds = \infty$. So we can argue similar to the preceding subsection, arriving at the asymptotic landing distribution (which appears very similar to Equation (28)):

$$\mathbb{L}(x,m) = N\phi\Big(H + |v|t^*(x), C^{-1}(m, t^*(x))\Big), \tag{42}$$

but here the landing time t^* is given by Equation (37). Again, as shown in Figure 1 (b), we obtain a map from launching ϕ to landing $\mathbb{L}$. Notice that we set $\mathbb{L}(x,m) = 0$ if $C^{-1}(m, t^*(x)) < 0$ (*i.e.*, we allow only nonnegative masses).

3.3. Case (W2,V1): Logarithmic Profile for w, Constant Vertical Velocity

In this subsection, we again assume a constant vertical velocity $v < 0$, and we introduce the logarithmic horizontal wind profile (first employed by Albini in spot fire modelling [1]):

$$w(z) = \frac{u_*}{\kappa} \ln\left(\frac{z-d}{z_0}\right). \tag{43}$$

where $z \geq H + d \approx H$, u_* is the friction velocity, κ is von Karman's constant, z_0 is the zero-datum displacement, where H is the canopy height and d is the distance above the canopy where horizontal winds begin [49]. We will approximate $d \approx 0$.

As in the preceding subsections, we may determine the landing time $t^*(x)$. After some work (see [29] for details), we arrive at an implicit expression for the landing time:

$$\frac{u_*}{\kappa} H \ln H - x + \frac{t^* u_*}{\kappa}(\ln H - 1) = \frac{u_*}{v\kappa}\left((H + |v|t^*) \ln (H + |v|t^*)\right). \tag{44}$$

For given values of the parameters (u_*, H, v) we can then use a numerical method, like Newton's iterative root-finding method, to compute the landing time as a function of x to any desired precision, since we cannot obtain an explicit expression for the landing time. Because the landing time t^* must be increasing with x by uniqueness of firebrand trajectories, with some more work we obtain once more a very similar expression for the the asymptotic landing distribution,

$$\mathbb{L}(x,m) = N\phi\Big(H + |v|t^*(x), C^{-1}(m, t^*(x))\Big), \tag{45}$$

though the formula is slightly less attractice since t^* must be solved implicitly from Equation (44).

3.4. The Spotting Distribution $\mathbb{S}(x)$ Determined from $\mathbb{L}(x,m)$

The uniqueness of firebrand trajectories for the constant wind, power law and logarithmic wind profiles, and the assumption that the ignition probability $E(m)$ depends only on the landed mass, will allow us to use the asymptotic landing distributions $\mathbb{L}(x,m)$ obtained in the previous section, to obtain the spotting distribution $\mathbb{S}(x)$. Recall our Formula (23) for the spotting distribution; having computed several asymptotic landing distributions, we now have several examples of the spotting distribution. Our assumption that the continuous ignition operator $E(m) \in [0,1]$ depends only on the landed mass implies that the *spotting distribution* $\mathbb{S}(x) \in [0,1]$, which characterizes the probability of a spotfire igniting due to an impulse release from $x = 0$, is given by:

$$\mathbb{S}(x) \;=\; E\left[N \int_0^{\bar{m}} m\, \phi\Big(H + |v|t^*(x), C^{-1}(m, t^*(x))\Big)\, dm\right]. \tag{46}$$

An explicit example for constant wind is given in Section 3.6.

We can extend this concept to firebrands released at location $x - y$, at time $t - t^*(x-y)$, to determine:

$$\mathbb{S}(x-y) = E\left[N \int_0^{\bar{m}} m\, \phi\Big(H + |v|t^*(x-y), C^{-1}(m, t^*(x-y))\Big)\, dm\right]. \tag{47}$$

The Formula (47) describes the kernel for a firebrand release at $x - y$, at time $t - t^*(x-y)$. Notice that our expression is very general, in that we are free to employ a variety of ignition or combustion models, through inclusion of specific functional forms for $E(m)$ or C respectively.

3.5. Case (W1,V1, F0, L3, I3): A Family of Simple Spotting Kernels

Let us consider the threshold ignition law (I3) presented in the Appendix, in the case where $\underline{m} \to 0$ (see Equation (A35)). Then any firebrand landing on a location which is not burning, will instantly generate a fire. We again have constant w and v, and further we will suppose that no mass is lost during transport corresponding to the constant burning law $f = const$ with zero rate of combustion (the transport process being assumed very rapid).

Let's assume that the firebrand vertical launching distribution $\phi(z, m)$ satisfies the assumption (L3), so that

$$\phi(z, m) = \mathbb{Z}(z)\ \mu(m). \tag{48}$$

This says the lofting heights z are independent of the masses m. This may occur, for example, in the case of well-mixed line thermals. We can choose the mass distribution $\mu(m)$ in accordance with experiments such as in the experiments by Manzello [5], or out of mathematical curiosity we could consider any other probability distribution. What is most important is to assume that $\mathbb{Z}(z)$ is not exponentially bounded, in order to obtain a fat-tailed kernel. For example, one could assume:

$$\mathbb{Z}(z) = e^{-z^{\beta}}, \tag{49}$$

for $\beta \in (0, 1)$. This kernel is of particular interest here, since it decays sub-exponentially, and has been shown in integro-difference models [26] to give rise to accelerating propagation of the corresponding solution in space [26]. Physically this could correspond to extreme spotting conditions like in the presence of fire whirls.

From Equation (46), with the form Equation (49) appearing in the landed mass distribution $M(x)$, we find that for a release of N firebrands the landed distribution $M(x)$ of masses at location x is:

$$\begin{aligned} M(x) &= N \int_0^{\bar{m}} m\ \mathbb{Z}\left(|v|\frac{x}{w}\right)\ \mu(m)\ dm && (50)\\ &= N \int_0^{\bar{m}} m\ e^{-(|v|\frac{x}{w})^{\beta}}\ \mu(m)\ dm && (51)\\ &= e^{-(|v|\frac{x}{w})^{\beta}}\ N \int_0^{\bar{m}} m\ \mu(m)\ dm && (52)\\ &= he^{-(|v|\frac{x}{w})^{\beta}}, && (53) \end{aligned}$$

where $h > 0$ is the total landed mass at x, namely

$$h\colon = N \int_0^{\bar{m}} m\ \mu(m)\ dm < \infty. \tag{54}$$

Since we are assuming instant ignition, the spotting distribution is the same as the landed mass distribution. Referring to Equation (50), we can then write the spotting kernel in this case:

$$\mathbb{S}(x) = he^{-(\frac{|v|x}{w})^{\beta}}. \tag{55}$$

This whole procedure can be generalized. If instead of the kernel Equation (49) we use an arbitrary kernel $\mathbb{Z}(z)$, fat-failed or not, repeating the steps leading to Equation (55), we obtain the spotting distribution:

$$\mathbb{S}(x) = h\mathbb{Z}(|v|\ \frac{x}{w}). \tag{56}$$

We can also extend the latter formula to include our other combustion models, by altering h defined in Equation (54) to read

$$h := N \int_0^{\bar{m}} m\ \mu\left(C^{-1}(m, \frac{x}{w})\right)\ dm, \tag{57}$$

however the resulting kernel Equation (56) will now have compact support, since we define $C^{-1}(m, \frac{x}{w})$ only for values of x where $C^{-1}(m, \frac{x}{w}) \geq 0$.

3.6. Applications: Examples of the Spotting Distribution

In this subsection we show some explicit examples for the landed mass and for the spotting distribution. First we consider an explicit version for the cases (L3, G1, W1, V1, F0, I2), which is a special case of case (W1, V1) that was studied in Section 3.1. We consider

$$\phi(z,m) = Z(z)\mu(m), \tag{58}$$
$$Z(z) = \lambda e^{-\lambda z}, \tag{59}$$
$$\mu(m) = am^{-0.5}, \qquad 0 \leq m \leq \bar{m}, \tag{60}$$
$$\frac{dm}{dt} = -\kappa, \tag{61}$$
$$C^{-1}\left(m, \frac{x}{w}\right) = m + \kappa\frac{x}{w}. \tag{62}$$

According to Formula (28) we have a landing distribution of

$$\mathbb{L}(x,m) = N\phi\left(|v|\frac{x}{w}, C^{-1}\left(m, \frac{x}{w}\right)\right). \tag{63}$$

Substituting the explicit forms from Equations (58)–(62) into the landing distribution, we get an explicit formula for $m \geq 0$:

$$\mathbb{L}(x,m) = \begin{cases} N\lambda e^{-\lambda|v|\frac{x}{w}} a\left(m + \kappa\frac{x}{w}\right)^{-0.5} & 0 \leq m + \kappa\frac{x}{w} \leq \bar{m} \\ 0 & m + \kappa\frac{x}{w} > \bar{m} \end{cases}, \tag{64}$$
$$= \begin{cases} N\lambda e^{-\lambda|v|\frac{x}{w}} a\left(m + \kappa\frac{x}{w}\right)^{-0.5} & 0 \leq m \leq \bar{m} - \kappa\frac{x}{w} \\ 0 & m > \bar{m} - \kappa\frac{x}{w} \end{cases}. \tag{65}$$

The condition $0 \leq m \leq \bar{m} - \kappa\frac{x}{w}$ implies that

$$x \leq x_{max} := \frac{\bar{m}w}{\kappa}, \tag{66}$$

and

$$\mathbb{L}(x,m) = 0 \qquad for \quad x > x_{max}. \tag{67}$$

To obtain the total landed mass, we multiply $\mathbb{L}$ with m and integrate from 0 to $\bar{m}$:

$$M(x) = \int_0^{\bar{m}} m\mathbb{L}(x,m)dm = \int_0^{\bar{m}-\kappa\frac{x}{w}} mN\lambda e^{-\lambda|v|\frac{x}{w}} a\left(m + \kappa\frac{x}{w}\right)^{-0.5}. \tag{68}$$

Now we use the integral

$$\int_0^{\bar{m}-A} m(m+A)^{-0.5}dm = \frac{2}{3}\bar{m}^{3/2} - 2A\sqrt{\bar{m}} + \frac{4}{3}A^{3/2} \tag{69}$$

and we find and explicit formula for the landed mass

$$M(x) = N\lambda e^{-\lambda|v|\frac{x}{w}} a\left(\frac{2}{3}\bar{m}^{3/2} - \frac{2\kappa x\sqrt{\bar{m}}}{w} + \frac{4}{3}\left(\frac{\kappa x}{w}\right)^{3/2}\right). \tag{70}$$

To obtain the spotting distribution, we need an ignition law. Here we use the piecewise linear ignition function with lower ignition threshold $\underline{m}$ and we obtain

$$\mathbb{S}(x) = E(M(x)) = \begin{cases} \frac{M(x)}{\underline{m}} & M(x) \leq \underline{m} \\ 1 & M(x) > \underline{m} \end{cases} \tag{71}$$

with $M(x)$ given by Equation (70). Hence we find an explicit formula for the spotting distribution. As examples we chose the parameters as in Table 1 and we plot the landed mass $M(x)$ and the spotting distributions $\mathbb{S}(x)$ for these cases in Figure 3.

Table 1. Summary of the model parameters for the examples (in metric units).

Parameter	w	v	$\bar{m}$	$\underline{m}$	a	κ	N	λ	x_{max}
base case (thick solid)	2	−1	0.004	0.001	7.91	0.00005	1000	0.01	160
slow burner (dotted)	2	−1	0.004	0.001	7.91	0.00003	1000	0.01	266.66
lower release height (dashed)	2	−1	0.004	0.001	7.91	0.00005	1000	0.05	160
Parameter	w	v	$\bar{m}$	$\underline{m}$	a	η	N	λ	x_{max}
Tarifa's case (thin blue)	2	−1	0.004	0.001	7.91	0.000286	1000	0.01	∞

The parameters in Table 1 have been chosen to model realistic physical scenarios, at the lower limit where spotting might begin to have an impact. The windspeed w of two metres per second is on the low end of observed values [3]. Terminal speed of one metre per second is also low, but on the same order of magnitude as firebrands with diameter, mass and length as described in Manzello's study. The upper and lower bounds on the mass, measured in kilograms, correspond to the values found in Manzello's combustion experiments [5]. The rate of combustion, chosen between 0.03 and 0.05 g per second, results in flaming burnout in under two minutes. Choosing 1000 firebrands, corresponds to combustion of about ten trees—a moderate estimate for the width of our front. The decay rates λ were chosen so that firebrand launching would drop off appreciably beyond several hundred metres. Finally the parameter a is a normalization constant for our mass distribution.

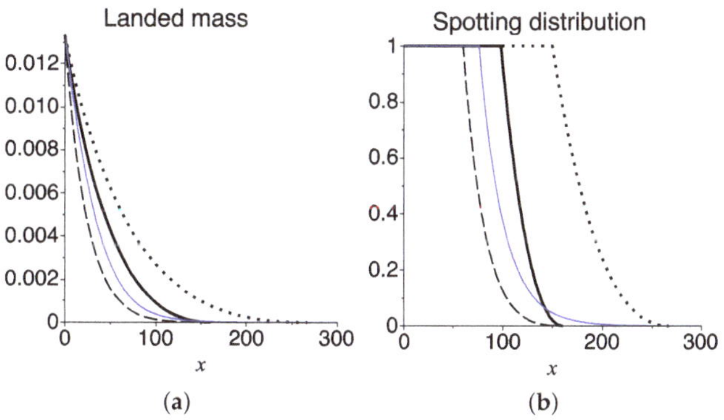

Figure 3. (**a**) Examples of landed mass distributions, with mass in kilograms along the y-axis and distance in metres along the x-axis; (**b**) Corresponding spotting distributions, dimensionless values on the y-axis and distance in metres along the x-axis. The parameters are from Table 1, with the base case in thick solid black, the slow burning case in dotted, the lower release height case in dashed and Tarifa's model in thin solid blue.

Next we consider the effect of changing combustion model, choosing instead our modified version of Tarifa's model (F2), while keeping all other parameters fixed (see Table 1). It is thus important to reconsider the inverse combustion operator. According to the model (F2), we have

$$C^{-1}(m, t^*(x)) = m\left(1 + \eta \frac{x^2}{w^2}\right), \tag{72}$$

since this is the initial mass, which travelled through the air after launching during a time of $t^* = \frac{x}{w}$. We can start with Equation (65) and replace the term $m + \kappa\frac{x}{w}$ by $m\left(1 + \eta\frac{x^2}{w^2}\right)$. Then the landed mass is computed to be

$$M(x) = \frac{2}{3} N \lambda e^{-\lambda |v| \frac{x}{w}} a \left(1 + \eta \frac{x^2}{w^2}\right)^{-2} \bar{m}^{\frac{3}{2}} \tag{73}$$

and the spotting distribution $S(x)$ is again explicitly given by Equation (71). As an example we use parameters as in Table 1 and plot the landed mass and the spotting distribution as thin blue lines in Figure 3. Notice that the inverse combustion operator in Equation (72) is always positive for all $x > 0$. Hence in this case we have no maximum spotting range and $x_{max} = \infty$, even though it is effectively zero beyond 200 m.

It is important to determine the minimum spatial extent of the spotting distribution required for spotting to be important. Below this extent, the main fire will outrun spot fires before they can develop into a separate front. As a specific example, Alexander has estimated such an effective distance for crown fires in North America to be approximately 300 m (in the most extreme ignition conditions [3]). In Figure 3 we consider Alexander's crown fire situation. In comparing the Tarifa-case with the dotted curve (base case with slower burning), both distributions extend at least approximately to the 300 m cutoff—though in the Tarifa-case the distribution is essentially zero there. We further note that, while not shown, just halving the combustion rate in the base case pushes the distribution past the 300 m mark. By changing the combustion model or parameters, we obtain very different outcomes for the importance of spotting.

4. Discussion

4.1. Usage of the Spotting Distribution

One of our primary motives in determining the spotting distribution, is to employ the latter as a redistribution kernel in fire spread models. For example, the model in [35] uses an indicator function approach to describe the interface between burned and unburned regions. They include spotting in the form of a log-normal distribution, but other spotting distributions, such as computed here, could be included as well.

A somewhat different approach was taken in [29,30], where an integro-PDE equation approach was used to model the probability of fire at a certain location. The time evolution of the likelihood $u(t, x)$ to observe a fire a time t at location x is given by the integro-PDE

$$u_t = D u_{xx} + \int_{-\infty}^{\infty} \mathbb{S}(x - g)\, u(g, t)\, dg + c(u)u - \delta(u)u, \tag{74}$$

where $\mathbb{S}(x)$ is the spotting distribution, $c(u)$ describes combustion, $\delta(u)$ denotes the heat loss term and the diffusion Du_{xx} describes local fire spread. This model was used in [30] to investigate invading fire fronts and their asymptotic invasion speeds. For example, we could there show that spotting is able to increase the fire invasion speed. A model of the above type is quite versatile and it can include various spotting distributions as well as different combustion and fuel dynamics.

A more standard approach starts from conservation laws for physical quantities like energy of chemical species, leading to coupled reaction-diffusion systems for temperature and mass evolution [31,52,53] and knowledge of the spotting distribution might be a useful addition to these models.

Changing gears from our discussion of the "spread problem", knowledge of the spotting distribution is also important for the "breaching problem"—of great practical importance for wildfire management, but poorly understood. For example, suppose a crowning forest fire reaches a wide river. On the opposite side is a continuous stretch of dense coniferous forest. The likelihood of a spot fire occurring on the non-burning side of the river equals the integral of the spotting distribution along the opposing side of the river. It gives a direct quantitative measurement for fire risk beyond fuel breaks. A similar problem arises at the wildland-urban interface and the spotting distribution can tell us how far the spotting is likely to reach.

4.2. Measurements of Spotting Distributions

The primary challenge for the spotting experimentalist is the paucity of quantitative observations from real fires. The landing fire brands could be visualized by visual recording or infrared recording. In the paper "Monitoring Insect Dispersal", by J.L. Osborne *et al.* in the collection [27], the authors discuss "vertically looking radar" (VLR). Such radar has been used to study insect dispersal, and consists of a series of gates capable of monitoring the skies from 100 m to a kilometre above ground. In addition, the VLR system is capable of determining flying mass, direction and magnitude of the velocity vector of a flying brand. One could imagine field experiments where such instrumentation is employed, in order to count how many, to which height and with what mass are the firebrands being released.

In field situations, the ignition probability can be estimated by the number of spot fires which result per unit of landed firebrand mass, or ideally extrapolated from laboratory experiments. Satelite data, or other forms of radar collection could be of use here as well; while individual spotting events may only be observable through the appearance of a new fire, if we had detailed spotting information for a particular fire situation, such information could help inform the likelihood of long-distance dispersal.

There are some field experiments, in particular from Australia's Project Vesta [54,55], where attempts have been made to directly measure the spotting distribution. In particular, in a series of experimental fires, firebrand distributions were measured by catching brands on plastic sheets. The goal was to validate earlier firebrand modelling employing the CSIRO wind tunnel by Ellis [9], though the results were mostly qualitative and the need for model improvement was a primary finding.

In controlled lab experiments, several trees could be alighted in a wind stream and all landing embers can be first put out, then counted and weighed, as has been done in experiments by Manzello [44,45]. Both of these approaches (field and lab) give us the landing distribution $\mathbb{L}(x,m)$. To get the spotting distribution $\mathbb{S}(x)$ we need a second ingredient, which is the probability of ignition $E(m)$. Several lab experiments have been done already, where burning material is thrown into various fuel beds and the ignition probability has been measured using regression analysis in quite some depth ([46,47]). In the very heterogeneous natural spotting environment, understanding the variability in spotting ignition probabilities over space and time is very important.

We believe that many of this data is already available in various fire management and research centers, but, to our knowledge, they have not been systematically examined to estimate a spotting distribution. Promising data are available for example for the 1961 Basin Fire in the Sierra National Forest (USA) [56], the 1994 South Canyon Fire in Colorado [57], the 1994 Butte City Fire in Idaho [58] and the Oakland/Berkeley Hills Fire [59], which in particular was described in the review paper by Koo *et al.* [12]. We leave a detailed exploration of these data to future research and we write this paper in the hope of generating increased interest in better quantifying the lesser known model components.

We have seen that the derivation of specific spotting distributions depends on many physical details, which we outline in the Appendix. However, direct measurement might enable us to skip the detailed physical modelling and rather use an empirically measured spotting distribution.

4.3. Future Studies

The key idea behind the spotting distribution is a separation of time scales for the relevant processes, namely the fast wind-distribution of firebrands versus the relatively slow crawling combustion of the main front. In real wildfires, the total flight time of burning material is of the order of seconds to minutes, while we may assume that the overall fire front progression is of the order of minutes or hours. The maximum possible flight time of firebrands has been established through wind-tunnel experiments, which confirm that combustion of firebrands results in extinction after at most several minutes, and flaming combustion even prior to that, though there is the possibility for re-flaming if travel is fast enough (e.g., [5,9,31]). Ignition is highly dependent on firebrand state upon landing (e.g., flaming vs. glowing). For forest fires, the required separation of timescales between spotting and local spread is valid—wind transport is much more rapid. For grass fires, however, the time scale of local spread and spotting is comparable and our scaling argument is invalid for grassfires.

We only consider horizontal winds perpendicular to the main fire front. However, dispersal of firebrands happens in all directions, often leading to a "V"-shaped spread downwind—similar to the wake left by a boat travelling through water. In our case, any spread parallel to the front will not change the front's shape. However, increased travel times due to horizontal movement parallel to the front's axis may lead to decreased support for the downwind landing distribution, due to earlier burnout. Further, there may be a greater net accumulation of firebrands than is described in our distributions, due to cross-wind contributions from launches further down the front. The latter point is less concerning, since there is already uncertainty in the number N of firebrands launched. Since our idealized distributions will be translation-invariant parallel to the front, extra accumulation could be accounted for by increasing N.

From a mathematical perspective, integro-differential equations, which employ redistribution kernels to describe long-distance dispersal, such as Equation (74), have become of much greater interest of late. There is considerable overlap with research on plant seed dispersal [26,28] and we expect that analysis methods that are used in seed dispersal to become useful for fire spotting as well. Employing our spotting distribution as such a redistribution kernel, we hope to be able to provide more complete answers to the questions posed in Section 1.3. Including topography, spatial variation in fuel and weather leads to heterogeneous and nonlinear mathematical models, which would be further complicated—but made more realistic—by adding an additional spatial dimension.

Many avenues of research are opened by considering the analysis of such models, since such models are mathematically complex and at the forefront of current applied analysis and numerical modelling research. Hence another way our models could be improved is a better understanding of the analytical aspects of nonlocal models in heterogeneous media, which is another avenue of research underway by the authors and others [60].

Acknowledgments: This work resulted from a MITACS (Mathematics of Information Technology and Complex Systems) Full Project on Forest Fire Modelling. The authors are grateful for feedback and support through the researchers involved in this MITACS project. Details of this work were discussed during Journal Club meetings at the University of Alberta, and we are grateful for the lively interest and feedback from the Journal Club participants. Parts of J.M.'s work was supported by the PIMS (Pacific Institute for Mathematical Sciences) International Graduate Student Training Centre, two MITACS summer internships, and part of this work has been carried out in the framework of the project NONLOCAL (ANR-14-CE25-0013), funded by the French National Research Agency (ANR). T.H. is grateful to support through NSERC (Natural Sciences and Engineering Research Council of Canada).

Author Contributions: The models and results of this paper are part of Jonathan Martin's PhD thesis [29] under supervision of Thomas Hillen. The model derivation and analysis, as well as the writing of the manuscript has been done in equal parts. The extensive appendix was developed by Jonathan Martin.

Conflicts of Interest: The authors declare no conflict of interest.

Appendix. Ember Release, Burning, Flying and Fuel Ignition

In this Appendix we will discuss the modelling of physical subprocesses that are involved in the spotting problem. We will review the extensive literature in this area, and derive mathematical models for the firebrand mass distribution, plume models and the vertical launch distribution, firebrand combustion and temperature, vertical and horizontal transport speeds, and ignition probabilities. We summarize all models and their references in Table A1.

Table A1. Physical and empirical models for spotting subprocesses.

Process	Number, Description	Reference
launching $\phi(z,m)$	L1, Unique launching height $z(m)$ L2 , Normally distributed L3, Heights and masses independent: $\phi(z,m)=\mathbb{Z}(z)\mu(m)$	[1,21,22]. New. New.
launched mass $\mu(m)$	G1, Power law G2, Slash burning	New; [44] [32].
Wind transport w	W1, Constant horizontal wind W2, Logarithmic wind profile W3, Power-law wind profile	New. [1,15]. [1,13].
Terminal vertical velocity v.	V1, Constant v V2, Experiments on cylindrical firebrands.	[9,68]. [16].
Combustion models f	F0, Constant burn rate F1, Tarifa's model F2, Simplified Tarifa's model F3, Negligible combustion F4, Fernandez-Pello model F5, Refinements to Fernandez-Pello model F6, Albini's line thermal model	New. [1] New. New. [69]. [22,70]. [3].
Ignition probability $I(m)$	I1, Piecewise linear I2, Heaviside step function I3, Smoothed step function	[32,46,47]. New. New.
Temperature $T(t)$	T1, Newton's Law of Cooling, T2, Stefan-Boltzmann law	 [22].

Appendix A.1. The Launching Distribution $\phi(z,m)$

The greatest challenge in modelling spotting is to determine how many firebrands, distributed according to their various characteristics, are both generated and subsequently launched into the atmosphere. The latter process takes place in the fire plume, which is a high-velocity, buoyancy-driven flow induced by the combustion at the surface [61]. The plume of a wildland fire is often called its *convection column* [32]. Any fire plume is more turbulent than laminar, and our knowledge about plumes is mostly experimental [61]. There is a complicated interaction between the atmosphere and the fire, hereafter referred to as *fire-atmosphere interactions*, which has been extensively studied from experiments, and physical modelling [1,2,17,18,22–24,31,62–67].

The most widely used fire plume model in spotfire modelling is the model developed by Baum and McCaffery [18,21], as used for example in [1,7,13,19,20,22,67]. The model consists of three burning regions, illustrated in Figure A1. Our discussion here closely follows that of the book [61] and the paper [22]. Region I lies at the base, and is the continuous burning region. Here the flow is pulsating and unsteady. Region II is an intermittent zone, in which flame patches break off from the below-anchored flame, while at the top of Region II all combustion ceases. Finally we have region III, the non-combusting plume, where we assume that the time-averaged upward velocity and temperature drop off radially in a Gaussian manner.

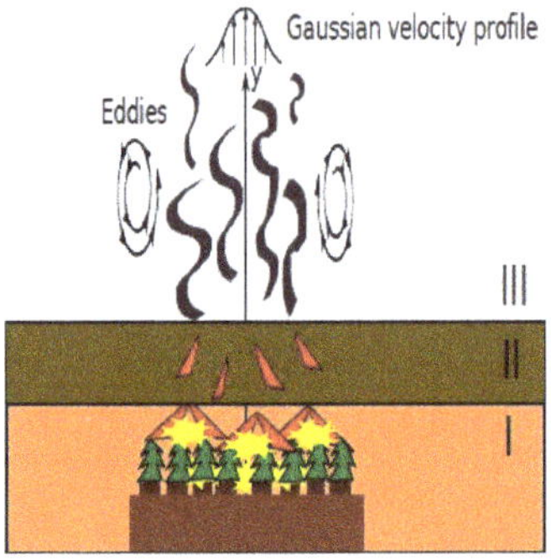

Figure A1. Sketch of the Baum and McCaffery plume. Region I is the continuous (canopy) burning region. Region II is a transition zone over which the plume velocity is constant. The buoyant upward motion in region III is reinforced by large ambient eddies, which cause entrainment of air into the plume.

The relevant parameters are height z, plume velocity U_p, and temperature T. These are made dimensionless by the scaling $\frac{z}{z_c} := z^*$, $\frac{U_p}{U_c} := U^*$, $\frac{T-T_a}{T_a} := T^*$, where:

$$z_c := \left(\frac{Q}{\rho_a c_p T_0 \sqrt{g}}\right)^{2/5}, \quad U_c := \sqrt{g z_c}. \tag{A1}$$

The parameters appearing in Equation (A1) are the heat release rate Q, the density of air $\rho_a = 1.2$ kg·m^{-3}, $c_p = 1$ kg^{-1}·K $^{-1}$ is the specific heat capacity of air at constant pressure, T_a is the temperature of the ambient air, and g is the gravitational constant.

The analysis in [71] then leads us to the mean plume-centerline velocity and temperature as a function of the rescaled height z^*:

$$U^* = 3.64(z^*)^{-1/3} \quad \text{and} \quad T^* = 8.41(z^*)^{-4/3}. \tag{A2}$$

For a given height z in the plume region there is associated a unique mass $m(z)$, such that this mass attains terminal velocity exactly at height z. In other words, the drag induced by the upward plume velocity U_p is balanced by the weight of the mass at this height.

In the idealized case of a spherical particle, we have a direct connection between the cross-sectional area A, the diameter d, the drag coefficient C_D and the density ρ_s. Employing the relation Equation (A2), we obtain the unique lofting height:

$$z = \left(\frac{40 C_D \rho_a}{4 d \rho_s g}\right)^{3/2} z_c, \tag{A3}$$

Employing Equation (A1) in the latter equation, we can re-write the lofting height as:

$$z = \left(\frac{40 C_D \rho_a}{4 d \rho_s g}\right)^{3/2} \left(\frac{Q}{\rho_a c_p T_0 \sqrt{g}}\right)^{2/5}. \tag{A4}$$

We will introduce a constant $\gamma = [40 C_D \rho_a (4 d \rho_s g)^{-1} (\rho_a c_p T_0 \sqrt{g})^{-4/15}]^{3/2}$ which absorbs all the constants in Equation (A4), and re-write the Equation:

$$z(m, Q) = \gamma m^{-3/2} Q^{2/5}. \tag{A5}$$

Recent work suggests that the fire-atmosphere interaction results in distinctly non-Gaussian distributions [22]. In the physically realistic computer simulations of grass fire plumes of [22], which employs a Large Eddy Simulator to model the atmospheric winds, the time-averaged velocity profiles are not observed to be Gaussian and the Baum and McCaffery plume model needs to be extended.

Based on these observations, we study three launching models:

Model L1. We assume that each firebrand of mass m is lofted to a unique height $z = z(m)$, as for example in Equation (A5). We can then define

$$\phi(z, m) = \delta(z - z(m))\mu(m), \tag{A6}$$

where $\mu(m)$ is a given mass distribution, and δ represents the Dirac delta functional.

Model L2. Instead of assuming that each mass is lofted to a unique height, we might instead suppose that it is launched randomly about the standard lofting height $z(m)$. For example, if heights are normally distributed about the lofting height $z(m)$, we can write $\phi(z, m) = \mathbb{N}(z(m), \sigma)\mu(m)$, a one-sided normal distribution where $\mathbb{N}$ has mean $z(m)$ and variance σ, with

$$\mathbb{N}(z(m), \sigma) := \frac{A}{\sigma\sqrt{2\pi}} \exp\left(-\frac{(z - z(m))^2}{2\sigma^2}\right), \tag{A7}$$

where the constant A is chosen so that the distribution is normalized. We allow values of z to lie in $[H, z_{max})$, where H is the canopy height and z_{max} is the maximum lofting height predicted by the Baum and McCaffery plume.

Model L3. Finally, we consider the case where the launching heights z, and masses m, are independent of each other, so we can write

$$\phi(z, m) = \mathbb{Z}(z)\mu(m), \tag{A8}$$

where $\mathbb{Z}(z)$ is a probability distribution which describes how firebrands are distributed with height z, and $\mu(m)$ is the mass distribution.

A similar sort of model was employed by Albini in the context of firebrand transport by line thermals [13], where $\mathbb{Z}$ is a uniform distribution .

$$\mathbb{Z}(z) = U[z_{min}, z_{max}]. \tag{A9}$$

Appendix A.2. Distribution of Launched Masses

A recent series of studies by Manzello and colleagues [5,44,45] investigated firebrands emitted from the controlled burning of either pine or fir trees. In the Manzello experiments, trees both 2.6 m and 5.2 m tall were investigated. For each tree, more than 70 firebrands were collected. These were all cylindrical in shape. The average firebrand length and diameter for the 2.6 m class was 40 mm in length and 3 mm in diameter, while for the 5.2 m class the average was 53 mm in length and 4 mm in diameter. The most recent of the experiments, on Korean pine [44], confirmed that the distribution was approximately the same. The total number of firebrands collected numbered almost 1000.

With respect to mass, all three of Manzello's studies indicate that between 60 and 80 percent of the firebrands have masses less than 0.1 g. Further, for both pine and fir taller trees produce larger firebrands, with the largest found at about 5 g. In addition, between 58 and 65 percent of the firebrand mass is needles, which are insignificant in long-range spotting, but may be effective at igniting local fuel only in short-range spotting [46]. We remove the needles from Manzello's mass distribution, to obtain an 'effective mass' distribution.

Model G1: Regression analysis on Manzello's data. In order to determine a functional form for the effective mass distribution, we reproduced the histograms from [44,45], an example of which is shown in Figure A2. Between 60 to 70 percent of the mass is needles which are negligible for long-range spotting. Hence we first removed the needles lying in the 0.1 g mass class, to obtain an effective firebrand distribution. We use non-linear regression of the functional form:

$$\mu(m) = am^{-b}, \qquad 0 \leq m \leq \bar{m}, \tag{A10}$$

where $\bar{m} = 4$, corresponding to the maximum firebrand mass. Regression analysis gives $b = 1/2$, and $a = 1/4$, which gave a better fit than an exponential form.

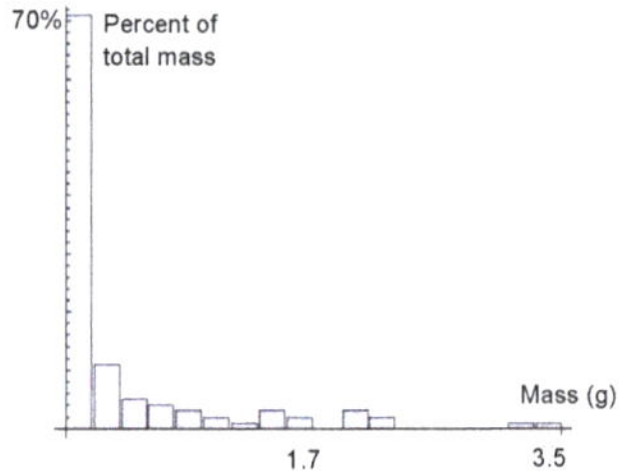

Figure A2. The mass distribution for the 5.2 m Douglas fir firebrands plotted as histogram from the data from [5]. The histograms for the other taller specimens for each species studied are similar. Along the x-axis we plot firebrand mass in grams.

Model G2: Models obtained from burning slash. Another firebrand distribution was suggested in [32], which relates the possible radius r of a firebrand to the mass consumption rate f, in the form:

$$p(r) = \frac{\alpha f}{r\sigma\sqrt{2\pi}} \exp\left(\frac{-1}{2\sigma^2}\log{(r/r_0)^2}\right), \tag{A11}$$

where r_0 and σ were parameters determined by regression analysis, and α represents the number of firebrands generated per unit mass. If we assume a relationship of the form $m = m(r) = \rho V(r)$, where ρ is the density and $V(r)$ is the firebrand volume, then from Equation (A11) we find the mass density:

$$\mu(m) = \int_0^\infty \delta(m - m(r))p(r)dr, \tag{A12}$$

where δ denotes the Dirac delta distribution.

From the experiments of Manzello described in this chapter, it was found that approximately one percent of the total mass lost during combustion appeared as firebrands. This could inform our parameter α, and in turn determine the total mass and total number of firebrands released when a given number of trees begin to spot.

Appendix A.3. The Atmospheric Boundary Layer

The atmospheric boundary layer (ABM) is the lowest portion of the atmosphere, extending on average about one kilometre, and ranging up to at most about three kilometres above the Earth's surface [72]. It consists of a number of distinct sublayers, and it is of utmost importance since the ABM is where firebrand transport occurs. At the bottom is the laminar sublayer, which has a thickness of only a few millimetres. This is a region where high viscosity, induced by the "roughness" of the surface, results in molecular diffusion being the basis for transport of momentum and heat [73]. Above the laminar layer is a transition region, leading into the Prandtl layer, in which turbulent convective motion is the dominant transport process [73]. The lower boundary of the Prandtl layer is called the *roughness height* y_0. At the top of the ABM is the Ekman layer, throughout which the effects of turbulent convection lessen with height, decreasing to zero near the top of the Ekman layer [73].

Firebrands are transported by convection, and hence are subject to the turbulent fluctuations in wind velocity present in the ABM. Turbulence is a dissipative process which converts kinetic energy in a fluid into heat energy, and it is essentially three-dimensional and rotational [74].

Turbulent eddies, which may be envisioned as large sheets of wind rolling over one another, exist on length scales from 10^{-3} m to 10^3 m. The largest eddies can therefore extend up to the height of the ABM [74]. In the case of a fire's convection column, the eddies swirling parallel to the column result in the entrainment of ambient air into the column [22].

Because of the inherent variability in the transport process, we introduce the standard Reynolds decomposition for the velocity components [74]. This means we decompose the horizontal velocity w into a slowly-varying mean component $\bar{w}$ and a rapidly-varying component w', so that:

$$w := \bar{w} + w'. \tag{A13}$$

In general the mean windspeed increases with height, though exactly how this happens is effected by surface roughness and variable topography, to say nothing of the fire-ABM interactions. In our transport model we will be interested in the time-mean behavior of the stochastic flight process, so we will focus exclusively on the mean velocity $\bar{w}$. We drop the bar in what follows for notational simplicity.

Model W1: Constant horizontal wind. The simplest assumption for the windspeed w is that it does not vary with height z, so that

$$w(z) = w > 0. \tag{A14}$$

Model W2: Logarithmic horizontal wind. Another commonly used wind model is the logarithmic profile, introduced in the context of spotting first by Albini [1]:

$$w(z) = \frac{u_*}{\kappa} \log\left(\frac{z-d}{y_0}\right). \tag{A15}$$

Appearing on the right hand side of Equation (A15) is the von Karman constant $\kappa = 0.41$, the roughness height y_0, the zero-datum displacement d, and the friction velocity u_*. The friction velocity is generally defined by $u_* := \sqrt{T/\rho}$, where T is the time-mean flux of tangential momentum towards the surface, borne by turbulence outside the lower ABM and by viscosity within it [15]. Typical values for u_* in a strongly upward-convective atmosphere are around $u_* = 2$ m/s, while for a roll-dominated atmosphere $u_* = 0.7$ m/s [24]. The roughness height y_0 corresponds to the lower boundary of the Prandtl layer. At the upper end examples include $y_0 = 0.5 - 1.0$ m for dense forest or shrubs, $y_0 = 0.1 - 0.5$ m for low crops or bushes, and flat grassland has $y_0 = 0.03$ m. When there is significant roughness or dense forests, the zero-datum displacement d is employed in Equation (A15) to offset the height at which the windspeed aloft vanishes. The value of d is usually about 2/3 the average height of the obstacles. We show velocity versus height for three different cases in Figure A3.

Model W3: Power-law wind profile. A third wind model was also first introduced in the context of spotting by Albini [1]. It assumes a power-law profile for the horizontal velocity versus height,

$$w(z) = w_H \left(\frac{z}{H}\right)^{\beta}, \tag{A16}$$

where H is the canopy height, w_H is the windspeed at the canopy's base, and $\beta \in [0,1]$. In Albini's work, he chose the constant $\beta = 1/7$ [13]. Model (A16) is a better approximation to the windflow when it is over terrain which is not covered by tall vegetation. Further, this model may be seen as an approximation to the logarithmic profile, and is consistent with the constant-wind model (to see this, set $\beta = 0$ in Equation (A16)).

A comparison of our three functional forms is presented in Figure A3.

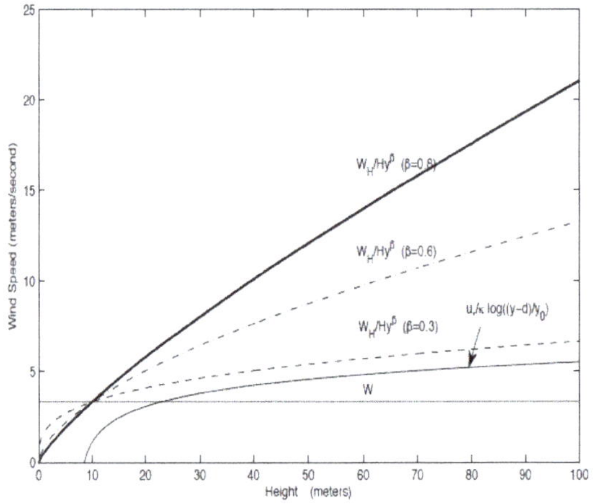

Figure A3. A comparison of the wind profiles discussed in this Section. Shown are three power-law models for different values of the parameter β, together with a logarithmic wind profile and a constant wind profile. We have chosen $w_H = 5$ m/s, and the canopy height to be 10 m.

Appendix A.4. Drag, Gravity and Terminal Velocity

In order to accurately model the falling of firebrands in the atmosphere, it is important to model the drag experienced by the firebrand. Often, the drag is proportional to the object's speed, or speed squared, depending on the Reynolds number of the flow. For firebrands, it is more accurate to model drag as proportional to the speed squared, since we have a relatively high Reynolds number flow [22].

Let us denote the drag force by D. Then the speed-squared assumption is generally written as:

$$D = \frac{1}{2} C_d \rho_a A v^2, \tag{A17}$$

where ρ_a refers to the mass density of the ambient fluid, and the object is assumed to have constant cross-sectional area A. The parameter C_d appearing in Equation (A17) is called the drag coefficient. This is a dimensionless number, with values typically ranging between 0.001 and 2, assumed to vary with shape. This constant is typically determined by experiment.

Newton's Second Law then provides us with an expression for the terminal velocity. If we focus on the vertical direction, the balancing of the drag force D with the weight W implies:

$$0 = \text{Drag} - \text{Weight} = \frac{1}{2} C_d \rho_a A v^2 - mg. \tag{A18}$$

Solving for v in Equation (A18), we obtain an expression for the terminal velocity:

$$v = \sqrt{\frac{2mg}{C_D \rho_a A}}. \tag{A19}$$

Model V1: Constant terminal velocity. The simplest assumption (V1) is to assume the terminal velocity does not change during transport.

Model V2: Experimentally determined model. Experimental analysis on the Aerospace Corporation's experiments, appearing in [16], revealed that for the cylindrical firebrands in the study the mass $m(t)$ is related to the terminal velocity $v(t)$ according to:

$$v(t) = v(0) \left(\frac{m(t)}{m(0)} \right)^{1/2}, \qquad v(0) < 0. \tag{A20}$$

Appendix A.5. Firebrand Combustion

Several combustion models have been proposed in the literature, and the interested reader is invited to peruse the review in [7]. They are usually modeled as continuous-time processes, so we will present functional forms for the density, effective radius, or simply the mass as functions of time.

Suppose the mass $m(t; m_0)$ is the unique solution of the initial value problem

$$\frac{dm}{dt} = f, \qquad m(0) = m_0. \tag{A21}$$

where f is the rate of combustion. We introduce the *combustion operator* C as the unique solution

$$C(m_0, t) := m(t; m_0). \tag{A22}$$

In our spotting problem, we need to compute backwards. If a fire brand of mass m lands, then we like to know its initial mass $m(-t)$. Hence we define the inverse combustion operator $C^{-1}(m, t)$, which is the unique solution to the initial value problem $\frac{dm}{dt} = -f, m(0) = m$ at time $t \geq 0$. Before continuing, it is important to note that the combustion process does not continue past the point where the firebrands have masses less than or equal to zero. Hence the combustion operator is only defined up to the burnout time t_b, and similarly for its inverse.

Model F0: Constant burning rate. Suppose the burning rate is a constant $f = -\kappa < 0$, then we have $m(t) = m(0) - \kappa t < m(0)$, and $C(m_0, t) = m_0 - \kappa t$. The inverse satisfies $C^{-1}(m, s) = m + \kappa s$ for $s > 0$.

Model F1: Tarifa's original experiments and models. The first experiments on the density and shape changes in combusting firebrands were carried out by Tarifa and collaborators at the Aerospace Corporation [1]. This data was fit by regression analysis obtained from windtunnel experiments, where both spherical and cylindrical firebrands were examined under a variety of ambient windspeeds.

Suppose $\rho(t)$ represents the density of a firebrand at time t, and $\rho(0)$ is the initial density. Then Tarifa found the density varied as:

$$\rho(t) = \frac{\rho(0)}{1 + \eta t^2}, \tag{A23}$$

where the constant $\eta = 2.86 \times 10^{-4}$ was determined from regression analysis.

Models F2: Caricature of Tarifa's model. We employ an analogue of Tarifa's density evolution for the mass $m(t)$, namely

$$m(t) = \frac{m(0)}{1 + \eta t^2}. \tag{A24}$$

Model F3: Constant mass. If we take case F2 to its extreme, sending $\eta \to 0$, we get

$$m(t) = m(0). \tag{A25}$$

Model F4: Tse and Fernandez-Pello's improvements. Tse and Fernandez-Pello revisited Tarifa's data set [75], and determined the model which best fit the data for the effective radius evolution is:

$$r(t)^4 = r(0)^4 - \frac{\chi \beta^2 t^2}{16}. \tag{A26}$$

Clearly the latter is only defined while $r(t) > 0$, which implies a finite burnout time. Here the parameter χ depends on the wood species and moisture content of the firebrand, and β is described below in the derivation of Equation (A31). Based on Tarifa's data, the parameter $\chi = 3.5$ gave the best fit [69].

To determine the mass $m(t)$, we first employ the simple relation:

$$m(t) = \rho(t)\mathrm{Vol}(r(t)), \tag{A27}$$

where $\mathrm{Vol}(r(t))$ is the volume of the firebrand, and hence dependent on the radius $r(t)$.

As a simple example, suppose the firebrand is approximately spherical, so that $\mathrm{Vol}(r(t)) = \frac{4}{3}\pi r(t)^3$. Then we combine Equations (A24) and (A26), so that the mass evolution Equation (A27) is given by

$$m(t) = \rho(t)\frac{4}{3}\pi r(t)^3 = \frac{4}{3}\frac{\rho(0)}{1+\eta t^2}\left(r(0)^4 - \frac{\chi\beta^2}{16}t^2\right)^{3/4}. \tag{A28}$$

Taking for example cedar wood as representative of forest fuel, as did the authors in [22], the initial density $\rho(0)$ of the firebrands would be 513 $\mathrm{kg/m^3}$, while in the experiments of Manzello [44] he found radii ranging from one half to five centimetres.

Another important result following the analysis in [69], is that one can reasonably approximate extinction of the firebrand to occur once:

$$\frac{m(t)}{m_0} = m_c, \tag{A29}$$

where m_c is a critical mass ratio which depends on the wood species. Beyond this point all that remains is non-flaming char, whose temperature then decreases according to Equation (A42) described in the next subsection. From regression analysis in [69], we find $m_c = 0.24$ for maple and pine.

Model F5: Including more physical realism. Another model has been derived in [22], and is based in part on experimental fitting of data by Tse and Fernandez-Pello [69]. It is based on Nusselt's physically-motivated combustion theory, known as 'shrinking drop theory' [22,70]. We include a discussion of the model here because, as we will see later, the firebrand's burning temperature may influence its flight path significantly [22]. In Nusselt's combustion theory, the firebrand's surface is assumed to be held at constant temperature, maintaining its geometrical shape while a pyrolysis wave propagates inward. One employs the so-called Frossling relation, in which we define the *effective mass diameter* d_{eff}, and an experimental constant β, such that:

$$\frac{d}{dt}\left(d_{\mathrm{eff}}\right)^2 = -\beta. \tag{A30}$$

The constant β, introduced earlier in Equations (A26) and (A28), is determined physically from the equation:

$$\beta = \beta^0\left(1 + 0.276 R_e^{1/2} S_c^{1/3}\right), \tag{A31}$$

where on the right hand side we have the Reynolds number R_e, and the Schmidt number S_c, and an experimentally determined constant β^0. The Reynolds number $R_e = \frac{2|v|r}{\nu_{\mathrm{air}}}$, where $|v|$ is the firebrand's speed relative to the surrounding air, r is the firebrand radius, and ν_{air} is the kinematic viscosity of the surrounding air [22]. R_e is a dimensionless constant which measures the ratio of the effects of inertia to viscosity for the firebrand. The Schmidt number S_c is a dimensionless constant which gives the ratio for viscosity to mass density, for mass-transfer problems. For air S_c is approximately constant, *i.e.*, $S_c \approx 0.7$ for a wide range of temperatures. Based on the data from [69], which was computed experimentally for firebrands with temperatures at about 993 K, we find the constant β^0 which appears in Equation (A31) to be $\beta^0 = 4.8 \times 10^{-7}\ m^2/s$.

Employing Tarifa's model for the density evolution ρ Equation (A23), and the solution to the differential equation for d_{eff}, the firebrand mass m is then approximated by:

$$m = \rho\pi d_{\mathrm{eff}}^3/6. \tag{A32}$$

Model F6: Albini's combustion model within line thermals. A final model was derived by Albini [13] in the context of firebrand transport by line thermals. Line thermals are well-mixed horizontal columns of warm air which rise above large forest fires, and are subsequently transported in a coherent manner downwind. Albini modeled line thermals as well-mixed cylinders of air rising above a fire.

If we assume that a firebrand has mass density $\rho(t)$, and terminal velocity v relative to the line thermal, Albini's model reads:

$$\frac{d}{dt}\rho(t) = -k\rho(t)v, \tag{A33}$$

where the constant $k = 8.15 \times 10^{-3}$ was chosen to match wind-tunnel data [13].

Appendix A.6. Ignition Models

Based on the observations of fuel bed ignition mentioned before, we assume that the ignition probability depends on the flaming mass that lands in a given location, where a mass larger than $\underline{m}$ will always ignite a spot fire. We consider three cases of a linear increase (I1), a step function (I2), and a smoothed out step function (I3).

Model I1:

$$E(m) := \begin{cases} m/\underline{m}, & m < \underline{m}; \\ 1, & m \geq \underline{m} \end{cases} \tag{A34}$$

Model I2:

$$E(m) = H(m - \underline{m}). \tag{A35}$$

Model I3:

$$E(m) = \left(1 + \text{erf}(Am - 2)\right)) /2. \tag{A36}$$

Here the function $\text{erf}(m)$ denotes the error function, related to the standard normal distribution. The parameter A measures the steepness of the transition region; the larger A, the steeper the transition.

In the most extreme fire burning situations, where the fuel beds are so dry that almost any landed mass will cause ignition, we can send $\underline{m} \to 0$ as an approximation.

Appendix A.7. Models for Firebrand Temperature

A model for firebrand temperature based on energy conservation is given in [22]. Let us define the density ρ, volume V, specific heat capacity at constant pressure c_p, and surface area A for a given firebrand. Then the temperature $T(t)$ appears in the energy conservation equation,

$$(\rho V c)\frac{DT}{Dt} = -A(q_{\text{conv}} + q_{\text{rad}}) \tag{A37}$$

where the terms q_{conv} and q_{rad}, appearing on the right hand side in the latter equation, represent the heat loss due to convection and radiation; the notation $\frac{DT}{Dt}$ represents the material derivative of T along the firebrand's trajectory. The convective heat flux can be modelled by Newton's law of cooling [22],

$$q_{\text{conv}} = h(T - T_a), \tag{A38}$$

where h is an average heat-transfer coefficient, and T_a is the ambient temperature. To determine the heat-transfer coefficient, we first introduce the average *Nusselt number* Nu,

$$Nu = 2 + 0.6Re^{1/2}Pr^{1/3}, \tag{A39}$$

where Pr the *Prandtl number* for air, and Re is the Reynolds number as described following Equation (A31) [22]. The Prandtl number is a dimensionless constant which measures the ratio of viscosity to thermal diffusivity. For example, when $P_r << 1$, heat diffuses quickly compared to momentum. For air, we have $Pr \approx 0.70$ [22]. Then the average heat transfer coefficient is given by:

$$h = \frac{k_{\text{air}} Nu}{2r}, \tag{A40}$$

where $k_{\text{air}} \approx 2.7 \times 10^{-2}\ m^2/s$ is the thermal conductivity of air, and r is the firebrand's radius.

The radiative heat flux can be approximated by the Stefan-Boltzmann law, which states:

$$q_{\text{rad}} = \sigma\epsilon(T^4 - T_a^4), \tag{A41}$$

where σ is the Stefan-Boltzmann constant, and ϵ is the emissivity. From experiment, $\epsilon = 0.9$.

It follows that the energy balance equation for the temperature evolution can be written:

$$\frac{dT}{dt} = -\frac{6}{\rho V c2\ r(t)}\left[h(T - T_a) + \sigma\epsilon(T^4 - T_a^4)\right]. \tag{A42}$$

This can be solved assuming the firebrand is initially at flaming temperature, e.g., $T(0) = 993$ K, to determine the temperature evolution once flaming combustion has halted when $m = m_c$.

References

1. Albini, F.A. *Spot Fire Distance From Burning Trees—A Predictive Model*; Technical Report; Intermountain Forest and Range Experiment Station, Forest Service, US Department of Agriculture: Ogden, UT, USA, 1979.
2. Clark, T.; Jenkins, M.A.; Coen, J.; Packham, D. A coupled atmospheric-fire model: Convective feedback on fire line dynamics. *J. Appl. Meteorol.* **1996**, *35*, 875–901.
3. Alexander, M.E.; Cruz, M.G. Evaluating a model for predicting active crown fire rate of spread using wildfire observations. *Can. J. For. Res.* **2006**, *36*, 3015–3028.
4. Cruz, M.G.; Alexander, M.E. Assessing crown fire potential in coniferous forests of western north america: A critique of current approaches and recent simulation studies. *Int. J. Wildland Fire* **2010**, *19*, 377–398.
5. Manzello, S.L.; Cleary, T.G.; Shields, J.R.; Yang, J.C. On the ignition of fuel beds by firebrands. *Fire Mater.* **2006**, *30*, 77–87.
6. University of California. *Man, Fire and Chapparal: A Conference on Southern California Wildland Research Problems*; University of California Conference Center at Lake Arrowhead, Lake Arrowhead, CA, USA, 1961.
7. Woycheese, J.P.; Pagni, P.J.; Liepmann, D. Brand propagation from large-scale fires. *J. Fire Protect. Eng.* **1999**, *10*, 32–44.
8. Schroeder, M.; Buck, C. *Fire Weather: A Guide for Application of Meteorological Information to Forest Fire Control Operations*; Forest Service, US Department of Agriculture: Washington, DC, USA, 1970.
9. Ellis, P.F. The Aerodynamic and Combustion Characteristics of Eucalypt Bark—A Firebrand Study. Ph.D. Thesis, Australian National University, Canberra, Australia, 2000.
10. Luke, R.H.; McArthur, A.G. *Bushfires in Australia*; Technical Report; Australian Government Publishing Service: Canberra, Australia, 1978.
11. McArthur, A.G. *Fire Behaviour in Eucalypt Forests*; Technical Report; Forestry and Timber Bureau: Canberra, Australia, 1967.
12. Koo, E.; Pagni, P.J.; Weise, D.R.; Woycheese, J.P. Firebrands and spotting ignition in large-scale fires. *Int. J. Wildland Fire* **2010**, *19*, 818–843.
13. Albini, F.A. Transport of firebrands by line thermals. *Combust. Sci. Technol.* **1983**, *32*, 277–288.
14. Johnson, E.A.; Miyanishi, K. *Forest Fires*; Academic Press: Waltham, MA, USA, 2001.
15. McIlveen, R. *Fundamentals of Weather and Climate*, 2nd ed.; Oxford University Press: Oxford, UK, 2010.
16. Muraszew, A.; Fedele, J.B.; Kuby, W.C. Trajectory of firebrands in and out of fire whirls. *Combust. Flame* **1977**, *30*, 321–324.
17. Clark, T.; Coen, J.; Latham, D. Description of a coupled atmosphere-fire model. *Int. J. Wildland Fire* **2004**, *13*, 49–63.

18. Clark, T.; Jenkins, M.A.; Coen, J.; Packham, D. A coupled atmospheric-fire model: Convective Froude number and dynamic fingering. *Int. J. Wildland Fire* **1996**, *6*, 177–190.
19. Finney, M.A. *Farsite: Fire Area Simulator—Model Development and Evaluation*; Technical Report; USDA Forest Service: Washington, DC, USA, 2004.
20. Tymstra, C. *Prometheus—The Canadian Wildland Fire Growth Model: Development and Assessment*; Technical Report; Canadian Forest Service, Northern Forestry Centre: Edmonton, AB, Canada, 2007.
21. Baum, H.R.; McCaffrey, B.J. *Fire Induced Flow Field—Theory and Experiment, Fire Safety Science*; Technical Report; Hemisphere: Washington, DC, USA, 1983.
22. Bhutia, S.; Jenkins, M.A.; Sun, R. Comparison of firebrand propagation prediction by a plume model and a coupled-fire/atmosphere large-eddy simulator. *J. Adv. Model. Earth Syst.* **2010**, doi:10.3894/JAMES.2010.2.4.
23. Sun, R.; Jenkins, M.A.; Krueger, S.K.; Mell, W.; Charney, J.J. An evaluation of fireplume properties simulated with the fire dynamics simulator (FDS) and the clark coupled wildfire model. *Can. J. For. Res.* **2006**, *36*, 2894–2908.
24. Sun, R.; Krueger, S.K.; Jenkins, M.A.; Zulauf, M.A.; Charney, J.J. The importance of fire-atmosphere coupling and boundary-layer turbulence to wildfire spread. *Int. J. Wildland Fire* **2009**, *18*, 50–60.
25. Sofiev, M.; Ermakova, T.; Vankevich, R. Evaluation of the smoke-injection height from wild-land fires using remote-sensing data. *Atmos. Chem. Phys.* **2012**, *12*, 1995–2006.
26. Kot, M.; Lewis, M.; van den Driessche, P. Dispersal data and the spread of invading organisms. *Ecology* **1996**, *77*, 2027–2042.
27. Bullock, J.M.; Kenward, R.E.; Hails, E.R.S. *Dispersal Ecology*, 1st ed.; British Ecological Society: London, UK, 2002.
28. Robbins, T. Seed Dispersal and Biological Invasion: A Mathematical Analysis. Ph.D. Thesis, University of Utah, Salt Lake City, UT, USA, 2003.
29. Martin, J. Mathematical Modelling and Analysis of a Model for Wildfire Spread Including Spotting. Ph.D. Thesis, University of Alberta, Edmonton, AB, Canada, 2013.
30. Hillen, T.; Greese, B.; Martin, J.; de Vries, G. Birth-jump processes, with applications to wildfire spotting. *J. Theor. Biol.* **2015**, *9* (Suppl. S1), 104–127.
31. Sardoy, N.; Consalvi, J.-L.; Porterie, B.; Fernandez-Pello, A.C. Modeling transport and combustion of firebrands from burning trees. *Combust. Flame* **2007**, *150*, 151–169.
32. Wang, H.H. Analysis on downwind distribution of firebrands sourced from a wildland fire. *Fire Technol.* **2011**, *47*, 321–340.
33. Boychuk, D.; Braun, W.J.; Kulperger, R.J.; Krougly, Z.L.; Stanford, D.A. A stochastic model for forest fire growth. *INFOR* **2008**, *45*, 9–16.
34. Forestry Canada Danger Group. *Development and Structure of the Canadian Forest Fire Behaviour Prediction System*; Technical Report; Forestry Canada: Edmonton, AB, Canada, 1992.
35. Kaur, I.; Mentrelli, A.; Bosseur, F.; Filippi, J.B.; Pagnini, G. Turbulence and fire-spotting effects into wild-land fire simulators. *Commun. Nonlinear Sci. Numer. Simul.* **2016**, *39*, 300–320.
36. Alexander, M.E.; Tymstra, C.; Frederick, K.W. *Incorporating Breaching and Spotting Considerations into Prometheus, the Canadian Wildland Fire Growth Model*; Technical Report; Foothills Model Forest: Hinton, AB, Canada, 2004.
37. Pagnini, G.; Mentrelli, A. Modelling wildland fire propagation by tracking random fronts. *Nat. Hazards Earth Syst. Sci.* **2014**, *14*, 2249–2263.
38. Mandel, J.; Beezley, J.D.; Kochanski, A.K. Coupled atmosphere-wildland fire modeling with WRF 3.3 and SFIRE 2011. *Geosci. Model Dev.* **2011**, *4*, 591–610.
39. Bova, A.S.; Mell, W.E.; Hoffman, C.M. A comparison of level set and marker methods for the simulation of wildland fire front propagation. *J. Appl. Meteorol.* **2015**, 25, 229–241.
40. Alexander, M.E. Crown Fire Thresholds in Exotic Pine Plantations of Australasia. Ph.D. Thesis, Australian National University, Canberra, Australia, 1998.
41. Ryu, S.D. (University of Alberta, Edmonton, AB, Canada). The seven types of spotting. Private communication, 2013.
42. Blackmarr, W.H. *Moisture Content Influences Ignitability of Slash Pine Litter*; Technical Report; USDA Forest Service South-Eastern Forest Experimental Station: Asheville, NC, USA, 1972.

43. Muraszew, A.; Fedele, J.B. *Firebrand Investigation*; Technical Report; The Aerospace Corporation: El Segundo, CA, USA, 1975.
44. Manzello, S.L.; Maranghides, A.; Mell, W.E. Firebrand generation from burning vegetation. *Int. J. Wildland Fire* **2007**, *16*, 458–462.
45. Manzello, S.L.; Maranghides, A.; Shields, J.R.; Mell, W.E.; Hayashi, Y. Mass and size distribution of firebrands generated from burning Korean pine trees. *Fire Mater.* **2009**, *33*, 21–31.
46. Gentaume, A.; Lampin-Maillet, C.; Guijarro, M.; Hernando, C.; Jappiot, M.; Fonturbel, T.; Perez-Gorostiaga, P.; Vega, J. Spot fires: Fuel ammability and capability of firebrands to ignite fuel beds. *Int. J. Wildland Fire* **2009**, *18*, 951–969.
47. Gentaume, A. Laboratory characterization of firebrands involved in spot fires. *Ann. For. Sci.* **2011**, *68*, 531–541
48. Ellis, P.F. Fuelbed ignition potential and bark morophology explain the notoriety of eucalypt messmate 'stringybark' for intense spotting. *Int. J. Wildland Fire* **2011**, *20*, 897–907.
49. Okubo, A.; Levin, S.A. *Diffusion and Ecological Problems: Modern Perspectives*, 2nd ed.; Springer-Verlag: New York, NY, USA, 2002.
50. Evans, L.C. *Partial Differential Equations*; AMS Press: New York, NY, USA, 1998.
51. Muraszew, A. *Firebrand Phenomena*; Technical Report; The Aerospace Corporation: El Segundo, CA, USA, 1974.
52. Asensio, M.I.; Ferragut, L. On a wildland fire model with radiation. *Int. J. Numer. Meth. Eng.* **2002**, *54*, 135–157.
53. Mandel, J.; Bennethum, L.S.; Beezley, J.D.; Coen, J.L.; Douglas, C.C.; Kim, M.; Vodacek, A. A wildland fire model with data assimilation. *Math. Comput. Simul.* **2008**, *79*, 584–606.
54. Ellis, P.F. Spotting and firebrand behaviour in dry eucalypt forest and the implications for fuel management in relation to fire suppression and to "ember" (firebrand) attack on houses. In Proceedings of the 3rd International Wildland Conference, Sydney, Australia, 3–6 October 2003.
55. Gould, J.S.; Cheney, N.P.; McCaw, L. Project Vesta-Research into the effects of fuel structure and fuel load on behaviour of moderate to high-intensity fires in dry eucalypt forest: Progress Report. In Proceedings of the Australasian Bushfire Conference, Christchurch, New Zealand, 3–6 July 2001; pp. 104–127.
56. Chandler, C.C. *Fire Behaviour of the Basin Fire, Sierra National Forest*; Technical Report; USDA Forest Service Pacific Southwest Forest and Range Experiment Station: Berkeley, CA, USA, 1961.
57. Butler, B.W.; Bartlette, R.A.; Bradshaw, L.S.; Cohen, J.D.; Andrews, P.L.; Putnam, T.; Mangan, R.J. *Fire Behaviour Associated with the 1994 South Canyon Fire on Storm King Mountain, Colorado*; Technical Report; USDA Forest Service Rocky Mountain Research Station: Asheville, NC, USA, 1998.
58. Butler, B.W.; Reynolds, T.D. *Wildfire Case Study: Butte City Fire, Southeastern Idaho, July 1, 1994.* Technical Report; USDA Forest Service Intermountain Research Station: Asheville, NC, USA, 1997.
59. Baden, W.; Klem, T.; Teague, P.E. *The Oakland/Berkeley Hills Fire*; Technical Report, National Wildland/Urban Interface Fire Protection Initiative: Quincy, MA, USA, 1991.
60. Martin, J.; Coville, J.; Hamel, F. Generalized transition waves for nonlocal equations with periodic heterogeneous dispersal. Unpublished work, 2016.
61. Quintiere, J.G. *Fundamentals of Fire Phenomena*; John Wiley and Sons Ltd.: Hoboken, NJ, USA, 2006.
62. Ai, S.; Huang, W. Travelling wavefronts in combustion and chemical reaction models. *Proc. R. Soc. Edinb.* **2007**, *137A*, 671–700.
63. Cunningham, P.; Goodrick, S.L.; Hussaini, M.Y.; Linn, R.R. Coherent vertical structures in numerical simulations of buoyant plumes from wildland fires. *Int. J. Wildland Fire* **2005**, *14*, 61–75.
64. Cunningham, P.; Linn, R.R. Numerical simulations of grass fires using a coupled atmosphere-fire model: Dynamics of fire spread. *J. Geophys. Res.* **2007**, *112*, doi:10.1029/2006JD007638.
65. Mercer, G.N.; Weber, R.O. Plumes above line fires in a cross wind. *Int. J. Wildland Fire* **1994**, *4*, 201–207.
66. Nmira, F.; Consalvi, J.J.; Boulet, P.; Porterie, B. Numerical study of wind effects on the characteristics of flames from non-propagating vegetation fires. *Fire Saf. J.* **2010**, *45*, 129–141.
67. Porterie, P.; Louraud, J.C.; Morvan, D.; Larini, M. A numerical study of buoyant plumes in cross-flow conditions. *Int. J. Wildland Fire* **1999**, *9*, 101–108.
68. Sanchez, C.T.; del Notario, P.P.; Moreno, F.G. On the flight paths and lifetimes of burning particles of wood. *Symp. Int. Combust.* **1967**, *10*, 1021–1037.

69. Tse, S.D.; Fernandez-Pello, A.C. On the flight paths of metal particles and embers generated by power lines in high winds—A potential source of wildland fires. *Fire Saf. J.* **1998**, *30*, 333–356.
70. Kanuri, A.M. *Introduction to Combustion Phenomena*; Gordon and Breach Science: New York, NY, USA, 1995.
71. McCaffrey, B. Momentum implications for buoyant dffusion flames. *Combust. Flames* **1983**, *52*, 149–167.
72. Linn, R.; Winterkamp, J.; Edminster, C.; Colman, J.J.; Smith, W.S. Coupled influences of topography and wind on wildland fire behaviour. *Int. J. Wildland Fire* **2007**, *16*, 183–195.
73. Zdunkowski, W.; Bott, A. *Dynamics of the Atmosphere: A Course in Theoretical Meteorology*; Cambridge University Press: Cambridge, UK, 2003.
74. Lynch, A.H.; Cassano, J.J. *Applied Atmospheric Dynamics*; John Wiley and Sons: Hoboken, NJ, USA, 2006.
75. Taylor, M.E. *Partial Differential Equations I: Basic Theory*; Springer-Verlag: New York, NY, USA, 1996.

applied sciences

MDPI

Article

Validation of a Mathematical Model for Green Algae (*Raphidocelis Subcapitata*) Growth and Implications for a Coupled Dynamical System with *Daphnia Magna*

Michael Stemkovski [1], Robert Baraldi [1], Kevin B. Flores [1,2,3,*] and H.T. Banks [1,3]

1 Center for Research in Scientific Computation, North Carolina State University, Raleigh, NC 27695-8212, USA; mstemko@ncsu.edu (M.S.); rjbarald@ncsu.edu (R.B.); htbanks@ncsu.edu (H.T.B.)
2 Center for Quantitative Sciences in Biomedicine, North Carolina State University, Raleigh, NC 27695-8212, USA
3 Department of Mathematics, North Carolina State University, Raleigh, NC 27695-8212, USA
* Correspondence: kbflores@ncsu.edu; Tel.: +1-919-513-0821

Academic Editor: Yang Kuang
Received: 31 March 2016; Accepted: 5 May 2016; Published: 18 May 2016

Abstract: Toxicity testing in populations probes for responses in demographic variables to anthropogenic or natural chemical changes in the environment. Importantly, these tests are primarily performed on species in isolation of adjacent tropic levels in their ecosystem. The development and validation of coupled species models may aid in predicting adverse outcomes at the ecosystems level. Here, we aim to validate a model for the population dynamics of the green algae *Raphidocelis subcapitata*, a planktonic species that is often used as a primary food source in toxicity experiments for the fresh water crustacean *Daphnia magna*. We collected longitudinal data from three replicate population experiments of *R. subcapitata*. We used this data with statistical model comparison tests and uncertainty quantification techniques to compare the performance of four models: the Logistic model, the Bernoulli model, the Gompertz model, and a discretization of the Logistic model. Overall, our results suggest that the logistic model is the most accurate continuous model for *R. subcapitata* population growth. We then implement the numerical discretization showing how the continuous logistic model for algae can be coupled to a previously validated discrete-time population model for *D. magna*.

Keywords: algae growth models; uncertainty quantification; asymptotic theory; bootstrapping; model comparison tests; *Raphidocelis subcapitata*; *Daphnia magna*

1. Introduction

Studies of the population dynamics of phytoplankton and their zooplankton predators in lentic habitats have found a variety of patterns. Plankton communities have been observed to either oscillate in low or high amplitude cycles or to remain relatively stable throughout the year [1]. The same lake may exhibit stability on a given year but switch to oscillation during the following year, and vice versa. A variety of explanations have been proposed for this behavior in the field, including predator-prey interactions, temperature fluctuations, and external influences on nutrient content [2–4]. Fewer studies have attempted to answer the question of what drives these population dynamics in the laboratory setting. Of the predator-prey models that have been proposed for plankton communities, most do not consider certain elements of zooplankton biology such as density-dependent mortality or age-specific fecundity. These traits are crucial for describing the population growth of zooplankton such as *Daphnia magna* [5] and may be an important factor when modeling population dynamics that are observed in lakes and other ecosystems.

The aim of the present study is to validate and parametrize continuous models for the growth of green algae (*Raphidocelis subcapitata*) in the absence of predation by zooplankton. The broader goal is to couple a validated green algae model with a validated discrete-time population model for *Daphnia magna* [5], and ultimately with an analogous continuous-time model, in order to investigate the possibility of oscillations in the laboratory setting similar to those found in lentic environments. In our previous study of *D. magna* population dynamics [5], we carried out laboratory experiments in which green algae were fed to *D. magna* populations on a daily basis, and populations were reared in media optimized for daphnia survival, but not necessarily ideal for algae growth. In particular, it was not known whether green algae could proliferate in daphnia media to an extent that would affect daphnia population dynamics. Thus, it is of central importance to quantitate the rate at which green algae grow in daphnia media and whether this growth has the potential to significantly alter the fecundity and survival of daphnia. In theory, such changes would thereby affect the quantification of population level risk assessments involving experimental exposure of daphnia populations to environmental perturbations, e.g., toxins or temperature change.

We tested several commonly used growth models for organisms with a limiting nutrient: the Gompertz, the Logistic (continuous and discretized), and the Bernoulli population models. We note that the Bernoulli model is a generalization of the continuous Logistic model, which allows for nested model comparison. Each model has been used to describe populations across many scenarios associated with saturated growth processes in biology. We collected experimental data from three replicates of green algae grown in isolation of predation. We describe goodness of fit of several mathematical models for green algae growth in the context of asymptotic theory and bootstrapping techniques. We provide estimated parameter values and computed confidence intervals for the model predictions. Finally, we implement a numerical scheme that can be used to approximate the concentration of green algae on a daily time scale in order to combine the continuous green algae model that we had the most confidence in with a discrete time population model for *Daphnia magna*. We performed simulations of an unvalidated coupled green algae and daphnia model in order to explore the possible effects of green algae growth on daphnia population dynamics.

2. Data and Methods

2.1. Data

To observe the growth of *Raphidocelis subcapitata* populations (previously known as *Pseudokirchneriella subcapitata* and *Selenastrum capricornutum*), we seeded three beakers containing 1 L of media reconstituted from deionized water for *Daphnia magna* culturing (previously described in [6]) and recorded the population density for eight days. Each population was kept in an incubator (Thermo Fisher Scientific, Waltham, MA, USA) at 20 °C on a 16/8 h light/dark cycle (6 AM–10 PM light, 10 PM–6 AM dark). Media lost to evaporation was replaced daily with deionized water in order to retain a 1 L volume and avoid replenishing nutrients. The 1 L algae cultures were uncovered and inspected for contamination during measurements. We selected a seeding concentration of 7×10^7 cells based on previous studies of algal growth in order to observe both the early (growth) and late (saturation) stage dynamics of the population [7–9]. We measured the density of each population replicate twice using a hemocytometer (Hausser Scientific, Horsham, PA, USA) at 9 AM, 3 PM, and 9 PM daily in order to obtain sufficient data points for parameter estimation and uncertainty quantification. The two measurements of density at each time point for each replicate were averaged to minimize human measurement error. This yielded a total of 23 data points for each of the three replicates.

2.2. Asymptotic Theory

The goal of this paper is to determine the most accurate model for algae growth in the absence of consumption. The uncertainty in parameters for each model can be quantified via asymptotic theory. In this section we provide the theory behind our asymptotic theory methodology. The estimation of parameters using asymptotic theory requires mathematical models of the form

$$\begin{aligned} \frac{dy}{dt} &= g(t, y(t), q), \\ y(t_0) &= y_0, \end{aligned} \tag{1}$$

and the corresponding observation process

$$f(t, \theta) = Cy(t, \theta), \tag{2}$$

where $\theta = (q, \tilde{y_0}) \in \mathbb{R}^{p+\tilde{p}}$ is the vector of unknown parameters, q is a vector of p model parameters, $\tilde{y_0}$ is the number $\tilde{p}$ of initial conditions that is unknown, and C maps the model solution $y(t, \theta)$ in $\mathbb{R}^l$ to the observed states $f(t, \theta)$. We consider the initial condition to be unknown because of measurement error. In this investigation, the observation operator will always produce a scalar, and thus C maps $\mathbb{R}^l$ to $\mathbb{R}$. In fact, in all our considerations we have $\tilde{p} = l = 1$, *i.e.*, the models are scalar and $C = I$.

Due to the discrete nature of our experimental data, the observations for our statistical error model occur at $n = 23$ discrete times t_i. Thus, the observations will be

$$f(t_i, \theta) = Cy(t_i, \theta), i = 1, \ldots, n. \tag{3}$$

To account for measurement error, we use the statistical model

$$Y_i = f(t_i, \theta_0) + \mathcal{E}_i, i = 1, \ldots, n \tag{4}$$

for our observations, where $\mathcal{E}_i$ is a zero mean random variable representing identically, independently distributed (*i.i.d.*) noise that causes our observed data to deviate from our model solution, and θ_0 is the hypothesized "true" or "nominal" parameter vector that generates the observations $\{Y_i\}_{i=1}^n$. The existence of this "true" parameter vector θ_0 is a standard assumption in frequentist statistical formulations. The *i.i.d.* nature of the error in our model implies that $\mathbb{E}(\mathcal{E}_i) = 0$ for each i, and that $\mathcal{E}_i = 1, \ldots, n$, are identically distributed with variance σ_0^2.

Since $\mathcal{E}_i$ is a random variable, Y_i is a random variable with corresponding realizations y_i. Asymptotic theory seeks to estimate θ_0 by creating a random variable Θ whose realizations for a given data set y_i will be estimates $\hat{\theta}$ of θ_0. These estimates $\hat{\theta}$ will approximate the true parameters θ_0, and are obtained by minimizing the ordinary least squares (OLS) cost functional [10,11]

$$S(Y; \theta) = \sum_{i=1}^{n} [Y_i - f(t_i, \theta)]^2, \tag{5}$$

where $\mathbf{Y} = (Y_1, Y_2, \ldots, Y_n)^T$. Thus, with Ω being the space of admissible parameters and y_i being the realizations of the random variable Y_i,

$$\theta_0 \approx \hat{\theta}_{OLS}^n = \underset{\theta \in \Omega}{\operatorname{argmin}} \sum_{i=1}^{n} [y_i - f(t_i, \theta)]^2 \tag{6}$$

provides an estimate for θ_0. The process of estimating parameters from data is known as an inverse problem, and all inverse problems in this experiment are computed using `fmincon` in MATLAB (Mathworks, 2015b, Natick, MA, USA, 2015) with function and step tolerances of 10^{-20} and 1000 iterations.

Once we have an estimate $\hat{\theta}^n_{OLS}$, we wish to ascertain the statistical properties of the estimator Θ. Although we do not know the distribution of the estimator Θ^n_{OLS}, we can approximate it under *asymptotic theory (as $n \to \infty$)* by the multivariate Gaussian distribution [10–12]

$$\Theta^n_{OLS} \sim N(\theta_0, \Sigma^n_0) \tag{7}$$

where, based on previous assumptions, the covariance matrix Σ^n_0 is approximated by

$$\Sigma^n_0 \approx \hat{\Sigma}^n = \hat{\sigma}^2_{OLS}[\chi^{nT}(\hat{\theta})\chi^n(\hat{\theta})]^{-1}. \tag{8}$$

Here χ^n is the sensitivity matrix

$$\chi^n_{jk}(\theta) = \frac{\partial f(t_i, \theta)}{\partial \theta_k}, i = 1, \dots, n; k = 1, \dots, p, \tag{9}$$

where θ_k is the kth component of the vector $\theta \in \mathbb{R}^{1\times p}$. The unbiased estimator for σ^2_0 is

$$\hat{\sigma}^2_{OLS} = \frac{1}{n-p}\sum_{i=1}^{n}[y_i - f(t_i, \hat{\theta}^n_{OLS})]^2 \tag{10}$$

where, for our own examples, $n = 23$ and $p = 3$ or 4 is the number of model parameters. Both $\hat{\theta}^n_{OLS}$ and $\hat{\sigma}^2_{OLS}$ are then used in Equation (8) (*i.e.*, $\hat{\theta} = \hat{\theta}_{OLS}$ and $\hat{\sigma}^2 = \hat{\sigma}^2_{OLS}$).

In our calculations, the sensitivity equations are calculated analytically by solving the differential equation at $\hat{\theta}$

$$\frac{d}{dt}\left(\frac{\partial y}{\partial \theta}\right) = \frac{\partial g}{\partial y}\frac{\partial y}{\partial \theta} + \frac{\partial g}{\partial \theta}. \tag{11}$$

Note that $\frac{dy}{dt} = g(t, y(t), \hat{\theta})$ is the differential equation for the green algae model and $f(t_j, \hat{\theta}) = y(t_j, \hat{\theta})$ is the forward solution of each model. Because we know analytical formulas that provide solutions for $\frac{dy}{dt} = g(t, y(t), \hat{\theta})$, we can solve Equation (11) by setting up a differential equation in terms of the sensitivity [11].

The χ^n matrix provides a measure for how sensitive the mathematical model is to each of its parameters. This can be used to estimate the $p \times p$ covariance matrix, Σ^n_0,

$$\Sigma^n_0 \approx \hat{\Sigma}^n = \hat{\sigma}^2[\chi^{nT}(\hat{\theta})\chi^n(\hat{\theta})]^{-1}. \tag{12}$$

In order to determine the confidence we have in the parameter estimates, we also compute the asymptotic theory based standard error $SE(\hat{\theta}_k) = \sqrt{\hat{\Sigma}^n_{kk}}$ for the kth parameter.

2.3. Boostrapping

We implemented bootstrapping techniques to complement our asymptotic theory approach with regards to estimating parameter uncertainty. We again assume that we are have experimental data for a dynamical system from an underlying observation process in Equation (4) where $\tilde{\mathcal{E}}_i$ are *i.i.d.* with mean zero and constant variance σ^2_0 and θ_0 is the "true value" hypothesized to exist in statistical treatments of data [10]. The random variable also has realizations

$$y_i = f(t_i, \theta_0) + \tilde{\epsilon}_i. \tag{13}$$

We use the following algorithm [10] to compute the bootstrapping estimate $\hat{\theta}_{BOOT}$ of θ_0 and its empirical distribution.

1. First estimate $\hat{\theta}^0$ from the entire sample $\{y_i\}_{i=1}^n$ using OLS.
2. Using this estimate, define the standardized residuals

$$\bar{r}_i = \sqrt{\frac{n}{n-p}}(y_i - f(t_i, \hat{\theta}^0)) \tag{14}$$

for $i = 1, \ldots, n$, where n is the number of data points, and p are the number of model parameters. Set $m = 0$, which will represent the total number of artificial samples we will create.
3. Create a bootstrapping sample of size n using random sampling with replacement from the data (realizations) $\{\bar{r}_1, \ldots, \bar{r}_n\}$ to form a bootstrapping sample $\{\bar{r}_1^m, \ldots, \bar{r}_n^m\}$.
4. Create bootstrap sample points

$$y_i^m = f(t_i, \hat{\theta}^0) + r_i^m \tag{15}$$

for $i = 1, \ldots, n$.
5. Obtain a new estimate $\hat{\theta}^{m+1}$ from the bootstrapping sample $\{y_i^m\}$ using OLS.
6. Set $m = m + 1$ and repeat steps 3–5 until $m \geq 1000$ (this can be any large value, but for these experiments we used $M = 1000$).

We then calculate mean, standard error, and confidence intervals using the formulas:

$$\hat{\theta}_{BOOT} = \frac{1}{M}\sum_{m=1}^{M} \hat{\theta}^m, \tag{16}$$

$$Var(\theta_{BOOT}) = \frac{1}{M-1}\sum_{m=1}^{M} (\hat{\theta}^m - \hat{\theta}_{BOOT})^T(\hat{\theta}^m - \hat{\theta}_{BOOT}), \tag{17}$$

$$SE_k(\hat{\theta}_{BOOT}) = \sqrt{Var(\theta_{BOOT})_{kk}}, \tag{18}$$

where θ_{BOOT} denotes the bootstrapping estimator. We present the results of these techniques as standard errors about the mean of the parameter estimates, as well as the parameter distributions created. This procedure is performed for each replicate in our experiments.

2.4. Model Comparison: Nested Restraint Tests

We used nested model comparison tests to determine the loss in accuracy by constraining certain models, *i.e.*, holding some parameters constant. In general, we assume that we have an inverse problem for the model observations $f(t, \theta)$ and are given n observations with the cost function described above in Equation (5). We are interested in using data to question whether the "true" parameter θ_0 can be found in a subset $\Omega_H \subset \Omega$, for which we make the same assumptions as Banks, Hu, and Thompson [10]. Thus, we want to test the null hypothesis H_0:$\theta_0 \in \Omega_H$, or that the constrained model provides an adequate fit to the data. We then define

$$\Theta_H^n(Y) = \underset{\theta \in \Omega_H}{\operatorname{argmin}}\, S^n(Y; \theta) \tag{19}$$

and

$$\hat{\theta}_H^n(y) = \underset{\theta \in \Omega_H}{\operatorname{argmin}}\, S^n(y; \theta), \tag{20}$$

where y is a realization of Y. It is important to note that $S^n(y; \hat{\theta}_H^n) \geq S^n(y; \hat{\theta}^n)$. We define the nonnegative test statistics and their realizations, respectively, by

$$T_n(Y) = S^n(Y; \hat{\theta}_H^n) - S^n(Y; \hat{\theta}^n) \tag{21}$$

and

$$\hat{T}_n = T_n(y) = S^n(y; \hat{\theta}_H^n) - S^n(y; \hat{\theta}^n). \tag{22}$$

We refer to [10] for a description of asymptotic convergence as $n \to \infty$, which yields the model comparison result

$$U_n(Y) = \frac{nT_n(Y)}{S^n(Y;\theta^n)} \tag{23}$$

with the corresponding realizations

$$\hat{u}_n = U_n(y) \tag{24}$$

which can be compared to a χ^2 distribution with r degrees of freedom. In this project we use a $\chi^2(1)$ table when comparing the results from the Logistic model to those from the Bernoulli model.

2.5. Akaike Information Criterion

In some cases (such as comparison between the Logistic and the Gompertz), the models are not nested (although they are related through a limiting process–see below) and hence we cannot use the model comparison tests outlined above. However, we can use an alternative model evaluation framework and implement the Akaike Information Criterion (AIC_c) with a small size sample correction [10] in the context of an ordinary least squares framework

$$AIC_c = n[1 + \ln(2\pi)] + n\ln\left(\frac{\sum_{i=1}^{n}(y_i - f(t_i, \hat{\theta}^n_{OLS}))^2}{n}\right) + 2(p+1) + \frac{2p(p+1)}{n-p-1} \tag{25}$$

where n is the sample size and p is the number of unknowns (parameters). This will allow us to suggest which model provides a better fit for the data (models with smaller AIC_c values provide better fits). While other goodness of fit tests may be useful for selecting models, we chose to use AIC_c, since it is a widely adopted measure of model accuracy (see Sections 4.2, 4.3 of [10]).

3. Models

3.1. Logistic Model

The first model we consider is the widely used logistic model for bounded growth of a population $P(t)$, given by the differential equation

$$\frac{dP}{dt} = RP(t)\left(1 - \frac{P(t)}{K}\right) \tag{26}$$

where R is the intrinsic growth rate, and K is the carrying capacity for the population under consideration.

3.2. Bernoulli Model

We also analyze the data within the context of a Bernoulli model due to Richards [13], given by the differential equation

$$\frac{dP}{dt} = RP(t)(1 - (\frac{P(t)}{K})^{\beta}). \tag{27}$$

The Bernoulli model has three model parameters, R, K, and β. The parameter β extends the logistic model to allow flexibility in the growth dynamics by allowing the inflection point to change while keeping the carrying capacity approximately the same [13]. Setting $\beta = 1$ yields the logistic model in Equation (26); hence, the logistic model is a special case of the Bernoulli model, thereby enabling us to use nested model comparison tests described above.

3.3. Gompertz Growth Model

The next model we consider is the Gompertz growth model, which is widely used for biological and economic phenomena where population growth is not symmetric about the point of inflection, *i.e.*, growth rates are time dependent. The differential equation form of this model is

$$\frac{dP}{dt} = \kappa P(t)(\log(K) - \log(P(t))) = \kappa P(t)\log(\frac{K}{P(t)}), \tag{28}$$

where K is the carrying capacity and κ scales the time. For both the Logistic and Gompertz models, we let X_0 represent the initial condition, *i.e.*, $P(t_0) = X_0$.

The Logistic and Gompertz models, while not nested, are related through a limiting process. Since

$$\lim_{\nu\to\infty} \nu(1 - (\frac{P(t)}{K})^{\frac{1}{\nu}})) = -\log(\frac{P(t)}{K}), \tag{29}$$

we find that the Gompertz model is the limit as $\nu \to \infty$ of the *generalized* logistic model for $\nu > 0$

$$\frac{dP}{dt} = \nu\kappa P(t)(1 - (\frac{P(t)}{K})^{\frac{1}{\nu}}). \tag{30}$$

3.4. Logistic Model: Numerical Discretization

Another model we consider is a discrete numerical approximation of the Logistic model. We note that the continuous models described above were simulated using the `ode45` algorithm in Matlab. In order to ensure that the logistic model can be coupled to a discrete time model for a *D. magna* population in which the population size is updated once per day [5], we investigated the logistic model using a forward Euler scheme that was discretized on an hour basis. In this paper, we refer to this discrete Euler-method logistic model as the DEL model. This numerically discretized logistic model is given by the difference equation

$$P_{t+1} = P_t + RP_t\left(1 - \frac{P_t}{K}\right) \tag{31}$$

where P_t is the population at time t hours and P_{t+1} is the population at the next time step. The parameters R and K are analogous to those of the continuous Logistic model and can be interpreted as the intrinsic population growth rate and carrying capacity, respectively. We refer to the initial population, $P_{t=0}$, as X_0 in our results and data fitting procedure.

4. Results

4.1. Data fitting and Model Comparisons

Overall, we found that all models provide a reasonable fit to the data. Figures 1–4 show results of the least squares estimation for the three different replicates of the Logistic, Bernoulli, DEL, and Gompertz models. These figures contain 68% and 95% confidence bands around the fits to data. These were constructed by generating 1000 random parameter sets from a normal distribution described by the mean and standard error obtained by the asymptotic theory results, computing the model for each of these parameter sets, and then calculating the respective confidence intervals from model generated points $f(t_i, \theta^k)$, where $k = 1, \ldots, 1000$ [5]. One primary difference between the fits to data that we found was that each model tends to underestimate the initial data and the Bernoulli model provided the closest fit (Figure 4). We note that we chose to estimate the initial condition due to measurement error associated with a low cell density as well as how much influence these discrepancies in error would affect the outcome of the model.

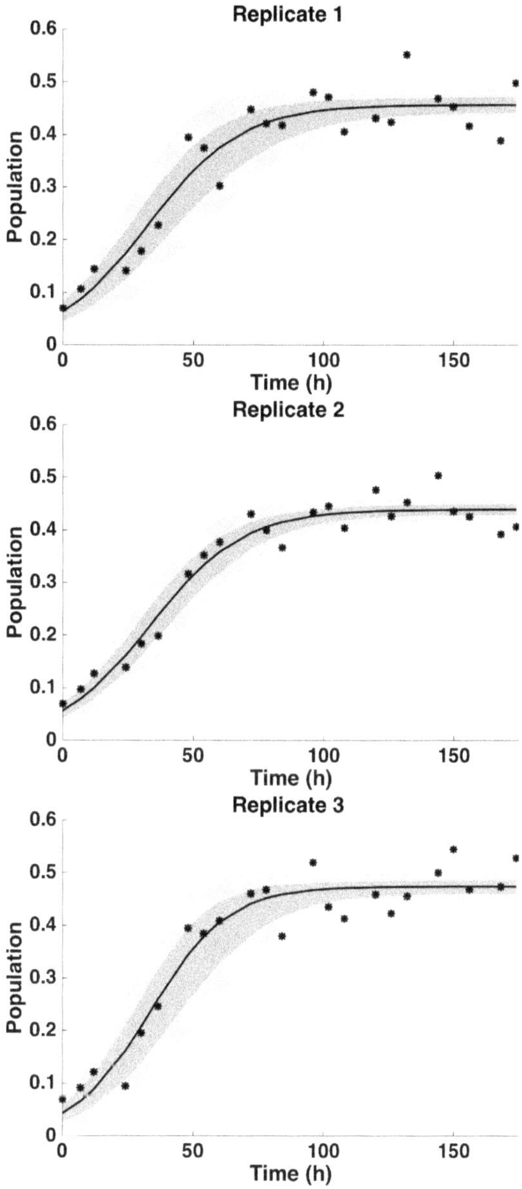

Figure 1. Plots of forward solutions for the Logistic curve for the three replicates of the data. Replicate one is on top and three is on the bottom. The lighter and darker shades of grey represent the 95% and 68% confidence bars on the model solution, respectively. The algae population is represented as $cells \times 10^7/\text{L}$. Data points are shown as "*".

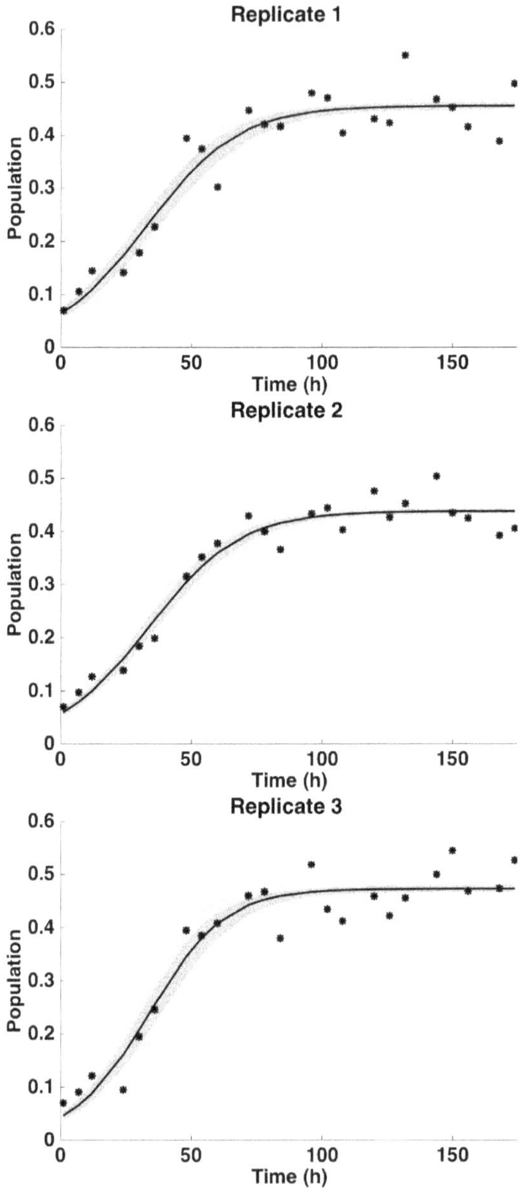

Figure 2. Plots of forward solutions for the discrete Euler-method logistic (DEL) please confirm. model for the three replicates of the data from left to right. Replicate one is on top and three is on the bottom. The lighter and darker shades of grey represent the 95% and 68% confidence bars on the model solution, respectively. The algae population is represented as $cells \times 10^7/\mathrm{L}$. Data points are shown as "*".

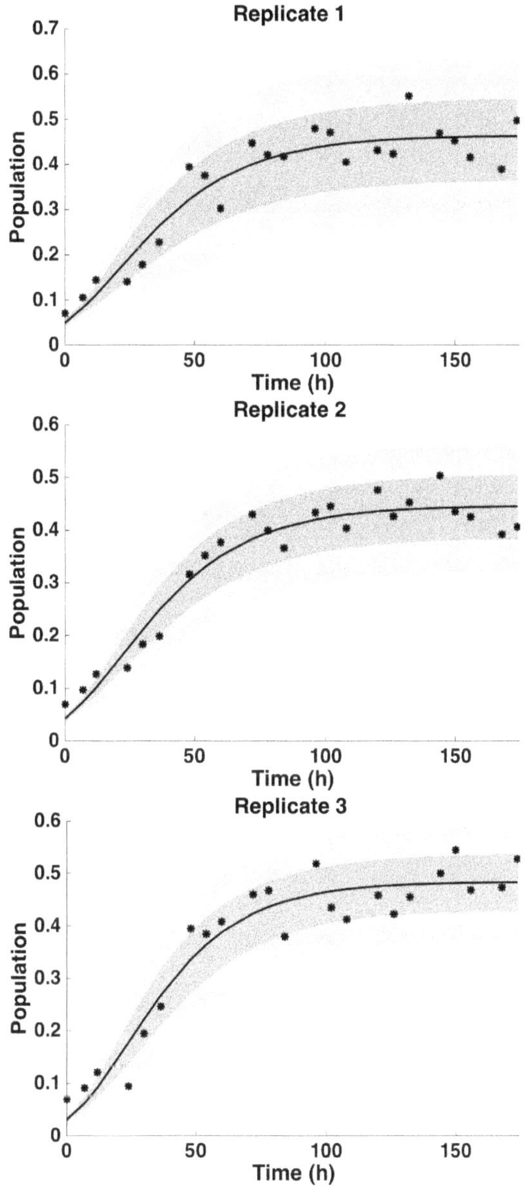

Figure 3. Plots of forward solutions for the Gompertz curve for the three replicates of the data. Replicate one is on top and three is on the bottom. The lighter and darker shades of grey represent the 95% and 68% confidence bars on the model solution, respectively. The algae population is represented as $cells \times 10^7/\text{L}$. Data points are shown as "*".

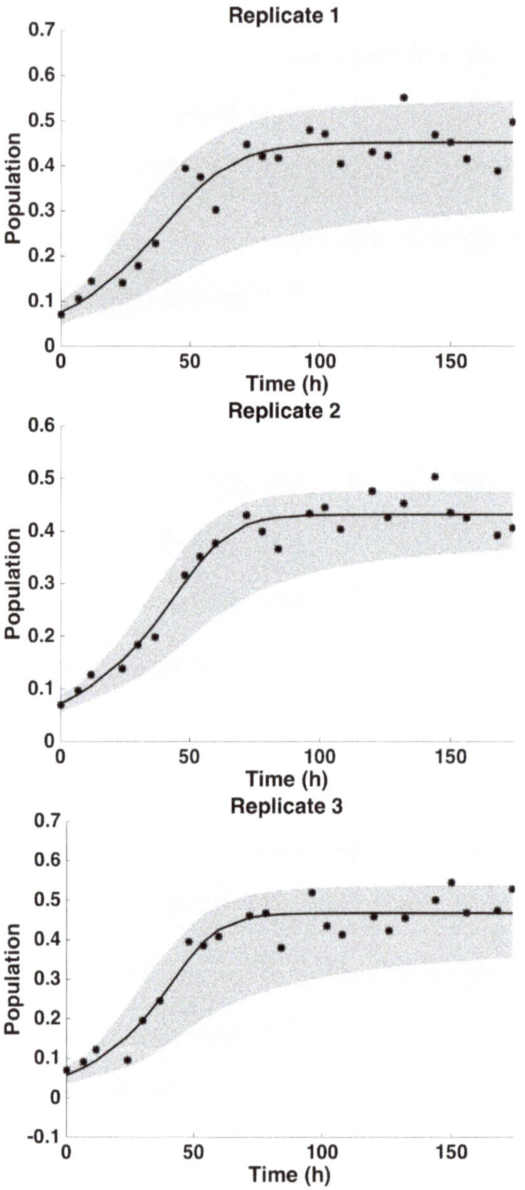

Figure 4. Plots of forward solutions for the Bernoulli curve for the three replicates of the data. Replicate one is on top and three is on the bottom.The lighter and darker shades of grey represent the 95% and 68% confidence bars on the model solution, respectively. The algae population is represented as $cells \times 10^7/\text{L}$. Data points are shown as "*".

The results in Table 1 show the small sample corrected Akaike Information Criterion (AIC_c) scores based on Equation (25) for each replicate and each model. These results suggest that the discrete and continuous logistic population models are better able to describe the green algae growth data than either the Gompertz or Bernoulli models, although the differences in AIC_c values were small between all models. The results of the model comparison test performed between the continuous Logistic model and the Bernoulli model are given in Table 2. In addition to fixing β to 1 to reduce the Bernoulli model to the Logistic model, we also fixed the initial condition X_0 at the seed value to enunciate the model improvement with an unalterable initial condition. Of all comparisons, only the Logistic model with X_0 fixed on replicates 2 and 3 yielded a significant result at the $\alpha = 0.9$ confidence level (or $\chi^2(1)_{.9} = 2.706$). This indicates that, in general, the OLS cost associated with the Bernoulli model was significantly improved by fixing β and reducing it to the Logistic model. However, restricting the Logistic model further by fixing X_0 does not significantly affect the cost of the Bernoulli model. The Logistic model also has benefits with regards to identifiability, which will be seen in subsequent passages of this document.

Table 1. Corrected Akaike Information Criterion scores for each model and replicate.

Replicate	Gompertz	Logistic	Bernoulli	Discrete Euler-Method Logistic (DEL)
1	−69.4203	−71.5919	−69.2189	−72.6155
2	−84.2435	−89.0016	−89.3905	−90.4114
3	−71.3972	−74.2560	−72.4414	−75.3515

Table 2. Model comparison $\hat{u}_n$ scores for the continuous Logistic model compared to Bernoulli model for each replicate. We also chose to fix the initial condition X_0 at the seed population value to enunciate model improvement if X_0 was unalterable. Note that values less than 2.706 indicate the restricted model is better.

	Bernoulli Restricted to:	Bernoulli Restricted to:
Replicate	Logistic	Logistic with X_0 fixed
1	0.5935	0.7233
2	2.4718	3.6216
3	1.1733	3.4118

4.2. Uncertainty Analysis

We compared results of parameter estimation and multiple uncertainty quantification techniques (asymptotic theory and bootstrapping) for the Logistic, Gompertz, and Bernoulli growth models, as well as the numerically discretized version of the Logistic model (DEL model), since each of them provided reasonable fits to the data. We first note that the usual assumption of *i.i.d.* residuals required for uncertainty analysis held for all models investigated (Supplementary Figures S1–S4) Although methods involving autocorrelation on residuals may be used to investigate the *i.i.d.* assumption, we investigated this assumption by visually inspecting residual plots, since there were not enough data to perform autocorrelation tests. Visual inspection of residual plots is a commonly used procedure when not enough longitudinal data are available (see [10], Section 3.6). We also note that the normality assumption for the parameter distributions in asymptotic theory was confirmed by our bootstrapping results in all but the Bernoulli model (Supplementary Figures S5–S17). We divide our analysis of the results from uncertainty quantification among the sets of parameters with similar interpretations for each model. For example, each model has an initial condition, a growth rate, and a parameter governing the saturation of growth due to population density.

4.2.1. Uncertainty Analysis: Initial Condition

The bootstrapping distribution results of X_0 estimation are presented in Supplementary Figures S5–S8. These appear to be approximately normally distributed, with some exceptions occurring where the estimates are close to the zero boundary. Bootstrapping estimates of uncertainty for X_0

are compared to asymptotic theory in Supplementary Tables S1–S4. We observed that the parameter estimates for X_0 for all replicates were lowest in the Gompertz model and highest in the Bernoulli model. The standard errors also varied between the asymptotic and bootstrapping techniques depending on the model. For example, the order of magnitude for the standard errors from bootstrapping was greater than asymptotic theory for the Gompertz model and the DEL model.

4.2.2. Uncertainty Analysis: Growth Rate

Each model that we investigated has a parameter that describes the population growth rate (Logistic and DEL: R, Gompertz κ, Bernoulli R). Because numerical estimates for the growth rate parameters will not be equal across models, we analyzed their consistency and uncertainty across replicate data sets within the same model. The bootstrapping distributions for the growth rates for each model were normally distributed except for the Bernoulli model, which was skewed to the right (Supplementary Figures S9–S12). We postulate that one reason for this skewness may be that the Bernoulli model is over-parameterized. Similarly, the standard errors computed for the growth rate within each model were of reasonable size and of the same order of magnitude except for the Bernoulli model (Supplementary Tables S5–S8). The growth rate estimates for the logsitic model differed between the continuous version and its numerical discretization using the euler method (DEL model). Specifically, the growth rate estimates for the DEL model were consistently higher. In addition, the asymptotic standard errors for the DEL model were lower than for the continuous logistic model.

4.2.3. Uncertainty Analysis: Saturation Parameter

The saturation parameter K has the same interpretation for all models we investigated, it is the carrying capacity of the green algae population. The estimates for K were remarkably similar across all models (Supplementary Tables S9–S12). The standard errors were inconsistent between asymptotic theory and bootstrapping for the Bernoulli and Gompertz models. For example, the asymptotic standard error for the estimate of K in replicate 1 for the Bernoulli model was 0.1138, whereas the bootstrapping error was 0.0158 (Supplementary Table S12). These results are important, because the asymptotic standard error would result in a much wider confidence band around the model fit to the data, which is indeed the case for replicate 1 of the Bernoulli model (Figure 4, Top). Since the bootstrapping distributions for K for all of the models are normally distributed, this indicates that the bootstrapping standard errors are accurate (Supplementary Figures S13–S16).

4.2.4. Uncertainty Analysis: Bernoulli Model Parameter β

The parameter β is unique to the Bernoulli model, and scales the rate at which the green algae population reaches carrying capacity. In particular, when $\beta = 1$, the Bernoulli model reduces to the logistic model. We found that the estimates for β with asymptotic theory and bootstrapping were >1. We can not confidently say that these estimates are accurate, since the corresponding standard errors are unreasonably high for both uncertainty techniques and for all three replicates (Table 3). Moreover, the bootstrapping distributions for β were not normally distributed and heavily skewed to the right for all three replicates (Supplementary Figure S17), indicating the possible presence of correlations with other model parameters. We speculate that the parameters β and K may be correlated for the Bernoulli model.

Table 3. β estimate and standard error for the Bernoulli model.

Asymptotic Results: β	Replicate	Estimate	SE
	1	2.1646	2.5440
	2	3.4574	2.8118
	3	2.8188	2.8758
Bootstrapping Results: β	**Replicate**	**Estimate**	**SE**
	1	38.31	113.72
	2	29.78	92.89
	3	43.27	113.81

4.3. Coupling to the Discrete-Time Daphnia magna Population Model

Our uncertainty analysis results indicate that the Logistic model and its numerically discretized counterpart, the DEL model, are the most accurate models among those we investigated. We forgo a summary of the evidence for this conclusion until the discussion section. Our ultimate aim of validating a model for green algae growth was to couple it to a model for *D. magna* population dynamics. In this section, we couple two validated models for algae and *Daphnia* population growth to create a theoretical, unvalidated model to explore potentially complex predator-prey dynamics.

The *D. magna* model we use is one that we recently validated with experimental population data [5]. The validated *D. magna* model is a specification of the following discrete-time discrete-age structured population model:

$$\begin{bmatrix} p(t+1,1) \\ p(t+1,2) \\ p(t+1,3) \\ \vdots \\ p(t+1,i_{max}) \end{bmatrix} = \begin{bmatrix} a(t,1) & a(t,2) & a(t,3) & \dots & a(t,i_{max}) \\ b(t,1) & 0 & 0 & \dots & 0 \\ 0 & b(t,2) & 0 & \dots & 0 \\ \vdots & & \ddots & \dots & \vdots \\ 0 & 0 & 0 & \dots & b(t,i_{max}-1) \end{bmatrix} \begin{bmatrix} p(t,1) \\ p(t,2) \\ p(t,3) \\ \vdots \\ p(t,i_{max}) \end{bmatrix}. \tag{32}$$

The population is divided into one-day age classes up to some maximum age i_{max} and the number of daphnids of age i at a time t is $p(t,i)$. The average fecundity of each age class i is given by $a(t,i)$ and the survival rate for daphnids of age i is given by $b(t,i)$. The validated functional forms are $a(t,i) = \alpha(i)(1-q)^{M(t-\tau)}$ and $b(t,i)$ is defined piecewise as $\mu(1-c)^{M(t)}$ if $i \leq 4$ and μ if $i \geq 5$. A summary of the parameters and variables in the model are listed in Table 4 (see [5] for further details).

We coupled the *D. magna* population model to the DEL green algae model by assuming that *D. magna* consumes green algae and that the density-dependent survival and fecundity rates of *D. magna* are influenced directly by the algae concentration. We modeled the algae population with predation as

$$P_{t+1} = P_t + RP_t\left(1 - \frac{P_t}{K}\right) - \delta M_t P_t \tag{33}$$

where δ is a predation coefficient, and M_t is the total *Daphnia* biomass at time t. We chose this functional form based on the assumption that *Daphnia* consume aglae at a rate proportional to the density of aglae and the total biomass of the *Daphnia* population, as adult daphnids consume food at a higher rate than younger ones. We used a 24 h time discretization to model algae growth for our simulation study and transformed parameters accordingly, setting $\delta = 0.001$, $K = 0.4559$, $R = 1.34$, and the initial algae population $P_0 = 0.0633$. We modeled the algal influence on *Daphnia* fecundity and survivorship by setting $a(t,i) = \alpha(i)(1-q)^{1/P_{t-\tau}}$ and $b(t,i) = \mu(1-c)^{1/P_t}$ if $i \leq 4$. In this model, we changed the functional form of the *Daphnia* model based on the assumption that the negative density-dependent fecundity and survivorship effects that daphnids experience are driven by a lack of food in the form of algae, represented as $\frac{1}{P_t}$ and its time-delayed analogue $\frac{1}{P_{t-\tau}}$. The *Daphnia* matrix model is otherwise unchanged.

Table 4. Summary of *Daphnia magna* and algae model parameters and variables.

Parameter/Variable	Description	Units
$p(t,i)$	Number of daphnids of age i	# of daphnids
$N(t)$	Total population size at time $t := \sum_{i=1}^{i_{max}} p(t,i)$	# of daphnids
q	Density-dependent fecundity constant	dimensionless
$\alpha(i)$	Density-independent fecundity rates	# neonates·daphnid^{-1}·day^{-1}
μ	Density-independent survival rate	day^{-1}
τ	Delay for density-dependent fecundity	days
c	Density-dependent survival constant	dimensionless
$M(t)$	Total biomass at time $t := \sum_{i=1}^{i_{max}} p(t,i) \frac{kZ_0 e^{ri}}{k+Z_0(e^{ri}-1)}$	mm
k	Average maximum daphnid size (major axis)	mm
r	Average daphnid growth rate	mm/hour
Z_0	Average neonate size (major axis)	mm
R	Intrinsic growth rate of algae	cells $\cdot 10^7 \cdot$L$^{-1}\cdot$day^{-1}
K	Algal population carrying capacity	cells $\cdot 10^7 \cdot$L^{-1}
δ	Density dependent predation constant	mm$^{-1}\cdot$cells$\cdot 10^{-7}$

We found that a coupled model could result in both steady state dynamics as well as oscillatory behavior for different choices of parameter values in the *Daphnia* model (Figure 5). The deterministic simulations in Figure 5 left show steady state behavior with small deviations relative to the population size. The seemingly random perturbations are due to the age-dependency of the density-indpendent fecundity rate $\alpha(i)$. We found that lowering the density-dependent survival competition parameter c yielded sustained oscillations, and increasing it led to both populations reaching a steady state. Other parameter combinations may also yield similar dynamics, but detailing those values is not the aim of the present study.

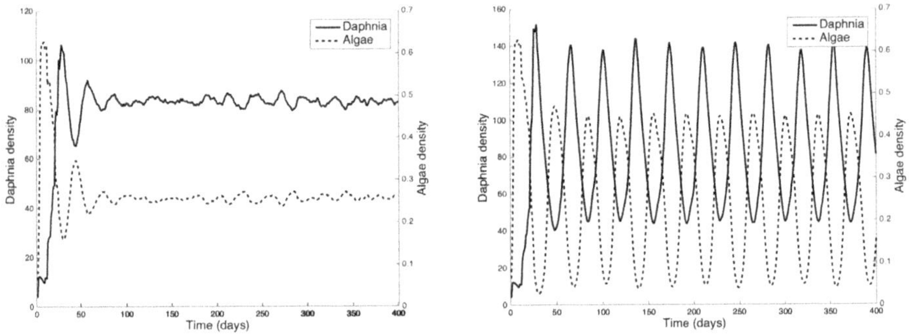

Figure 5. Simulations of the coupled daphnia and green algae dynamics model resulting in steady state behavior (**Left**, $c = 0.01$) and sustained oscillations (**Right**, $c = 0.04$).

5. Discussion

Our results highlight the importance of performing uncertainty quantification in validated biological models, even in the simple case of saturating growth dynamics encountered for green algae. For example, the ordinary least squares regression seemed to indicate that each of the models we investigated provide reasonable fits to the algae growth data. In addition, parameter estimates were consistent between replicate data sets for each model. From this information alone, one might conclude that the Bernoulli model was the best performing model, since it best fit the initial data X_0 (Figure 4). However, a deeper investigation with uncertainty analysis allowed us to generate confidence bands around the fits to the data, showing that the Bernoulli model was the model for which we could have the least confidence. This result was corroborated by a thorough examination of the standard errors for each of the parameters with similar interpretations across all models. For example, the growth rates for the Bernoulli model had unreasonably high standard errors, while the uncertainty in the corresponding growth rates for the other models were relatively low. Our results also indicated that the Gompertz model had inconsistent standard errors between asymptotic theory and bootstrapping for the initial condition and saturation parameter, emphasizing the importance of using multiple uncertainty quantification techniques to ascertain the best validated model. We observed that the logistic and DEL models have different confidence regions for the model solution in Figures 1 and 2. We attribute the change in computed confidence regions to the differences in numerical discretization, the time step for the DEL model was equal to one hour while the logistic model had a coarser mesh.

We collected replicate data and our results had strong agreement across the three replicates. Although methodology exists to fit all three data sets simultaneously to the same model, we chose to fit them separately to test whether the model validation results were consistent across several repeated experiments. We noticed a slight trend in the residuals resulting from fits of the model to data for each replicate: the fit sometimes either over- or under-estimates the data in groups of threes (Supplementary Figures S1–S4). We surmise that this phenomenon may be explained by human measurement error; some people tend to over or under count the algae when using a hemocytometer. Since all models were confounded with this possibility for human error, we can assume that human error did not influence the analysis by favoring one model over any other. In future work, more accurate and consistent cell counts may be performed with a flow cytometer. Alternatively, a spectrophotometer may be used to approximate algae concentrations.

Overall, we suggest that the population growth of *Raphidocelis subcapitata* is most accurately modeled using the Logistic equation among the simple growth models we investigated. It is important to note that our findings are limited to controlled laboratory settings with unchanging temperature, constant photoperiod, and no change in nutrient availability. For example, we did not consider the possible influence of photoperiod (light/dark) conditions on algae growth. We also did not consider the affect of limiting nutrients such as carbon, nitrogen, phosphorous, or sulfur. The models we investigated here represent our first approximation of algae growth and seemed to fit the data well even without considering how light affects algae growth and that the affect of limiting nutrients could be described by saturating algae growth. The simplifying assumption we made that growth parameters are independent of light conditions may be investigated in future work to yield a closer fit to the data. Our work here serves the purpose of coupling our green algae model with one of zooplankton (*D. magna*) population growth in a laboratory setting, e.g., for toxicity testing, but should not be directly extrapolated to populations in lakes. In order to develop an accurate model of community fluctuations in the field, we will need to consider predation by various zooplankton and microbes, competition with other algae, nutrient fluctuation, abiotic drivers, and habitat heterogeneity. This study is, however, a useful step toward developing a more comprehensive model. In particular, our results showing that a coupling a validated green algae model with a validated daphnia model are important because it exemplifies the possibility of using a mathematical model to recapitulate the oscillatory dynamics seen in nature. In contrast, the previously validated daphnia model that did not include algae dynamics was not able to produce oscillations and only resulted in steady state behavior [5], a result that did not account for the broader range of plankton dynamics seen in natural systems [1]. We note that our ultimate goal is to validate a coupled continuous time daphnia/algae model since continuous time models, such as the Sinko-Streifer model, are described by partial differential equations (PDEs) with a continuously structured variable and can be more amenable to the estimation of age-dependent parameters than a discretely structured model [14–17]. In this work, we investigated the dynamics of a coupled algae/daphnia discrete-time model as a coarse approximation to understand the qualitative impact of including a dynamic food source on daphnia populations. Our finding that increasing the density-dependent survival constant (c) in a coupled predator-prey model yields oscillatory dynamics compliments previous work that has predicted limit cycles based on increased mortality [18]. In an ecological setting, changes in the parameter c could reflect differing nutrient (algae) requirements on the density-dependent survival of daphnia due to differences in species size (e.g., *Daphnia pulex vs. Daphnia magna*) or other increased sources of density-dependent mortality such as predation on daphnia. Changes in c may also be induced toxicologically. For example, endocrine toxins are known to alter the molt cycle of adult daphnids through incomplete ecdysis [19], which may have an indirect affect on density-dependent survival by lowering competition for algae. Together, these results suggest that a structural change in the validated daphnia model, *i.e.*, including predation, and not just a change in parameter values is required to reproduce population oscillations observed in laboratory and natural settings. This finding is important in the context of our previous and current ongoing efforts, since oscillations were not observed in our previous daphnia population experiments [5]. Although the experiments performed in this work did not involve toxins, the species investigated are commonly used in toxicity assessments. Thus, our results can be used to provide a baseline to compare effects in a toxicity setting in future work.

Supplementary Materials: The following are available online at www.mdpi.com/2076-3417/6/5/155/s1.

Acknowledgments: This research was supported in part by the National Science Foundation under NSF Undergraduate Biomathematics grant number DBI-1129214, NSF grant number DMS-0946431, NSF grant number DMS-1514929 and in part by the Air Force Office of Scientific Research under grant numbers AFOSR FA9550-12-1-0188 and AFOSR FA9550-15-1-0298. The authors are grateful to Amanda Laubmeier and Kaska Adoteye for their assistance in collection of the data used in our efforts.

Author Contributions: Michael Stemkovski conceived and designed the experiments and performed simulations; Michael Stemkovski and Robert Baraldi performed the experiments, wrote the code, and performed the uncertainty analysis; Michael Stemkovski, Robert Baraldi, H. T. Banks, and Kevin B. Flores analyzed the results and wrote the paper.

Conflicts of Interest: The authors declare no conflict of interest.

References

1. McCauley, E.; Murdoch, W.W. Cyclic and stable populations: Plankton as paradigm. *Am. Nat.* **1987**, *129*, 97–121.
2. Benincà, E.; Dakos, V.; van Nes, E.H.; Huisman, J.; Scheffer, M. Resonance of plankton communities with temperature fluctuations. *Am. Nat.* **2011**, *178*, E85–E95.
3. McCauley, E.; Nisbet, R.M.; Murdoch, W.W.; de Roos, A.M.; Gurney, W.S. Large-amplitude cycles of Daphnia and its algal prey in enriched environments. *Nature* **1999**, *402*, 653–656.
4. Scheffer, M.; Rinaldi, S.; Kuznetsov, Y.A.; van Nes, E.H. Seasonal dynamics of Daphnia and algae explained as a periodically forced predator-prey system. *Oikos* **1997**, *80*, 519–532.
5. Adoteye, K.; Banks, H.T.; Cross, K.; Eytcheson, S.; Flores, K.B.; LeBlanc, G.A.; Nguyen, T.; Ross, C.; Smith, E.; Stemkovski, M. Statistical validation of structured population models for *Daphnia magna*. *Math. Biosci.* **2015**, *266*, 73–84.
6. Wang, Y.H.; Wang, G.; LeBlanc, G.A. Cloning and characterization of the retinoid X receptor from a primitive crustacean Daphnia magna. *Gen. Comp. Endocrinol.* **2007**, *150*, 309–318.
7. Caperon, J. Population growth in micro-organisms limited by food supply. *Ecology* **1967**, *48*, 715–722.
8. Schanz, F.; Zahler, U. Prediction of algal growth in batch cultures. *Schw. Zeit. Hydr.* **1981**, *43*, 103–113.
9. Thompson, P.-A.; Couture, P. Short-and long-term changes in growth and biochemical composition of *Selenastrum capricornutum* populations exposed to cadmium. *Aqua. Toxic.* **1991**, *21*, 135–143.
10. Banks, H.T.; Thompson, W.C.; Hu, S. *Modeling and Inverse Problems in the Presence of Uncertainty*; CRC Press: Boca Raton, FL, USA, 2014.
11. Banks, H.T.; Tran, H.T. *Mathematical and Experimental Modeling of Physical and Biological Processes*; CRC Press: Boca Raton, FL, USA, 2009.
12. Seber, G.A.; Wild, C.J. *Nonlinear Regression*; Wiley: Hoboken, NJ, USA, 2003.
13. Richards, F.J. A flexible growth function for empirical use. *J. Exp. Bot.* **1959**, *10*, 290–300.
14. Banks, H.T.; Banks, J.E.; Dick, L.K.; Stark, J.D. Estimation of dynamic rate parameters in insect populations undergoing sublethal exposure to pesticides. *Bull. Math. Biol.* **2007**, *69*, 2139–2180.

15. Banks, H.T.; Davis, J.L.; Ernstberger, S.L.; Hu, S.; Artimovich, E.; Dhar, A.K. Experimental design and estimation of growth rate distributions in size-structured shrimp populations. *Inv. Probl.* **2009**, *25*, 095003, doi:10.1088/0266-5611/25/9/095003.
16. Banks, J.E.; Dick, L.K.; Banks, H.T.; Stark, J.D. Time-varying vital rates in ecotoxicology: Selective pesticides and aphid population dynamics. *Ecol. Model.* **2008**, *210*, 155–160.
17. Wood, S.N. Obtaining birth and mortality patterns from structured population trajectories. *Ecol. Monogr.* **1994**, *64*, 23–44.
18. May, R.M.; Conway, G.R.; Hassell, M.P.; Southwood, T.R.E. Time delays, density-dependence and single-species oscillations. *J. Anim. Ecol.* **1974**, *43*, 747–770.
19. Wang, Y.H.; Kwon, G.; Li, H.; LeBlanc, G.A. Tributyltin synergizes with 20-Hydroxyecdysone to produce endocrine toxicity. *Toxicol. Sci.* **2011**, *123*, 71–79.

applied sciences

MDPI

Article

A Simple Predator-Prey Population Model with Rich Dynamics

Bing Li [1,2], Shengqiang Liu [1,*], Jing'an Cui [3] and Jia Li [4,*]

1 Academy of Fundamental and Interdisciplinary Science, Harbin Institute of Technology, Harbin 150080, China; leeicer@126.com
2 School of Mathematical Science, Harbin Normal University, Harbin 150025, China
3 College of Science, Beijing University of Civil Engineering and Architecture, Beijing 100044, China; cuijingan@bucea.edu.cn
4 Department of Mathematical Sciences, The University of Alabama in Huntsville, Huntsville, AL 35899, USA
* Correspondence: sqliu@hit.edu.cn (S.L.); li@math.uah.edu (J.L.); Tel.: +86-451-8640-2559 (S.L.); +1-256-824-6470 (J.L.)

Academic Editor: Yang Kuang
Received: 31 March 2016; Accepted: 10 May 2016; Published: 16 May 2016

Abstract: A non-smooth switched harvest on predators is introduced into a simple predator-prey model with logistical growth of the prey and a bilinear functional response. If the density of the predator is below a switched value, the harvesting rate is linear; otherwise, it is constant. The model links the well studied predator-prey model with constant harvesting to that with a proportional harvesting rate. It is shown that when the net reproductive number for the predator is greater than unity, the system is permanent and there may exist multiple positive equilibria due to the effects of the switched harvest, a saddle-node bifurcation, a limit cycle, and the coexistence of a stable equilibrium and a unstable circled inside limit cycle and a stable circled outside limit cycle. When the net reproductive number is less than unity, a backward bifurcation from a positive equilibrium occurs, which implies that the stable predator-extinct equilibrium may coexist with two coexistence equilibria. In this situation, reducing the net reproductive number to less than unity is not enough to enable the predator to go extinct. Numerical simulations are provided to illustrate the theoretical results. It seems that the model possesses new complex dynamics compared to the existing harvesting models.

Keywords: predator-prey model; switched harvest; limit cycle; rich dynamics

MSC: Primary: 92D25, 34K60; Secondary: 34K18

1. Introduction

Mathematical modeling of predator-prey interactions have attracted wide attention since the original work by Lotka and Volterra in 1920s, and there have been extensively studied for their rich dynamics [1–3]. Since the rich and complex dynamics for interactive species are common in the real world, many researchers have investigated the processes that affect the dynamics of prey-predator models and wanted to know what models can best represent species interactions.

As a simplest form, the interaction between a predator and prey may be modeled by a pair of differential equations [1,3–5],

$$\begin{aligned} \frac{dN}{dt} &= rN(1-\frac{N}{K}) - aNP \\ \frac{dP}{dt} &= caNP - dP \end{aligned} \tag{1}$$

where N and P represent the prey and predator species, respectively; $r, K, a, c,$ and d are positive constants. In the absence of the predation, the prey grows logistically with intrinsic growth rate r and carrying capacity K. In the presence of the predator, the prey species decreases at a rate proportional to the functional response aN, where a presents the rate of predation. The factor c denotes the efficiency of predation which divides a maximum per capita birth rate of the predators into a maximum per capita consumption rate. Without the prey, no predation occurs and the predator species decreases exponentially with mortality rate d.

To consider the dynamics of model Equation (1), it is shown that the origin $(0,0)$ is a saddle point. Define the net reproductive number of the predator population n_0, *i.e.*, the expected number of a predator individual producing as the predator population is introduced into a stable prey population [6,7], as

$$n_0 = \frac{caK}{d} \tag{2}$$

Then if $n_0 < 1$, the boundary equilibrium $(K,0)$ with the predator going extinct is globally asymptotically stable and there exists no positive equilibrium for the prey-predator interaction. If $n_0 > 1$, the boundary equilibrium $(K,0)$ is a saddle point and there exists a positive coexistence equilibrium $(N_*, P_*) = \left(\frac{K}{n_0}, \frac{r}{a}\left(1 - \frac{1}{n_0}\right)\right)$, which is globally asymptotically stable. The model dynamics are relatively simple. Using n_0 as bifurcation parameter, we have a transcritical bifurcation at $n_0 = 1$ as (N_*, P_*) is bifurcated. Notice that model Equation (1) is a special case of the prey-predator model in [8].

Apparently, model Equation (1) fails to show the complicated dynamics of the predator-prey interactions in the real world. Later, many researchers improve and enrich model Equation (1) by incorporating some other elements, for example, stage-structure [9–13], nonlinear functional response function [14–18], dispersal among patchers [19], delays [9–11,20], or impulsive effects [18,21].

In the real world, from the point of view of predators' needs, the exploitation of biological resources and harvest are commonly practiced in fishery, forestry, and wildlife management. There is an interest in the use of bioeconomic models to gain insight into the scientific management of renewable resources [16]. Moreover, harvesting is an important and effective method to prevent and control the explosive growth of predators or prey when they are enough. So, generally speaking, it is reasonable and necessary for one to introduce the harvest of populations into models. Taking the above reasons into a consideration, we focus on the predator-prey model with harvest [12,13,16,17,20–28].

Normally, harvesting has several forms in predator-prey models. The most common one of these harvesting forms is a nonzero constant [16,17,20,22–24,26] or a linear harvesting rate [12,13,22,25–28]. In Ref. [22], a two-prey-one-predator model with predator harvested was studied. The authors are particularly interested in the stability properties of different harvest strategies. Two types of harvest strategies are: with a nonzero constant and a linear harvesting rate. The choice of idealized harvest strategies will contribute to a qualitative understanding of the properties of different harvesting strategies. Xiao and Jennings [16] considered the dynamical properties of the ratio dependent predator-prey model with constant prey harvesting. There existed numerous kinds of bifurcations, such as the saddle-node bifurcation, the subcritical and supercritical Hopf bifurcations. There also existed a limit cycle, a homoclinic or heteroclinic orbit satisfying different parameter values. In Ref. [17], the ratio-dependent predator-prey model with constant predator harvesting was focused on. Philip *et al.* [26] also discussed two predator-prey models with linear or nonzero constant predator harvesting.

The above two types of harvesting rates seemingly have their own advantages as well as disadvantages in fitting the harvest in the real world. When the density of the predator or prey is rather low, the nonzero constant harvesting rate is not as reasonable as of the proportional type [13,22,26]; while if the predator or prey is abundant, linear harvesting rate is less possible than the constant harvesting rate [16]. In Ref. [22], to compare the stability properties of the system with two harvest strategies, they applied linear or constant harvesting rate, respectively. In that comparative study, the authors demonstrated that switching from linear to constant harvesting rate may turn a stable stationary state to a periodic or chaotic oscillatory mode from a mathematical perspective. However, when deciding the constant level of harvesting, the instability of the constant harvest strategy calls for great care. In Ref. [29], Beddington *et al.* introduced a more realistic smooth harvesting function in which the fishing effort is limited upwards because the constant harvest cannot be achieved for small populations of the fish. The adjustment of a harvesting function can prevent extinction and increase the stability to some extent. Moreover, the dramatic increase of the predator or prey challenges the normal ecological balance and capacity of harvest. Thus, it is interesting to construct a new kind of harvesting rate that combines the advantages from both linear and constant harvesting rates.

Motivated by these ideas, in this paper, we consider a predator-prey model with a novel harvesting rate. Our ideas to develop the harvesting rate are derived from the capacity of treatments of diseases that had been well studied in the dynamical epidemic models [30–32].

Using model Equation (1) as our baseline model, we assume that harvesting takes place, but only the predator population is under harvesting and introduce harvesting function $H(P)$ of the predator to prey-predator model Equation (1) for discussing its dynamical features. The interactive dynamics are governed by the following system

$$\begin{aligned} \frac{dN}{dt} &= rN\left(1-\tfrac{N}{K}\right) - aNP \\ \frac{dP}{dt} &= caNP - dP - H(P) \end{aligned} \tag{3}$$

Following the methods in [14–17,30], we investigate the existence and stability of multiple equilibria, bifurcations, and limit cycles, and study the effects of switched harvest on the dynamics of the predator-prey model.

This paper is organized as follow. Sections 2 and 3.1 represent the boundedness of model Equation (3) and existence of multiple equilibria. In Section 3.2, we study the stability of equilibria, bifurcations, and the existence and stability of a limit cycle. In Section 4, we give numerical simulations to verify our results. Brief discussions are presented finally in Section 5.

2. Model Formulation

Now, we consider model Equation (3). Firstly, we describe harvesting function $H(P)$ of the predators in model Equation (3), which has the following form

$$H(P) = \begin{cases} mP, & 0 \le P \le P_0 \\ h, & P_0 < P \end{cases} \tag{4}$$

We assume that the harvesting rate is proportional to the predator population size until it reaches a threshold value due to limited facilities of harvesting or resource protection. The harvesting rate will then be kept as a constant. Denote the harvesting threshold value as $h = mP_0$.

When $0 \le P \le P_0$, model Equation (3) is

$$\begin{aligned} \frac{dN}{dt} &= rN\left(1-\tfrac{N}{K}\right) - aNP \\ \frac{dP}{dt} &= caNP - dP - mP \end{aligned} \tag{5}$$

When $P > P_0$, model Equation (3) becomes

$$\begin{aligned} \frac{dN}{dt} &= rN\left(1-\tfrac{N}{K}\right) - aNP \\ \frac{dP}{dt} &= caNP - dP - h \end{aligned} \tag{6}$$

It is straightforward to verify that solutions of Equation (3) with positive initial conditions are all positive for $t \geq 0$ and ultimately bounded. Thus the following set

$$D = \left\{(N,P) : N \geq 0, P \geq 0, cN + P \leq \frac{ck(r+d)^2}{4rd}\right\}$$

is positive invariant for system Equation (3).

3. Preliminary Results

3.1. Existence of Equilibria

In this section, we explore the existence of all nonnegative equilibria. First, the origin $(0,0)$ is still a trivial equilibrium and the predator-free equilibrium $(K,0)$ exists. Moreover, it is easy to see that there exists no positive equilibrium in region D if $n_0 \leq 1$. We thus assume $n_0 > 1$ hereafter, and present our results of the existence of positive equilibria as follows.

Theorem 1. *System Equation (3) has a positive coexistence equilibrium $E^*(N^*, P^*)$, in the subregion of D with $0 < P \leq P_0$,*

$$N^* = \frac{K}{\hat{n}}, \quad P^* = \frac{r}{a}\left(1 - \frac{1}{\hat{n}}\right) \tag{7}$$

if and only if $P_0 \geq \frac{r}{a}$ and $\hat{n} > 1$, or $P_0 < \frac{r}{a}$ and $1 < \hat{n} \leq \frac{r}{r - aP_0}$, where $\frac{r}{a}$ represents the maximum predator density for which the prey population can establish itself from a small initial population and $\hat{n}$ is the net reproductive number of the predator under harvesting defined by

$$\hat{n} = \frac{caK}{d+m} \tag{8}$$

Proof. In the subregion of D with $0 < P \leq P_0$, a positive equilibrium of Equation (3) satisfies

$$\begin{cases} rN^*\left(1 - \frac{N^*}{K}\right) - aN^*P^* = 0 \\ caN^*P^* - (d+m)P^* = 0 \end{cases} \tag{9}$$

Then it follows that

$$N^* = \frac{d+m}{ca} = \frac{K}{\hat{n}}$$

and then

$$P^* = \frac{r}{a}\left(1 - \frac{1}{\hat{n}}\right) > 0$$

if $\hat{n} > 1$. In the mean time, it follows from

$$P^* = \frac{r}{a}\left(1 - \frac{1}{\hat{n}}\right) \leq P_0$$

and then

$$\frac{1}{\hat{n}} \geq \frac{r - aP_0}{r}$$

that there exists a positive equilibrium in this subregion if

$$\frac{r - aP_0}{r} \leq 0$$

or if $\frac{r - aP_0}{r} > 0$ and

$$\hat{n} \leq \frac{1}{1 - \frac{aP_0}{r}} = \frac{r}{r - aP_0} \tag{10}$$

□

To investigate the existence of positive equilibria in the subregion of D with $P > P_0$, we first give the relation of roots and coefficients for a quadratic equation.

Quadratic equation $(x - A)(B - x) = C$, with constants A, B, and C positive, has two positive roots

$$x_1 = \frac{A + B - \sqrt{(B - A)^2 - 4C}}{2} < x_2 = \frac{A + B + \sqrt{(A - B)^2 - 4C}}{2}$$

a unique positive root

$$x = \frac{A + B}{2}$$

or no positive root, if

$$C < \frac{(B - A)^2}{4}, \; C = \frac{(B - A)^2}{4}, \text{ or } C > \frac{(B - A)^2}{4}$$

respectively.

The results for the existence of positive equilibria of Equation (3) in the subregion of D with $P > P_0$ are provided as follows.

Theorem 2. *We assume* $P_0 < \frac{r}{a}$ *and define*

$$\hat{P} = \frac{r(n_0 - 1)}{2an_0}, \; h_1 = \frac{crK}{4}\left(1 - \frac{1}{n_0}\right)^2, \; h_2 = \frac{rm}{a}\left(1 - \frac{1}{\hat{n}}\right) \tag{11}$$

System Equation (3), in the subregion of D with $P > P_0$*, has*

(a) No positive equilibrium if

$$h > h_1, \quad \text{or} \quad \begin{cases} P_0 > \hat{P} \\ h_2 \leq h < h_1 \end{cases}$$

(b) A unique positive equilibrium with

$$N = \frac{K(1 + n_0)}{2n_0}, \quad P = \frac{r(K - N)}{aK} = \frac{r(n_0 - 1)}{2an_0} \tag{12}$$

if

$$h = h_1, \quad \begin{cases} P_0 > \hat{P}, \\ h < h_1 \text{ and } h < h_2 \end{cases} \quad \text{or} \quad \begin{cases} P_0 < \hat{P}, \\ h < h_1 \text{ and } h \leq h_2 \end{cases}$$

(c) Two positive equilibria $E_i(N_i, P_i)$, $i = 1, 2$, *where*

$$\begin{aligned} N_1 &= \frac{crK(1 + n_0) - n_0\sqrt{4crK\,(h_1 - h)}}{2crn_0} \\ &< \frac{crK(1 + n_0) + n_0\sqrt{4crK\,(h_1 - h)}}{2crn_0} = N_2 \end{aligned} \tag{13}$$

and

$$P_2 = \frac{r(K - N_2)}{aK} < P_1 = \frac{r(K - N_1)}{aK}$$

if $h < h_1$ *and*

$$\begin{cases} P_0 < \hat{P} \\ \hat{n} < 1 \end{cases} \quad or \quad \begin{cases} P_0 < \hat{P} \\ h_2 < h \end{cases}$$

Proof. In the subregion of D with $P > P_0$, a positive equilibrium satisfies

$$rN(1 - \frac{N}{K}) - aNP = 0 \tag{14a}$$

$$caNP - dP - h = 0 \tag{14b}$$

It follows from Equation (14a) that

$$P = \frac{r(K - N)}{aK}$$

Substituting it into Equation (14b) yields

$$\left(N - \frac{K}{n_0}\right)(K - N) = \frac{K}{cr}h \tag{15}$$

It follows from the relation of roots and coefficients for the quadratic equation $(x - A)(B - x) = c$ shown above that equation Equation (15) has no, unique, or two positive solutions if $h > h_1$, $h = h_1$, or $h < h_1$.

To have $P_i > P_0$, $i = 1, 2$, we need

$$P_i = \frac{r(K - N_i)}{aK} > P_0$$

that is,

$$N_i < K - \frac{aK}{r}P_0 = \frac{aK}{r}\left(\frac{r}{a} - P_0\right), \quad i = 1, 2 \tag{16}$$

Thus, if $P_0 \geq r/a$, there is no positive equilibrium of Equation (15) in the subregion of D with $P > P_0$. We assume $P_0 < r/a$.

Suppose there are two positive solutions, $N_1 < N_2$, to Equation (15) and hence $P_2 < P_1$. Then $P_2 > P_0$ if and only if

$$N_2 < K - \frac{aK}{r}P_0 \tag{17}$$

Substituting N_2 in Equation (13) into Equation (17) yields

$$n_0\sqrt{4crK(h_1 - h)} < crK(n_0 - 1) - 2caKn_0P_0$$

that is,

$$\frac{\sqrt{4crK}}{2caK}\sqrt{(h_1 - h)} < \frac{r(n_0 - 1)}{2an_0} - P_0 = \hat{P} - P_0 \tag{18}$$

Then if $P_0 > \hat{P}$, we have $P_2 \leq P_0$.

Assume $P_0 < \hat{P}$. Squaring both sides of Equation (18) yields

$$\frac{r}{ca^2K}(h_1 - h) < \frac{r^2(n_0 - 1)^2}{4a^2n_0^2} - \frac{r(n_0 - 1)}{an_0}P_0 + P_0^2 \tag{19}$$

It follows from the definition of h_1 that

$$\frac{rh_1}{ca^2K} = \frac{r}{ca^2K}\frac{crK(n_0 - 1)^2}{4n_0^2} = \frac{r^2(n_0 - 1)^2}{4a^2n_0^2}$$

Then Equation (19) becomes

$$\begin{aligned} 0 &< h^2 - \frac{rm}{a}\left(\frac{n_0 - 1}{n_0} - \frac{m}{caK}\right)h = \left(h - \frac{rm}{a}\left(1 - \frac{d+m}{caK}\right)\right)h \\ &= \left(h - \frac{rm}{a}\left(1 - \frac{1}{\hat{n}}\right)\right)h \end{aligned} \tag{20}$$

If $\hat{n} < 1$, we have $P_2 > P_0$. Otherwise, Equation (20) is equivalent to

$$0 < (h - h_2)h$$

Thus, if $h_2 < h < h_1$, $P_2 > P_0$, and if $h \leq h_2$, $P_2 < P_0$.

We now consider $P_1 > P_0$, that is,

$$N_1 < \frac{aK}{r}\left(\frac{r}{a} - P_0\right) \tag{21}$$

Substituting N_1 in Equation (13) into Equation (21), we have

$$\frac{\sqrt{4crK}}{2caK}\sqrt{(h_1 - h)} > P_0 - \hat{P} \tag{22}$$

If $P_0 \leq \hat{P}$, inequality Equation (21) is satisfied which implies $P_1 > P_0$.

Suppose $P_0 > \hat{P}$. Similarly as above, inequality Equation (22) is equivalent to

$$0 > (h - h_2)h$$

Thus, if $h < h_2$, $P_1 > P_0$, and if $h \geq h_2$, $P_1 < P_0$.

By putting all together, the proof is completed. □

Remark 1. *System Equation (1) with linear predator harvest strategy leads to the predator extinct if the net reproductive number $\hat{n} < 1$. However, by Theorem 2 (c), we find that for system Equation (1) with switched predator harvest strategy; that is, when the density of predator is below harvest level P_0, the linear harvesting rate is applied to the system, whereas when the density of predators is higher than harvest level, the system adopts nonzero constant harvesting rate, even if $\hat{n} < 1$, the prey and predator may coexist.*

3.2. Stability of Equilibria

In this section, we discuss the stability of equilibria of model Equation (3).

Theorem 3. *Equilibrium $E_0(K, 0)$ is locally asymptotically stable if $\hat{n} < 1$, and unstable if $\hat{n} > 1$. Moreover, $E_0(K, 0)$ is globally asymptotically stable in $D \backslash \{(0,0)\}$ if $\hat{n} < 1$ and $h > h_1$. If $\hat{n} > 1$, system Equation (3) is permanent.*

Proof. It is easy to obtain that the characteristic roots to the linearized equation of system Equation (3) at $E_0(K, 0)$ are $\lambda_1 = -r < 0$ and $\lambda_2 = caK - d - m = (d+m)(\hat{n} - 1)$. Thus, E_0 is locally asymptotically stable if $\hat{n} < 1$ and unstable if $\hat{n} > 1$.

Next, note that $(0, 0)$ is always unstable. If $\hat{n} < 1$, by Theorems 1 and 2, there is not any other equilibrium of system Equation (3) than E_0 in $D \backslash \{(0,0)\}$. Since D is the invariant set of system Equation (3) and E_0 is locally asymptotically stable, it follows from the Bendixson Theorem that every solution of system Equation (3) in D approaches E_0 when t tends to positive infinity.

Since E_0 is unstable as $\hat{n} > 1$, following the similar arguments to Cantrell and Cosner ([33] Theorem 3.1) (see also ([34] Theorems 3,4)), which is based on the uniform persistence theory introduced by Hale and Waltman [35], we are able to conclude that system Equation (3) is permanent if $\hat{n} > 1$.

The proof is complete. □

Remark 2. *It follows from Cantrell and Cosner ([33] Theorem 3.1) that system Equation (5) is permanent if and only if $\hat{n} > 1$. Thus Theorem 3 suggests that if system Equation (5) is permanent, then so is system Equation (3).*

For the three positive equilibria E^*, E_1, E_2, we have the following results. Firstly, we consider the stability of E^*.

Theorem 4. *The positive equilibrium $E^*(N^*, P^*)$ of system Equation (3) is globally asymptotically stable if $P_0 < \frac{r}{a}$ and $1 < \hat{n} \leq \frac{r}{r - aP_0}$.*

Proof. According to Theorem 1, positive equilibrium $E^*(N^*, P^*)$ exists if and only if $P_0 < \frac{r}{a}$ and $1 < \hat{n} \leq \frac{r}{r - aP_0}$. The Jacobian matrix of system Equation (3) at $E^*(N^*, P^*)$ is

$$J^* = \begin{pmatrix} r(1 - \frac{N^*}{K}) + rN^*(-\frac{1}{K}) - aP^* & -aN^* \\ caP^* & caN^* - d - m \end{pmatrix} \tag{23}$$

Because N^* and P^* satisfy Equation (9), by means of Equation (9), the trace and determinant of J^* are simplified as

$$\begin{aligned} tr(J^*) &= r - \frac{2rN^*}{K} - aP^* + caN^* - d - m = -\frac{rN^*}{K} < 0 \\ det(J^*) &= (r - \frac{2rN^*}{K} - aP^*)(caN^* - d - m) + ca^2N^*P^* = ca^2N^*P^* > 0 \end{aligned} \tag{24}$$

Therefore, all eigenvalues of matrix J^* have negative real parts when $P_0 < \frac{r}{a}$ and $1 < \hat{n} \leq \frac{r}{r - aP_0}$. It follows that $E^*(N^*, P^*)$ is locally asymptotically stable.

Then, it suffices for us to prove the global attractiveness of $E^*(N^*, P^*)$. Inspired by the work of McCluskey [36], we define a Lyapunov function

$$M(t) = cN^*(\frac{N(t)}{N^*} - \ln \frac{N(t)}{N^*} - 1) + P^*(\frac{P(t)}{P^*} - \ln \frac{P(t)}{P^*} - 1)$$

where $N^* = \frac{K}{\hat{n}}$, and $P^* = \frac{r}{a}(1 - \frac{1}{\hat{n}})$.

We know that $\frac{N(t)}{N^*} - \ln \frac{N(t)}{N^*} - 1$ and $\frac{P(t)}{P^*} - \ln \frac{P(t)}{P^*} - 1 \geq 0$ for all $N(t), P(t) > 0$. (The equality holds if and only if $N(t) = N^*, P(t) = P^*$.) From the definition of $M(t)$, we know that $M(t)$ is well-defined and $M(t) \geq 0$. The equality holds if and only if $N(t) = N^*$ and $P(t) = P^*$.

Differentiating $M(t)$ along the solutions of system Equation (3), we obtain

$$\begin{aligned} \frac{dM(t)}{dt} &= c(1 - \frac{N^*}{N})\frac{dN}{dt} + (1 - \frac{P^*}{P})\frac{dP}{dt}. \\ &= -\frac{cr}{K}N^2 + \frac{2rN(d+m)}{aK} - \frac{r(d+m)^2}{ca^2K}. \\ &= -\frac{cr}{K}(N - \frac{d+m}{ca})^2 \leq 0 \end{aligned}$$

It follows that $M(t)$ is bounded and non-increasing. Thus $\lim_{t\to\infty} M(t)$ exists. Note that $\frac{dM}{dt} = 0$ if and only if $N = N^*$. Substituting $N = N^*$ into the first equation of Equation (3), one can directly get $P = P^*$. Therefore, the maximal compact invariant set in $\frac{dM}{dt} = 0$ is the singleton E^*. By the LaSalle invariance principle (see, for example, Theorem 5.3.1 in Hale and Verduyn Lunel [37]), positive equilibrium E^* is globally attracting. Further, E^* is globally asymptotically stable. □

Now we concentrate on the stability of coexistence equilibria $E_i(N_i, P_i)$, $i = 1, 2$. The Jacobian matrix of system Equation (3), at $E_i(N_i, P_i)$, is

$$J_i = \begin{pmatrix} r - \frac{2rN_i}{K} - aP_i & -aN_i \\ caP_i & caN_i - d \end{pmatrix}, \quad i = 1,2 \tag{25}$$

Theorem 5. *If the coexistence equilibrium $E_2(N_2, P_2)$ of system Equation (3) exists, it is unstable.*

Proof. For the coexistence equilibrium $E_2(N_2, P_2)$, N_2 and P_2 satisfy Equation (14). By Equations (13)–(15), after direct calculations, the determinant of matrix J_2 is

$$det(J_2) = -\frac{2car}{K}(N_2 - \frac{d + caK}{4ca})^2 + \frac{2car(d + caK)^2}{16c^2a^2K} \tag{26}$$

Because

$$N_2 = \frac{crK(1 + n_0) + n_0\sqrt{4crK(h_1 - h)}}{2crn_0} > \frac{crK(1 + n_0)}{2crn_0} = \frac{caK + d}{2ca}$$

we obtain that $det(J_2) < 0$. Thus $E_2(N_2, P_2)$ is a saddle point and unstable. □

For the coexistence equilibrium $E_1(N_1, P_1)$, similarly, the determinant of J_1 is

$$det(J_1) = -\frac{2car}{K}(N_1 - \frac{d + caK}{4ca})^2 + \frac{2car(d + caK)^2}{16c^2a^2K} \tag{27}$$

but since

$$N_1 = \frac{crK(1 + n_0) - n_0\sqrt{4crK(h_1 - h)}}{2crn_0} < \frac{crK(1 + n_0)}{2crn_0} = \frac{caK + d}{2ca}$$

we obtain $det(J_1) > 0$, and $E_1(N_1, P_1)$ may be node, focus or center.

By Equations (13)–(15), the sign of the trace of matrix J_1 is determined by

$$\varphi = r^2d(d - caK) + 2ca^2K(caK - r)h - ard\sqrt{4crK(h_1 - h)} \tag{28}$$

According to Equation (28), we obtain

(a) If $caK - r \leq 0$, then $\varphi < 0$.
(b) If $caK - r > 0$, and $h \leq \frac{r^2d(caK-d)}{2ca^2K(caK-r)} = \frac{r^2d^2(n_0-1)}{2ca^2K(caK-r)} = h_3$, then $\varphi < 0$.
(c) If $caK - r > 0$, $h > h_3$ and $\eta = 4crK(h_1 - h) - \frac{[2ca^2K(caK-r)h - r^2d^2(n_0-1)]^2}{a^2r^2d^2} > 0$, then $\varphi < 0$.

Summarizing the above discussions, we have the following results on the stability of equilibrium $E_1(N_1, P_1)$.

Theorem 6. *Suppose that the coexistence equilibrium $E_1(N_1, P_1)$ exists and if one of following conditions is satisfied:*

(a) $caK \leq r$;
(b) $caK > r$ and $h \leq h_3$;
(c) $caK > r$ and $h > h_3$ and $\eta > 0$,

then $E_1(N_1, P_1)$ is locally asymptotically stable. It is unstable if $caK > r, h > h_3$, and $\eta < 0$.

Combining the existence and stability results of equilibria of model Equation (3), by following similar arguments to ([31] Corollary 2.3), we present the following corollary to give conditions for bifurcation.

Corollary 1. *If* $\hat{n} < 1$ *and* $h < min\{h_1, m\hat{P}\}$, *then system Equation (3) has a backward bifurcation of positive equilibria.*

Next, we give examples to demonstrate the bifurcation of multiple equilibria.

For various parameter values, model Equation (3) has a forward bifurcation from one positive equilibrium to another positive equilibrium (see Example 3.1.) and a backward bifurcation with a predator-extinct equilibrium and two positive equilibria (Example 3.2.). Note that the conditions in Theorem 2 *a*) and Theorem 2 *c*) guarantee the existence of three positive equilibria E^* and E_1, E_2 (Example 3.3.).

Example 3.1. Using the following parameter values $r = 0.1, K = 0.5, a = 0.25, c = 0.8, d = 0.01$, and $m = 0.03$, we obtain $\hat{n} = 2.5$, and $h_2 = 0.0072$. When $h \leq h_2$, a bifurcation diagram is shown in Figure 1. When the parameter h decreases, the bifurcation at $h = h_2$ is forward, and model Equation (3) has a unique positive equilibrium for $h > 0$, which is similar to ([30] Example 2.4).

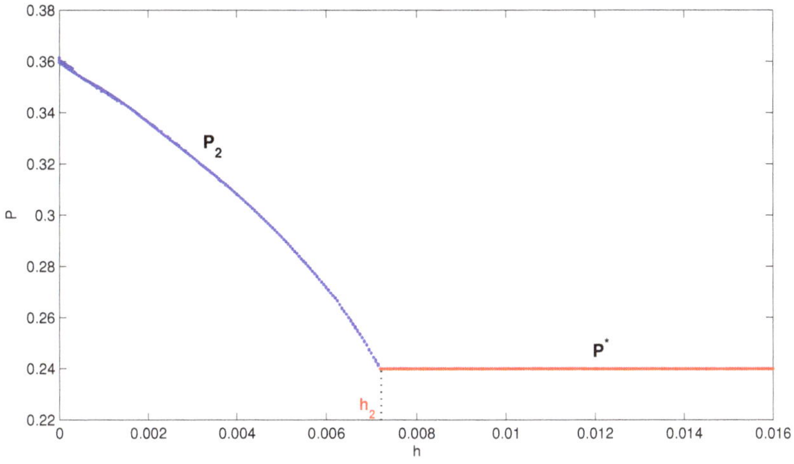

Figure 1. The forward bifurcation diagram from P^* to P_1 *versus u* for Equation (3). The line with P_1 indicates the curve of the predator with coexistent equilibrium E_1, and the line with P^* indicates the curve of the predator with coexistent equilibrium E^*.

Example 3.2. Choosing $r = 0.1, K = 0.4, a = 0.1, c = 0.8, d = 0.01$, and $m = 0.03$, we obtain $\hat{n} = 0.8 < 1$, $h_1 = 0.00378$, and $\hat{P} = 0.34367$. A backward bifurcation diagram is given in Figure 2, where the horizontal line denotes the curve of the predator with predator-extinct equilibrium E_0. Two positive equilibria E_1 and E_2 arise simultaneously at $h = h_1$ when the parameter h decreases.

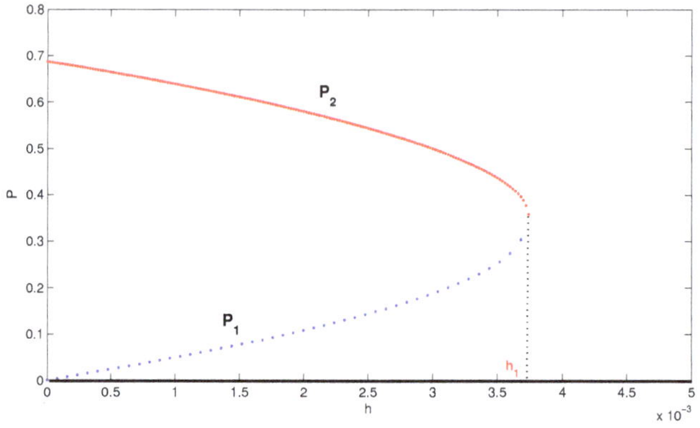

Figure 2. The backward bifurcation diagram of P_1 and P_2 *versus* h for Equation (3). The solid line with P_1 and dotted line with P_2 represent the curves of the predator with coexistent equilibrium E_1 and E_2, respectively.

Remark 3. *In Figure 2, we consider how to set up the harvesting threshold value h. We find that h_1 is an important harvesting amount. If the harvesting threshold value exceeds h_1, the system does not have a positive equilibrium; that is, the predator eventually tends to extinction. If the harvesting threshold value is less than h_1, the system has two positive equilibria among which one is unstable and the other may be stable,* i.e., *the predator and prey may coexist.*

Example 3.3. For model Equation (3), we choose $r = 0.1, K = 0.25, a = 0.25, c = 0.8, d = 0.01$, $m = 0.03$ and $P_0 = 0.1$. Thus we obtain $\hat{n} = 1.25$, $h_1 = 0.0032$, $h_2 = 0.0024$, and $\hat{P} = 0.16$. A bifurcation diagram is illustrated in Figure 3, where the horizontal blue line presents the curve of the predator with the positive equilibrium E^*. It displays that there is a bifurcation at $h = h_1$ when the parameter h reduces, which produces three equilibria E^* ,E_1 and E_2.

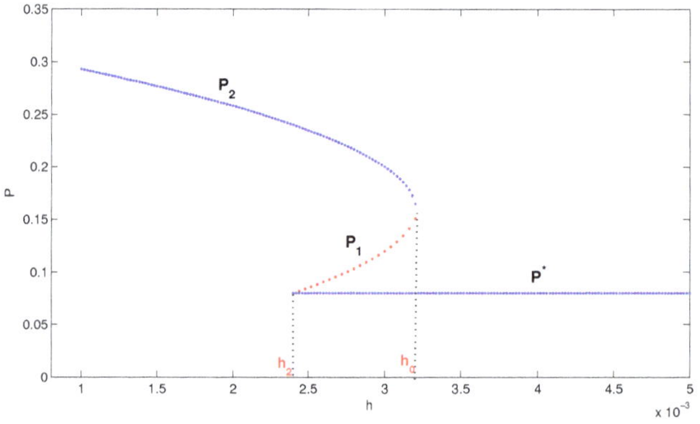

Figure 3. The bifurcation diagram with P_1, P_2 and P^* *versus* h. The lines with P_1 and P_2 indicate the curves of the predator with coexistent equilibria E_1 and E_2, respectively, and the line with P^* indicates the curve of the predator with coexistent equilibrium E^*.

The existence of limit cycles plays an important role in determining the dynamical behavior of the system. For example, if there is no limit cycle in system Equation (3) and its positive equilibrium is unique and locally asymptotically stable, then it must be globally stable. Now, we explore the existence of limit cycles in system Equation (3).

Theorem 7. *Suppose $\hat{n} > 1$ and $h < min\{h_1, h_2\}$. If $\varphi > 0$, then system Equation (3) has at least a stable limit cycle which encircles E_1.*

Proof. For $\hat{n} > 1$ and $h < h_2$, it is known from Theorem 2 $a)$ that the equilibrium E^* of system Equation (3) does not exist. Furthermore, because $\hat{n} > 1$, $h < h_1$, and $h < h_2$, it follows from Theorem 2 $b)$ that the equilibrium E_2 of system Equation (3) does not exist, but the equilibrium E_1 exists.

It follows from $\varphi > 0$ that E_1 is an unstable focus or node. It is easy to see that the unstable manifold at the saddle point $E_0(K, 0)$ is in the first quadrant. As the set D is positively invariant for system Equation (3), and system Equation (3) does not have any equilibrium in the interior of $D \setminus \{E_1\}$. It follows from the Poincaré-Bendixson theorem that system Equation (3) has at least a stable limit cycle which encircles E_1. □

In general, Dulac functions are only applied to smooth vector fields in the study of nonexistence of limit cycles. Since the right-hand sides of Equation (3), denoted by f_1 and f_2, are not smooth, following the similar arguments as in Wang ([31] Lemma 3.2), which is based on *Green's* Theorem, we are able to obtain sufficient conditions for the nonexistence of limit cycles in system Equation (3).

Theorem 8. *System Equation (3) does not have a limit cycle if $caK < d + r$.*

Proof. By the first equation of Equation (3), it is easy to see that the positive solutions of Equation (3) eventually enter and remain in the region

$$C = \{(N, P) : N \leq K\}$$

Thus, if a limit cycle exists, it must lie in the region C. Take a Dulac function $F = \frac{1}{N}$. Then we have

$$\frac{\partial(Ff_1)}{\partial N} + \frac{\partial(Ff_2)}{\partial P} = -\frac{r}{K} + ca - \frac{d+m}{N} \leq ca - \frac{d+m+r}{K} < 0$$

if $0 < P < P_0$. If $P > P_0$, it is easy to see that

$$\frac{\partial(Ff_1)}{\partial N} + \frac{\partial(Ff_2)}{\partial P} = -\frac{r}{K} + ca - \frac{d}{N} \leq ca - \frac{d+r}{K} < 0$$

Hence, system Equation (3) does not have a limit cycle. □

4. Numerical Simulation

In this section, we present numerical examples for system Equation (3).

Example 4.1. (**Example 3.1.** continued) The parameters values r, K, a, c, d and m are the same as in Example 3.1. We obtain $\hat{n} > 1$, $h_1 = 0.0081$, and $h_2 = 0.0072$. A forward bifurcation diagram is given in Figure 1.

Selecting $P_0 = 0.25$, we get $h = 0.0075 > h_2$. The equilibrium $E^*(0.2, 0.24)$ exists, but E_1 and E_2 do not exist (Theorem 2 $b)$). Its phase portrait is given in Figure 4, which shows that the unique positive equilibrium E^* is globally asymptotically stable.

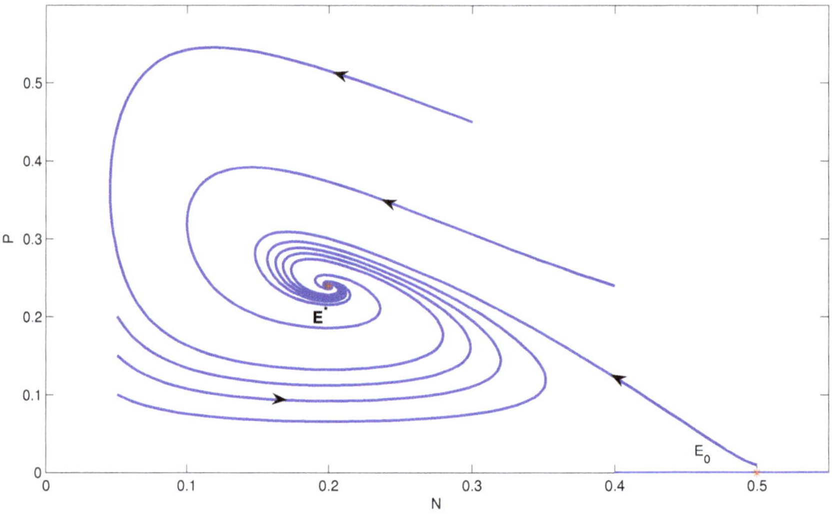

Figure 4. The phase portrait of model Equation (3) when E^* is globally asymptotically stable and E_0 is unstable.

If we choose $P_0 = 0.2$, then $h = 0.006 < h_2$. Equilibrium $E_1(0.16044, 0.27165)$ exists, but E_2 and E^* do not exist (Theorem 2 *b*)). Its phase portrait is given in Figure 5. The unique positive equilibrium E_1 is globally asymptotically stable in D.

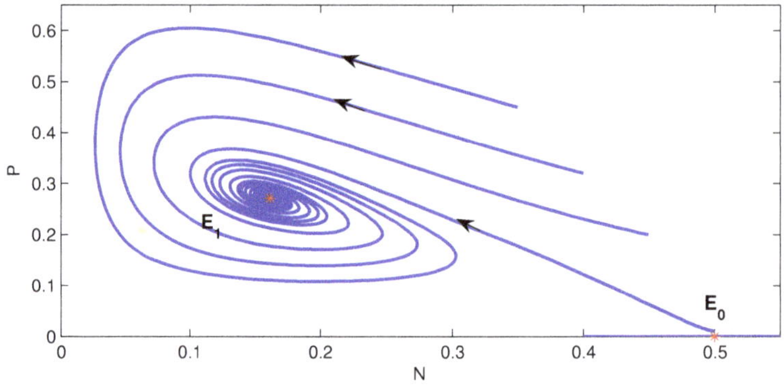

Figure 5. The phase portrait of model Equation (3) when E_1 is locally asymptotically stable and E_0 is unstable.

The equilibrium E^* in Figure 4 corresponds to some point on the curve of P^* in Figure 1 ($h > h_2$), and the equilibrium E_1 in Figure 5 corresponds to some point on the curve of P_1 in Figure 1 ($h < h_2$).

Example 4.2. (Example 3.2. continued) Choosing the same parameters values as in Example 3.2, we have $h_1 = 0.00378$ and $\hat{P} = 0.34367$. A backward bifurcation diagram is given in Figure 2.

If we choose $P_0 = 0.1$, then $P_0 < \hat{P}$. Equilibria $E_1(0.2, 0.5)$ and $E_2(0.325, 0.1875)$ exist, but equilibrium E^* does not exist. Its phase portrait is illustrated in Figure 6. It shows that equilibria E_0 and E_1 are asymptotically stable.

The equilibria E_1 and E_2 in Figure 6 correspond to some points on the curves of P_1 and P_2 , respectively, in Figure 2 ($h < h_1$).

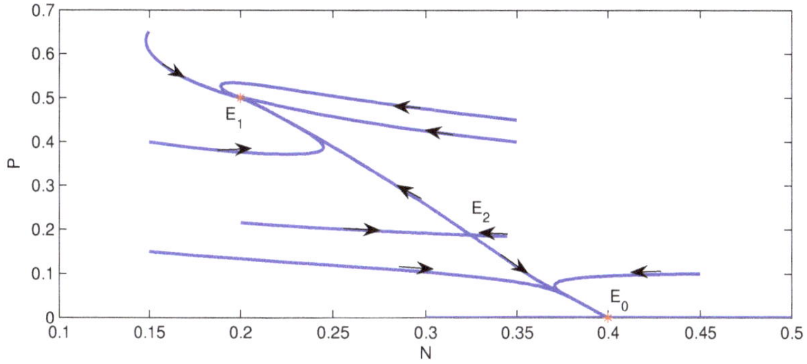

Figure 6. The phase portrait of model Equation (3) when E_0 and E_1 are locally asymptotically stable and E_2 is unstable.

Example 4.3 (I). (**Example 3.3.** continued) We identically select parameter values as in Example 3.3, and set $P_0 = 0.1$. Thus we have $\hat{n} > 1$, $h_1 = 0.0032$, $h_2 = 0.0024$, $\hat{P} = 0.16$, and $h = 0.003$. The above parameter values satisfy conditions $a)$ and $c)$ in Theorem 2, and condition $a)$ in Theorem 6.

Obviously, equilibria $E_0(0.25, 0)$, $E^*(0.2, 0.08)$, $E_1(0.125, 0.2)$, and $E_2(0.175, 0.12)$ all exist. The phase portrait of model Equation (3) is shown in Figure 7. Equilibria E^* and E_1 are asymptotically stable, and E_0 and E_2 are unstable. So model Equation (3) has bistable positive equilibria E^* and E_1.

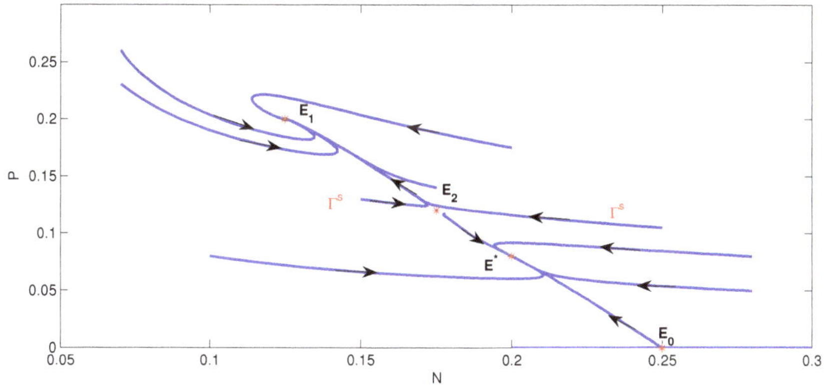

Figure 7. The phase portrait of model Equation (3) with bistable positive equilibria E^* and E_1, and unstable equilibria E_0 and E_2.

It follows from Figure 7 that the stable region Γ^s of the saddle point E_2 divides the positive invariant set into two regions. The attractive basin for the stable equilibrium E_1 is the region above Γ^s and the region below Γ^s is the basin of attraction for the stable equilibrium E^*.

The equilibria E_1 and E_2 in Figure 7 correspond to some points on the curves of P_1 and P_2, respectively, and E^* corresponds to some point on the curve of P^* in Figure 3 ($h_2 < h < h_1$).

Example 4.3 (II). We set parameter values $r = 0.03, K = 0.25, a = 0.5, c = 0.4, d = 0.01, m = 0.03$, and $P_0 = 0.0133$. Thus we have $\hat{n} = 1.25$, $uh_1 = 0.00048$, $h_2 = 0.00036$, $\hat{P} = 0.024$, and $h = 0.000399$. These parameter values satisfy conditions $a)$ and $c)$ in Theorem 2 and the condition $c)$ in Theorem 6. Thus, equilibria $E_0(0.25, 0)$, $E^*(0.2, 0.012)$, $E_1(0.1092, 0.0338)$, and $E_2(0.1908, 0.0142)$ all exist. The phase portrait of model Equation (3) is shown in Figure 8. Equilibria E^* and E_1 are stable, and E_0 and E_2 are unstable. So model Equation (3) has bistable positive equilibria E^* and E_1.

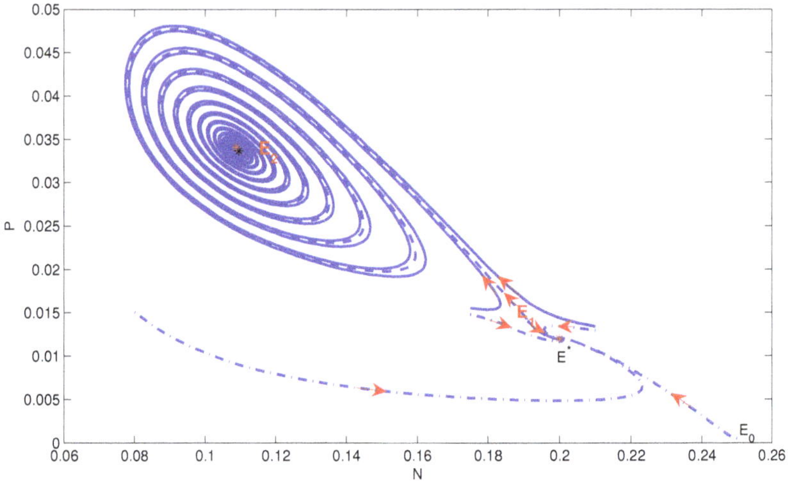

Figure 8. The phase portrait of model Equation (3) with positive stable equilibria E^* and E_1, and unstable equilibria E_0 and E_2.

The model with the parameter values in Example 4.3 (II) has a similar bifurcation diagram as that in Example 4.3 (I), and the stability of equilibria in Figure 8 is identical with that in Figure 7.

Example 4.4 (I). We select $r = 0.004, K = 0.2, a = 0.1, c = 0.4, d = 0.001$ and $m = 0.005$. Then $\hat{n} = 1.333$, $h_1 = 6.125 * 10^{-5}$, $\hat{P} = 1.75 * 10^{-2}$, and $h_2 = 5 * 10^{-5}$.

Choosing $P_0 = 0.008$, we have $h = 2 * 10^{-5}$. Thus, $h < h_2 < h_1$, $P_0 < \hat{P}$, $caK > r$, $h > h_3$, and $\eta < 0$. Equilibrium $E_1(0.04069, 0.03186)$ exists but is unstable, and E^* and E_2 do not exist. The parameter values satisfy the conditions of Theorem 7. Its phase portrait is given in Figure 9, which shows that model Equation (3) has a stable limit cycle which encircles E_1.

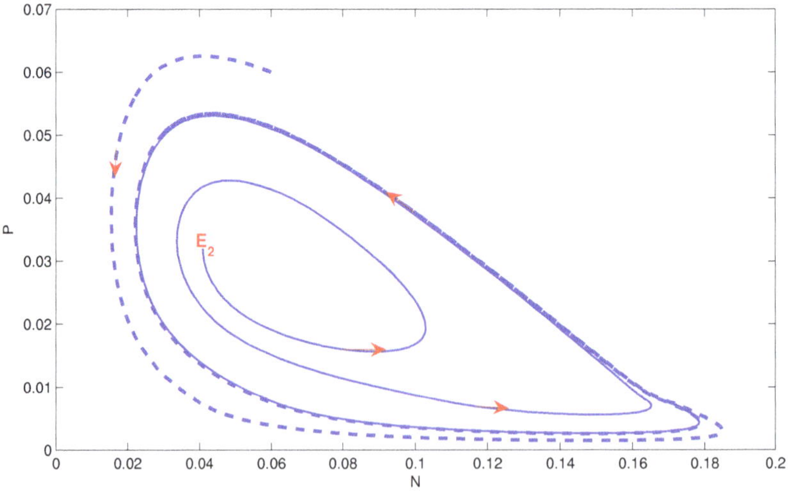

Figure 9. A stable limit cycle of model Equation (3) encircling the unstable equilibrium E_1.

Example 4.4 (II). Set $r = 0.03, K = 0.264, a = 0.5, c = 0.4, d = 0.01$, and $m = 0.031$. We obtain $\hat{n} = 1.2878, h_1 = 5.2041 * 10^{-4}, \hat{P} = 2.4318 * 10^{-2}$, and $h_2 = 4.1568 * 10^{-4}$.

Choosing $P_0 = 0.0133$, we obtain $h = 4.123 * 10^{-4}$. The parameter values satisfy Theorem 2 (b) and Theorem 6 (c), but do not satisfy Theorem 2 (a). Thus, equilibria E^* and E_2 do not exist and $E_1(0.1082, 0.0354)$ exists. The phase portrait of model Equation (3) is shown in Figure 10. In Figure 10, equilibrium E_1 is stable and two periodic orbits encircle E_1. We can see that the outside periodic orbit is stable. However, the inside periodic orbit is unstable. A trajectory (the dotted line) between the outside and inside periodic orbits ultimately tends to the outside periodic orbit, but a trajectory (the thin black line) starting from within the unstable periodic orbit finally tends to equilibrium E_1. So the initial state is important for ultimate trends of trajectories.

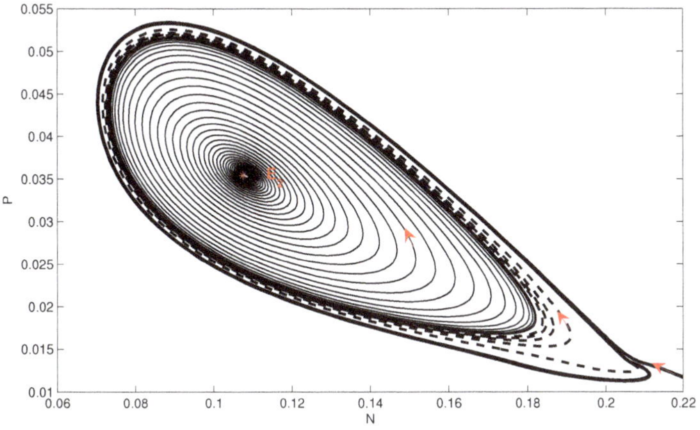

Figure 10. The phase portrait of model Equation (3) when E_1 is stable.

5. Conclusions

In this paper, we proposed and studied a new predator-prey model with non-smooth switched harvest on the predator. If the density of the predator is below a switched value, the harvest has a linear harvesting rate. Otherwise, the harvesting rate is constant. Our model exhibits new dynamical features compared to those with a linear harvesting rate or a constant harvesting rate.

According to the Kolmogorov Theorem [38], under certain assumptions, the model with a linear predator harvesting rate has either a stable equilibrium or a stable limit cycle, whereas the model with a constant harvesting rate on the predator has richer dynamics [16,23,23]. For example, for a class of predator-prey systems, Brauer and Soudack [23,23] obtained different types of dynamics for which the harvesting was in prey or a predator; Xiao and Jennings [16] further studied a ratio-dependent predator-prey model with a constant harvest on prey. They proved that the model could exhibit complicated bifurcation phenomena, including the Bogdanov-Takens bifurcation of cusp type, the heteroclinic bifurcation, or a separatrix connecting a saddle-node and a saddle bifurcation.

For the model studied in this paper, we showed that (see Theorem 2 and Corollary 1) a backward bifurcation from the predator-prey coexistence equilibrium may occur, which shows that reducing the net reproductive number of the predator to less than unity is not enough to eradicate the predator. On the other hand, when the net reproductive number of the predator is greater than unity, we showed that the predator always coexists with the prey permanently (Theorem 3), and the model may exhibit the following dynamics: (i) a unique globally asymptotically stable coexistence equilibrium; (ii) the coexistence of positive saddle equilibria connecting with either a locally asymptotically stable positive equilibrium (biostable) or a limit cycle; (iii) two stable positive equilibria coexisting with a saddle point. By numerical examples, we also showed that the model could exhibit more new dynamical features:

(a) a limit cycle encircling a unique positive equilibrium (see Figure 9); (b) two cycles surround an identical positive equilibrium, with one stable and one unstable (see Figure 10).

We would like to point out that we have assumed the simple functional response of the bilinear type in our current model Equation (3). We may also consider other types of functional responses. The dynamics may be richer and more complex. Further investigations are planned in our future studies.

Acknowledgments: Bing Li is supported by Natural Science Foundation of Heilongjiang A201411. Shengqiang Liu is supported by the NNSF of China (No. 11471089, 11301453) and the Fundamental Research Funds for the Central Universities (No.HIT.IBRSEM.A.201401). Jing'an Cui is supported by the National Natural Science Foundation of China (11371048, 11071011) and Funding Project for Academic Human Resources Development in Institutions of Higher Learning Under the Jurisdiction of Beijing Municipality(PHR201107123). Jia Li is supported partially by U.S. National Science Foundation grant DMS-1118150.

Author Contributions: Jia Li and Shengqiang Liu conceived and designed the study; Bing Li performed the simulation; Jing'an Cui participated in the analysis of the results; Bing Li, Jia Li and Shengqiang Liu wrote the paper. All authors read and approved the manuscript.

Conflicts of Interest: The authors declare no conflict of interest.

References

1. Murdoch, W.W.; Briggs, C.J.; Nisbet, R.M. *Consumer-Resource Dynamics*; Princeton University Press: Princeton, NJ, USA, 2003.
2. Seo, G.; DeAngelis, D.L. A predator-prey model with a Holling type I functional response including a predator mutual interference. *J. Nonlinear Sci.* **2011**, *21*, 811–833.
3. Turchin, P. *Complex Population Dynamics: A Theoretical/empirical Synthesis*; Princeton University Press: Princeton, NJ, USA, 2003.
4. Gutierrez, A.P. *Applied Population Ecology: A Supply-Demand Approach*; John Wiley and Sons: New York, NY, USA, 1996.
5. Seo, G.; Kot, M. A comparison of two predator-prey models with Hollingąŕs type I functional response. *Math. Biosci.* **2008**, *212*, 161–179.
6. Cushing, J.M.; Zou, Y. The net reproductive value and stability in matrix population models. *Nat. Resour. Model.* **1994**, *8*, 297–333.
7. Cushing, J.M. *An Introduction to Structured Population Dynamics*; SIAM: Philadelphia, PA, USA, 1998.
8. Kar, T.K. Stability analysis of a prey-predator model incorporating a prey refuge. *Commun. Nonlinear Sci. Numer. Simul.* **2005**, *10*, 681–691.
9. Chakraborty, K.; Haldar, S.; Kar, T.K. Global stability and bifurcation analysis of a delay induced prey-predator system with stage structure. *Nonlinear Dyn.* **2013**, *73*, 1307–1325.
10. Liu, S.Q.; Beretta, E. A stage-structured predator-prey model of Beddington-Deangelis type. *SIAM J. Appl. Math.* **2006**, *66*, 1101–1129.
11. Qu, Y.; Wei, J.J. Bifurcation analysis in a time-delay model for preyÍCpredator growth with stage-structure. *Nonlinear Dyn.* **2007**, *49*, 285–294.
12. Zhang, X.A.; Chen, L.S.; Neumann, A.U. The stage-structured predator-prey model and optimal harvesting policy. *Math. Biosci.* **2000**, *168*, 201–210.
13. Zhang, Y.; Zhang, Q.L. Dynamic behavior in a delayed stage-structured population model with stochastic fluctuation and harvesting. *Nonlinear Dyn.* **2011**, *66*, 231–245.
14. Lai, X.H.; Liu, S.Q.; Lin, R.Z. Rich dynamical behaviours for predator-prey model with weak Allee effect. *Appl. Anal.* **2010**, *89*, 1271–1292.
15. Lin, R.Z.; Liu, S.Q.; Lai, X.H. Bifurcations of a predator-prey system with weak Allee effects. *J. Korean Math. Soc.* **2013**, *50*, 695–713.
16. Xiao, D.M.; Jennings, L.S. Bifurcations of a ratio-dependent predator-prey system with constant rate harvesting. *SIAM J. Appl. Math.* **2005**, *65*, 737–753.
17. Xiao, D.M.; Li, W.X.; Han, M.A. Dynamics in a ratio-dependent predator-prey model with predator harvesting. *J. Math. Anal. Appl.* **2006**, *324*, 14–29.

18. Zhang, Y.; Zhang, Q.L.; Zhang, X. Dynamical behavior of a class of prey-predator system with impulsive state feedback control and Beddington-DeAngelis functional response. *Nonlinear Dyn.* **2012**, *70*, 1511–1522.
19. Gao, Y.; Liu, S.Q. Global stability for a predator-prey model with dispersal among patches. *Abstr. Appl. Anal.* **2014**, *2014*, 176493.
20. Martin, A.; Ruan, S.G. Predator-prey models with delay and prey harvesting. *J. Math. Biol.* **2001**, *43*, 247–267.
21. Wei, C.J.; Chen, L.S. Periodic solution and heteroclinic bifurcation in a predatorÍCprey system with Allee effect and impulsive harvesting. *Nonlinear Dyn.* **2014**, *76*, 1109–1117.
22. Azar, C.; Holmberg, J.; Lindgren, K. Stability analysis of harvesting in a predator-prey model. *J. Theor. Biol.* **1995**, *174*, 13–19.
23. Brauer, F.; Soudack, A.C. Stability regions and transition phenomena for harvested predator-prey systems. *J. Math. Biol.* **1979**, *7*, 319–337.
24. Brauer, F.; Soudack, A.C. Stability regions in predator-prey systems with constant-rate prey harvesting. *J. Math. Biol.* **1979**, *8*, 55–71.
25. Kar, T.K. Selective harvesting in a prey-predator fishery with time delay. *Math. Comput. Model.* **2003**, *38*, 449–458.
26. Lenzini, P.; Rebaza, J. Nonconstant predator harvesting on ratio-dependent predator-prey models. *Appl. Math. Sci.* **2010**, *4*, 791–803.
27. Negi, K.; Gakkhar, S. Dynamics in a Beddington-DeAngelis prey-predator system with impulsive harvesting. *Ecol. Model.* **2007**, *206*, 421–430.
28. Xiao, M.; Cao, J.D. Hopf bifurcation and non-hyperbolic equilibrium in a ratio-dependent predator-prey model with linear harvesting rate: Analysis and computation. *Math. Comput. Model.* **2009**, *50*, 360–379.
29. Beddington, J.R.; Cooke, J.K. Harvesting from a prey-predator complex. *Ecol. Model.* **1982**, *14*, 155–177.
30. Hu, Z.X.; Ma, W.B.; Ruan, S.G. Analysis of an SIR epidemic model with nonlinear incidence rate and treatment. *Math. Biosci.* **2012**, *238*, 12–20.
31. Wang, W.D. Backward bifurcation of an epidemic model with treatment. *Math. Biosci.* **2006**, *201*, 58–71.
32. Zhang, X.; Liu, X.N. Backward bifurcation and global dynamics of an SIS epidemic model with general incidence rate and treatment. *Nonlinear Anal. Real World Appl.* **2009**, *10*, 565–575.
33. Cantrell, R.S.; Cosner, C. On the dynamics of predator-prey models with the Beddington—DeAngelis functional response. *J. Math. Anal. Appl.* **2001**, *257*, 206–222.
34. Lu, Y.; Li, D.; Liu, S.Q. Modeling of hunting strategies of the predators in susceptible and infected prey. *Appl. Math. Comput.* **2016**, *284*, 268–285.
35. Hale, J.; Waltman, P. Persistence in infinite-dimensional systems. *SIAM J. Math. Anal.* **1989**, *20*, 388–395.
36. McCluskey, C. Lyapunov Functions for Tuberculosis Models with Fast and Slow Progression. *Math. Biol. Eng.* **2006**, *3*, 603–614.
37. Hale, J.; Lunel, S.V. *Introduction to Functional Differential Equations*; Springer-Verlag: New York, NY, USA, 1993.
38. Freedman, H.I. *Deterministic Mathematical Models in Population Ecology*; Marcel Dekker: New York, NY, USA, 1980.
39. Cui, J. ; Mu, X.; Wan, H. Saturation recovery leads to multiple endemic equilibria and backward bifurcation. *J. Theor. Biol.* **2008**, *254*, 275–283.
40. Van den Driessche, P.; Watmough, J. Reproduction numbers and sub-threshold endemic equilibria for compartmental models of disease transmission. *Math. Biosci.* **2002**, *180*, 29–48.
41. McQuaid, C.F.; Britton, N.F. Trophic structure, stability, and parasite persistence threshold in food webs. *Bull. Math. Biol.* **2013**, *75*, 2196–2207.
42. Zhang, X.; Liu, X.N. Backward bifurcation of an epidemic model with saturated treatment function. *J. Math. Anal. Appl.* **2008**, *348*, 433-443.
43. Zhou, L.H.; Fan, M. Dynamics of an SIR epidemic model with limited medical resources revisited. *Nonlinear Anal. Real World Appl.* **2012**, *13*, 312–324.

applied sciences

MDPI

Article

Mathematical Modeling of Bacteria Communication in Continuous Cultures

Maria Vittoria Barbarossa [1,*] and Christina Kuttler [2]

1 Faculty for Mathematics and Informatics, Universität Heidelberg, Im Neuenheimer Feld 205, D-69120 Heidelberg, Germany

2 Faculty for Mathematics, Technische Universität München, Boltzmannstraße 3, D-85748 Garching bei München, Germany; kuttler@ma.tum.de

* Correspondence: barbarossa@uni-heidelberg.de; Tel.: +49-6221-54-14134

Academic Editor: Yang Kuang
Received: 30 March 2016; Accepted: 3 May 2016; Published: 16 May 2016

Abstract: Quorum sensing is a bacterial cell-to-cell communication mechanism and is based on gene regulatory networks, which control and regulate the production of signaling molecules in the environment. In the past years, mathematical modeling of quorum sensing has provided an understanding of key components of such networks, including several feedback loops involved. This paper presents a simple system of delay differential equations (DDEs) for quorum sensing of *Pseudomonas putida* with one positive feedback plus one (delayed) negative feedback mechanism. Results are shown concerning fundamental properties of solutions, such as existence, uniqueness, and non-negativity; the last feature is crucial for mathematical models in biology and is often violated when working with DDEs. The qualitative behavior of solutions is investigated, especially the stationary states and their stability. It is shown that for a certain choice of parameter values, the system presents stability switches with respect to the delay. On the other hand, when the delay is set to zero, a Hopf bifurcation might occur with respect to one of the negative feedback parameters. Model parameters are fitted to experimental data, indicating that the delay system is sufficient to explain and predict the biological observations.

Keywords: quorum sensing; chemostat; mathematical model; differential equations; delay; bifurcations; dynamical system; numerical simulation

MSC: 92C40; 34K17; 34K60; 34C60

1. Background

More than twenty years ago it was first discovered that even primitive single-celled organisms such as bacteria are able to communicate with each other and coordinate their behavior [1,2]. Bacterial communication is based on the exchange of signaling molecules, or autoinducers, which are produced and released in the surrounding space. At the same time, bacteria are able to measure the autoinducer concentration in the environment, and according to this, they can coordinate and even switch their behavior, adapting to environmental changes. The term "quorum sensing" was coined to summarize the cell-to-cell communication mechanism thanks to which single bacteria cells are able to measure ("sense") the whole population density [3]. Quorum sensing was first observed for the species *Vibrio fischeri* [2], which uses such a mechanism to regulate its bioluminescence. Nowadays, it is known that many bacterial species are able to use similar regulation systems, controlling biofilm formation, swarming motility, and the production of antibiotics or virulence factors [4–6].

The basis for cell-to-cell communication is a gene regulatory network that not only controls certain target genes, but often also their own production, resulting in a positive feedback loop. Gram-positive bacteria use so-called two-component systems (see e.g., [7]), whereas Gram-negative bacteria produce autoinducers directly in the cells, release them to and take them up from the extracellular space without any further modification or transformation.

In the following, we focus on the architecture of a quorum sensing system in Gram-negative bacteria, which mainly communicate via *N*-Acyl homoserine lactones (AHLs) [3,8], typically produced by a synthase. AHL molecules bind to receptors, which control the transcription of target genes. The receptor–AHL complex usually induces the expression of AHL synthases in a positive feedback loop.

We restrict our considerations to the bacteria species *Pseudomonas putida*, a root colonizing, plant growth-promoting organism [9]. Nevertheless, these basic principles may be easily transferred to related bacterial species.

Mathematical modeling of quorum sensing systems has developed in the last decade. Basic principles for a mathematical approach can be found, for example, in [10], where quasi-steady state assumptions for mRNA and corresponding protein in *Pseudomonas aeruginosa* were introduced, or in [11], which focuses on the basic feedback system of *Vibrio fischeri* and the resulting bistability. Alternative approaches for Gram-negative bacteria can be found in [12] (focussing on population dynamics) and [13] (including a further feedback loop). Classical mathematical models for Gram-positive bacteria were introduced, for example, for *Staphylococcus aureus* in [7,14,15].

Several model approaches have also been proposed for *Pseudomonas putida*, in closed systems (batch) as well as in continuous cultures (chemostat) [16–18]. The goal of this manuscript is to review such models, investigating mathematical properties and principles underlying the equations. The interesting component of quorum sensing models of *Pseudomonas putida* is that beside a positive feedback for the autoinducer one also finds a negative feedback via an autoinducer-degrading enzyme, a Lactonase. This is initialized with a certain time lag, leading to a system of delay differential equations (DDEs).

The paper is organized as follows. In Section 2 we provide a short overview of previous modeling approaches for quorum sensing of *Pseudomonas putida*. Starting from ordinary differential equations (ODEs) for the regulatory network in one single cell, in a second step we extend to quorum sensing in populations, including signal exchange among cells and Lactonase activity. The latter component introduces delays into the system. The delay represents the activation time of the Lactonase-dependent negative feedback. Bacteria population might be considered in batch as well as in continuous cultures. It is our purpose to investigate the long term behavior of the presented dynamical systems, and this can be achieved via a reduced model of two delay equations. We explain in great detail how to obtain the two-equation system, maintaining key properties of the gene regulatory network.

In Section 3 we present results concerning the existence and uniqueness of solutions to the reduced model. Moreover, we show that non-negative initial data yield non-negative solutions, a fundamental property of models in biology that is often violated when working with delay differential equations (*cf.* [19]). We compute stationary states of the dynamical system and investigate local stability properties. To this purpose, we compare the DDE system (with a constant delay, $\tau > 0$) to the associated ODE system ($\tau = 0$), studying delay-induced stability switches. In the last part of Section 3, model parameters are fitted to experimental data from [18], indicating that in the long run the reduced model is sufficient to explain and predict the general behavior of the system.

Everywhere in this manuscript, if not otherwise specified, we shall denote variables dependent on time by x or $x(t)$. First derivatives with respect to time are denoted by $\dot{x}$, respectively by $\dot{x}(t)$.

2. Methods

2.1. Compartmental Models

We present in the following compartmental models for quorum sensing of bacteria in a continuous culture. One compartment represents either bacterial population density, the nutrient concentration in the medium, or the concentration of a certain protein/enzyme/signaling substance in a single cell or in the medium.

2.1.1. Regulatory Pathway in One Cell

Let us start to consider the gene regulatory system for a single *Pseudomonas putida* cell. We follow a standard approach for modeling the quorum sensing system in *Pseudomonas putida* (*ppu*), analogous to the *lux* system in *Vibrio fischeri* [11], where polymers of the receptor–AHL complex initiate a positive feedback loop. The autoinducer concentration in *Pseudomonas putida* is regulated by a (self-induced) positive feedback as well as by a negative feedback via the AHL-degrading enzyme Lactonase. Transcriptional activators PpuR bind to AHLs, forming a PpuR-AHL complex which polymerizes. PpuR-AHL n-mers bind to the AHL-dependent quorum sensing locus (ppu-box) and synthesize PpuI. This protein is finally responsible for AHL synthesis. We neglect possible feedbacks (*cf.* [13,20]) on the transcription of PpuR, as these seem to be of minor influence [16]. Thus, just a constant basic production of the receptor PpuR is considered, as in [10,11,17]. Further, we assume as in [16–18] that PpuR-AHL n-mers induce synthesis of Lactonase molecules. A schematic representation of this regulatory pathway is given in Figure 1.

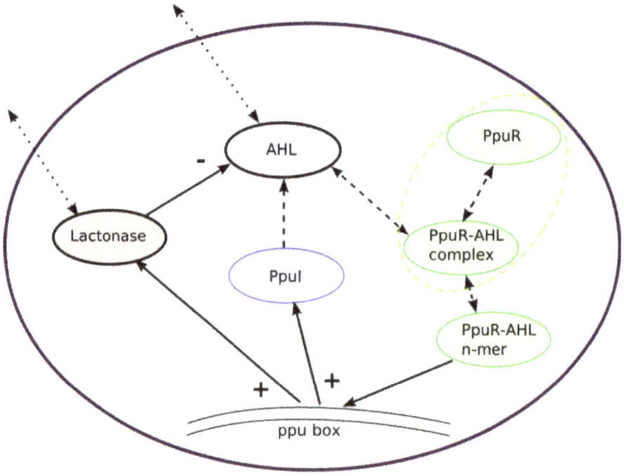

Figure 1. Model structure for the quorum sensing system in one *Pseudomonas putida* cell. *N*-Acyl homoserine lactone (AHL) concentration is regulated by a (self-induced) positive feedback (+) as well as by a negative feedback (−) via the AHL-degrading enzyme Lactonase. The transcriptional activator PpuR binds to AHL forming a PpuR–AHL complex, which polymerizes. PpuR–AHL n-mers bind to the AHL-dependent quorum sensing locus (ppu-box) and synthesize PpuI. This protein is finally responsible for AHL synthesis. Similarly, PpuR-AHL n-mers induce synthesis of Lactonase molecules. Feedbacks on the transcription of PpuR are neglected. Solid arrows represent activations and inhibitions. Dashed arrows indicate reactions and processes which are partially assumed to be in quasi-steady state. Dotted arrows represent the possible exchange of substances between intracellular and extracellular space. The dashed green ellipse refers to the special case in model version (4), where it is assumed that the total amount of PpuR in one cell (consisting of PpuR and the PpuR-AHL complex) is constant whereas in the other models, PpuR and the PpuR-AHL complex follow their own dynamics.

Everywhere in this work, mRNA equations are assumed to be in quasi-steady state. This assumption is justified by the evidence that many proteins are more stable than their own mRNA code (*cf.* [10] and references thereof). Let us denote the intracellular concentrations of PpuI and PpuR at time t by $I(t)$ and $R(t)$, respectively. The variables C and C_i indicate the concentration of the PpuR–AHL complex and of its i-mer in one bacterial cell, respectively. It is assumed that the formation of complex i-mers takes place via the combination of an $(i-1)$-mer with a single PpuR-AHL complex, *cf.* [11]. AHL concentration will be denoted by x; in some cases, it might be convenient to distinguish between intracellular (x_{int}) and extracellular concentration (x_{ext}). The variable y denotes Lactonase concentration.

To begin with, we consider only the positive feedback which regulates AHL. The Lactonase-degrading activity shall be included in a separate step. The positive feedback loop of the regulatory pathway on the protein level described in Figure 1 can be written in the form of an ODE system (*cf.* [21]):

$$
\begin{aligned}
\dot{I} &= \underbrace{\alpha_I}_{\substack{\text{basic}\\\text{production}}} + \underbrace{\beta_I \frac{C_n}{I_{th}+C_n}}_{\substack{\text{feedback-regulated}\\\text{production}}} - \underbrace{\gamma_I I}_{\substack{\text{natural}\\\text{decay}}} \\
\dot{x}_{int} &= \underbrace{\hat{\alpha} I}_{\text{production}} \underbrace{-\gamma_A x_{int}}_{\substack{\text{natural}\\\text{decay}}} \underbrace{-\pi_1^+ x_{int} R}_{\substack{\text{complex}\\\text{formation}}} + \underbrace{\pi_1^- C}_{\substack{\text{complex}\\\text{degradation}}} + \underbrace{d(x_{ext}-x_{int})}_{\substack{\text{exchange}\\\text{with medium}}} \\
\dot{x}_{ext} &= \underbrace{d(x_{int}-x_{ext})}_{\substack{\text{exchange}\\\text{with medium}}} - \underbrace{\gamma_A x_{ext}}_{\substack{\text{natural}\\\text{decay}}} \\
\dot{R} &= \underbrace{\alpha_R}_{\substack{\text{basic}\\\text{production}}} \underbrace{-\pi_1^+ x_{int} R}_{\substack{\text{complex}\\\text{formation}}} + \underbrace{\pi_1^- C}_{\substack{\text{complex}\\\text{degradation}}} - \underbrace{\gamma_R R}_{\substack{\text{natural}\\\text{decay}}} \\
\dot{C} &= \underbrace{\pi_1^+ x_{int} R}_{\substack{\text{complex}\\\text{formation}}} - \underbrace{\pi_1^- C}_{\substack{\text{complex}\\\text{degradation}}} + \underbrace{2\pi_2^- C_2}_{\substack{\text{dimer}\\\text{degradation}}} - \underbrace{2\pi_2^+ C^2}_{\substack{\text{dimer}\\\text{formation}}} + \underbrace{\sum_{j=3}^{n} \pi_j^- C_j}_{\substack{\text{j-mer}\\\text{degradation}}} - \underbrace{\sum_{j=3}^{n} \pi_j^+ C C_{j-1}}_{\substack{\text{j-mer}\\\text{formation}}} \\
\dot{C}_i &= \underbrace{\pi_i^+ C C_{i-1}}_{\substack{\text{i-mer}\\\text{formation}}} - \underbrace{\pi_i^- C_i}_{\substack{\text{i-mer}\\\text{degradation}}} + \underbrace{\pi_{i+1}^- C_{i+1}}_{\substack{\text{(i+1)-mer}\\\text{degradation}}} - \underbrace{\pi_{i+1}^+ C C_i}_{\substack{\text{(i+1)-mer}\\\text{formation}}} \qquad \text{for } 2 \le i \le n-1 \\
\dot{C}_n &= \underbrace{\pi_n^+ C C_{n-1}}_{\substack{\text{n-mer}\\\text{formation}}} - \underbrace{\pi_n^- C_n}_{\substack{\text{n-mer}\\\text{degradation}}} \,.
\end{aligned}
\tag{1}
$$

Although this regulatory pathway seems to be well understood, experimental settings cannot provide information on the dynamics of all components described in system (1). Typically only data for the time course of AHL (and for the population dynamics of the bacteria, which will be introduced in the next step) are available. For this reason, one is interested in a model reduction, decreasing the number of variables and parameters in the system of equations. In a first step, we assume the formation of complexes and its polymers to take place on a fast time scale. Quasi steady state assumptions ($\varepsilon \to 0$) yield for the n-mer,

$$
\varepsilon \dot{C}_n = \pi_n^+ C C_{n-1} - \pi_n^- C_n \quad \underbrace{\to}_{\varepsilon \to 0} \quad C_n = \left(\frac{\pi_n^+}{\pi_n^-}\right) C C_{n-1}
$$

Consider now the $(n-1)$-mer for $\varepsilon \to 0$ and substitute the last expression. We find

$$
0 = \pi_{n-1}^+ C C_{n-2} - \pi_{n-1}^- C_{n-1} + \underbrace{\pi_n^- C_n - \pi_n^+ C C_n}_{=0}
$$

It follows that

$$
C_n = \left(\frac{\pi_n^+ \pi_{n-1}^+}{\pi_n^- \pi_{n-1}^-}\right) C^2 C_{n-2}
$$

and, recursively,

$$C_n = \left(\prod_{j=2}^{n} \frac{\pi_j^+}{\pi_j^-}\right) C^n$$

We denote $p_I := \prod_{j=2}^{n} \frac{\pi_j^+}{\pi_j^-}$ and substitute the result of the quasi-steady state assumption into the I-equation in (1), obtaining

$$\dot{I} = \alpha_I + \beta_I \frac{p_I C^n}{I_{th} + p_I C^n} - \gamma_I I.$$

Observe that the Hill coefficient n covers the fact that polymers (n-mers) of the complex PpuR-AHL are relevant for the positive feedback loop (see also [17]).

To reduce the system further, we also assume that PpuI is in quasi steady state, as in [11,13,17], for example, resulting in

$$I = \frac{\alpha_I}{\gamma_I} + \frac{\beta_I}{\gamma_I} \frac{C^n}{I_{th}/p_I + C^n}$$

Let $C_{th} := \sqrt[n]{I_{th}/p_I}$, then the modified equation for x_{int} reads

$$\dot{x}_{int} = \alpha_A + \beta_A \frac{C^n}{C_{th}^n + C^n} - \gamma_A x_{int} - \pi_1^+ x_{int} R + \pi_1^- C + d(x_{ext} - x_{int}),$$

where $\alpha_A := \hat{\alpha}\alpha_I/\gamma_I$ and $\beta_A := \hat{\alpha}\beta_I/\gamma_I$

Diffusion through the cell membrane plays an important role in regulation processes. Nevertheless, AHL diffusion into and out of the cytoplasm does not require any transport mechanisms and the whole diffusion process goes rather fast, compared to the time scale chosen for the experimental measurements (1 h) [17,22]. This allows us to assume that x_{int} and x_{ext} are in equilibrium. Via steady state assumption, we get

$$x_{ext} = \frac{d x_{int}}{d + \gamma_A} \approx x_{int}$$

as $d \gg \gamma_A$. Taken together, the resulting AHL concentration (now simply denoted by x) follows

$$\dot{x} = \alpha_A + \beta_A \frac{C^n}{C_{th}^n + C^n} - \gamma_A x - \pi_1^+ x R + \pi_i^- C$$

and the simplified version of the single cell model (1) reads

$$\begin{aligned} \dot{x} &= \alpha_A + \beta_A \frac{C^n}{C_{th}^n + C^n} - \gamma_A x - \pi_1^+ x R + \pi_i^- C \\ \dot{R} &= \alpha_R - \pi_1^+ x R + \pi_1^- C - \gamma_R R \\ \dot{C} &= \pi_1^+ x R - \pi_1^- C \end{aligned} \tag{2}$$

2.1.2. Population Dynamics

In the next step, the model is adapted for a bacterial population, including its growth in the classical experimental situation of a batch culture [17]. We denote the bacteria density in the medium at time t by $N(t)$. The dynamics of the bacterial population is classically described by logistic growth,

$$\dot{N} = rN\left(1 - \frac{N}{K}\right)$$

where r is the bacterial growth rate and K the carrying capacity of the batch culture system.

Alternatively, one can consider the situation in a continuous culture, also called chemostat, with a continuous inflow of water and nutrient substrate for the bacteria and an outflow for all extracellular players. In this setting, one introduces a separate variable (S) for the available substrate concentration, which limits the bacterial growth. Consumption of nutrients is usually assumed to lead directly to a proportional increase of the biomass (N). The consumption term includes a saturation with the possibility of a further nonlinearity via the Hill coefficient n_s. Standard equations for nutrient–bacteria dynamics in a chemostat, with dilution rate $D > 0$, are given by [23]

$$\begin{aligned} \dot{S} &= \underbrace{DS_0}_{\text{inflow}} \underbrace{-\gamma_S N \frac{S^{n_s}}{K_m^{n_s} + S^{n_s}}}_{\text{consumption}} - \underbrace{DS}_{\text{washout}} \\ \dot{N} &= \underbrace{aN \frac{S^{n_s}}{K_m^{n_s} + S^{n_s}}}_{\text{growth}} - \underbrace{DN}_{\text{washout}} \end{aligned} \tag{3}$$

2.1.3. Lactonase Regulates AHL Degradation

It turned out by experimental observations [16,18] that a further process plays a major role in the AHL dynamics. In both the batch [16] and the continuous culture experiments [18], maximum concentrations of detected AHLs were followed by a rapid degradation of AHLs to Homoserines, indicating the presence of extracellular enzymatic activity. It is reasonable to assume that the AHL-degrading enzyme is a Lactonase [16], whose production or activation could also be initiated by polymers of the PpuR–AHL complex. Experiments in [16] suggested that Lactonases are activated with a certain delay (about 2 h) compared to the up-regulation of AHL production. From a mathematical point of view, this time lag can be included in the model via a delay differential equation [17].

2.1.4. Full Model

Let us now see how the regulatory pathway model (1), respectively the simplified system (2), can be adapted for a bacteria population. It can be convenient to distinguish between intracellular and extracellular components, and different assumptions are reasonable. For example, whereas in [17] the PpuR concentration was thought for the whole population, we consider here a system where the intracellular components (like PpuR) are interpreted per single (typical) cell.

In [18], to keep the model simple and at the same time to cover some details in the dynamics, equations for the concentrations of AHL (x) and Lactonase (y) in the medium, as well as one equation for the intracellular concentration of PpuR–AHL (C) were added to (3). At the same time, the total amount of PpuR (either free or in the PpuR–AHL complex) in one cell was assumed to be constant.

This does not correspond exactly to reality, but covers the idea that a cell typically maintains the number of receptors within a certain range. This simplification is justified by the still realistic resulting AHL-dynamics (see [18] for details).

The result is the following system of equations:

$$
\begin{aligned}
\dot{S}(t) &= DS_0 - \gamma_S N(t)\frac{S(t)^{n_s}}{K_m^{n_s} + S(t)^{n_s}} - DS(t) \\
\dot{N}(t) &= aN(t)\frac{S(t)^{n_s}}{K_m^{n_s} + S(t)^{n_s}} - DN(t) \\
\dot{x}(t) &= \underbrace{\left(\alpha_A + \beta_A \frac{C(t)^n}{C_{th}^n + C(t)^n}\right) N(t)}_{\substack{\text{total} \\ \text{AHL production}}} - \underbrace{\gamma_A x(t)}_{\substack{\text{natural} \\ \text{decay}}} - \underbrace{\pi_1^+ (R_{const} - C(t))x(t)}_{\substack{\text{complex} \\ \text{formation}}} \\
&\quad + \underbrace{\pi_1^- C(t)}_{\substack{\text{complex} \\ \text{degradation}}} - \underbrace{Dx(t)}_{\text{washout}} - \underbrace{\delta x(t)y(t)}_{\substack{\text{Lactonase-regulated} \\ \text{degradation}}} \\
\dot{C}(t) &= \pi_1^+ (R_{const} - C(t))x(t) - \pi_1^- C(t) \\
\dot{y}(t) &= \underbrace{\alpha_L \frac{C(t-\tau)^m}{C_{th2}^m + C(t-\tau)^m} N(t)}_{\substack{\text{total} \\ \text{Lactonase production}}} - \underbrace{\gamma_L y(t)}_{\substack{\text{natural} \\ \text{decay}}} - \underbrace{Dy(t)}_{\text{washout}}
\end{aligned}
\tag{4}
$$

where m, C_{th2} are the Hill coefficient and the threshold for Lactonase activation, respectively, and δ is the Lactonase-dependent degradation rate of AHLs. Observe that there is no outflow term in the complex equation, as PpuR–AHL is considered to be intracellular.

The model (4) can be extended by adding one equation for PpuR dynamics in one cell, as in [17] or in system (1). Then the system reads

$$
\begin{aligned}
\dot{S}(t) &= DS_0 - \gamma_S N(t)\frac{S(t)^{n_s}}{K_m^{n_s} + S(t)^{n_s}} - DS(t) \\
\dot{N}(t) &= aN(t)\frac{S(t)^{n_s}}{K_m^{n_s} + S(t)^{n_s}} - DN(t) \\
\dot{x}(t) &= \left(\alpha_A + \beta_A \frac{C(t)^n}{C_{th}^n + C(t)^n}\right) N(t) - \gamma_A x(t) - \pi_1^+ R(t)x(t) + \pi_1^- C(t) - Dx(t) - \delta x(t)y(t) \\
\dot{C}(t) &= \pi_1^+ R(t)x(t) - \pi_1^- C(t) \\
\dot{R}(t) &= \alpha_R + \pi_1^- C(t) - \pi_1^+ R(t)x(t) - \gamma_R R(t) \\
\dot{y}(t) &= \alpha_L \frac{C(t-\tau)^m}{C_{th2}^m + C(t-\tau)^m} N(t) - \gamma_L y(t) - Dy(t)
\end{aligned}
\tag{5}
$$

2.1.5. Reduced Model

When being interested in the long term behavior of regulatory systems in the chemostat, one can assume that substrate concentration and bacterial density have approximately assumed a stationary state (N^*, S^*). We consider the system (5) for large values of t and impose quasi-steady state conditions for PpuR and complex. In other words, we assume that when bacteria stay at their saturation level, the dynamics of R and C is slow compared to those of AHL and Lactonase. The equilibrium conditions are given by

$$
R^* = \frac{\alpha_R}{\gamma_R}, \qquad C^* = \underbrace{\frac{\pi_1^+ \alpha_R}{\pi_1^- \gamma_R}}_{=:\tilde{\gamma}} x = \tilde{\gamma} x \tag{6}
$$

Define the parameters

$$\begin{aligned} \alpha = \alpha_A N^*, \quad \beta = \beta_A N^*, \quad x_{th} = C_{th}/\tilde{\gamma}, \quad \omega = \gamma_L + D \\ \gamma = \gamma_A + D, \quad \rho = \alpha_L N^*, \quad y_{th} = C_{th2}/\tilde{\gamma} \end{aligned} \tag{7}$$

Substituting the equilibrium conditions (6) into (5), we obtain the system

$$\begin{aligned} \dot{x}(t) &= \alpha - \gamma x(t) - \delta x(t)y(t) + \beta \frac{x(t)^n}{x_{th}^n + x(t)^n} \\ \dot{y}(t) &= \rho \frac{x(t-\tau)^m}{y_{th}^m + x(t-\tau)^m} - \omega y(t) \end{aligned} \tag{8}$$

Observe that all parameter values are non-negative. Their meaning is summarized in Table 1.

2.2. Experimental Data

We report experimental data as published in the previous publication [18]. *Pseudomonas putida* IsoF was cultivated and grown in a continuous culture with a working volume of 2 L, under controlled conditions at 30 °C, enabling the reproducible establishment of defined environmental conditions.

AHL molecules and their degradation products were identified and quantified via two different methods. The first one is the so-called ultra-high-performance liquid chromatography (UHPLC), a technique used to separate different components in a mixture. The second method, the enzyme-linked immunosorbent assay (ELISA), allows the rapid detection and quantification of AHLs and Homoserines directly in biological samples with the help of antibodies.

2.3. Parameter Estimation

In [18], the model (4) was fitted to a first set of experimental data using a mean square error algorithm and the simplex search algorithm in MATLAB® (Version 2013b, The Mathworks, Natick, MA, USA, 2013). Obtained parameter values were used to validate further data sets with minor adaptations for some initial values, which increased the quality of the fit.

Starting from these estimated parameter values, we fit the reduced system (8) to the same experimental data published in [18]. The fit was performed using curve fitting tools in MATLAB® and Wolfram Mathematica® (Version 10, Wolfram Research, Champaign, IL, USA, 2014). The reduced model (8) is obtained assuming the cell population to be in equilibrium; that is, it holds only for times $t > t_{ec}$, where t_{ec} is the time at which the cell population has reached its saturation level.

3. Results

In this section we present analytical results concerning qualitative properties of the solution of the reduced model (8), as well as numerical simulations and data fit.

3.1. Existence of Solutions

Theorem 1. *Let the system (8) hold for $t \geq t_0$, and let initial data $x(t) = x_0(t)$, $y(t) = y_0(t)$ be given for $t \in [t_0 - \tau, t_0]$, $\tau > 0$, with x_0, y_0 Lipschitz continuous. Then there is a unique solution to (8) in $[t_0, \infty)$. Moreover, if x_0, y_0 are non-negative, the solution is also non-negative.*

Proof. The proof follows from basic principles of DDE theory, *cf.* [19,24,25]. We provide here a sketch of the proof steps. For simplicity, we shall denote the right-hand side of the system (8) by $f(u, v)$, where $u = (x(t), y(t))$ and $v = (x(t-\tau), y(t-\tau))$.

Local existence. For the construction of a local (maximal) solution on an interval $[t_0, t_0 + \Delta)$, $\Delta > 0$, it is sufficient to guarantee Lipschitz continuity of the initial data, as well as of f with respect to both arguments, *cf.* [25] (Thm. 2.2.1). It is easy to verify that the right-hand side of (8) is continuously differentiable with respect to the delayed, as well as to the non-delayed argument, and that the partial derivatives are bounded (computation not shown).

Non-negativity. Preservation of positivity is due to the fact that the delay only appears in the positive feedback term. Indeed, if for some $\bar{t} > t_0$, $x(\bar{t}) = 0$ then $\dot{x}(\bar{t}) = \alpha > 0$, and $x(t)$ remains non-negative. With this result it follows that also y stays non-negative. If for some $\bar{t} > t_0$, $y(\bar{t}) = 0$, then $\dot{y}(\bar{t}) = \rho \frac{x(\bar{t}-\tau)^m}{y_{th}^m + x(\bar{t}-\tau)^m} \geq 0$.

Global existence. We show that the maximal solution is bounded. This follows with estimates on the right-hand side. Observe that

$$\begin{aligned} \dot{y}(t) &= \rho \frac{x(t-\tau)^m}{y_{th}^m + x(t-\tau)^m} - \omega y(t) \\ &\leq \rho - \omega y(t) \end{aligned}$$

hence for all $t \geq t_0$ we have $0 \leq y(t) \leq \hat{y}$, where $\hat{y} := \left(y_0(t_0) - \frac{\rho}{\omega}\right) e^{-\omega t} + \frac{\rho}{\omega}$.

Similarly,

$$\begin{aligned} \dot{x}(t) &= \alpha - \gamma x(t) - \delta x(t) y(t) + \beta \frac{x(t)^n}{x_{th}^n + x(t)^n} \\ &\leq (\alpha + \beta) - (\gamma + \delta y(t)) x(t) \\ &\leq (\alpha + \beta) - \gamma x(t) \end{aligned}$$

Thus for all $t \geq t_0$ we have $0 \leq x(t) \leq \hat{x}$, with $\hat{x} := \left(x_0(t_0) - \frac{\alpha+\beta}{\gamma}\right) e^{-\gamma t} + \frac{\alpha+\beta}{\gamma}$. The maximal solution is bounded, hence it exists on $[t_0, \infty)$, *cf.* [25] (Thm. 2.2.2).

3.2. Fixed Points

Fixed points of (8) are given by the solutions of

$$\begin{cases} 0 &= \alpha - \gamma \bar{x} - \delta \bar{x} \bar{y} + \beta \frac{\bar{x}^n}{x_{th}^n + \bar{x}^n} \\ 0 &= \rho \frac{\bar{x}^m}{y_{th}^m + \bar{x}^m} - \omega \bar{y}. \end{cases}$$

So we have

$$\bar{y} = \frac{\rho}{\omega} \frac{\bar{x}^m}{y_{th}^m + \bar{x}^m}$$

where $\bar{x}$ is given by the solutions of

$$\alpha - \gamma \bar{x} - \frac{\delta \rho}{\omega} \frac{\bar{x}^{m+1}}{y_{th}^m + \bar{x}^m} + \beta \frac{\bar{x}^n}{x_{th}^n + \bar{x}^n} = 0 \tag{9}$$

Recall that for the biological motivation of the model, we are only interested in non-negative $\bar{x}$. In the following, for simplicity of notation, we shall omit the bars from $\bar{x}$.

In the general case $n \neq m$ and $x_{th} \neq y_{th}$, solutions of (9) are the zeros of the polynomial

$$a_0 x^{n+m+1} + a_1 x^{n+m} + a_2 x^{n+1} + a_3 x^{m+1} + a_4 x^n + a_5 x^m + a_6 x + a_7 = 0$$

where

$$a_0 = -(\gamma\omega + \rho\delta) < 0, \qquad a_1 = \omega(\alpha + \beta) > 0,$$
$$a_2 = -\gamma y_{th}^m \omega < 0, \qquad a_3 = -x_{th}^n(\gamma\omega + \rho\delta) < 0,$$
$$a_4 = y_{th}^m \omega(\alpha + \beta) > 0, \qquad a_5 = \alpha\omega x_{th}^n > 0,$$
$$a_6 = -\omega\gamma x_{th}^n y_{th}^m < 0, \qquad a_7 = \alpha\omega x_{th}^n y_{th}^m < 0.$$

Let us consider a special case which is relevant for our application, and assume $n = m = 2$ and $x_{th} = y_{th}$. Then, fixed points $(\bar{x}, \bar{y})$ satisfy

$$\bar{y} = \frac{\rho}{\omega} \frac{\bar{x}^2}{x_{th}^2 + \bar{x}^2} \tag{10}$$

with $\bar{x}$ given by the solutions of a cubic equation

$$(\delta\rho + \omega\gamma)\bar{x}^3 - \omega(\beta + \alpha)\bar{x}^2 - \omega\gamma x_{th}^2 \bar{x} - \alpha\omega x_{th}^2 = 0 \tag{11}$$

which has either three real zeros, or one real and two complex solutions. Thus, we might have up to three biologically-relevant fixed points.

3.3. The case $\tau = 0$

Consider the ODE system obtained from (8) by setting $\tau = 0$:

$$\begin{aligned} \dot{x}(t) &= \alpha - \gamma x(t) - \delta x(t) y(t) + \beta \frac{x(t)^n}{x_{th}^n + x(t)^n} \\ \dot{y}(t) &= \rho \frac{x(t)^m}{y_{th}^m + x(t)^m} - \omega y(t), \end{aligned} \tag{12}$$

It is important to know the dynamics of (12), because for small delays ($\tau > 0$), the DDE system (8) will very likely behave as the ODE system (12), *cf.* [24].

Observe that the ODE system (12) and the DDE system (8) have exactly the same equilibrium points. In general, a DDE system and the associated ODE system have the same number of fixed points, but if the delay appears in the coefficients, the fixed points of the DDE system could be shifted with respect to those of the ODE system.

The presence of a negative feedback in (12) leads to the hypothesis that oscillatory solutions might show up. We investigate local properties of the steady states, looking for Hopf-bifurcations. For linear (local) stability analysis, we compute the Jacobian matrix of system (12),

$$J = \begin{pmatrix} -\gamma - \delta\bar{y} + \beta \frac{n x_{th}^n \bar{x}^{n-1}}{(x_{th}^n + \bar{x}^n)^2} & -\delta\bar{x} \\ \rho \frac{m y_{th}^m \bar{x}^{m-1}}{(y_{th}^m + \bar{x}^m)^2} & -\omega \end{pmatrix}.$$

In the special case $n = m = 2$ and $x_{th} = y_{th}$, we have

$$J = \begin{pmatrix} -\gamma - \delta\bar{y} + \beta \frac{2 x_{th}^2 \bar{x}}{(x_{th}^2 + \bar{x}^2)^2} & -\delta\bar{x} \\ \rho \frac{2 x_{th}^2 \bar{x}}{(x_{th}^2 + \bar{x}^2)^2} & -\omega \end{pmatrix} \tag{13}$$

The trace and the determinant of (13) at a stationary point at $(\bar{x}, \bar{y})$, with $\bar{y}$ in (10), are given by

$$\begin{aligned} \mathrm{Tr}(J) &= \frac{2\beta x_{th}^2 \bar{x}}{(x_{th}^2 + \bar{x}^2)^2} - \gamma - \delta \frac{\rho}{\omega} \frac{\bar{x}^2}{x_{th}^2 + \bar{x}^2} - \omega \\ \det(J) &= -\omega \left(\frac{2\beta x_{th}^2 \bar{x}}{(x_{th}^2 + \bar{x}^2)^2} - \gamma - \delta \frac{\rho}{\omega} \frac{\bar{x}^2}{x_{th}^2 + \bar{x}^2} \right) + \delta \bar{x} \frac{2\rho x_{th}^2 \bar{x}}{(x_{th}^2 + \bar{x}^2)^2}. \end{aligned}$$

If there was only one stationary point and this one is a repellor, one can use the fact that all solutions are bounded and stay positive, then the Poincare–Bendixson theorem yields the existence of periodic solutions. For a Hopf-bifurcation, necessary conditions are $\mathrm{Tr}(J) = 0$ and $\Delta(J) = \mathrm{Tr}(J)^2 - 4\det(J) < 0$. We choose δ, the Lactonase activity, as bifurcation parameter. From the trace condition, we get

$$\delta \left[\frac{\rho}{\omega} \bar{x}^2 (x_{th}^2 + \bar{x}^2) \right] - 2\beta x_{th}^2 \bar{x} + \gamma (x_{th}^2 + \bar{x}^2)^2 + \omega (x_{th}^2 + \bar{x}^2)^2 = 0.$$

We solve for δ and obtain

$$\begin{aligned} \delta = \delta(\bar{x}) &= \frac{2\beta x_{th}^2 \bar{x} - (\gamma + \omega)(x_{th}^2 + \bar{x}^2)^2}{\frac{\rho}{\omega} \bar{x}^2 (x_{th}^2 + \bar{x}^2)} \\ &= \frac{-\omega(\gamma + \omega)(x_{th}^2 + \bar{x}^2)}{\rho \bar{x}^2} + \frac{2\beta x_{th}^2 \omega}{\rho \bar{x} (x_{th}^2 + \bar{x}^2)}. \end{aligned} \tag{14}$$

Note that, in turn, $\bar{x}$ also depends on δ, *cf.* Equation (11). Neglecting this for a minute, we observe that $\lim_{\bar{x} \to \infty} \delta(\bar{x}) = -(\gamma + \omega)\frac{\omega}{\rho} < 0$, whereas $\lim_{\bar{x} \to 0} \delta(\bar{x}) \to +\infty$. Due to the intermediate value theorem, there exists a $\tilde{x} > 0$, such that $\delta(\bar{x}) > 0$ for $\bar{x} > \tilde{x}$. It is possible to choose $\tilde{x}$ as the smallest positive solution of $(\gamma + \omega)(x_{th}^2 + \bar{x}^2)^2 > 2\beta x_{th}^2 \bar{x}$. If a $\bar{x} = \bar{x}(\delta) > 0$ satisfies this condition, then $\mathrm{Tr}(J) = 0$ at $(\bar{x}, \bar{y})$.

In the next step, we check the discriminant condition ($\Delta(J) < 0$), or equivalently, $\det(J) > 0$, as for the Hopf-bifurcation we need simultaneously $\mathrm{Tr}(J) = 0$.

$$\begin{aligned} \det(J) &= -\omega(\mathrm{Tr}(J) + \omega) + \delta \frac{2\rho x_{th}^2 \bar{x}^2}{(x_{th}^2 + \bar{x}^2)^2} \\ &= -\omega^2 + \delta \frac{2\rho x_{th}^2 \bar{x}^2}{(x_{th}^2 + \bar{x}^2)^2} \\ &= \ldots \\ &= \omega^2 \left[\frac{-3x_{th}^2 - \bar{x}^2}{x_{th}^2 + \bar{x}^2} \right] + \omega \frac{2\bar{x}^2}{(x_{th}^2 + \bar{x}^2)^3} \left[2\beta \bar{x} x_{th}^2 - \gamma (x_{th}^2 + \bar{x}^2)^2 \right]. \end{aligned}$$

We solve $\det(J) = 0$ in dependence of the Lactonase decay rate $\omega > 0$, and find the roots $\omega_1 = 0$ (which does not provide further information), and

$$\omega_2 = \frac{\frac{2x_{th}^2}{(x_{th}^2 + \bar{x}^2)^2} \left[2\beta \bar{x} x_{th}^2 - \gamma (x_{th}^2 + \bar{x}^2)^2 \right]}{3x_{th}^2 + \bar{x}^2}.$$

Hence $\det(J) > 0$ when $\omega_2 > 0$. We need to distinguish between two cases. If $\bar{x} > x_{th}$, *i.e.*, a stationary state with high AHL concentration and activated bacteria, then we get

$$2\beta \bar{x} x_{th}^2 - \gamma (x_{th}^2 + \bar{x}^2)^2 > 2\beta x_{th}^3 - \gamma (2\bar{x}^2)^2 = 2\beta x_{th}^3 - 4\gamma \bar{x}^4.$$

Thus, if $2\beta x_{th}^3 - 4\gamma \bar{x}^4 > 0$, then $\omega_2 > 0$. Analogously, we get $\omega_2 > 0$ if $2\beta \bar{x}^3 - 4\gamma x_{th}^4 > 0$, in case of $\bar{x} < x_{th}$, *i.e.*, with bacteria in a non-activated quorum sensing state. All in all, if the model parameters and the resulting stationary point satisfy the last condition yielding $\omega_2 > 0$, and simultaneously $\delta > 0$ according to (14), then a Hopf-bifurcation takes place.

3.4. The Case $\tau > 0$

We are interested in stability switches due to the presence of a delay $\tau > 0$ in (8). Consider the case $n = m = 2$ and $x_{th} = y_{th}$, and let $(\bar{x}, \bar{y})$ be one equilibrium point of (8). The linearized system about $(\bar{x}, \bar{y})$ is given by

$$\dot{Z}(t) = AZ(t) + BZ(t - \tau), \tag{15}$$

with

$$Z(t) = \begin{pmatrix} z_1(t) \\ z_2(t) \end{pmatrix}, A = \begin{pmatrix} a & b \\ 0 & d \end{pmatrix}, B = \begin{pmatrix} 0 & 0 \\ c & 0 \end{pmatrix},$$

and

$$\begin{aligned} a &= -\gamma - \delta\bar{y} + 2\beta x_{th}^2 \frac{\bar{x}}{(x_{th}^2 + \bar{x}^2)}, \\ b &= -\delta\bar{x} \leq 0, \\ c &= 2\rho x_{th}^2 \frac{\bar{x}}{(x_{th}^2 + \bar{x}^2)} \geq 0 \\ d &= -\omega < 0. \end{aligned} \tag{16}$$

The characteristic equation corresponding to (15) is given by

$$det\left(\lambda \mathbb{I} - A - Be^{-\lambda\tau}\right) = 0,$$

or equivalently,

$$\lambda^2 - \lambda(a + d) + ad - bce^{-\lambda\tau} = 0. \tag{17}$$

Characteristic equations of this and more general type have been studied in [24]. In the following, we report results from [24], adapting them to our specific example. We apply standard methods for the analysis of characteristic equations and switches with respect to increasing delays, hence we consider purely imaginary roots, $\lambda = i\varphi, \varphi > 0$. Separating real and imaginary parts in (17) we obtain

$$\begin{aligned} \varphi^2 - ad &= -bc\cos(\varphi t) \\ \varphi(a + d) &= bc\sin(\varphi t). \end{aligned}$$

Now we square left- and right-hand sides and sum up the two equations, obtaining

$$\varphi^4 + \varphi^2(a^2 + d^2) + a^2d^2 = b^2c^2. \tag{18}$$

Its roots are

$$\varphi_{\pm}^2 = \frac{1}{2}\left(-(a^2 + d^2) \pm \sqrt{(a^2 - d^2)^2 + 4b^2c^2}\right).$$

It can be seen from (18) that the parabola in φ^2 is open upwards and it has:

- no positive intercept with the horizontal axis, if $a^2d^2 - b^2c^2 > 0$, *i.e.*, if $|ad| > -bc$;
- one positive intercept (φ_+) with the horizontal axis, if $|ad| < -bc$.

In the first case, there is no stability switch with respect to τ; that is, the stability of the equilibrium point $(\bar{x}, \bar{y})$ remains the same for any $\tau \geq 0$, and it is sufficient to study the ODE case (12). In the case $|ad| < -bc$, there is one root (φ_+) with positive imaginary part, hence one stability switch. In order to find out in which direction the stability switch occurs, we study the sign of the real part $\Re\lambda(\tau)$ in $\lambda = i\varphi_+$, for $\tau > 0$. From (17) we have

$$\left\{2\lambda - (a+d) + \tau bce^{-\lambda\tau}\right\}\frac{d\lambda(\tau)}{d\tau} = \lambda bce^{-\lambda\tau}.$$

It follows

$$\begin{aligned}
\operatorname{sign}\left\{\frac{d\Re\lambda(\tau)}{d\tau}\right\}_{\lambda=i\varphi_+} &= \operatorname{sign}\left\{\Re\left(\frac{d\lambda}{d\tau}\right)^{-1}\right\}_{\lambda=i\varphi_+} \\
&= \operatorname{sign}\left\{\Re\left(\frac{2\lambda-(a+d)}{\lambda(\lambda^2-(a+d)\lambda+ad)}\right)\right\}_{\lambda=i\varphi_+} \\
&= \operatorname{sign}\left\{\frac{2(\varphi_+^2-ad)+(a+d)^2}{(\varphi_+^2(a+d)^2+(ad-\varphi_+^2)^2}\right\} \\
&= \operatorname{sign}\left\{2(\varphi_+^2-ad)+(a+d)^2\right\} \\
&= \operatorname{sign}\left\{2\varphi_+^2+a^2+d^2\right\} \\
&= +1.
\end{aligned}$$

Roots cross the imaginary axis from the left to the right, indicating stability loss. If the solution $(\bar{x}, \bar{y})$ is asymptotically stable for $\tau = 0$, then it is uniformly asymptotically stable for all $\tau < \tau_c$ and unstable for $\tau > \tau_c$, where

$$\tau_c = \frac{\theta_c}{\varphi_+}, \tag{19}$$

with θ_c implicitly defined by

$$\arctan(\theta_c) = \frac{(a+d)\varphi_+}{ad - \varphi_+^2}. \tag{20}$$

All in all, we have shown the following result.

Theorem 2. *Let $(\bar{x}, \bar{y})$ be one equilibrium point of (8), with $\tau > 0$, $n = m = 2$ and $x_{th} = y_{th}$. Assume that $|ad| < bc$, with a, b, c, d given in (16). Then, the equilibrium point is uniformly asymptotically stable for all $0 < \tau < \tau_c$ and unstable for $\tau > \tau_c$, with τ_c defined by (19)–(20).*

3.5. Numerical Simulations and Data Fitting

We consider experimental data published in [18] and perform numerical simulations in MATLAB® and Wolfram Mathematica®. The reduced model (8) is obtained assuming the cell population to be in equilibrium, that is, for times $t > t_{ec}$, where t_{ec} is the time at which bacteria have reached the saturation level. In Figure 2, we read from experimental data that the cell population reaches the equilibrium after *ca.* 20 h from the beginning of the experiment. Hence, we take $t_{ec} = 20$ as the starting time point for numerical simulations of the reduced system (8), and define initial data on the time interval $[t_{ec} - \tau, t_{ec}]$.

For simplicity, we assume that initial data are constant functions on the definition interval, see also [17,18]. We fix the value of the delay, $\tau = 2$ h, as in [18]. Then we take $x(t) = \hat{x}_{19}$, for $t \in [18, 20]$, $\hat{x}_{19}$ being the mean value of ELISA and UHPLC measurements at 19 h from the beginning of the experiment. Initial data for the Lactonase are estimated from simulations of the full model (4) in [18]. To date, there is no experimental data available for Lactonase concentration, thus parameters associated with Lactonase production (ρ), decay (ω), and activity (δ) can be only estimated from AHL experimental data. This means in turn that there are several plausible solutions for the estimation of ρ, ω, and δ. We choose to maintain parameter values as estimated in [18].

It can be seen from the model reduction assumptions, as well as from the simplified parameters (7) that we lack information on the receptor production (α_R) and decay (γ_R); indeed, there is no equation for PpuR in (4). These parameters play a role for the critical threshold value, x_{th}, in the complex-regulated processes. We fit x_{th} and $y(t) = y_0$, $t \in [18, 20]$, fixing all other parameter values as in [18]. The results are summed up in Table 1, with parameter values as provided by the fitting procedure, without rounding. In Figure 3 we show a comparison between the numerical solution of model (4), that of the simplified model (8) and experimental data for AHL time series.

With the estimated parameter values in Table 1, we consider the analytical results in Section 3.2 and Section 3.4. The system has three equilibrium points $\bar{x}_1, \bar{x}_2, \bar{x}_3$, but we only consider the stability properties of the largest one $(x_3, y_3) = (1.593 \times 10^{-7}, 4.809 \times 10^4)$, which corresponds to high AHL level and to an activated state of the bacteria population. Parameter values satisfy $a^2d^2 > b^2c^2$, thus there is no stability switch with respect to τ, and the system behaves as in the case $\tau = 0$. We go back to Section 3.3 and consider the Jacobian matrix (13), obtaining $tr(J) = -7.4243$ and $\det(J) = 0.7685$. Hence, with the estimated parameter values, the system (8) has a locally asymptotically stable equilibrium (x_3, y_3) in which bacteria are activated.

Table 1. Variables and parameters in model (8), with values used for data fit in Figure 3.

Symbol	Description	Value (Unit)	Comments/Source
N^*	Cell density at equilibrium	4.5929×10^{11} (cells/lit)	[18]
α	Basic AHL production rate	1.0564×10^{-7} (mol/(lit$^2 \cdot$ h))	$= \alpha_A * N_{equi}$, [18]
γ	AHL decay rate (includes washout)	0.105 (1/h)	$= \gamma_A + D$, [18]
δ	Lactonase-dependent degradation rate	1.5000×10^{-4} (lit/(mol $\cdot$ h))	[18]
β	Feedback-regulated AHL production rate	1.0564×10^{-6} (mol/(lit$^2 \cdot$ h))	$= \beta_A * N_{equi}$, [18]
n	Hill coefficient for x	2.3 (dimensionless)	[18]
x_{th}	Critical threshold for positive-feedback in x	3.597×10^{-13} (mol/lit)	estimated
ω	Lactonase decay rate (includes washout)	0.105 (1/h)	$= \gamma_e + D$, [18]
ρ	Lactonase production rate	5.0521×10^3 (mol/(lit$^2 \cdot$ h))	$= \alpha_e * N_{equi}$, [18]
τ	Delay in the release of y	2 (h)	[18]
m	Hill coefficient for x	2.5 (dimensionless)	[18]
y_{th}	Critical threshold for positive-feedback in y	3.597×10^{-13} (mol/lit)	estimated
$x_0(t)$	AHL concentration (initial data) $t \in [18, 20]$	5.4044×10^{-7} (mol/lit)	mean of exp. data
$y_0(t)$	Lactonase (initial data) $t \in [18, 20]$	5.2×10^3 (mol/lit)	estimated

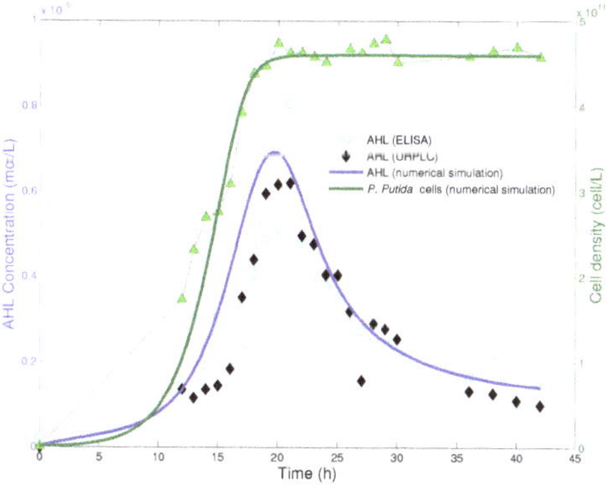

Figure 2. Experimental data and numerical solution of the mathematical model (4). Picture adapted from [18]. Copyright 2014, Springer-Verlag Berlin Heidelberg. The cell population reaches its equilibrium after approximatively 20 h from the beginning of the experiment.

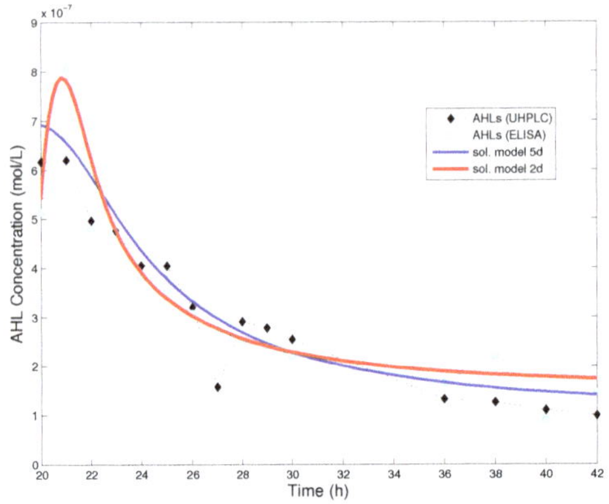

Figure 3. Comparison between the numerical solution of the dynamical systems and experimental data. Red curve: solution of the reduced system (8); Blue curve: solution of the full model (4) in [18]. Initial data for the reduced system are $x(t) = x_0(t)$, $y(t) = y_0(t)$, $t \in [18, 20]$, where $y_0(t) \equiv 5.2 \times 10^{-13}$ was fitted and $x_0(t) \equiv 5.4044 \times 10^{-7}$ is the mean value of ELISA and UHPLC measurements at 19 h from the beginning of the experiment. When the cell population has reached its stationary level, the reduced model provides a good approximation of the dynamics. Parameter values used for the reduced model are given in Table 1.

4. Discussion

In this paper we have introduced a system of two delay differential Equation (8) for quorum sensing of *Pseudomonas putida* in a continuous culture. Motivated by experimental data, a more detailed mathematical model (4) was previously proposed in [18]. Though the system (4) describes the regulatory network in greater detail, in the long run, bacteria reach a saturation level and the model can be reduced to two governing Equation (8), as we have shown in Section 2.1.

Surprisingly, even a simple model such as (8) can be used to explain experimental data (Figure 3), maintaining parameter values from a previous fit [18] for almost all model parameters. However, one should take into account that this is valid only from the moment the bacteria population has reached its saturation level. If one is interested in understanding quorum sensing in the initial phases (lag and exponential phase) of bacterial population, then it is convenient to use a more detailed model, such as (4) or (5).

The advantage of system (8) is that it can be investigated thoroughly thanks to well-established methods. We have shown existence and uniqueness of solutions and, more importantly, we could guarantee preservation of positivity. This property is often violated in systems of delay differential equations. We have studied linearized stability of non-negative equilibria and proved that the delay system (8) might show stability switches as the delay increases. On the other side, the Lactonase activity (δ) can induce Hopf-bifurcation in the associated ODE system (12).

For simplicity of computation in the analysis of the system, we have considered only the case of small Hill coefficients ($n = m = 2$), which corresponds to a maximum of three biologically-relevant stationary states. This assumption is, however, not as restrictive as it seems. Three stationary states, with an intermediate unstable one, are the basis for the bistability situation already discovered in analogous regulatory networks [11,17]. Moreover, similar small values for the Hill coefficients were found to fit experimental data (Table 1) and correspond well to the biological assumption of a dimer being relevant for the positive feedback in the quorum sensing system of *Pseudomonas putida* [16]. With the fitted parameter values, we do not find delay-induced stability switches. This is not a hint of the delay being not relevant. Though a positive time lag might not change the main qualitative behavior of the system, the DDE model still describes the experimentally-determined data and their time course quantitatively better than the associated ODE system, in particular when the bacteria population is in the lag or exponential phase. Stability switches and periodic oscillatory behavior might appear for a different choice of the parameters in system (8). As the main focus of this work was to provide a description for a real biological process, we decided to omit further numerical investigation on the qualitative behavior of (8).

Delay equations have been previously used in mathematical models of continuous cultures. Commonly, a time lag was included to describe the time necessary for the bacteria to convert nutrients in new biomass [26,27]. Being interested in the long term dynamics with bacteria being at an equilibrium, we have chosen not to consider such reproduction lags in our model. In our case, the time lag arises from the dynamics of the regulatory network, in particular from the initialization processes of the AHL-degrading enzyme.

Taken together, the presented simplified delay equation system is a good compromise between refined modeling for a well-known gene regulatory network with several players, and a system of equations which still allows explicit analysis of the basic qualitative behavior as well as parameter determination from few experimental data.

Acknowledgments: Maria Vittoria Barbarossa is supported by the European Social Fund and by the Ministry of Science, Research and Arts Baden-Württemberg.

Author Contributions: M.V.B. and C.K. wrote the paper. M.V.B. collected and analyzed data. C.K. collected literature. M.V.B. and C.K. conceived the study. M.V.B. and C.K. developed the model. M.V.B. and C.K. performed model analysis. M.V.B. performed numerical simulations and parameter fitting.

Conflicts of Interest: The authors declare no conflict of interest.

References

1. Nealson, K.H.; Platt, T.; Hastings, J.W. Cellular control of the synthesis and activity of the bacterial luminescent system. *J. Bacteriol.* **1970**, *104*, 313–322.
2. Fuqua, W.C.; Winans, S.C.; Greenberg, E.P. Quorum sensing in bacteria: The LuxR-LuxI family of cell density-responsive transcriptional regulators. *J. Bacteriol.* **1994**, *176*, 269–275.
3. Williams, P.; Winzer, K.; Chan, W.C.; Camara, M. Look who's talking: Communication and quorum sensing in the bacterial world. *Philos. Trans. R. Soc. Lond. Biol.* **2007**, *362*, 1119–1134.
4. Yarwood, J.M.; Bartels, D.J.; Volper, E.M.; Greenberg, E.P. Quorum sensing in *Staphylococcus aureus* biofilms. *J. Bacteriol.* **2004**, *186*, 1838–1850.
5. Whitehead, N.; Barnard, A.; Slater, H.; Simpson, N.; Salmond, G. Quorum-sensing in Gram-negative bacteria. *FEMS Microbiol. Rev.* **2001**, *25*, 365–404.
6. Rumbaugha, K.; Griswold, J.; Hamood, A. The role of quorum sensing in the *in vivo* virulence of *Pseudomonas aeruginosa*. *Microbes Infect.* **2000**, *2*, 1721–1731.
7. Gustafsson, E.; Nilsson, P.; Karlsson, S.; Arvidson, S. Characterizing the dynamics of the quorum sensing system in *Staphylococcus aureus*. *J. Mol. Microbiol. Biotechnol.* **2004**, *8*, 232–242.
8. Cooley, M.; Chhabra, S.R.; Williams, P. *N*-Acylhomoserine lactone-mediated quorum sensing: A twist in the tail and a blow for host immunity. *Chem. Biol.* **2008**, *15*, 1141–1147.
9. Steidle, A.; Sigl, K.; Schuhegger, R.; Ihring, A.; Schmid, M.; Gantner, S.; Stoffels, M.; Riedel, K.; Givskov, M.; Hartmann, A.; *et al.* Visualization of *N*-acylhomoserine lactone-mediated cell-cell communication between bacteria colonizing the tomato rhizosphere. *Appl. Environ. Microbiol.* **2001**, *67*, 5761–5770.
10. Dockery, J.D.; Keener, J. A mathematical model for quorum sensing in *Pseudomonas aeruginosa*. *Bull. Math. Biol.* **2001**, *63*, 95–116.
11. Müller, J.; Kuttler, C.; Hense, B.A.; Rothballer, M.; Hartmann, A. Cell-cell communication by quorum sensing and dimension-reduction. *J. Math. Biol.* **2006**, *53*, 672–702.
12. Ward, J.; King, J.; Koerber, A.; Williams, P.; Croft, J.; Sockett, R. Mathematical modelling of quorum sensing in bacteria. *Math Med. Biol.* **2001**, *18*, 263–292.
13. Williams, J.; Cui, X.; Levchenko, A.; Stevens, A. Robust and sensitive control of a quorum-sensing circuit by two interlocked feedback loops. *Mol. Syst. Biol.* **2008**, *4*, doi:10.1038/msb.2008.70.

14. Jabbari, S.; King, J.; Koerber, A.; Williams, P. Mathematical modelling of the *agr* operon in *Staphylococcus aureus*. *J. Math. Biol.* **2010**, *61*, 17–54.
15. Koerber, A.; King, J.; Williams, P. Deterministic and stochastic modelling of endosome escape by *Staphylococcus aureus*: Quorum sensing by a single bacterium. *J. Math. Biol.* **2005**, *50*, 440–488.
16. Fekete, A.; Kuttler, C.; Rothballer, M.; Hense, B.A.; Fischer, D.; Buddrus-Schiemann, K.; Lucio, M.; Müller, J.; Schmitt-Kopplin, P.; Hartmann, A. Dynamic regulation of *N*-acyl-homoserine lactone production and degradation in *Pseudomonas putida* IsoF. *FEMS Microbiol. Ecol.* **2010**, *72*, 22–34.
17. Barbarossa, M.V.; Kuttler, C.; Fekete, A.; Rothballer, M. A delay model for quorum sensing of *Pseudomonas putida*. *Biosystems* **2010**, *102*, 148–156.
18. Buddrus-Schiemann, K.; Rieger, M.; Mühlbauer, M.; Barbarossa, M.V.; Kuttler, C.; Hense, A.B.; Rothballer, M.; Uhl, J.; Fonseca, J.R.; Schmitt-Kopplin, P.; *et al.* Analysis of *N*-acylhomoserine lactone dynamics in continuous cultures of *Pseudomonas putida* IsoF by use of ELISA and UHPLC/qTOF-MS-derived measurements and mathematical models. *Anal. Bioanal. Chem.* **2014**, *406*, 6373–6383.
19. Smith, H.L. *An Introduction to Delay Differential Equations with Applications to the Life Sciences*; Springer: New York, NY, USA, 2011.
20. Goryachev, A.B.; Toh, D.J.; Lee, T. Systems analysis of a quorum sensing network: Design constraints imposed by the functional requirements, network topology and kinetic constants. *Biosystems* **2006**, *83*, 178–187.
21. Kuttler, C.; Hense, B.A. Interplay of two quorum sensing regulation systems of *Vibrio fischeri*. *J. Theor. Biol.* **2008**, *251*, 167–180.
22. Pearson, J.P.; van Delden, C.; Iglewski, B. Active efflux and diffusion are involved in transport or *Pseudomonas aeruginosa* cell-to-cell signals. *J. Bacteriol.* **1999**, *181*, 1203–1210.
23. Smith, H.L.; Waltman, P. *The Theory of the Chemostat: Dynamics of Microbial Competition*; Cambridge University Press: Cambridge, UK, 1995; Volume 13.
24. Kuang, Y. *Delay Differential Equations: With Applications in Population Dynamics*; Academic Press: Cambridge, MA, USA, 1993.
25. Bellen, A.; Zennaro, M. *Numerical Methods for Delay Differential Equations*; Oxford University Press: Oxford, UK, 2013.
26. Ellermeyer, S.F. Competition in the chemostat: Global asymptotic behavior of a model with delayed response in growth. *SIAM J. Appl. Math.* **1994**, *54*, 456–465.
27. Freedman, H.I.; So, J.W.H.; Waltman, P. Coexistence in a model of competition in the chemostat incorporating discrete delays. *SIAM J. Appl. Math.* **1989**, *49*, 859–870.

applied sciences

MDPI

Article

Dynamics of a Stochastic Intraguild Predation Model

Zejing Xing [1,2,†], Hongtao Cui [2,†] and Jimin Zhang [2,*,†]

1 College of Automation, Harbin Engineering University, Harbin, Heilongjiang 150001, China; xingzejing@hlju.edu.cn
2 School of Mathematical Sciences, Heilongjiang University, 74 Xuefu Street, Harbin, Heilongjiang 150080, China; hongtaocui1989@163.com
* Correspondence: zhangjimin@hlju.edu.cn; Tel.: +86-0451-8661-3762
† These authors contributed equally to this work.

Academic Editors: Yang Kuang, Meng Fan, Shengqiang Liu and Wanbiao Ma
Received: 25 December 2015; Accepted: 14 April 2016; Published: 22 April 2016

Abstract: Intraguild predation (IGP) is a widespread ecological phenomenon which occurs when one predator species attacks another predator species with which it competes for a shared prey species. The objective of this paper is to study the dynamical properties of a stochastic intraguild predation model. We analyze stochastic persistence and extinction of the stochastic IGP model containing five cases and establish the sufficient criteria for global asymptotic stability of the positive solutions. This study shows that it is possible for the coexistence of three species under the influence of environmental noise, and that the noise may have a positive effect for IGP species. A stationary distribution of the stochastic IGP model is established and it has the ergodic property, suggesting that the time average of population size with the development of time is equal to the stationary distribution in space. Finally, we show that our results may be extended to two well-known biological systems: food chains and exploitative competition.

Keywords: intraguild predation; random perturbations; persistence; stationary distribution; global asymptotic stability

MSC: 60H10, 92D25, 60H30

1. Introduction

Interactions among species can structure biological communities by affecting the identity, number and abundance of species present. Intraguild predation (IGP) has been playing an important role in structuring ecological communities, strongly influencing the structure and function of food webs. IGP describes an interaction in which one predator species consumes another predator species with whom it also competes for shared prey [1,2]. This suggests that IGP combines two important structuring forces in ecological communities: competition and predation. Accordingly, IGP is not only a taxonomically widespread interaction within communities which can occur at different trophic levels, but also a central force to forecast the stability of food webs and the maintenance of biodiversity.

The simplest form of IGP is depicted by a simple food web model in which IGP can occur: a top predator (IG predator P), an intermediate consumer (IG prey N), and a shared prey (R). The development of IGP model can be traced back to Holt and Polis [1] who initially introduce a three species model with the Lotka–Volterra type to study the species coexistence of IGP and point out that it is very difficult to achieve a stable three-species steady state. After that, there are some articles to consider an IGP model with different structures and forms, such as, IGP model with the Lotka–Volterra type [3–5], the IGP model with special forms of the functional and numerical responses [6–8], the IGP model with prey switching or adaptive prey behavior [9,10], and the IGP model with generalist predator or time delay [11–13].

The effect of the random variation of environment is an integral part of any realistic ecosystem. Stochastic models may be more important in characterizing population dynamics in contrast to the deterministic models. In essence, random factors can lead to complete extinction of populations even if the population size is relatively large. Previous studies have explored the dynamic properties for stochastic single species models [14–16], stochastic predator–prey models [17–23], stochastic competitive models [24–27], stochastic mutualism model [28–31]. Specially, Liu and Wang [32] investigated a two-prey one-predator model with random perturbations. However, there are few studies to investigate dynamics of a stochastic IGP model.

Motivated by the existing nice studies and the above considerations, we consider a following IGP model with the Lotka–Volterra type

$$\begin{aligned}
\frac{dR(t)}{dt} &= R(t)(r - a_{rr}R(t) - a_{rn}N(t) - a_{rp}P(t)), \\
\frac{dN(t)}{dt} &= N(t)(-d_n + e_{rn}a_{rn}R(t) - a_{nn}N(t) - a_{np}P(t)), \\
\frac{dP(t)}{dt} &= P(t)(-d_p + e_{rp}a_{rp}R(t) + e_{np}a_{np}N(t) - a_{pp}P(t)),
\end{aligned} \tag{1}$$

where $R(t), N(t)$ and $P(t)$ are the densities of the shared prey, IG prey and IG predator, respectively; r is the per capita growth rate of the shared prey and $d_i(i = n, p)$ is the death rate of species i; $a_{ii}(i = r, n, p)$ is the intraspecific competition rate of species i; a_{rp} and a_{np} are the predation rates of IG predator to the shared prey and IG prey; a_{rn} is the predation rate of IG prey to the shared prey; $e_{ij}(i = r, n, j = n, p)$ is the conversion rates of resource consumption into reproduction for IG prey and IG predator. Here, a_{rn}, a_{rp}, a_{np} is nonnegative constants and the remaining parameters are all positive constants. In view of the fact that the per capita growth rate and the death rate exhibit random fluctuation to a greater or lesser extent (see [33]), we assume that the environmental fluctuation mainly affects the parameters r, d_n and d_p and model these fluctuations by means of independent Gaussian white noises. Let $(B_r(t), B_n(t), B_p(t))^T$ be a three-dimensional Brownian motion defined on a complete probability space $(\Omega, \mathcal{F}, \mathcal{P})$ and

$$r \to r + \alpha_r \dot{B}_r(t),\ d_n \to d_n - \alpha_n \dot{B}_n(t),\ d_p \to d_p - \alpha_p \dot{B}_p(t), \tag{2}$$

where $\alpha_r^2, \alpha_n^2, \alpha_p^2$ are the intensity of the white noise. Thus we consider the Itô's stochastic IGP model as follows:

$$\begin{aligned}
dR(t) &= R(t)(r - a_{rr}R(t) - a_{rn}N(t) - a_{rp}P(t))dt + \alpha_r R(t)dB_r(t), \\
dN(t) &= N(t)(-d_n + e_{rn}a_{rn}R(t) - a_{nn}N(t) - a_{np}P(t))dt + \alpha_n N(t)dB_n(t), \\
dP(t) &= P(t)(-d_p + e_{rp}a_{rp}R(t) + e_{np}a_{np}N(t) - a_{pp}P(t))dt + \alpha_p P(t)dB_p(t).
\end{aligned} \tag{3}$$

The main aim of this paper is to study the dynamics of the model (3). Theoretical studies have suggested that it is very difficult to a achieve stable three-species steady state for the deterministic IGP model. Hence, the first interesting topic of the present paper is whether we can establish a criterion for three- species coexistence under the influence of environmental noise and give the sufficient conditions for global asymptotic stability of the positive solution of model (3). Another important and interesting problem is whether there is a stationary distribution of the stochastic IGP model (3) and if it has the ergodic property.

The rest of the paper is organized as follows. In the next section, we do some necessary preparations including some notations and several important lemmas. In Section 3, we explore stochastic persistence and the extinction of model (3) for five different cases and compare them with the corresponding results of the deterministic model (1).

Then, we establish global asymptotic stability of the positive solution of the model (3). In Section 4, we prove that there is a stationary distribution of model (3), and it has the ergodic property by using the theory of Has'minskii [34]. In the final section, according to the conclusions of previous sections, we first study dynamic properties of two well-known biological systems under random perturbations: food chains and exploitative competition. We state biological implications of our mathematical findings and present some figures to illustrate or complement our mathematical findings.

2. Preliminaries

In this section, we first introduce several important lemmas.

Lemma 1 (see [32]). *Let* $z \in C(\Omega \times [0, +\infty), \mathbb{R}_+)$, $[z]^* = \limsup_{t\to+\infty} \frac{1}{t}\int_0^t z(s)ds$ *and* $[z]_* = \liminf_{t\to+\infty} \frac{1}{t}\int_0^t z(s)ds$.

(i) *If there exist two positive constants* T *and* λ_0 *such that*

$$\ln z(t) \le \lambda t - \lambda_0 \int_0^t z(s)ds + \sum_{i=1}^{n} \sigma_i B_i(t) \tag{4}$$

for all $t \ge T$, *where* $B_i(t)$, $1 \le i \le n$, *are independent standard Brownian motions and* σ_i, $1 \le i \le n$, *are constants, then* $[z]^* \le \lambda/\lambda_0$ *a.s. if* $\lambda \ge 0$ *or* $\lim_{t\to+\infty} z(t) = 0$ *a.s. if* $\lambda < 0$.

(ii) *If there exist three positive constants* T, λ, *and* λ_0 *such that*

$$\ln z(t) \ge \lambda t - \lambda_0 \int_0^t z(s)ds + \sum_{i=1}^{n} \sigma_i B_i(t) \tag{5}$$

for all $t \ge T$, *where* $B_i(t)$, $1 \le i \le n$, *are independent standard Browniam motions and* σ_i, $1 \le i \le n$, *are constants, then* $[z]_* \ge \dfrac{\lambda}{\lambda_0}$ *a.s.*

Similar to Theorem 2.1, Lemma 3.1 and Lemma 3.4 in [25], we have the following lemma:

Lemma 2. *For any given initial value* $(R(0), N(0), P(0))^T \in \mathbb{R}^3_+$ *and any* $p > 0$, *model* (3) *has a unique solution* $(R(t), N(t), P(t))^T$ *on* $t \ge 0$ *which will remain in* $\mathbb{R}^3_+$ *with probability 1 and there is a constant* $K = K(p)$ *such that*

$$\limsup_{t\to+\infty} E(R(t)^p) \le K, \ \limsup_{t\to+\infty} E(N(t)^p) \le K, \ \limsup_{t\to+\infty} E(P(t)^p) \le K. \tag{6}$$

Moreover, the solution $(R(t), N(t), P(t))^T$ *of* (3) *has the properties that*

$$\limsup_{t\to+\infty} \frac{\ln R(t)}{\ln t} \le 1 \text{ a.s.}, \ \limsup_{t\to+\infty} \frac{\ln N(t)}{\ln t} \le 1 \text{ a.s.}, \ \limsup_{t\to+\infty} \frac{\ln P(t)}{\ln t} \le 1 \text{ a.s.} \tag{7}$$

In order to obtain the conditions of global asymptotic stability of solutions for the stochastic model (3), we need the following two lemmas.

Lemma 3 (see [35]). *If there exist positive constants* ω_1, ω_2 *and* κ *such that an* n*-dimensional stochastic process* $Y(t)$, $t \ge 0$ *satisfies*

$$E|Y(t) - Y(s)|^{\omega_1} \le \kappa |t-s|^{1+\omega_2} \tag{8}$$

for $0 \leq t, s < +\infty$, *then there exists a continuous modification* $\overline{Y}(t)$ *of* $Y(t)$ *such that for every* $\omega \in (0, \omega_1/\omega_2)$ *there is a positive random variable* $h(\omega)$ *such that*

$$P\left\{ \sup_{\substack{0<|t-s|<h(\omega) \\ 0\leq t,s<+\infty}} \frac{|\overline{Y}(t,\omega) - Y(s,\omega)|}{|t-s|^{\omega}} \leq \frac{2}{1-2^{-\omega}} \right\} = 1, \tag{9}$$

which implies that almost every sample path of $\overline{Y}(t)$ *is locally but uniformly Hölder continuous with exponent* ω.

Lemma 4 (see [36]). *If* g *is a non-negative function defined on* $[0, +\infty)$ *such that* g *is integrable and is uniformly continuous, then* $\lim_{t\to+\infty} g(t) = 0$.

To establish the existence of a stationary distribution of model (3) in Section 4, we introduce the theory of Has'minskii [34] and let $Y(t)$ be a homogeneous Markov process in E^l (E^l is an l-dimensional Euclidean space) described by the stochastic equation

$$dY(t) = b(Y)dt + \sum_{m=1}^{k} g_m(Y)dB_m(t). \tag{10}$$

Let the diffusion matrix be $\Lambda(x) = (a_{ij}(x))$, $a_{ij}(x) = \sum_{m=1}^{k} g_m^i(x) g_m^j(x)$.

Assumption 1. *There is a bounded domain* $U \subset E^l$ *with regular boundary* Γ *such that*

- (H_1) *In the domain* U *and some neighborhood thereof, the smallest eigenvalue of the diffusion matrix* $\Lambda(x)$ *is bounded away from zero;*
- (H_2) *If* $x \in E^l \setminus U$, *the mean time* τ *at which a path issuing from* x *reaches the set* U *is finite, and* $\sup_{x\in K} E_x\tau < +\infty$ *for every compact subset* $K \subset E^l$.

It is worth noting that we can use the following two stronger conditions to verify (H_1) and (H_2) in Assumption 1:

- (K_1) To obtain (H_1), we only need to show that T is uniformly elliptical in U, where $Tu = b(x)u_x + tr(\Lambda(x)u_{xx})/2$, that is, there exists a $c > 0$ such that $\sum_{i,j=1}^{k} a_{ij}(x)\xi_i\xi_j \geq c|\xi|^2$, $x \in U$, $\xi \in R^l$ (see [37,38]);
- (K_2) To obtain (H_2), we only need to prove that there exist a neighborhood U and a nonnegative C^2-function $V(x)$ such that for any $x \in E^l \setminus U$, $LV(x) < 0$ (see [39]).

Lemma 5 ([34]). *If Assumption 1 holds, then the Markov process* $Y(t)$ *has a stationary distribution* $\mu(\cdot)$. *Moreover, if* $f(\cdot)$ *is a function integrable with respect to the measure* μ, *then*

$$P\Big\{ \lim_{t\to+\infty} \frac{1}{t}\int_0^t f(Y(s))ds = \int_{E^l} f(x)\mu(dx) \Big\} = 1. \tag{11}$$

In order to study dynamic properties of model (3), we do the following notations:

$$[g(t)] = \frac{1}{t}\int_0^t g(s)ds,\ [g]^* = \limsup_{t\to+\infty} \frac{1}{t}\int_0^t g(s)ds,\ [g]_* = \liminf_{t\to+\infty} \frac{1}{t}\int_0^t g(s)ds; \tag{12}$$

$$
L = \begin{vmatrix} a_{rr} & a_{rn} & a_{rp} \\ -e_{rn}a_{rn} & a_{nn} & a_{np} \\ -e_{rp}a_{rp} & -e_{np}a_{np} & a_{pp} \end{vmatrix}, \quad M = \begin{vmatrix} a_{rr} & r & \alpha_r^2/2 \\ -e_{rn}a_{rn} & -d_n & \alpha_n^2/2 \\ -e_{rp}a_{rp} & -d_p & \alpha_p^2/2 \end{vmatrix};
$$
$$
L_1 = \begin{vmatrix} r & a_{rn} & a_{rp} \\ -d_n & a_{nn} & a_{np} \\ -d_p & -e_{np}a_{np} & a_{pp} \end{vmatrix}, \quad M_1 = \begin{vmatrix} \alpha_r^2/2 & a_{rn} & a_{rp} \\ \alpha_n^2/2 & a_{nn} & a_{np} \\ \alpha_p^2/2 & -e_{np}a_{np} & a_{pp} \end{vmatrix};
$$
$$
L_2 = \begin{vmatrix} a_{rr} & r & a_{rp} \\ -e_{rn}a_{rn} & -d_n & a_{np} \\ -e_{rp}a_{rp} & -d_p & a_{pp} \end{vmatrix}, \quad M_2 = \begin{vmatrix} a_{rr} & \alpha_r^2/2 & a_{rp} \\ -e_{rn}a_{rn} & \alpha_n^2/2 & a_{np} \\ -e_{rp}a_{rp} & \alpha_p^2/2 & a_{pp} \end{vmatrix}; \quad (13)
$$
$$
L_3 = \begin{vmatrix} a_{rr} & a_{rn} & r \\ -e_{rn}a_{rn} & a_{nn} & -d_n \\ -e_{rp}a_{rp} & -e_{np}a_{np} & -d_p \end{vmatrix}, \quad M_3 = \begin{vmatrix} a_{rr} & a_{rn} & \alpha_r^2/2 \\ -e_{rn}a_{rn} & a_{nn} & \alpha_n^2/2 \\ -e_{rp}a_{rp} & -e_{np}a_{np} & \alpha_p^2/2 \end{vmatrix}.
$$

3. Stochastic Persistence and Stochastic Extinction

To illuminate the effect of the stochastic perturbations for population and compare the stochastic IGP model (3) with the deterministic IGP model (1), we first explore the existence and local stability of boundary and positive equilibria for model (1). The summary of conditions for the existence and local stability of equilibria are listed in Table 1.

Table 1. Existence and local stability of equilibria for model (1).

Equilibria	Existence	Local Stability
$E_0(0,0,0)$	Always	Never
$E_r(r/a_{rr},0,0)$	Always	$\delta_5<0, \delta_6<0$
$E_{rn}(\delta_4/\delta_1, \delta_5/\delta_1, 0)$	$\delta_5>0$	$L_3<0$
$E_{rp}(\delta_3/\delta_2, 0, \delta_6/\delta_2)$	$\delta_6>0$	$L_2<0$
$E_{rnp}(L_1/L, L_2/L, L_3/L)$	$L>0, L_i>0, i=1,2,3$	$\delta_1\delta_2+\delta_7\delta_8>0$

Here, $\delta_1 = a_{rr}a_{nn} + e_{rn}a_{rn}^2; \delta_2 = a_{rr}a_{pp} + e_{rp}a_{rp}^2; \delta_3 = ra_{pp} + a_{rp}d_p; \delta_4 = ra_{nn} + a_{rn}d_n;$ $\delta_5 = -a_{rr}d_n + re_{rn}a_{rn}; \delta_6 = -a_{rr}d_p + re_{rp}a_{rp}; \delta_7 = -a_{rr}e_{np}a_{np} + a_{rn}e_{rp}a_{rp}; \delta_8 = a_{rr}a_{np} + a_{rp}e_{rn}a_{rn}.$

Now, we analyze stochastic persistence and stochastic extinction of model (3).

Definition 1 (see [32]). *Species $x(t)$ is said to be persistent in the mean if $[x]_* > 0$.*

Let

$$
\begin{aligned}
&\bar{\delta}_1 = \alpha_r^2 a_{nn}/2 - \alpha_n^2 a_{rn}/2, \quad \bar{\delta}_2 = \alpha_r^2 a_{pp}/2 - \alpha_p^2 a_{rp}/2, \\
&\bar{\delta}_3 = \alpha_n^2 a_{rr}/2 + \alpha_r^2 e_{rn}a_{rn}/2, \ \bar{\delta}_4 = \alpha_p^2 a_{rr}/2 + \alpha_r^2 e_{rp}a_{rp}/2.
\end{aligned} \quad (14)
$$

A direct calculation gives

$$
\begin{aligned}
2r/\alpha_r^2 - \delta_5/\bar{\delta}_3 &= a_{rr}(r\alpha_n^2 + d_n\alpha_r^2)/(\alpha_r^2\bar{\delta}_3) > 0, \\
2r/\alpha_r^2 - \delta_6/\bar{\delta}_4 &= a_{rr}(r\alpha_p^2 + d_p\alpha_r^2)/(\alpha_r^2\bar{\delta}_4) > 0, \\
\delta_4/\bar{\delta}_1 - 2r/\alpha_r^2 &= a_{rn}(r\alpha_n^2 + d_n\alpha_r^2)/(\bar{\delta}_1\alpha_r^2) > 0, \\
\delta_3/\bar{\delta}_2 - 2r/\alpha_r^2 &= a_{rp}(r\alpha_p^2 + d_p\alpha_p^2)/(\bar{\delta}_2\alpha_r^2) > 0.
\end{aligned} \quad (15)
$$

Theorem 1. *The following five cases hold:*

(i) *If $2r < \alpha_r^2$, then all the populations are extinction a.s.*

(ii) *If* $2r/\alpha_r^2 > 1 > \max\{\delta_5/\bar{\delta}_3, \delta_6/\bar{\delta}_4\}$, *then* $N(t)$ *and* $P(t)$ *are extinction a.s. and*

$$\lim_{t\to+\infty} \frac{1}{t}\int_0^t R(s)ds = \frac{r-\alpha_r^2/2}{a_{rr}} \quad a.s. \tag{16}$$

(iii) *If* $L > 0$, $\delta_5/\bar{\delta}_3 > 1$ *and* $L_3 < M_3$, *then* $P(t)$ *is extinction a.s. and*

$$\begin{aligned} \lim_{t\to+\infty} \frac{1}{t}\int_0^t R(s)ds &= \frac{\delta_4-\bar{\delta}_1}{\delta_1} \quad a.s., \\ \lim_{t\to+\infty} \frac{1}{t}\int_0^t N(s)ds &= \frac{\delta_5-\bar{\delta}_3}{\delta_1} \quad a.s. \end{aligned} \tag{17}$$

(iv) *If* $L > 0$, $\delta_6/\bar{\delta}_4 > 1$ *and* $L_2 < M_2$, *then* $N(t)$ *is extinction a.s. and*

$$\begin{aligned} \lim_{t\to+\infty} \frac{1}{t}\int_0^t R(s)ds &= \frac{\delta_3-\bar{\delta}_2}{\delta_2} \quad a.s., \\ \lim_{t\to+\infty} \frac{1}{t}\int_0^t P(s)ds &= \frac{\delta_6-\bar{\delta}_4}{\delta_2} \quad a.s. \end{aligned} \tag{18}$$

(v) *If* $L > 0$, $L_i > M_i$, $i = 1,2,3$, *then*

$$\begin{aligned} \lim_{t\to+\infty} \frac{1}{t}\int_0^t R(s)ds &= \frac{L_1-M_1}{L} \quad a.s., \\ \lim_{t\to+\infty} \frac{1}{t}\int_0^t N(s)ds &= \frac{L_2-M_2}{L} \quad a.s., \\ \lim_{t\to+\infty} \frac{1}{t}\int_0^t P(s)ds &= \frac{L_3-M_3}{L} \quad a.s. \end{aligned} \tag{19}$$

Proof. It follows from Itô's formula that

$$\begin{aligned} d\ln R &= (r-\alpha_r^2/2 - a_{rr}R(t) - a_{rn}N(t) - a_{rp}P(t))dt + \alpha_r dB_r(t), \\ d\ln N &= (-d_n - \alpha_n^2/2 + e_{rn}a_{rn}R(t) - a_{nn}N(t) - a_{np}P(t))dt + \alpha_n dB_n(t), \\ d\ln P &= (-d_p - \alpha_p^2/2 + e_{rp}a_{rp}R(t) + e_{np}a_{np}N(t) - a_{pp}P(t))dt + \alpha_p dB_p(t). \end{aligned} \tag{20}$$

By integrating from 0 to t on both sides of the above equation and dividing by t, we have

$$\begin{aligned} \frac{1}{t}\ln\frac{R(t)}{R(0)} &= r - \frac{\alpha_r^2}{2} - a_{rr}[R(t)] - a_{rn}[N(t)] - a_{rp}[P(t)] + \frac{\alpha_r B_r(t)}{t}, \\ \frac{1}{t}\ln\frac{N(t)}{N(0)} &= -d_n - \frac{\alpha_n^2}{2} + e_{rn}a_{rn}[R(t)] - a_{nn}[N(t)] - a_{np}[P(t)] + \frac{\alpha_n B_n(t)}{t}, \\ \frac{1}{t}\ln\frac{P(t)}{P(0)} &= -d_p - \frac{\alpha_p^2}{2} + e_{rp}a_{rp}[R(t)] + e_{np}a_{np}[N(t)] - a_{pp}[P(t)] + \frac{\alpha_p B_p(t)}{t}. \end{aligned} \tag{21}$$

(i) It follows from the first equality of Equation (21) that

$$\frac{1}{t}\ln\frac{R(t)}{R(0)} \le r - \frac{\alpha_r^2}{2} - a_{rr}[R(t)] + \frac{\alpha_r B_r(t)}{t}. \tag{22}$$

By Lemma 1, we have

$$\lim_{t\to+\infty} R(t) = 0 \quad a.s. \tag{23}$$

since $2r < \alpha_r^2$ holds. Substituting Equation (23) into the second equality of Equation (21) yields

$$\frac{1}{t}\ln\frac{N(t)}{N(0)} \le -d_n - \frac{\alpha_n^2}{2} + \varepsilon - a_{nn}[N(t)] + \frac{\alpha_n B_n(t)}{t} \tag{24}$$

for sufficiently large t and sufficiently small ε such that $-d_n - \alpha_n^2/2 + \varepsilon < 0$. Applying Lemma 1 to Equation (24), we get

$$\lim_{t\to+\infty} N(t) = 0 \ a.s. \tag{25}$$

Similarly, in view of the third equality of Equations (21), (23), (25) and Lemma 1, we can conclude that $\lim_{t\to+\infty} P(t) = 0\ a.s.$ This implies that (i) of Theorem 1 holds.

(ii) It follows from Equation (22) and Lemma 1 that

$$[R]^* \le \frac{r - \alpha_r^2/2}{a_{rr}} \ a.s. \tag{26}$$

Combining the second equality of Equation (21) with Equation (26) gives

$$\begin{aligned} \frac{1}{t}\ln\frac{N(t)}{N(0)} &\le -d_n - \frac{\alpha_n^2}{2} + e_{rn}a_{rn}[R]^* + \varepsilon - a_{nn}[N(t)] + \frac{\alpha_n B_n(t)}{t} \\ &\le \frac{\delta_5 - \bar{\delta}_3}{a_{rr}} + \varepsilon - a_{nn}[N(t)] + \frac{\alpha_n B_n(t)}{t} \end{aligned} \tag{27}$$

for sufficiently large t. Then

$$\lim_{t\to+\infty} N(t) = 0 \ a.s. \tag{28}$$

if $\delta_5/\bar{\delta}_3 < 1$ and ε is sufficiently small such that $\delta_5 - \bar{\delta}_3 + a_{rr}\varepsilon < 0$. It follows from the third equality of Equation (21), (26), (28) and Lemma 1 that

$$\lim_{t\to+\infty} P(t) = 0 \ a.s. \tag{29}$$

since $\delta_6/\bar{\delta}_4 < 1$. From Equation (28) and (29) and Lemma 1, we obtain

$$\frac{1}{t}\ln\frac{R(t)}{R(0)} \ge r - \frac{\alpha_r^2}{2} - a_{rr}[R(t)] - \varepsilon + \frac{\alpha_r B_r(t)}{t}$$

for sufficiently large t and

$$[R]_* \ge \frac{r - \alpha_r^2/2}{a_{rr}} \ a.s. \tag{30}$$

Combining Equation (26) with Equation (30) implies that (ii) holds.

(iii) Let

$$\mu_1 = (a_{rn}a_{pp} + a_{rp}e_{np}a_{np})/(a_{nn}a_{pp} + e_{np}a_{np}^2),\ \mu_2 = -(a_{rn}a_{np} - a_{rp}a_{nn})/(a_{nn}a_{pp} + e_{np}a_{np}^2).$$

A direct calculation gives $a_{nn}\mu_1 - e_{np}a_{np}\mu_2 - a_{rn} = 0$ and $a_{np}\mu_1 + a_{pp}\mu_2 - a_{rp} = 0$. Multiplying both sides of three equalities of Equation (21) by -1, μ_1 and μ_2, respectively, and then adding these three equalities, we have

$$\begin{aligned} \frac{1}{t}\ln\frac{R(t)}{R(0)} =& \frac{\mu_1}{t}\ln\frac{N(t)}{N(0)} + \frac{\mu_2}{t}\ln\frac{P(t)}{P(0)} + \frac{L_1 - M_1}{a_{nn}a_{pp} + e_{np}a_{np}^2} - \frac{L}{a_{nn}a_{pp} + e_{np}a_{np}^2}[R(t)] \\ &+ \frac{\alpha_r B_1(t) - \mu_1\alpha_n B_2(t) - \mu_2\alpha_p B_3(t)}{t}. \end{aligned} \tag{31}$$

We consider the following two cases:

Case 1: if $\limsup_{t\to+\infty}(\ln P(t)/\ln t) < 0\ a.s.$, then $\lim_{t\to+\infty} P(t) = 0\ a.s.$

Case 2: if $\limsup_{t\to+\infty}(\ln P(t)/\ln t) \geq 0$ *a.s.*, then by Equation (7), for sufficiently large t, we get

$$\begin{aligned}\frac{1}{t}\ln\frac{R(t)}{R(0)} \leq & \frac{L_1 - M_1}{a_{nn}a_{pp} + e_{np}a_{np}^2} + \varepsilon - \frac{L}{a_{nn}a_{pp} + e_{np}a_{np}^2}[R(t)] \\ & + \frac{\alpha_r B_r(t) - \mu_1\alpha_n B_n(t) - \mu_2\alpha_p B_p(t)}{t}.\end{aligned} \tag{32}$$

It follows from Lemma 1 and the arbitrariness of ε that

$$[R]^* \leq (L_1 - M_1)/L. \tag{33}$$

On the other hand, a direct calculation also shows that

$$a_{rr}\delta_9/\delta_2 + e_{rp}a_{rp}\delta_8/\delta_2 - e_{rn}a_{rn} = 0, \quad -a_{rp}\delta_9/\delta_2 + a_{pp}\delta_8/\delta_2 - a_{np} = 0,$$

where $\delta_9 = a_{rn}a_{np} - a_{rp}a_{nn}$. For sufficiently large t, multiplying both sides of three equalities of Equation (21) by $-\delta_9/\delta_2$, -1 and δ_8/δ_2, respectively, and then adding these three equalities, we obtain

$$\begin{aligned}\frac{1}{t}\ln\frac{N(t)}{N(0)} = & -\frac{\delta_9}{\delta_2 t}\ln\frac{R(t)}{R(0)} + \frac{\delta_8}{\delta_2 t}\ln\frac{P(t)}{P(0)} + \frac{L_2 - M_2}{\delta_2} - \frac{L}{\delta_2}[N(t)] \\ & + \frac{\alpha_n B_n(t) + \delta_{10}/\delta_2\alpha_r B_r(t) - \delta_8/\delta_2\alpha_p B_p(t)}{t}.\end{aligned} \tag{34}$$

Here, we have $\limsup_{t\to+\infty}(\ln R(t)/\ln t) \geq 0$ *a.s.* In fact, if $\limsup_{t\to+\infty}(\ln R(t)/\ln t) < 0$ *a.s.*, then $\lim_{t\to+\infty} R(t) = 0$ *a.s.*, which implies that $\lim_{t\to+\infty} N(t) = 0$ *a.s.* and $\lim_{t\to+\infty} P(t) = 0$ *a.s.* This is a contradiction. By Equation (7), for sufficiently large t, we obtain

$$\begin{aligned}\frac{1}{t}\ln\frac{N(t)}{N(0)} \leq & \frac{L_2 - M_2}{\delta_2} + \varepsilon - \frac{L}{\delta_2}[N(t)] \\ & + \frac{\alpha_n B_n(t) + \delta_{10}/\delta_2\alpha_r B_r(t) - \delta_8/\delta_2\alpha_p B_p(t)}{t}.\end{aligned} \tag{35}$$

From Lemma 1, we get

$$[N]^* \leq (L_2 - M_2)/L. \tag{36}$$

since ε is arbitrary. For the third equality of Equation (21) and sufficiently large t, combining Equation (33) with Equation (36) gives

$$\begin{aligned}\frac{1}{t}\ln\frac{P(t)}{P(0)} & \leq -d_p - \frac{\alpha_p^2}{2} + \varepsilon + e_{rp}a_{rp}[R]^* + e_{np}a_{np}[N]^* - a_{pp}[P(t)] + \frac{\alpha_p B_p(t)}{t} \\ & \leq \frac{a_{pp}(L_3 - M_3)}{L} + \varepsilon - a_{pp}[P(t)] + \frac{\alpha_p B_p(t)}{t}.\end{aligned} \tag{37}$$

Then, $\lim_{t\to+\infty} P(t) = 0$ *a.s.* if ε is sufficiently small.

Combining case 1 with case 2 gives $\lim_{t\to+\infty} P(t) = 0$ *a.s.* The first equality of Equation (21) multiplied by $e_{rn}a_{rn}$ plus the second equality of Equation (21) multiplied by a_{rr} gives

$$\begin{aligned}\frac{a_{rr}}{t}\ln\frac{N(t)}{N(0)} &= -\frac{e_{rn}a_{rn}}{t}\ln\frac{R(t)}{R(0)} + \delta_5 - \bar{\delta}_3 - \delta_1[N(t)] - (e_{rn}a_{rn}a_{rp} + a_{rr}a_{np})[P(t)] \\ &+ \frac{e_{rn}a_{rn}\alpha_r B_r(t) + a_{rr}\alpha_n B_n(t)}{t} \\ &\geq \delta_5 - \bar{\delta}_3 - 2\varepsilon - \delta_1[N(t)] + \frac{e_{rn}a_{rn}\alpha_r B_r(t) + a_{rr}\alpha_n B_n(t)}{t}\end{aligned} \tag{38}$$

for sufficiently large t and sufficiently small ε since Equation (7) and $\lim_{t\to+\infty} P(t) = 0$ *a.s.* hold. It follows from Lemma 1 and the arbitrariness of ε that

$$[N]_* \geq (\delta_5 - \bar{\delta}_3)/\delta_1 \ \ a.s. \tag{39}$$

By applying the above inequality and $\lim_{t\to+\infty} P(t) = 0$ *a.s.* into the first equality of Equation (21), we get

$$\begin{aligned}\frac{1}{t}\ln\frac{R(t)}{R(0)} &\leq r - \frac{\alpha_r^2}{2} + 2\varepsilon - a_{rr}[R(t)] - a_{rn}[N]_* + \frac{\alpha_r B_r(t)}{t} \\ &\leq r - \frac{\alpha_r^2}{2} - \frac{a_{rn}(\delta_5 - \bar{\delta}_3)}{\delta_1} + 2\varepsilon - a_{rr}[R(t)] + \frac{\alpha_r B_r(t)}{t} \\ &= \frac{a_{rr}(\delta_4 - \bar{\delta}_1)}{\delta_1} + 2\varepsilon - a_{rr}[R(t)] + \frac{\alpha_r B_r(t)}{t}.\end{aligned} \tag{40}$$

Then,

$$[R]^* \leq (\delta_4 - \bar{\delta}_1)/\delta_1 \ \ a.s. \tag{41}$$

On the other hand, for sufficiently large t, substituting Equation (41) to the second equality of Equation (21) gives

$$\begin{aligned}\frac{1}{t}\ln\frac{N(t)}{N(0)} &\leq -d_n - \frac{\alpha_n^2}{2} + 2\varepsilon + e_{rn}a_{rn}[R]^* - a_{nn}[N(t)] + \frac{\alpha_n B_n(t)}{t} \\ &\leq -d_n - \frac{\alpha_n^2}{2} + 2\varepsilon + \frac{e_{rn}a_{rn}(\delta_4 - \bar{\delta}_1)}{\delta_1} - a_{nn}[N(t)] + \frac{\alpha_n B_n(t)}{t} \\ &= \frac{a_{nn}(\delta_5 - \bar{\delta}_3)}{\delta_1} + 2\varepsilon - a_{nn}[N(t)] + \frac{\alpha_n B_n(t)}{t},\end{aligned} \tag{42}$$

which implies that

$$[N]^* \leq (\delta_5 - \bar{\delta}_3)/\delta_1 \ \ a.s. \tag{43}$$

Combining Equation (39) with Equation (43) gives

$$[N]^* = (\delta_5 - \bar{\delta}_3)/\delta_1 \ \ a.s. \tag{44}$$

It follows from Equation (43) and $\lim_{t\to+\infty} P(t) = 0$ *a.s.* that

$$\begin{aligned}\frac{1}{t}\ln\frac{R(t)}{R(0)} &\geq r - \frac{\alpha_r^2}{2} - 2\varepsilon - a_{rr}[R(t)] - a_{rn}[N]^* + \frac{\alpha_r B_r(t)}{t} \\ &\geq r - \frac{\alpha_r^2}{2} - \frac{a_{rn}(\delta_5 - \bar{\delta}_3)}{\delta_1} - 2\varepsilon - a_{rr}[R(t)] + \frac{\alpha_r B_r(t)}{t} \\ &= \frac{a_{rr}(\delta_4 - \bar{\delta}_1)}{\delta_1} - 2\varepsilon - a_{rr}[R(t)] + \frac{\alpha_r B_r(t)}{t}.\end{aligned} \tag{45}$$

Then,

$$[R]^* \geq (\delta_4 - \bar{\delta}_1)/\delta_1 \ \ a.s. \tag{46}$$

Combining Equation (41) with (46) gets

$$[R]^* = (\delta_4 - \bar{\delta}_1)/\delta_1 \ \ a.s. \tag{47}$$

It follows from Equation (44) and (47) that (iii) holds.

(iv) Similar to the arguments of (iii), it follows from Equation (36) that $\lim_{t\to+\infty} N(t) = 0 \ a.s.$ if $L_2 < M_2$. The first equality of Equation (21) multiplied by $e_{rp}a_{rp}$ plus the third equality of Equation (21) multiplied by a_{rr} gives

$$\begin{aligned} a_{rr}\frac{1}{t}\ln\frac{P(t)}{P(0)} &= -e_{rn}a_{rn}\frac{1}{t}\ln\frac{R(t)}{R(0)} + \delta_6 - \bar{\delta}_4 - \delta_2[P(t)] - (a_{rn}e_{rp}a_{rp} - a_{rr}e_{np}a_{np})[N(t)] \\ &+ \frac{a_{rr}\alpha_p B_p(t) + e_{rp}a_{rp}\alpha_r B_r(t)}{t} \\ &\geq \delta_6 - \bar{\delta}_4 - 2\varepsilon - \delta_2[P(t)] + \frac{a_{rr}\alpha_p B_p(t) + e_{rp}a_{rp}\alpha_r B_r(t)}{t} \end{aligned} \tag{48}$$

for sufficiently large t. By Lemma 1 and the arbitrariness of ε, we have

$$[P]_* \geq (\delta_6 - \bar{\delta}_4)/\delta_2 \ \ a.s. \tag{49}$$

This implies that

$$\begin{aligned} \frac{1}{t}\ln\frac{R(t)}{R(0)} &\leq r - \frac{\alpha_r^2}{2} + 2\varepsilon - a_{rr}[R(t)] - a_{rp}[P]_* + \frac{\alpha_r B_r(t)}{t} \\ &\leq r - \frac{\alpha_r^2}{2} - \frac{a_{rp}(\delta_6 - \bar{\delta}_4)}{\delta_2} + 2\varepsilon - a_{rr}[R(t)] + \frac{\alpha_r B_r(t)}{t} \\ &= \frac{a_{rr}(\delta_3 - \bar{\delta}_2)}{\delta_2} + 2\varepsilon - a_{rr}[R(t)] + \frac{\alpha_r B_r(t)}{t} \end{aligned} \tag{50}$$

for sufficiently large t. From Lemma 1, we get

$$[R]^* \leq (\delta_3 - \bar{\delta}_2)/\delta_2 \ \ a.s. \tag{51}$$

It follows from $\lim_{t\to+\infty} N(t) = 0 \ a.s.$ and Equation (49) that

$$\begin{aligned} \frac{1}{t}\ln\frac{P(t)}{P(0)} &\leq -d_p - \frac{\alpha_p^2}{2} + 2\varepsilon + e_{rp}a_{rp}[R]^* - a_{pp}[P(t)] + \frac{\alpha_p B_p(t)}{t} \\ &\leq -d_p - \frac{\alpha_p^2}{2} + \frac{e_{rp}a_{rp}(\delta_3 - \bar{\delta}_2)}{\delta_2} + 2\varepsilon - a_{pp}[P(t)] + \frac{\alpha_p B_p(t)}{t} \\ &= \frac{a_{pp}(\delta_6 - \bar{\delta}_4)}{\delta_2} + 2\varepsilon - a_{pp}[P(t)] + \frac{\alpha_p B_p(t)}{t} \end{aligned} \tag{52}$$

for sufficiently large t. Then,

$$[P]^* \leq (\delta_6 - \bar{\delta}_4)/\delta_2 \ \ a.s. \tag{53}$$

Using Equation (53), we have

$$\begin{aligned}\frac{1}{t}\ln\frac{R(t)}{R(0)} &\geq r-\frac{\alpha_r^2}{2}-2\varepsilon-a_{rr}[R(t)]-a_{rp}[P]^*+\frac{\alpha_r B_r(t)}{t}\\ &\geq r-\frac{\alpha_r^2}{2}-\frac{a_{rp}(\delta_6-\bar{\delta}_4)}{\delta_2}-2\varepsilon-a_{rr}[R(t)]+\frac{\alpha_r B_r(t)}{t}\\ &=\frac{a_{rr}(\delta_3-\bar{\delta}_2)}{\delta_2}-2\varepsilon-a_{rr}[R(t)]+\frac{\alpha_r B_r(t)}{t}\end{aligned} \tag{54}$$

for sufficiently large t. Hence,

$$[R]_* \geq (\delta_3-\bar{\delta}_2)/\delta_2 \ a.s. \tag{55}$$

It follows from Equations (49)–(53) and Equation (55) that (iv) holds.

(v) By using Equation (37), we obtain

$$[P]^* \leq (L_3-M_3)/L \tag{56}$$

since $L_3 > M_3$. For sufficiently large t, it follows from (36) and Equation (56) that

$$\begin{aligned}\frac{1}{t}\ln\frac{R(t)}{R(0)} &\geq r-\frac{\alpha_r^2}{2}-2\varepsilon-a_{rr}[R(t)]-a_{rn}[N]^*-a_{rp}[P]^*+\frac{\alpha_r B_r(t)}{t}\\ &\geq r-\frac{\alpha_r^2}{2}-2\varepsilon-a_{rr}[R(t)]-\frac{a_{rn}(L_2-M_2)}{L}-\frac{a_{rp}(L_3-M_3)}{L}+\frac{\alpha_r B_r(t)}{t}\\ &=\frac{a_{rr}(L_1-M_1)}{L}-2\varepsilon-a_{rr}[R(t)]+\frac{\alpha_r B_r(t)}{t},\end{aligned} \tag{57}$$

which means

$$[R]_* \geq (L_1-M_1)/L. \tag{58}$$

Similarly, we have

$$\begin{aligned}\frac{1}{t}\ln\frac{N(t)}{N(0)} &\geq -d_n-\frac{\alpha_n^2}{2}-2\varepsilon+e_{rn}a_{rn}[R]_*-a_{nn}[N(t)]-a_{np}[P]^*+\frac{\alpha_n B_n(t)}{t}\\ &\geq -d_n-\frac{\alpha_n^2}{2}-2\varepsilon+\frac{e_{rn}a_{rn}(L_1-M_1)}{L}-a_{nn}[N(t)]-\frac{a_{np}(L_3-M_3)}{L}+\frac{\alpha_n B_n(t)}{t}\\ &=\frac{a_{nn}(L_2-M_2)}{L}-2\varepsilon-a_{nn}[N(t)]+\frac{\alpha_n B_n(t)}{t}\end{aligned} \tag{59}$$

and

$$\begin{aligned}\frac{1}{t}\ln\frac{P(t)}{P(0)} &\geq -d_p-\frac{\alpha_p^2}{2}-2\varepsilon+e_{rp}a_{rp}[R]_*+e_{np}a_{np}[N]_*-a_{pp}[P(t)]+\frac{\alpha_p B_p(t)}{t}\\ &\geq -d_p-\frac{\alpha_p^2}{2}-2\varepsilon+\frac{e_{rp}a_{rp}(L_1-M_1)}{L}+\frac{e_{np}a_{np}(L_2-M_2)}{L}-a_{pp}[P(t)]+\frac{\alpha_p B_p(t)}{t}\\ &=\frac{a_{pp}(L_3-M_3)}{L}-2\varepsilon-a_{pp}[P(t)]+\frac{\alpha_p B_p(t)}{t}\end{aligned} \tag{60}$$

for sufficiently large t. Then,

$$[N]_* \geq (L_2-M_2)/L, \ \ [P]_* \geq (L_3-M_3)/L. \tag{61}$$

By Equation (33), (36), (56), (58) and (61), (v) holds. The proof of the theorem is complete. □

Now, we establish the sufficient criteria for global asymptotic stability of the positive solutions for the stochastic model (3). This stochastic model (3) is said to be globally asymptotically stable (or globally attractive) if $\lim_{t\to\infty}\max\{|R_1(t)-R_2(t)|,|N_1(t)-N_2(t)|,|P_1(t)-P_2(t)|\} = 0$, where $(R_i(t),N_i(t),P_i(t)), i=1,2$ are two arbitrary solutions of (3) with initial values $(R_i(0),N_i(0),P_i(0)) \in \mathbb{R}^3_+, i=1,2$. By Lemma 3, similar to arguments as those of Lemma 15 in [32], we have the following lemma.

Lemma 6. *If $(R(t),N(t),P(t))$ is a positive solution of (3), then almost every sample path of $R(t),N(t)$ and $P(t)$ are uniformly continuous.*

Theorem 2. *If there exist positive constants δ_1,δ_2 and δ_3 such that*

$$\delta_1 a_{rr} \geq \delta_2 e_{rn} a_{rn} + \delta_3 e_{rp} a_{rp}, \quad \delta_2 a_{nn} \geq \delta_1 a_{rn} + \delta_3 e_{np} a_{np}, \quad \delta_3 a_{pp} \geq \delta_1 a_{rp} + \delta_2 a_{np}, \tag{62}$$

then (3) is globally asymptotically stable.

Proof. We let

$$V(t) = \delta_1|\ln R_1(t) - \ln R_2(t)| + \delta_2|\ln N_1(t) - \ln N_2(t)| + \delta_3|\ln P_1(t) - \ln P_2(t)| \tag{63}$$

for $t \geq 0$, where $(R_i(t),N_i(t),P_i(t))$, $i=1,2$ are two arbitrary solutions of (3) with initial values $(R_i(0),N_i(0),P_i(0)) \in \mathbb{R}^3_+, i=1,2$. A direct calculation gives

$$\begin{aligned}
D^+V(t) =& \delta_1 \operatorname{sgn}(R_1(t)-R_2(t)) \\
&\times [-a_{rr}(R_1(t)-R_2(t)) - a_{rn}(N_1(t)-N_2(t)) - a_{rp}(P_1(t)-P_2(t))]dt \\
&+ \delta_2 \operatorname{sgn}(N_1(t)-N_2(t)) \\
&\times [e_{rn}a_{rn}(R_1(t)-R_2(t)) - a_{nn}(N_1(t)-N_2(t)) - a_{np}(P_1(t)-P_2(t))]dt \\
&+ \delta_3 \operatorname{sgn}(P_1(t)-P_2(t)) \\
&\times [e_{rp}a_{rp}(R_1(t)-R_2(t)) + e_{np}a_{np}(N_1(t)-N_2(t)) - a_{pp}(P_1(t)-P_2(t))]dt \\
\leq& -(\delta_1 a_{rr} - \delta_2 e_{rn}a_{rn} - \delta_3 e_{rp}a_{rp})|R_1(t)-R_2(t)|dt \\
&-(\delta_2 a_{nn} - \delta_1 a_{rn} - \delta_3 e_{np}a_{np})|N_1(t)-N_2(t)|dt \\
&-(\delta_3 a_{pp} - \delta_1 a_{rp} - \delta_2 a_{np})|P_1(t)-P_2(t)|dt := -\Delta(t)dt.
\end{aligned} \tag{64}$$

Then,

$$V(t) + \int_0^t \Delta(s)ds \leq V(0) < +\infty. \tag{65}$$

It follows from Lemmas 6 and 4 that (3) is globally asymptotically stable. □

4. Stationary Distribution and Ergodicity

In this section, we establish the stationary distribution of the stochastic IGP model (3) and show that it has the ergodic property. It is clear that the diffusion matrix of (3) is $\Lambda(x) = \operatorname{diag}(\alpha_r^2 R^2, \alpha_n^2 N^2, \alpha_p^2 P^2)$. Let

$$\begin{aligned}
\lambda_1 &= a_{rr} - (a_{rn} + e_{rn}a_{rn} + a_{rp} + e_{rp}a_{rp})/2, \\
\lambda_2 &= a_{nn} - (a_{np} + e_{np}a_{np} + a_{rn} + e_{rn}a_{rn})/2, \\
\lambda_3 &= a_{pp} - (a_{np} + e_{np}a_{np} + a_{rp} + e_{rp}a_{rp})/2.
\end{aligned} \tag{66}$$

Theorem 3. *If $\lambda_i > 0, i=1,2,3$ and (R^*,N^*,P^*) is the positive equilibrium point of the deterministic model (1) with*

$$(\alpha_r^2 R^* + \alpha_n^2 N^* + \alpha_p^2 P^*)/2 < \min\{\lambda_1 (R^*)^2, \lambda_2 (N^*)^2, \lambda_3 (P^*)^2\}, \tag{67}$$

then there is a stationary distribution $\mu(\cdot)$ for (3) and it has the ergodic property

$$
\begin{aligned}
P\Big\{\lim_{t\to+\infty}\frac{1}{t}\int_0^t R(s)ds &= \int_{\mathbb{R}^3_+}\omega_1\mu(d\omega_1,d\omega_2,d\omega_3)\Big\} = 1,\\
P\Big\{\lim_{t\to+\infty}\frac{1}{t}\int_0^t N(s)ds &= \int_{\mathbb{R}^3_+}\omega_2\mu(d\omega_1,d\omega_2,d\omega_3)\Big\} = 1,\\
P\Big\{\lim_{t\to+\infty}\frac{1}{t}\int_0^t P(s)ds &= \int_{\mathbb{R}^3_+}\omega_3\mu(d\omega_1,d\omega_2,d\omega_3)\Big\} = 1.
\end{aligned}
\tag{68}
$$

Proof. To obtain the conclusion, we need to show that (K_1) and (K_2) hold. It follows from (67) that the ellipsoid

$$
-\lambda_1(R-R^*)^2-\lambda_2(N-N^*)^2-\lambda_3(P-P^*)^2+\frac{\alpha_r^2R^*}{2}+\frac{\alpha_n^2N^*}{2}+\frac{\alpha_p^2P^*}{2}=0 \tag{69}
$$

lies entirely in $\mathbb{R}^3_+$. Let U be a neighborhood of the ellipsoid with $\overline{U}\subseteq\mathbb{R}^3_+$. It is not difficult to show that there exists a $\rho>0$ such that

$$
\sum_{i,j=1}^{3}a_{ij}(x)\omega_i\omega_j=\alpha_r^2R^2\omega_1^2+\alpha_n^2N^2\omega_2^2+\alpha_p^2P^2\omega_3^2\geq\rho|\omega|^2 \tag{70}
$$

for $x\in\overline{U}$ and $\omega\in\mathbb{R}^3$. This implies that (K_1) holds.

Let

$$
V(R,N,P)=R-R^*-R^*\ln\frac{R}{R^*}+N-N^*-N^*\ln\frac{N}{N^*}+P-P^*-P^*\ln\frac{P}{P^*}. \tag{71}
$$

Then,

$$
\begin{aligned}
dV(R,N,P)&=LV(R,N,P)dt\\
&\quad+(R-R^*)\alpha_r dB_r(t)+(N-N^*)\alpha_n dB_n(t)+(P-P^*)\alpha_p dB_p(t),
\end{aligned}
\tag{72}
$$

where

$$
\begin{aligned}
LV(R,N,P)=&(R-R^*)[r-a_{rr}R-a_{rn}N-a_{rp}P]+\alpha_r^2R^*/2\\
&+(N-N^*)[-d_n+e_{rn}a_{rn}R-a_{nn}N-a_{np}P]+\alpha_n^2N^*/2\\
&+(P-P^*)[-d_p+e_{rp}a_{rp}R+e_{np}a_{np}N-a_{pp}P]+\alpha_p^2P^*/2.
\end{aligned}
\tag{73}
$$

Since (R^*,N^*,P^*) is the positive equilibrium point of (1), we have

$$
\begin{aligned}
LV(R,N,P)=&(R-R^*)[-a_{rr}(R-R^*)-a_{rn}(N-N^*)-a_{rp}(P-P^*)]+\alpha_r^2R^*/2\\
&+(N-N^*)[e_{rn}a_{rn}(R-R^*)-a_{nn}(N-N^*)-a_{np}(P-P^*)]+\alpha_n^2N^*/2\\
&+(P-P^*)[e_{rp}a_{rp}(R-R^*)+e_{np}a_{np}(N-N^*)-a_{pp}(P-P^*)]+\alpha_p^2P^*/2\\
\leq&-\lambda_1(R-R^*)^2-\lambda_2(N-N^*)^2-\lambda_3(P-P^*)^2\\
&+(\alpha_r^2R^*+\alpha_n^2N^*+\alpha_p^2P^*)/2.
\end{aligned}
\tag{74}
$$

Then, for any $x\in\mathbb{R}^3_+\setminus U$, we get $LV(x)<0$, which means that (K_2) holds. It follows from Lemma 5 that (3) has a stationary distribution $\mu(\cdot)$, and it is ergodic.

On the other hand, for any $m>0$, it follows from the dominated convergence theorem and Lemma 2 that

$$
E\Big[\lim_{t\to+\infty}\frac{1}{t}\int_0^t(R(s)\wedge m)ds\Big]=\lim_{t\to+\infty}\frac{1}{t}\int_0^t E(R(s)\wedge m)ds\leq K. \tag{75}
$$

By the ergodic property, we have

$$\int_{\mathbb{R}^3_+} (\omega_1 \wedge m)\mu(d\omega_1, d\omega_2, d\omega_3) = E\Big[\lim_{t\to+\infty} \frac{1}{t}\int_0^t (R(s) \wedge m)ds\Big] \le K. \tag{76}$$

Then, $\int_{\mathbb{R}^3_+} \omega_1 \mu(d\omega_1, d\omega_2, d\omega_3) \le K$ as $m \to +\infty$. By Lemma 5, the first equality of Equation (68) holds. Similarly, we can conclude that the second and third equalities of Equation (68) hold. The proof of the theorem is completed. □

5. Conclusions

In this section, we first focus on the stochastic food chains model and the stochastic exploitative competition model. In the model (3), if we let $a_{rp} = 0$ or $a_{np} = 0$, then we get the stochastic food chains model

$$\begin{aligned} dR(t) &= R(t)(r - a_{rr}R(t) - a_{rn}N(t))dt + \alpha_r R(t)dB_r(t), \\ dN(t) &= N(t)(-d_n + e_{rn}a_{rn}R(t) - a_{nn}N(t) - a_{np}P(t))dt + \alpha_n N(t)dB_n(t), \\ dP(t) &= P(t)(-d_p + e_{np}a_{np}N(t) - a_{pp}P(t))dt + \alpha_p P(t)dB_p(t), \end{aligned} \tag{77}$$

and the stochastic exploitative competition model

$$\begin{aligned} dR(t) &= R(t)(r - a_{rr}R(t) - a_{rn}N(t) - a_{rp}P(t))dt + \alpha_r R(t)dB_r(t), \\ dN(t) &= N(t)(-d_n + e_{rn}a_{rn}R(t) - a_{nn}N(t))dt + \alpha_n N(t)dB_n(t), \\ dP(t) &= P(t)(-d_p + e_{rp}a_{rp}R(t) - a_{pp}P(t))dt + \alpha_p P(t)dB_p(t). \end{aligned} \tag{78}$$

In view of the stochastic IGP model (3), Theorems 1, 2, 3 reduce the corresponding results of models (77) and (78), that is, we get the stochastic persistence and stochastic extinction, stationary distribution and ergodicity, and globally asymptotically stability of the positive solution for the stochastic food chains model (77), and the stochastic exploitative competition model (78), in the case of $a_{rp} = 0$ or $a_{np} = 0$.

In this paper, we have developed a stochastic IGP model (3) describing the interactions among a top predator (IG predator P), an intermediate consumer (IG prey N), and a shared prey (R) under the influence of environmental noise. We have analyzed the dynamic properties for the stochastic IGP model (3) and the deterministic IGP model (1). As applications, we show that our results may be extended to two well-known biological systems: food chains and exploitative competition.

Comparing the stochastic IGP model (3) with the deterministic IGP model (1) (see Theorems 1, 2, 3 and Table 1), we obtain the following conclusions:

- In the deterministic model (1), the total extinction of three populations is impossible since E_0 is unstable. However, this situation is possible for the stochastic model (3) when the noise intensity α_r is large enough (see Figure 1a);
- The existence of the shared prey with the extinction of both IG prey and IG predators is a possible outcome of the stochastic model (3) (see Figure 1b). There is also evidence that the noise is a harmful factor for the shared prey population (see E_r of Table 1 and (ii) of Theorem 1);
- The existence of both the shared prey and IG prey with the extinction of IG predators, and the existence of both the shared prey and IG predators with the extinction of IG prey are both possible outcomes of the stochastic model (3) with different sets of parameters (see Figure 1c,d). Here, it is worth noting that the noise has a negative effect for IG prey and IG predators, and may also have a positive effect for the shared prey if the values of α_n and α_p grow larger (see (iii) and (iv) of Theorem 1). This also implies that stochastic fluctuation of N or P would help R to grow larger;
- This study suggests that the shared prey, IG prey and IG predators can coexist together for the stochastic model (3), which implies that it is possible for the coexistence of three species under the influence of environmental noise (see Figure 1e). There is recognition that the noise may be

favorable to three-species coexistence if $M_i < 0, i = 1, 2, 3$ (see (v) of Theorem 1). In addition, we also prove that three-species is stable coexistence for the influence of environmental noise (see Theorem 2 and Figure 1f);

- The study of Theorem 3 suggests that the time average of the population size of model (3) with the development of time is equal to the stationary distribution in space.

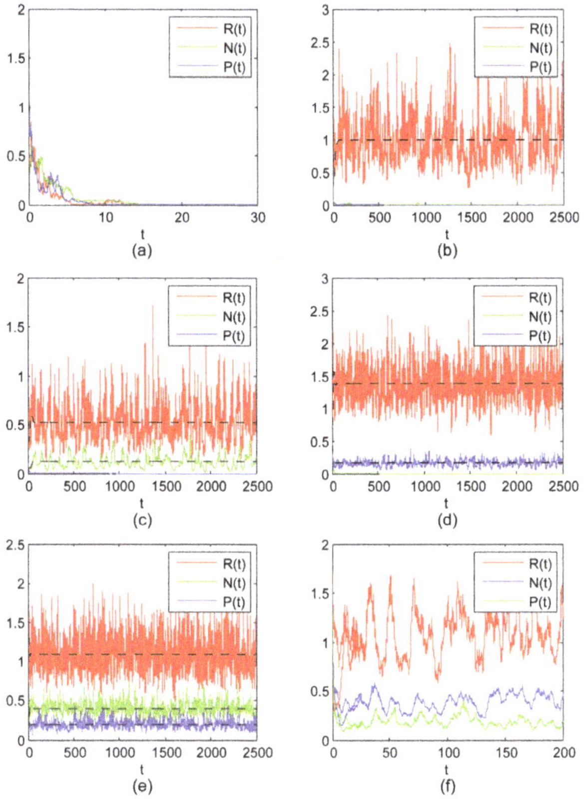

Figure 1. (**a**) $\alpha_r = 0.9177$, $\alpha_n = 0.4472$, $\alpha_p = 0.6325$, $0.8 = 2r < \alpha_r^2 = 0.8422$; (**b**) $\alpha_r = 0.7750$, $\alpha_n = 0.6325$, $\alpha_p = 0.5477$, $1.3333 = 2r/\alpha_r^2 > 1 > \max\{\delta_5/\bar{\delta}_3 = 0.9991, \delta_6/\bar{\delta}_4 = 0.5123\}$, $\lim_{t\to+\infty} \frac{1}{t}\int_0^t R(s)ds = \frac{r - \alpha_r^2/2}{a_{rr}} = 0.9970$; (**c**) $\alpha_r = 0.7746$, $\alpha_n = 0.1414$, $\alpha_p = 0.1414$, L= 0.1790, $\delta_5/\bar{\delta}_3 = 1.2088 > 1$ and $0.0445 = L_3 < M_3 = 0.0501$, $\lim_{t\to+\infty} \frac{1}{t}\int_0^t R(s)ds = \frac{\delta_4 - \bar{\delta}_1}{\delta_1} = 0.5250$ and $\lim_{t\to+\infty} \frac{1}{t}\int_0^t N(s)ds = \frac{\delta_5 - \bar{\delta}_3}{\delta_1} = 0.1187$; (**d**) $\alpha_r = 0.6$, $\alpha_n = 0.8$, $\alpha_p = 0.4$, $\delta_6/\bar{\delta}_4 = 1.2 > 1$ and $0.0730 = L_2 < M_2 = 0.0862$, $\lim_{t\to+\infty} \frac{1}{t}\int_0^t R(s)ds = \frac{\delta_3 - \bar{\delta}_2}{\delta_2} = 1.6565$ and $\lim_{t\to+\infty} \frac{1}{t}\int_0^t P(s)ds = \frac{\delta_6 - \bar{\delta}_4}{\delta_2} = 0.3826$; (**e**) $\alpha_r = 0.7746, \alpha_n = 0.1414, \alpha_p = 0.2449, 0.2015 = L_1 > M_1 = 0.0046, 0.0730 = L_2 > M_2 = 0.0014, 0.0445 = L_3 > M_3 = 0.0087$, $\lim_{t\to+\infty} \frac{1}{t}\int_0^t R(s)ds = \frac{L_1 - M_1}{L} = 1.1000$, $\lim_{t\to+\infty} \frac{1}{t}\int_0^t N(s)ds = \frac{L_2 - M_2}{L} = 0.4000$ and $\lim_{t\to+\infty} \frac{1}{t}\int_0^t P(s)ds = \frac{L_3 - M_3}{L} = 0.2000$; (**f**) $\alpha_r = 0.7746$, $\alpha_n = 0.1414$, $\alpha_p = 0.2449$, $R_1(0) = 1.7$, $N_1(0) = 0.6$, $P_1(0) = 0.3$, $R_2(0) = 0.7$, $N_2(0) = 1.6$, $P_2(0) = 1.3$. Here $r = 0.4$, $d_n = 0.1$, $d_p = 0.2$, $a_{rr} = 0.1$, $a_{rn} = 0.4$, $a_{rp} = 0.5$, $e_{rn} = 0.75$, $a_{nn} = 0.4$, $a_{np} = 0.3$, $e_{rp} = 0.6$, $e_{np} = 0.5$, $a_{pp} = 0.8$.

Acknowledgments: The authors are grateful to the anonymous referees for carefully reading the manuscript and for important suggestions and comments, which led to the improvement of their manuscript. This research is supported by National Natural Science Foundation of China (No.11201128,11302127) Natural Science Foundation of Heilongjiang Province (No.A201414,F2015032), The National High Technology Research and Development Program of China (No.2013AA122904), Science and Technology Innovation Team in Higher Education Institutions of Heilongjiang Province (No.2014TD005), and The Heilongjiang University Fund for Distinguished Young Scholars (No.201203).

Author Contributions: Zejing Xing and Jimin Zhang conceived the study and drafted the manuscript. Hongtao Cui participated in the design of the study and analysis of the results.

Conflicts of Interest: The authors declare no conflict of interest.

References

1. Holt, R.D.; Polis, G.A. A theoretical framework for intraguild predation. *Am. Nat.* **1997**, *149*, 745–764.
2. Polis, G.A.; Holt, R.D. Intraguild predation: The dynamics of complex trophic interactions. *Trends Ecol. Evol.* **1992**, *7*, 151–154.
3. Hsu, S.B.; Ruan, S.G.; Yang, T.H. Analysis of three species Lotka–Volterra food web models with omnivory. *J. Math. Anal. Appl.* **2015**, *426*, 659–687.
4. Shchekinova, E.Y.; Löder, M.G.J.; Boersma, M.; Wiltshire, K.H. Facilitation of intraguild prey by its intraguild predator in a three-species Lotka–Volterra model. *Theor. Popul. Biol.* **2014**, *92*, 55–61.
5. Velazquez, I.; Kaplan, D.; Velasco-Hernandez, J.X.; Navarrete, S.A. Multistability in an open recruitment food web model. *Appl. Math. Comput.* **2005**, *163*, 275–294.
6. Abrams, P.A.; Fung, S.R. Prey persistence and abundance in systems with intraguild predation and type-2 functional responses. *J. Theor. Biol.* **2010**, *264*, 1033–1042.
7. Freeze, M.; Chang, Y.; Feng, W. Analysis of dynamics in a complex food chain with ratio-dependent functional response. *J. Appl. Anal. Comput.* **2014**, *4*, 69–87.
8. Verdy, A.; Amarasekare, P. Alternative stable states in communities with intraguild predation. *J. Theor. Biol.* **2010**, *262*, 116–128.
9. Urbani, P.; Ramos-Jiliberto, R. Adaptive prey behavior and the dynamics of intraguild predation systems. *Ecol. Model.* **2010**, *221*, 2628–2633.
10. Zabalo, J. Permanence in an intraguild predation model with prey switching. *Bull. Math. Biol.* **2012**, *74*, 1957–1984.
11. Fan, M.; Kuang, Y.; Feng, Z.L. Cats protecting birds revisited. *Bull. Math. Biol.* **2005**, *67*, 1081–1106.
12. Kang, Y.; Wedekin, L. Dynamics of a intraguild predation model with generalist or specialist predator. *J. Math. Biol.* **2013**, *67*, 1227–1259.
13. Shu, H.Y.; Hu, X.; Wang, L.; Watmough, J. Delay induced stability switch, multitype bistability and chaos in an intraguild predation model. *J. Math. Biol.* **2015**, *71*, 1269–1298.
14. Golec, J.; Sathananthan, S. Stability analysis of a stochastic logistic model. *Math. Comput. Model.* **2003**, *38*, 585–593.
15. Jiang, D.Q.; Shi, N.Z.; Li, X.Y. Global stability and stochastic permanence of a non-autonomous logistic equation with random perturbation. *J. Math. Anal. Appl.* **2008**, *340*, 588–597.
16. Liu, M.; Wang, K.; Hong, Q. Stability of a stochastic logistic model with distributed delay. *Math. Comput. Model.* **2013**, *57*, 1112–1121.
17. Aguirre, P.; González-Olivares, E.; Torres, S. Stochastic predator–prey model with Allee effect on prey. *Nonlinear Anal. RWA* **2013**, *14*, 768–779.
18. Ji, C.Y.; Jiang, D.Q. Dynamics of a stochastic density dependent predator–prey system with Beddington-DeAngelis functional response. *J. Math. Anal. Appl.* **2011**, *381*, 441–453.
19. Liu, M.; Wang, K. Global stability of a nonlinear stochastic predator–prey system with Beddington-DeAngelis functional response. *Commun. Nonlinear Sci.* **2011**, *16*, 1114–1121.
20. Mandal, P.S.; Banerjee, M. Stochastic persistence and stability analysis of a modified Holling-Tanner model. *Math. Method. Appl. Sci.* **2013**, *36*, 1263–1280.
21. Saha, T.; Chakrabarti, C. Stochastic analysis of prey-predator model with stage structure for prey. *J. Appl. Math. Comput.* **2011**, *35*, 195–209.

22. Vasilova, M. Asymptotic behavior of a stochastic Gilpin-Ayala predator–prey system with time-dependent delay. *Math. Comput. Model.* **2013**, *57*, 764–781.
23. Yagi, A.; Ton, T.V. Dynamic of a stochastic predator–prey population. *Appl. Math. Comput.* **2011**, *218*, 3100–3109.
24. Jovanović, M.; Vasilova, M. Dynamics of non-autonomous stochastic Gilpin-Ayala competition model with time-varying delays. *Appl. Math. Comput.* **2013**, *219*, 6946–6964.
25. Li, X.; Mao, X. Population dynamical behavior of non-autonomous Lotka–Volterra competitive system with random perturbation. *Discrete Cont. Dyn. Syst. Ser.* **2009**, *24*, 523–593.
26. Lian, B.S.; Hu, S.G. Asymptotic behaviour of the stochastic Gilpin-Ayala competition models. *J. Math. Anal. Appl.* **2008**, *339*, 419–428.
27. Zhu, C.; Yin, G. On competitive Lotka–Volterra model in random environments. *J. Math. Anal. Appl.* **2009**, *357*, 154–170.
28. Ji, C.Y.; Jiang, D.Q.; Liu, H.; Yang, Q.S. Existence, uniqueness and ergodicity of positive solution of mutualism system with stochastic perturbation. *Math. Probl. Eng.* **2010**, *2010*, doi:10.1155/2010/684926.
29. Liu, M.; Wang, K. Analysis of a stochastic autonomous mutualism model. *J. Math. Anal. Appl.* **2013**, *402*, 392–403.
30. Liu, M.; Wang, K. Population dynamical behavior of Lotka–Volterra cooperative systems with random perturbations. *Discrete Cont. Dyn. Syst. Ser.* **2013**, *33*, 2495–2522.
31. Liu, Q. Analysis of a stochastic non-autonomous food-limited Lotka–Volterra cooperative model. *Appl. Math. Comput.* **2015**, *254*, 1–8.
32. Liu, M.; Wang, K. Dynamics of a two-prey one-predator system in random environments. *J. Nonlinear Sci.* **2013**, *23*, 751–775.
33. May, R.M. *Stability and Complexity in Model Ecosystems*. Princeton University Press: Princeton, NJ, USA, 1973.
34. Has'minskii, R.Z. *Stochastic Stability of Differential Equations*; Sijthoff & Noordhoff: Alphen aan den Rijn, the Netherlands, 1980.
35. Karatzas, I.; Shreve, S. *Brownian Motion and Stochastic Calculus*; Springer: Berlin, Germany, 1991.
36. Barbalat, I. Systems d'equations differentielles d'oscillations nonlineaires. *Rev. Roum. Math. Pures Appl.* **1959**, *4*, 267–270.
37. Gard, T.C. *Introduction to Stochastic Differential Equations*; Marcel Dekker, Inc.: New York, NY, USA, 1988.
38. Strang, G. *Linear Algebra and Its Applications*; Wellesley-Cambridge Press: London, UK, 1988.
39. Zhu, C.; Yin, G. Asymptotic properties of hybrid diffusion systems. *SIAM J. Control Optim.* **2007**, *46*, 1155–1179.

applied sciences

MDPI

Article

Novel Graphical Representation and Numerical Characterization of DNA Sequences

Chun Li [1,2,*], Wenchao Fei [1], Yan Zhao [1] and Xiaoqing Yu [3]

[1] Department of Mathematics, Bohai University, Jinzhou 121013, China; feiwenchao90@163.com (W.F.); zhaoyan_jinzh@126.com (Y.Z.)
[2] Research Institute of Food Science, Bohai University, Jinzhou 121013, China
[3] Department of Applied Mathematics, Shanghai Institute of Technology, Shanghai 201418, China; xqyu@sit.edu.cn
* Correspondence: lichwun@163.com; Tel.: +86-416-3402166

Academic Editor: Yang Kuang
Received: 10 December 2015; Accepted: 14 February 2016; Published: 24 February 2016

Abstract: Modern sequencing technique has provided a wealth of data on DNA sequences, which has made the analysis and comparison of sequences a very important but difficult task. In this paper, by regarding the dinucleotide as a 2-combination of the multiset $\{\infty \cdot A, \infty \cdot G, \infty \cdot C, \infty \cdot T\}$, a novel 3-D graphical representation of a DNA sequence is proposed, and its projections on planes (x,y), (y,z) and (x,z) are also discussed. In addition, based on the idea of "piecewise function", a cell-based descriptor vector is constructed to numerically characterize the DNA sequence. The utility of our approach is illustrated by the examination of phylogenetic analysis on four datasets.

Keywords: 2-combination; graphical representation; cell-based vector; numerical characterization; phylogenetic analysis

1. Introduction

The rapid development of DNA sequencing techniques has resulted in explosive growth in the number of DNA primary sequences, and the analysis and comparison of biological sequences has become a topic of considerable interest in Computational Biology and Bioinformatics. The traditional measure for similarity analysis of DNA sequences is based on multiple sequence alignment, which uses dynamic programming techniques to identify the globally optimal alignment solution. However, the sequence alignment problem is NP-hard (non-deterministic polynomial-time hard), making it infeasible for dealing with large datasets [1]. To overcome the limitation, a lot of alignment-free approaches for sequence comparison have been proposed.

The basic idea behind most alignment-free methods is to characterize DNA by certain mathematical models derived for DNA sequence, rather than by a direct comparison of DNA sequences themselves. Graphical representation is deemed to be a simple and powerful tool for the visualization and analysis of bio-sequences. The earliest attempts at the graphical representation of DNA sequences were made by Hamori and Ruskin in 1983 [2]. Afterwards, a number of graphical representations were well developed by researchers. For instance, by assigning four directions defined by the positive/negative x and y coordinate axes to the four nucleic acid bases, Gates [3], Nandy [4,5], and Leong and Morgenthaler [6] introduced three different 2-D graphical representations, respectively. While Jeffrey [7] proposed a chaos game representation (CGR) of DNA sequences, in which the four corners of a selected square are associated with the four bases respectively. In 2000, Randic *et al.* [8] generalized these 2-D graphical representations to a 3-D graphical representation, in which the center of a cube is chosen as the origin of the Cartesian (x,y,z) coordinate system, and the four corners with coordinates (+1,−1,−1), (−1,+1,−1), (−1,−1,+1), and (+1,+1,+1) are assigned to the four bases. Some other graphical representations of bio-sequences and their applications in the field of biological science and technology can be found in [9–24].

Numerical characterization is another useful tool for sequence comparison. One way to arrive at the numerical characterization of a DNA sequence is to associate the sequence with a vector whose components are related to *k*-words, including the single nucleotide, dinucleotide, trinucleotide, and so on [25–30]. In addition, the numerical characterization can be accomplished by associating with a graphical representation given by a curve in the space (or a plane) structural matrices, such as the Euclidean-distance matrix (ED), the graph theoretical distance matrix (GD), the quotient matrix (D/D, M/M, L/L), and their "higher order" matrices [8–18,31–33]. Once a matrix representation of a DNA sequence is given, some matrix invariants, e.g. the leading eigenvalues, can be used as descriptors of the sequence. This technique has been widely used in the field of biological science and medicine, and different types of matrices are defined to construct various invariants of DNA sequences. However, the order of these matrices is equal to n, the length of the DNA sequence considered. A problem we must face is that the calculation of these matrix invariants will become more and more difficult with larger n values [17,24,32].

In this paper, based on all of the 2-combinations of the multiset $\{\infty\cdot A, \infty\cdot G, \infty\cdot C, \infty\cdot T\}$, we propose a novel graphical representation of DNA sequences. Then, according to the idea of "piecewise function", we describe a particular scheme that transforms the graphical representation of DNA into a cell-based descriptor vector. The introduced vector leads to more simple characterizations and comparisons of DNA sequences.

2. Methods

2.1. The 3-D Graphical Representation

As we know, the four nucleic acid bases A, G, C, and T can be classified into three categories:

$$R = \{A,G\}/Y = \{C,T\};\ M = \{A,C\}/K = \{G,T\};\ W = \{A,T\}/S = \{G,C\}.$$

In fact, these groups are just all of the non-repetition 2-combinations of set {A,G,C,T}. If repetition is allowed, in other words, if we consider multiset $\{\infty\cdot A, \infty\cdot G, \infty\cdot C, \infty\cdot T\}$ instead of the set {A,G,C,T}, then the number of 2-combinations equals 10 (see Table 1).

Table 1. The 2-combinations of multiset $\{\infty\cdot A, \infty\cdot G, \infty\cdot C, \infty\cdot T\}$.

Base	A	G	C	T
A	{A,A}	{A,G}	{A,C}	{A,T}
G	-	{G,G}	{G,C}	{G,T}
C	-	-	{C,C}	{C,T}
T	-	-	-	{T,T}

Let V be a regular tetrahedron whose center is at the origin $O = (0,0,0)$. V_1 = (+1,+1,+1), V_2 = (−1,−1,+1), V_3 = (+1,−1,−1), and V_4 = (−1,+1,−1) are its four vertices. To each of the vertices we assign one of the four nucleic acid bases A, C, G and T. Moreover, to the midpoint of the line segment AC we assign M, and K to the midpoint of the line segment GT, R to that of the line segment AG, Y to that of the line segment CT, W to that of the line segment AT, and S to that of the line segment CG. We thus obtain ten fixed directions: $\vec{OA}, \vec{OC}, \vec{OG}, \vec{OT}, \vec{OM}, \vec{OK}, \vec{OR}, \vec{OY}, \vec{OW}, \vec{OS}$, based on which we can derive ten unit vectors:

$$r_A = \frac{1}{||\vec{OA}||}\cdot\vec{OA},\ r_C = \frac{1}{||\vec{OC}||}\cdot\vec{OC}, \ldots,\ r_S = \frac{1}{||\vec{OS}||}\cdot\vec{OS} \tag{1}$$

Obviously, the ten unit vectors are ten points on a unit sphere.

An idea arises naturally: each of the ten 2-combinations can be associated with one of the ten unit vectors. In detail, we have

$$\begin{gathered}\{A,A\} \leftarrow r_A,\ \{A,G\} \leftarrow r_R,\ \{A,C\} \leftarrow r_M, \{A,T\} \leftarrow r_W,\\ \{G,G\} \leftarrow r_G, \{G,C\} \leftarrow r_S, \{G,T\} \leftarrow r_K,\\ \{C,C\} \leftarrow r_C,\ \{C,T\} \leftarrow r_Y,\ \{T,T\} \leftarrow r_T.\end{gathered} \quad (2)$$

To obtain the spatial curve of a DNA sequence, we move a unit length in the direction that the above assignment dictates. Taking sequence segment ATGGTGCACCTGACTCCTGATCTGGTA as an example, we inspect it by stepping two nucleotides at a time. Starting from the origin $O = (0,0,0)$, we move in the direction dictated by the first dinucleotide AT, r_W, and arrive at P_1, the first point of the 3-D curve. From this point, we move in the direction dictated by the second dinucleotide TG, r_K, and arrive at the second point P_2. From here we move in the direction dictated by the third dinucleotide GG, r_G, and come to the third point P_3. Continuation of this process is illustrated in Table 2, and the corresponding 3-D graphical representation is shown in Figure 1.

Table 2. Cartesian 3-D coordinates for the sequence ATGGTGCACCTGACTCCTGATCTGGTA.

Point	Dinucleotide	x	y	z
1	AT	0	1	0
2	TG	0	1	−1
3	GG	0.5774	0.4226	−1.5774
4	GT	0.5774	0.4226	−2.5774
5	TG	0.5774	0.4226	−3.5774
6	GC	0.5774	−0.5774	−3.5774
7	CA	0.5774	−0.5774	−2.5774
8	AC	0.5774	−0.5774	−1.5774
9	CC	0	−1.1547	−1
10	CT	−1	−1.1547	−1
...	...	...	...	...

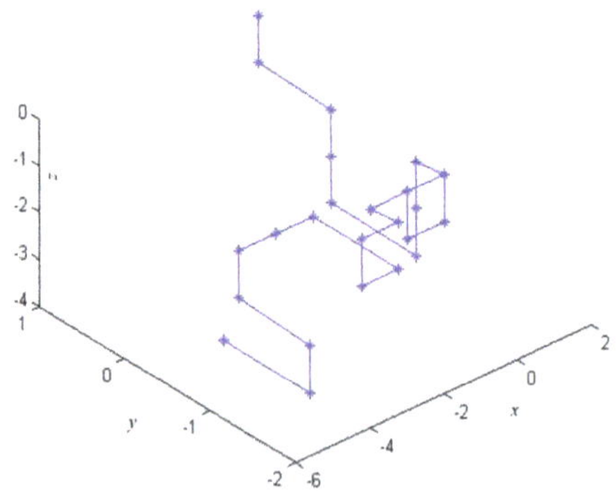

Figure 1. 3-D graphical representation of the sequence ATGGTGCACCTGACTCCTGATCTGGTA.

As the characterization of a research object, a good visualization representation should allow us to see a pattern that may be difficult or impossible to see when the same data is presented in its original form. In order to provide a direct insight into the local and global characteristics of a DNA sequence, the proposed 3-D curve can be projected on planes (x,y), (y,z) or (x,z), and thus three different 2-D graphical representations will be yielded. Figure 2 shows the projections of 3-D curves of 18 different DNA sequences listed in Table 3.

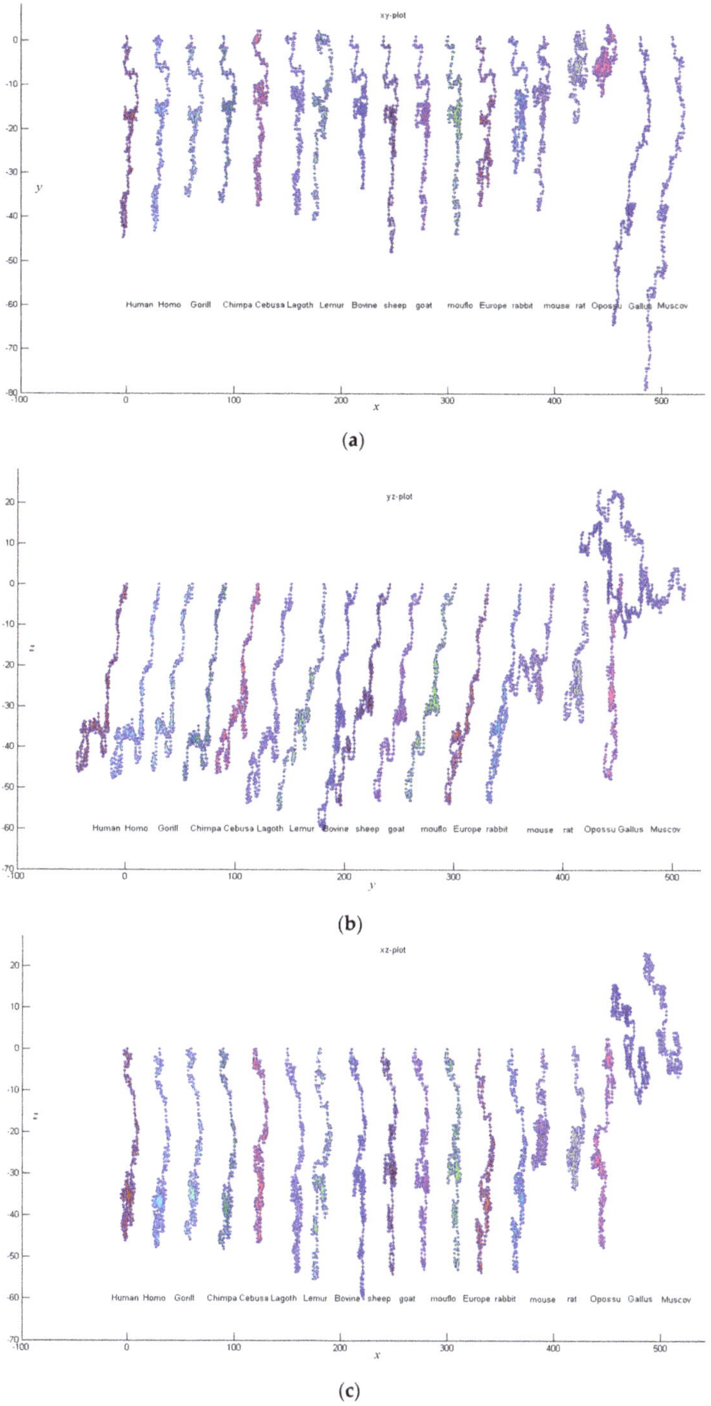

Figure 2. (**a**) The projection on the xy-plane of 3-D curves of 18 DNA sequences; (**b**) The projection on the yz-plane of 3-D curves of 18 DNA sequences; (**c**) The projection on the xz-plane of 3-D curves of 18 DNA sequences.

Table 3. The CDS (Coding DNA Sequence) of β-globin gene of 18 species.

No.	Species	AC (GenBank)	Location
1	*Human*	U01317	join(62187..62278, 62409..62631, 63482..63610)
2	*Homo*	AF007546	join(180..271,402..624,1475..1603)
3	*Gorilla*	X61109	join(4538..4630, 4761..4982, 5833..>5881)
4	*Chimpanzee*	X02345	join(4189..4293, 4412..4633, 5484..>5532)
5	*Lemur*	M15734	join(154..245, 376..598, 1467..1595)
6	*CebusaPella*	AY279115	join(946..1037, 1168..1390, 2218..2346)
7	*LagothrixLagotricha*	AY279114	join(952..1043, 1174..1396, 2227..2355)
8	*Bovine*	X00376	join(278..363, 492..714, 1613..1741)
9	*Goat*	M15387	join(279..364, 493..715, 1621..1749)
10	*Sheep*	DQ352470	join(238..323, 452..674, 1580..1708)
11	*Mouflon*	DQ352468	join(238..323, 452..674, 1578..1706)
12	*European hare*	Y00347	join(1485..1576, 1703..1925, 2492..2620)
13	*Rabbit*	V00882	join(277..368, 495..717, 1291..1419)
14	*Mouse*	V00722	join(275..367, 484..705, 1334..1462)
15	*Rat*	X06701	join(310..401, 517..739, 1377..>1505)
16	*Opossum*	J03643	join(467..558, 672..894, 2360..2488)
17	*Gallus*	V00409	join(465..556, 649..871, 1682..1810)
18	*Muscovy duck*	X15739	join(291..382, 495..717, 1742..1870)

It is easy to see that, in each projection, the trend of curves of the two non-mammals (*Gallus, Muscovy duck*) is distinguished from that of the mammals. On the other hand, the Primates species are similar to one another, so it is with the curves of *bovine*, *sheep*, *goat*, and *mouflon*. Also, the curves of *rabbit* and *European hare* show their great similarity. In addition, both Figure 2b, the projection on *yz*-plane, and Figure 2c, the projection on *xz*-plane, show *opossum* has relatively low similarity with the remaining mammals, while *mouse* and *rat* look similar to each other because both of their curves wind themselves into a mass and need a relatively small space.

2.2. Numerical Characterization of DNA Sequences

The graphical representations not only offer the visual inspection of data, helping in recognizing major differences among DNA sequences, but also provide with the numerical characterization that facilitates quantitative comparisons of DNA sequences. One way to arrive at the numerical characterization of a DNA sequence is to convert its graphical representation into some structural matrices, and use matrix invariants, e.g., the leading eigenvalues, as descriptors of the DNA sequence [8–18,31,32]. It is expected that effective invariants will emerge and enable to uniquely characterize the sequences considered. However, the difficulties associated with computing various parameters for very large matrices that are natural for long sequences have restricted the numerical characterizations, for instance, leading eigenvalues and the like [17,24]. The search for novel descriptors may be an endless project. The art is in finding useful descriptors, and those that have plausible structural interpretation, at least within the model considered [8]. In this section, we bypass the difficulty of calculating the invariants like the leading eigenvalue and propose a novel descriptor to numerically characterize a DNA sequence.

As described above, the pattern, including shape and trend, of curves for the 18 DNA sequences provides useful information in an efficient way. This inspires us to numerically characterize a DNA sequence with an idea of "piecewise function" as below.

For a given 3-D graphical representation with n vertices, by the order in which these vertices appear in the curve, we partition it into K parts, each of which is called a cell. All the cells contain $m = \left\lfloor \frac{n}{K} \right\rfloor$ vertices except the last one. For the i-th cell, $i = 1,2,\ldots,K$, the geometric center $U_i = (x_i, y_i, z_i)$ is viewed as its respective. Then we have

$$\overrightarrow{U_{i-1}U_i} = (x_i - x_{i-1}, y_i - y_{i-1}, z_i - z_{i-1}) \quad (3)$$

where $U_0 = (0,0,0)$. It is not difficult to find that $\overrightarrow{U_{i-1}U_i}$ reflects a certain "growing trend" of these cells. For convenience, we call $\overrightarrow{U_{i-1}U_i}$ the trend-point. On the basis of the K trend-points, a DNA sequence can be characterized by a $3K$-dimensional vector V_{tp}:

$$\begin{aligned} V_{tp} = \; & (x_1 - x_0, x_2 - x_1, \cdots, x_k - x_{k-1}, \\ & y_1 - y_0, y_2 - y_1, \cdots, y_k - y_{k-1}, \\ & z_1 - z_0, z_2 - z_1, \cdots, z_k - z_{k-1}) \end{aligned} \tag{4}$$

In this paper, K is determined by $round\left(\log_4 \frac{\overline{L}}{2\sqrt{2}}\right)$, where $\overline{L} = \frac{1}{N}\sum_{j=1}^{N} |s_j|$, N is the cardinality of the dataset Ω considered, and $|s_j|$ stands for the length of sequence $s_j \in \Omega$. Taking for example the two non-mammals of the 18 species, the corresponding vectors can be calculated as

$$\begin{aligned} V_{\text{Gallus}} = \; & (4.524, -9.588, -5.546, -10.962, -9.234, -20.304, \\ & -9.824, -12.093, -4.087, -0.450, 10.255, 5.615), \end{aligned} \tag{5}$$

$$\begin{aligned} V_{\text{MDuck}} = \; & (6.186, -10.593, -3.440, -12.511, -10.639, -21.519, \\ & -12.987, -18.351, -1.244, 0.498, 10.478, 9.325). \end{aligned} \tag{6}$$

3. Results and Discussion

In this section, we will illustrate the use of the proposed cell-based descriptor V_{tp} of a DNA sequence. For any two sequences S_a and S_b, suppose their descriptor vectors are $a = (a_1, a_2, \cdots, a_{3k})$ and $b = (b_1, b_2, \cdots, b_{3k})$, respectively. Then, their similarity can be examined by the following Euclidean distance. Clearly, the smaller the Euclidean distance is, the more similar the two DNA sequences are.

$$d(a,b) = \sqrt{\sum_{j=1}^{3k} (a_j - b_j)^2} \tag{7}$$

Firstly, we give a comparison for CDS (Coding DNA Sequence) of β-globin gene of 18 species listed in Table 3. The lengths of the 18 sequences are about 434 bp. Thus K is taken to be 4, and each of these sequences is converted into a 12-D vector. According to Equation (7), we calculate the distance between any two of the 18 DNA sequences. Then an 18×18 real symmetric matrix D_{18} is obtained. On the basis of D_{18}, a phylogenetic tree (see Figure 3) is constructed using UPGMA (Unweighted Pair Group Method with Arithmetic Mean) program included in MEGA4 [34]. Observing Figure 3, we find that the CDS are more similar for Primate group {*Gorilla, Chimpanzee, Human, Homo, CebusaPella, LagothrixLagotricha, Lemur*}, Cetartiodactyla group {*bovine, sheep, goat, mouflon*}, Lagomorpha group {*Rabbit, European hare*}, and Rodentia group {*mouse, rat*}, respectively. On the other hand, CDS of the two kinds of non-mammals {*Gallus, Muscovy duck*} are very dissimilar to the mammals because they are grouped into an independent branch. This is analogous to that reported in the literature [8,12,14,31], and the relationship of these species detected by their graphical representations as well. From this result, a conclusion one can draw is that the cell-based descriptors of the new graphical representation may suffice to characterize DNA sequences.

In order to further illustrate the effectiveness of our method, we test it by phylogenetic analysis on other three datasets: one consists of mitochondrial cytochrome oxidase subunit I (COI) genes of nine butterflies, another includes S segments of 32 hantaviruses (HVs), and the last is composed of 70 complete mitogenomes (mitochondrial genomes). For convenience, we denote the three datasets by COI, HV and mitogenome, respectively. In the COI dataset (see Table 4), which is taken from Yang *et al.* [12], eight belong to the *Catopsilia* genus and one belongs to *Appias* genus, which is used as the out-group. The average length of these COI gene sequences is 661 bp, and thus K, the number of cells, is calculated as 4. According to the method mentioned above, a distance matrix is constructed, and then a phylogenetic tree (see Figure 4) is generated. Figure 4 shows that the five *pomona* sub-species have relatively high similarity with each other, while the two *pyranthe* sub-species cluster together. In addition, *scylla* sub-species is situated at an independent branch, whereas the *Appias lyncida* stays outside of all the *Catopsilia*. This result is consistent with that reported in [12,35].

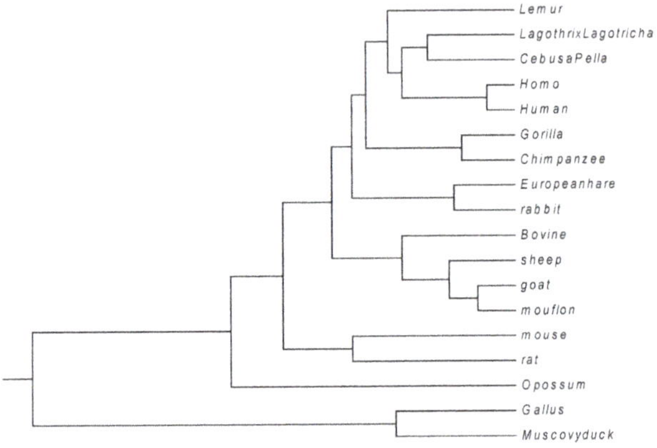

Figure 3. The relationship tree of 18 species.

Table 4. The COI (cytochrome oxidase subunit I) genes of nine butterflies.

NO.	Species	Code	AC (GenBank)	Region
1	*C.pomona pomona f.pomona*	PA	GU446662	Yexianggu, Yunnan
2	*C.pomona pomona f.hilaria*	HI	GU446664	Yexianggu, Yunnan
3	*C.pomona pomona f.crocale*	CR	GU446663	Menglun, Yunnan
4	*C.pomona pomona f.catilla*	CA	GU446666	Daluo, Yunnan
5	*C.pomona pomona f.jugurtha*	JU	GU446665	Daluo, Yunnan
6	*C.scylla scylla*	CS	GU446667	Yinggeling, Hainan
7	*C.pyranthe pyranthe*	CP	GU446668	Daluo, Yunnan
8	*C.pyranthe chryseis*	CH	GU446669	Yinggeling, Hainan
9	*Appias lyncida*	-	GU446670	Bawangling, Hainan

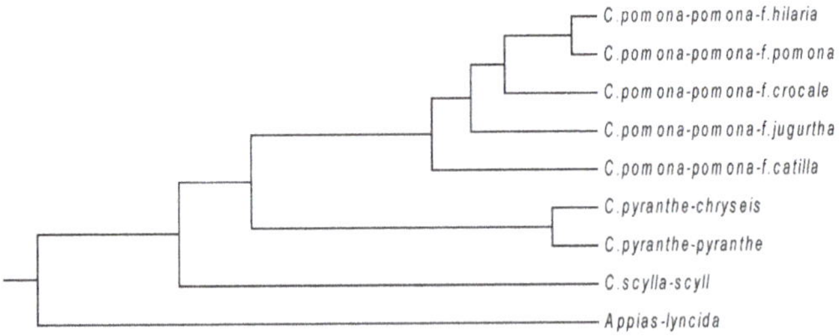

Figure 4. The relationship tree of nine COI (cytochrome oxidase subunit I) gene sequences.

The hantavirus (HV), which is named for the Hantan River area in South Korea, is a relatively newly discovered RNA virus in the family *Bunyaviridae*. This kind of virus normally infects rodents and does not cause disease in these hosts. Humans may be infected with HV, and some HV strains could cause severe, sometimes fatal, diseases in humans, such as HFRS (hantavirus hemorrhagic fever with renal syndrome) and HPS (hantavirus pulmonary syndrome). The later occurred in North and

South America, while the former mainly in Eurasia [12,36]. In Eastern Asia, particularly in China and Korea, the viruses that cause HFRS mainly include Hantaan (HTN) and Seoul (SEO) viruses, while Puumala (PUU) virus is found in Western Europe, Russia and northeastern China. The HV dataset analyzed in this paper includes 32 HV sequences. Phlebovirus (PV) is another genus of the family *Bunyaviridae*. Here, two PV strains KF297911 and KF297914 are used as the out-group. The name, accession number, type, and region of the 34 sequences are described in Table 5. The lengths of these sequences are in the range of 1.30–1.88 kbp. Thus K is calculated as 5, and each of the 34 viruses is converted into a 15-D vector. The phylogenetic tree constructed by our method is shown in Figure 5.

Table 5. Sequence information of S segment of hantavirus.

No.	Strain	AC (GenBank)	Type	Region
1	CGRn53	EF990907	HTNV	Guizhou
2	CGRn5310	EF990906	HTNV	Guizhou
3	CGRn93MP8	EF990905	HTNV	Guizhou
4	CGRn8316	EF990903	HTNV	Guizhou
5	CGRn9415	EF990902	HTNV	Guizhou
6	CGRn93P8	EF990904	HTNV	Guizhou
7	CGHu3612	EF990909	HTNV	Guizhou
8	CGHu3614	EF990908	HTNV	Guizhou
9	Z10	AF184987	HTNV	Shengzhou
10	Z5	EF103195	HTNV	Shengzhou
11	NC167	AB027523	HTNV	Anhui
12	CGAa4MP9	EF990915	HTNV	Guizhou
13	CGAa4P15	EF990914	HTNV	Guizhou
14	CGAa1011	EF990913	HTNV	Guizhou
15	CGAa1015	EF990912	HTNV	Guizhou
16	H5	AB127996	HTNV	Heilongjiang
17	76-118	M14626	HTNV	South Korea
18	Gou3	AF184988	SEOV	Jiande
19	ZJ5	FJ753400	SEOV	Jiande
20	80-39	AY273791	SEOV	South Korea
21	SR11	M34881	SEOV	Japan
22	K24-e7	AF288653	SEOV	Xinchang
23	K24-v2	AF288655	SEOV	Xinchang
24	Z37	AF187082	SEOV	Wenzhou
25	ZT10	AY766368	SEOV	Tiantai
26	ZT71	AY750171	SEOV	Tiantai
27	K27	L08804	PUUV	Russia
28	P360	L11347	PUUV	Russia
29	Sotkamo	X61035	PUUV	Finland
30	Fusong843-06	EF488805	PUUV	Jilin
31	Fusong199-05	EF488803	PUUV	Jilin
32	Fusong900-06	EF488806	PUUV	Jilin
33	91045-AG	KF297911	PV	Iran
34	I-58	KF297914	PV	Iran

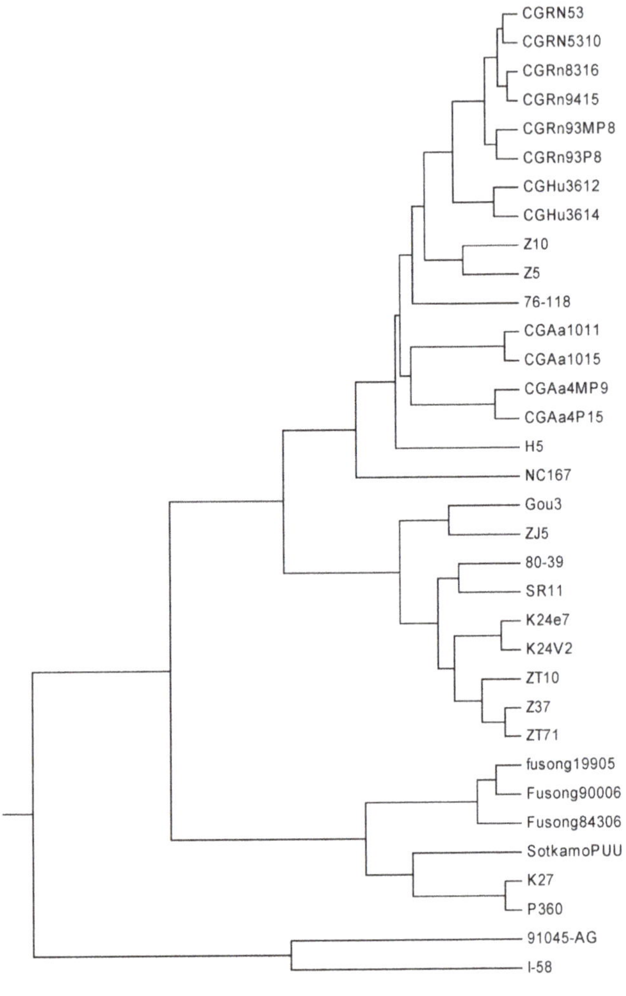

Figure 5. The relationship tree of 34 viruses.

From Figure 5, we find that the two PV strains form an independent branch, which can be distinguished easily from the HV strains, while the 32 HVs are grouped into three separate branches: the strains belonging to PUUV are clearly clustered together, the strains belonging to SEOV appear to cluster together, and so do the ones belonging to HTNV. A closer look at the subtree of HTNV, all CGRn strains whose host is *Rattus norvegicus* tend to cluster together, so it is with the CGHu strains whose host is Homo sapiens. In addition, all the four CGAa strains whose host is *Apodemus agrarius* are grouped closely. Needless to say, the phylogeny is not only closely related to the isolated regions, but also has certain relationship with the host. This result is similar to that reported in [12,37].

The mitogenome dataset comprises 70 complete mitochondrial genomes of Eukaryota. The name, accession number, and genome length are listed in Table 6. Among them, two species (*Argopecten irradians irradians* and *Argopecten purpuratus*) belong to family Pectinidae are used as the out-group. Four species belong to the Order Caudata under the Class Amphibia, while four species belong to the Order Anura under the same Class. The remaining belongs to the Class Actinopterygii. The average length of the 70 genome sequences is about 16817 bp. Thus, K is calculated as 6, and each of these genome sequences is converted into an 18-D vector. The phylogenetic tree constructed by our method is shown in Figure 6. It is easy to see from Figure 6 that the two *Pectinidae* species stay outside of the others, while the four *Hynobiidae* species and four *Ranidae* species form an independent branch. In the subtree of the Class Actinopterygii, the 60 genomes are separated into six groups: group 1 corresponds to genus *Anguilla* under family Anguillidae; group 2 includes genera *Bangana* and *Acrossocheilus* under family Cyprinidae; group 3 includes genera *Brachymystax* and *Hucho* under family Salmonidae; group 4 is genus *Alepocephalus* under family Alepocephalidae; group 5 is the family of Clupeidae; group 6 includes genera *Trichiurus, Amphiprion* and *Apolemichthys* under Acanthomorphata. This result agrees well with the established taxonomic groups. In addition, we make a comparison for the 70 genome sequences by using ClustalX2.1 [38], and the corresponding tree is shown in Figure 7. Observing Figure 7, we find that the tree includes four branches: the outside is the *Argopecten* branch, the following is *Babina*, then *Batrachuperus*, and the subtree consisting of the other 60 species. A closer look at the subtree shows that *Trichiurus* is distinguished from the remaining, which seems to be a disappointing phenomenon in the evolutionary sense.

Table 6. Sequence information of 70 complete mitogenomes.

No.	Genome	AC (GenBank)	Length
1	*Acrossocheilus barbodon*	NC_022184	16596
2	*Acrossocheilus beijiangensis*	NC_028206	16600
3	*Acrossocheilus fasciatus*	NC_023378	16589
4	*Acrossocheilus hemispinus*	NC_022183	16590
5	*Acrossocheilus kreyenbergii*	NC_024844	16849
6	*Acrossocheilus monticola*	NC_022145	16599
7	*Acrossocheilus parallens*	NC_026973	16592
8	*Acrossocheilus stenotaeniatus*	NC_024934	16594
9	*Acrossocheilus wenchowensis*	NC_020145	16591
10	*Alepocephalus agassizii*	NC_013564	16657
11	*Alepocephalus australis*	NC_013566	16640
12	*Alepocephalus bairdii*	NC_013567	16637
13	*Alepocephalus bicolor*	NC_011012	16829
14	*Alepocephalus productus*	NC_013570	16636
15	*Alepocephalus tenebrosus*	NC_004590	16644
16	*Alepocephalus umbriceps*	NC_013572	16640
17	*Alosa alabamae*	NC_028275	16708
18	*Alosa alosa*	NC_009575	16698
19	*Alosa pseudoharengus*	NC_009576	16646
20	*Alosa sapidissima*	NC_014690	16697
21	*Amphiprion bicinctus*	NC_016701	16645
22	*Amphiprion clarkia*	NC_023967	16976
23	*Amphiprion frenatus*	NC_024840	16774
24	*Amphiprion ocellaris*	NC_009065	16649
25	*Amphiprion percula*	NC_023966	16645
26	*Amphiprion perideraion*	NC_024841	16579
27	*Amphiprion polymnus*	NC_023826	16804
28	*Anguilla anguilla*	NC_006531	16683
29	*Anguilla australis*	NC_006532	16686
30	*Anguilla australis schmidti*	NC_006533	16682
31	*Anguilla bengalensis labiata*	NC_006543	16833
32	*Anguilla bicolor bicolor*	NC_006534	16700
33	*Anguilla bicolor pacifica*	NC_006535	16693
34	*Anguilla celebesensis*	NC_006537	16700
35	*Anguilla dieffenbachia*	NC_006538	16687
36	*Anguilla interioris*	NC_006539	16713
37	*Anguilla japonica*	NC_002707	16685
38	*Anguilla luzonensis* (Philippine eel)	NC_011575	16635

Table 6. *Cont.*

No.	Genome	AC (GenBank)	Length
39	*Anguilla luzonensis* (freshwater eel)	NC_013435	16632
40	*Anguilla malgumora*	NC_006536	16550
41	*Anguilla marmorata*	NC_006540	16745
42	*Anguilla megastoma*	NC_006541	16714
43	*Anguilla mossambica*	NC_006542	16694
44	*Anguilla nebulosa nebulosa*	NC_006544	16707
45	*Anguilla obscura*	NC_006545	16704
46	*Anguilla reinhardtii*	NC_006546	16690
47	*Anguilla rostrata*	NC_006547	16678
48	*Apolemichthys armitagei*	NC_027857	16551
49	*Apolemichthys griffisi*	NC_027592	16528
50	*Apolemichthys kingi*	NC_026520	16816
51	*Argopecten irradians irradians*	NC_012977	16211
52	*Argopecten purpuratus*	NC_027943	16270
53	*Babina adenopleura*	NC_018771	18982
54	*Babina holsti*	NC_022870	19113
55	*Babina okinavana*	NC_022872	19959
56	*Babina subaspera*	NC_022871	18525
57	*Bangana decora*	NC_026221	16607
58	*Bangana tungting*	NC_027069	16543
59	*Batrachuperus londongensis*	NC_008077	16379
60	*Batrachuperus pinchonii*	NC_008083	16390
61	*Batrachuperus tibetanus*	NC_008085	16379
62	*Batrachuperus yenyuanensis*	NC_012430	16394
63	*Brachymystax lenok*	NC_018341	16832
64	*Brachymystax lenok tsinlingensis*	NC_018342	16669
65	*Brachymystax tumensis*	NC_024674	16836
66	*Hucho bleekeri*	NC_015995	16997
67	*Hucho hucho*	NC_025589	16751
68	*Hucho taimen*	NC_016426	16833
69	*Trichiurus lepturus nanhaiensis*	NC_018791	17060
70	*Trichiurus japonicus*	NC_011719	16796

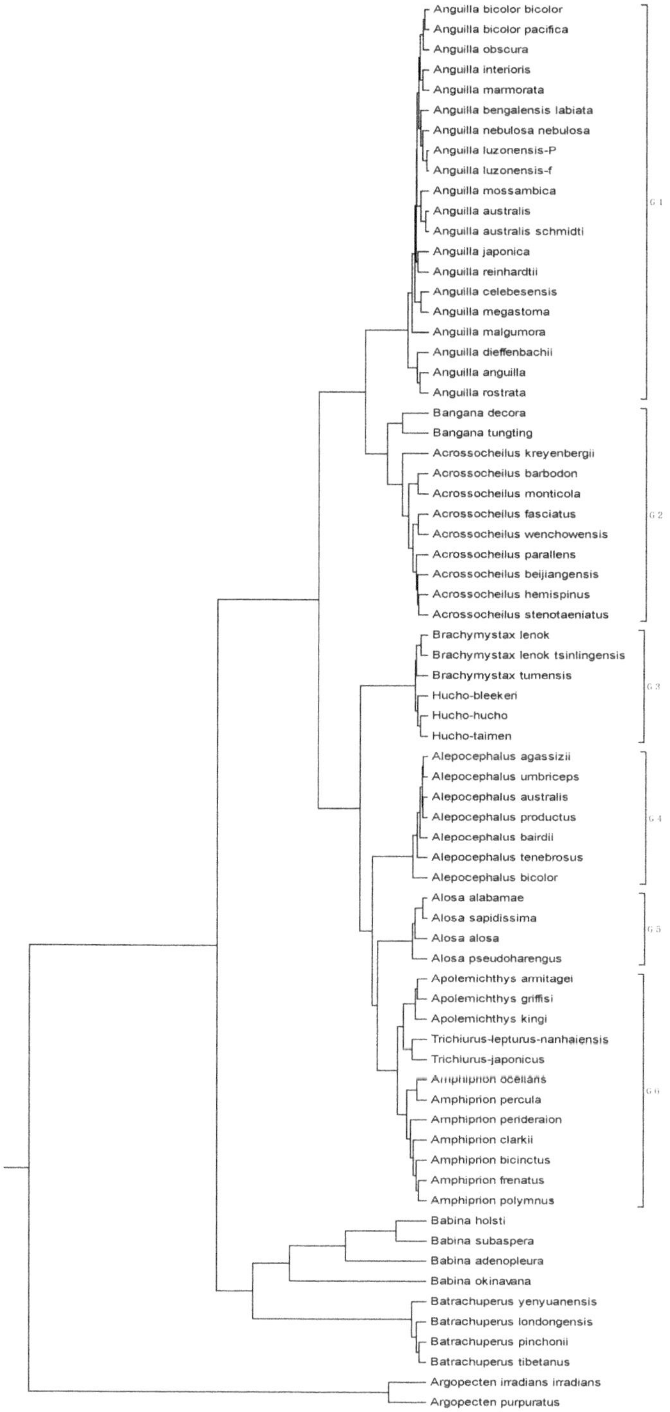

Figure 6. The tree of 70 genome sequences constructed with the current method.

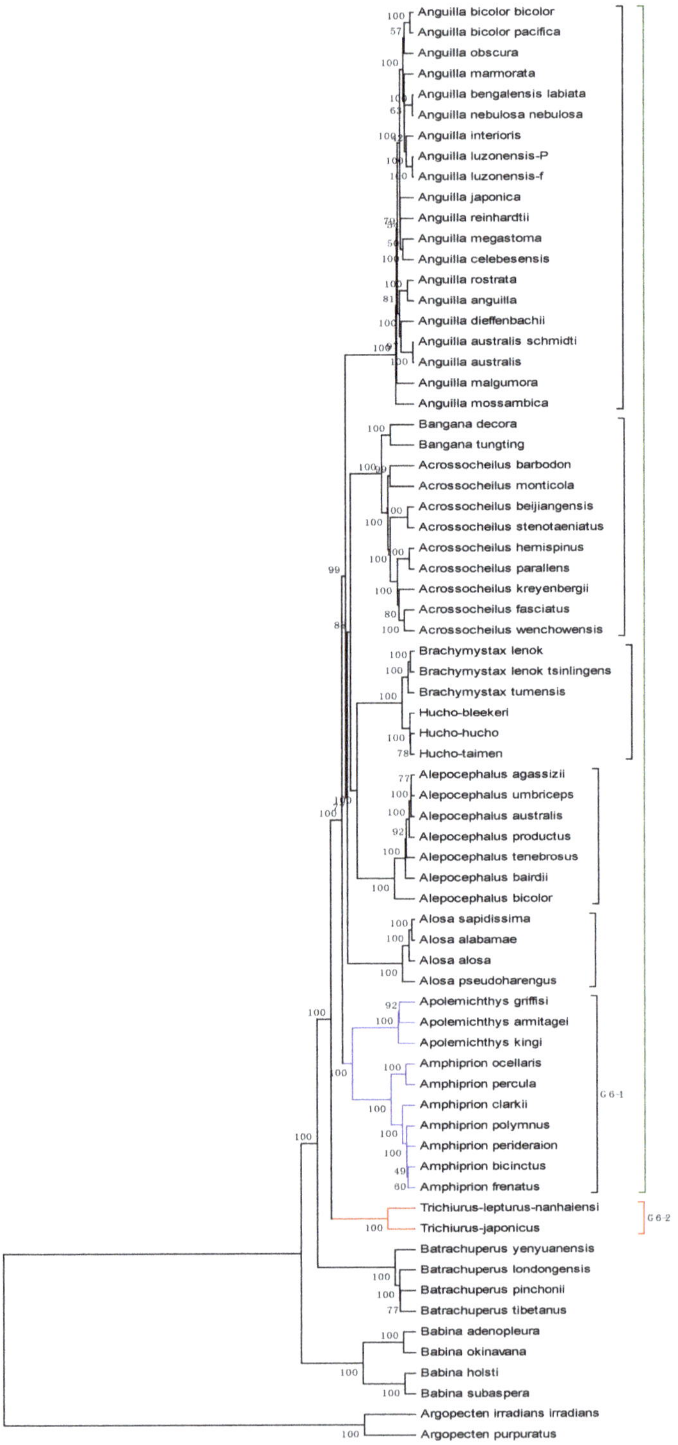

Figure 7. The tree of 70 genome sequences constructed with multiple alignment.

4. Concluding Remarks

By means of a regular tetrahedron whose center is at the origin, we associate the ten 2-combinations of multiset $\{\infty\cdot A, \infty\cdot G, \infty\cdot C, \infty\cdot T\}$ with ten unit vectors (points on a unit sphere), and then a novel 3-D graphical representation of a DNA sequence is proposed. Moreover, we partition the graph into K cells, and then a $3K$-dimensional cell-based vector is used to numerically characterize a DNA sequence. The proposed method is tested by phylogenetic analysis on four datasets. In comparison with other methods, our approach does not depend on multiple sequence alignment, and avoids the complex calculation as in the calculation of invariants for higher order matrices. Nevertheless, K, the number of cells, is dataset specific, which may restrict our approach. We will make efforts in our future work to find a possible formula for K that is independent of the dataset.

Acknowledgments: Acknowledgments: The authors wish to thank the three anonymous referees for their valuable suggestions and support. This work was partially supported by the National Natural Science Foundation of China (No. 11171042), the Program for Liaoning Innovative Research Team in University (LT2014024), the Liaoning BaiQianWan Talents Program (2012921060), and the Open Project Program of Food Safety Key Lab of Liaoning Province (LNSAKF2011034).

Author Contributions: Author Contributions: Chun Li and Xiaoqing Yu conceived the study and drafted the manuscript. Wenchao Fei and Yan Zhao participated in the design of the study and analysis of the results.

Conflicts of Interest: Conflicts of Interest: The authors declare no conflict of interest.

References

1. Tian, K.; Yang, X.Q.; Kong, Q.; Yin, C.C.; He, R.L.; Yau, S.S.T. Two dimensional Yau-hausdorff distance with applications on comparison of DNA and protein sequences. *PLoS ONE* **2015**, *10*. [CrossRef] [PubMed]
2. Hamori, E.; Ruskin, J. H curves, a novel method of representation of nucleotide series especially suited for long DNA sequences. *J. Biol. Chem.* **1983**, *258*, 1318–1327. [PubMed]
3. Gates, M.A. Simpler DNA sequence representations. *Nature* **1985**, *316*. [CrossRef]
4. Nandy, A. A new graphical representation and analysis of DNA sequence structure: I methodology and application to globin genes. *Curr. Sci.* **1994**, *66*, 309–314.
5. Nandy, A. Graphical representation of long DNA sequences. *Curr. Sci.* **1994**, *66*, 821.
6. Leong, P.M.; Morgenthaler, S. Random walk and gap plots of DNA sequences. *Comput. Appl. Biosci.* **1995**, *11*, 503–507. [CrossRef] [PubMed]
7. Jeffrey, H.J. Chaos game representation of gene structure. *Nucleic Acids Res.* **1990**, *18*, 2163–2170. [CrossRef] [PubMed]
8. Randic, M.; Vracko, M.; Nandy, A.; Basak, S.C. On 3-D graphical representation of DNA primary sequences and their numerical characterization. *J. Chem. Inf. Comput. Sci.* **2000**, *40*, 1235–1244. [CrossRef] [PubMed]
9. Randic, M.; Novic, M.; Plavsic, D. Milestones in graphical bioinformatics. *Int. J. Quantum Chem.* **2013**, *113*, 2413–2446. [CrossRef]
10. Randic, M.; Zupan, J.; Balaban, A.T.; Vikic-Topic, D.; Plavsic, D. Graphical representation of proteins. *Chem. Rev.* **2011**, *111*, 790–862. [CrossRef] [PubMed]
11. Li, C.; Tang, N.N.; Wang, J. Directed graphs of DNA sequences and their numerical characterization. *J. Theor. Biol.* **2006**, *241*, 173–177. [CrossRef] [PubMed]
12. Yang, Y.; Zhang, Y.Y.; Jia, M.D.; Li, C.; Meng, L.Y. Non-degenerate graphical representation of DNA sequences and its applications to phylogenetic analysis. *Comb. Chem. High Throughput Screen.* **2013**, *16*, 585–589. [CrossRef] [PubMed]
13. Gonzzlez-Diaz, H.; Perez-Montoto, L.G.; Duardo-Sanchez, A.; Paniagua, E.; Vazquez-Prieto, S.; Vilas, R.; Dea-Ayuela, M.A.; Bolas-Fernandez, F.; Munteanu, C.R.; Dorado, J.; *et al.* Generalized lattice graphs for 2D-visualization of biological information. *J. Theor. Biol.* **2009**, *261*, 136–147. [CrossRef] [PubMed]
14. Zhang, Z.J. DV-Curve: A novel intuitive tool for visualizing and analyzing DNA sequences. *Bioinformatics* **2009**, *25*, 1112–1117. [CrossRef] [PubMed]
15. Qi, Z.H.; Jin, M.Z.; Li, S.L.; Feng, J. A protein mapping method based on physicochemical properties and dimension reduction. *Comput. Biol. Med.* **2015**, *57*, 1–7. [CrossRef] [PubMed]

16. Waz, P.; Bielinska-Waz, D. 3D-dynamic representation of DNA sequences. *J. Mol. Model.* **2014**, *20*. [CrossRef] [PubMed]
17. Yao, Y.H.; Yan, S.; Han, J.; Dai, Q.; He, P.A. A novel descriptor of protein sequences and its application. *J. Theor. Biol.* **2014**, *347*, 109–117. [CrossRef] [PubMed]
18. Ma, T.T.; Liu, Y.X.; Dai, Q.; Yao, Y.H.; He, P.A. A graphical representation of protein based on a novel iterated function system. *Phys. A* **2014**, *403*, 21–28. [CrossRef]
19. Zhang, R.; Zhang, C.T. A brief review: The Z curve theory and its application in genome analysis. *Curr. Genom.* **2014**, *15*, 78–94. [CrossRef] [PubMed]
20. Zhang, C.T.; Zhang, R.; Ou, H.Y. The Z curve database: A graphic representation of genome sequences. *Bioinformatics* **2003**, *19*, 593–599. [CrossRef] [PubMed]
21. Zhang, R.; Zhang, C.T. Z curves, an intuitive tool for visualizing and analyzing DNA sequences. *J. Biomol. Struct. Dyn.* **1994**, *11*, 767–782. [CrossRef] [PubMed]
22. Herisson, J.; Payen, G.; Gherbi, R. A 3D pattern matching algorithm for DNA sequences. *Bioinformatics* **2007**, *23*, 680–686. [CrossRef] [PubMed]
23. Bianciardi, G.; Borruso, L. Nonlinear analysis of tRNAs squences by random walks: Randomness and order in the primitive information polymers. *J. Mol. Evol.* **2015**, *80*, 81–85. [CrossRef] [PubMed]
24. Ghosh, A.; Nandy, A. Graphical representation and mathematical characterization of protein sequences and applications to viral proteins. *Adv. Protein Chem. Struct. Biol.* **2011**, *83*. [CrossRef]
25. Karlin, S.; Burge, C. Dinucleotide relative abundance extremes: A genomic signature. *Trends Genet.* **1995**, *11*, 283–290. [PubMed]
26. Karlin, S. Global dinucleotide signatures and analysis of genomic heterogeneity. *Curr. Opin. Microbiol.* **1998**, *1*, 598–610. [CrossRef]
27. Yang, X.W.; Wang, T.M. Linear regression model of short *k*-word: A similarity distance suitable for biological sequences with various lengths. *J. Theor. Biol.* **2013**, *337*, 61–70. [CrossRef] [PubMed]
28. Li, C.; Ma, H.; Zhou, Y.; Wang, X.; Zheng, X. Similarity analysis of DNA sequences based on the weighted pseudo-entropy. *J. Comput. Chem.* **2011**, *32*, 675–680. [CrossRef] [PubMed]
29. Rocha, E.P.; Viari, A.; Danchin, A. Oligonucleotide bias in *Bacillus subtilis*: General trends and taxonomic comparisons. *Nucleic Acids Res.* **1998**, *26*, 2971–2980. [CrossRef] [PubMed]
30. Pride, D.T.; Meineramann, R.J.; Wassenaar, T.M.; Blaser, M.J. Evolutionary implications of microbial genome tetranucleotide frequency biases. *Genome Res.* **2003**, *13*, 145–158. [CrossRef] [PubMed]
31. Li, C.; Wang, J. Numerical characterization and similarity analysis of DNA sequences based on 2-D graphical representation of the characteristic sequences. *Comb. Chem. High. Throughput Screen.* **2003**, *6*, 795–799. [CrossRef] [PubMed]
32. Li, C.; Wang, J. New invariant of DNA sequences. *J. Chem. Inf. Model.* **2005**, *36*, 115–120. [CrossRef] [PubMed]
33. Bai, F.; Zhang, J.; Zheng, J.; Li, C.; Liu, L. Vector representation and its application of DNA sequences based on nucleotide triplet codons. *J. Mol. Graph. Model.* **2015**, *62*, 150–156. [CrossRef] [PubMed]
34. MEGA, Molecular Evolutionary Genetics Analysis. Available online: http://www.megasoftware.net (accessed on 15 January 2014).
35. Wang, J.; Shang, S.Q.; Zhang, Y.L. Phylogenetic relationship of genus catopsilia (Lepidoptera: Pieridae) based on partial sequences of NDI and COI genes from China. *Acta. Zootaxon. Sin.* **2010**, *35*, 776–781.
36. Zhang, Y.Z.; Dong, X.; Li, X.; Ma, C.; Xiong, H.P.; Yan, G.J.; Gao, N.; Jiang, D.M.; Li, M.H.; Li, L.P.; *et al.* Seoul virus and hantavirus disease, Shenyang, People's Republic of China. *Emerg. Infect. Dis.* **2009**, *15*, 200–206. [CrossRef] [PubMed]
37. Yao, P.P.; Zhu, H.P.; Deng, X.Z.; Xu, F.; Xie, R.H.; Yao, C.H.; Weng, J.Q.; Zhang, Y.; Yang, Z.Q.; Zhu, Z.Y. Molecular evolution analysis of hantaviruses in Zhejiang province. *Chin. J. Virol.* **2010**, *26*, 465–470.
38. Clustal: Multiple Sequence Alignment. Available online: http://www.clustal.org (accessed on 31 August 2012).

applied sciences

MDPI

Article

Numerical Characterization of Protein Sequences Based on the Generalized Chou's Pseudo Amino Acid Composition

Chun Li [1,2,*], Xueqin Li [1] and Yan-Xia Lin [2]

[1] Department of Mathematics, Bohai University, Jinzhou 121013, China; 18841655169@163.com
[2] NIASRA—National Institute for Applied Statistics Research Australia, School of Mathematics and Applied Statistics, University of Wollongong, Wollongong 2522, Australia; yanxia@uow.edu.au
* Correspondence: lichwun@163.com; Tel.: +86-416-3400-145

Academic Editors: Yang Kuang and Wanbiao Ma
Received: 18 September 2016; Accepted: 29 November 2016; Published: 6 December 2016

Abstract: The technique of comparison and analysis of biological sequences is playing an increasingly important role in the field of Computational Biology and Bioinformatics. One of the key steps in developing the technique is to identify an appropriate manner to represent a biological sequence. In this paper, on the basis of three physical–chemical properties of amino acids, a protein primary sequence is reduced into a six-letter sequence, and then a set of elements which reflect the global and local sequence-order information is extracted. Combining these elements with the frequencies of 20 native amino acids, a $(21 + \lambda)$ dimensional vector is constructed to characterize the protein sequence. The utility of the proposed approach is illustrated by phylogenetic analysis and identification of DNA-binding proteins.

Keywords: generalized pseudo amino acid composition; numerical characterization; phylogenetic analysis; identification of DNA-binding proteins

1. Introduction

In the task of comparison and analysis of biological sequences, choosing a type of DNA/protein representation is an important step. The usual representation of the primary structure of DNA is a string of four letters: A (adenine); G (guanine); C (cytosine); and T (thymine). This expression is called a letter sequence representation (LSR) or a DNA primary sequence. Similarly, a protein primary sequence is usually expressed in terms of a series of 20 letters, which denote 20 different amino acids. The sequence encodes information of the corresponding structure and function in a living organism. However, it is difficult to obtain the information from the representation of a primary sequence directly. Therefore, various sequence representation techniques have been developed for encoding bio-sequences and extracting the hidden information.

Graphical representation of DNA is a useful tool for visualizing and analyzing DNA sequences. By using the tool, one can obtain a route to condense the information coded by DNA primary sequences into a set of invariants [1,2]. Early attempts towards graphical representations of DNA were made by Hamori and Ruskin in 1983 [3], Hamori in 1985 [4], and Gates in 1985 [5]. Afterwards, more graphical representations of DNA sequences were well developed by researchers [1,2,6–15]. In comparison with DNA, graphical representations of proteins emerged only very recently [2,16–27]. As a matter of fact, most of the graphical representations of DNA involve some degree of arbitrariness, such as the selection of directions to be assigned to individual bases. For a string like DNA sequence over an alphabet with size 4, there are $4! = 24$ possible ways of assigning 4 directions to 4 nucleic acid bases. If these methods are directly extended to protein sequences, the corresponding figure is

$20! \approx 2.433 \times 10^{18}$. It is impracticable to represent one protein sequence by such an enormous number of graphs. This is probably the most important reason why protein graphical representations have not been advanced [19,23]. It is found that reducing the alphabet or fixing the directions assigned to amino acid residues plays an important role in addressing this problem. For details, we refer to some recent publications [2,16,21,23,24,28].

Matrix representation of a biological sequence is another powerful tool for characterization and comparison of sequences. These matrices include: The frequency matrix; Euclidean-distance matrix (*ED*); graph theoretical distance matrix (*GD*); line distance matrix (*LD*); quotient matrix (*D/D*, *M/M*, *L/L*); and their "higher order" matrices [1,2,12,13,20,21,27,29,30]. Among them, *ED*, *GD*, *L/L*, etc., are derived from a graphical representation. For example, *L/L* is a symmetric matrix whose diagonal entries are zero, while other entries are defined as the quotient of the Euclidean distance between two points of the graph and the sum of geometrical lengths of edges between the two points. Once the matrix is given, some of matrix invariants can be used as descriptors of the sequence. Eigenvalues of a matrix are one of the best-known matrix invariants [31]. In fact, two graphs are isomorphic if and only if their adjacency matrices are similar. It is of interest to note that similar matrices have the same eigenvalues. Among all the eigenvalues, the leading eigenvalue often plays a special role and has been widely used in the field of biological science and chemistry. However, a problem we must face is that the calculation of the eigenvalue will become more and more difficult with the order of the matrix large. ALE-index is an alternative invariant we proposed in 2005 [32]. The ALE-index can be viewed as an Approximation of the Leading Eigenvalue (ALE) of the corresponding matrix (it is just in this sense that it is called 'ALE'-index), while it is much simpler for calculation than the latter. Therefore, it may be more economical to adopt the ALE-index when one is interested only in the leading eigenvalue.

The third method for formulating a protein sequence is the pseudo amino acid composition (PseAAC), with the advantage of avoiding loss of the sequence-order information. Ever since the concept of PseAAC [33,34] or Chou's PseAAC [35,36] was proposed, it has rapidly penetrated into nearly all fields of computational proteomics (see a long list papers cited in [36,37]). Stimulated by the great successes of PseAAC in dealing with protein/peptide sequences, the concept of PseAAC has been extended [38–42] to cover DNA/RNA sequences as well via the form of PseKNC (pseudo K-tuple nucleotide composition) [43,44], which has been proven very useful in studying many important genome analysis problems, as summarized in a recent review paper [45]. Also, because the concept of PseAAC has been increasingly and widely used in both computational proteomics and genomics, a very powerful web-server called "Pse-in-One" [46] was established that can be used to generate the pseudo components for both protein/peptide and DNA/RNA sequences.

In this paper, we modify the method of Chou's PseAAC and propose a novel approach for numerically characterizing a protein sequence. We characterize a protein sequence by a $(21 + \lambda)$ dimensional vector, whose first 20 components are the occurrence frequencies of 20 native amino acids, while the last $\lambda + 1$ components are based on a six-letter sequence derived from the protein primary sequence. The former is used to reflect the effect of the amino acid composition, and the latter is used to reflect the effect of sequence order and property of the residues. It is well known that a sequence naturally contains two pieces of information: the elements of the sequence; and the orders of the elements. Any methodologies based on the amino acid composition alone are worthy of further investigation. However, as pointed out by Chou [33,34], it is not feasible to completely include all sequence order patterns. It was stirring to see that Chou creatively developed an approach as mentioned above to extract the important feature beyond amino acid composition. Our scheme is similar to, but different from, that of Chou. Experiments about phylogenetic analysis on two datasets and identification of DNA-binding proteins illustrate the utility of the proposed method.

2. Methods

A protein sequence can be viewed as a string of 20 amino acids. Without loss of generality, by the numerical indices 1,2, ... ,20, we represent the 20 native amino acids according to the alphabetical order of their single-letter codes: A,C,D,E,F,G,H,I,K,L,M,N,P,Q,R,S,T,V,W and Y. Then the frequencies of appearance of the 20 amino acids in a protein sequence are often used to construct a vector

$$[f_1, f_2, \ldots, f_{20}]$$

This is the conventional amino acid composition. The advantage of such a vector representation is that it is easy in statistical treatment, but it cannot reflect the effect regarding sequence order and property. In what follows, we will take this effect into account through a set of elements in addition to the 20 components.

Hydrophobicity, isoelectric point (pI), and relative distance (RD) are three important physicochemical properties of the 20 native amino acids. Here RD can be viewed as an integration of the information on three side chain properties: composition; polarity; and molecular volume—where composition is defined as the atomic weight ratio of hetero (noncarbon) elements in end groups or rings to carbons in the side chain (for details, see [47]). Listed in Table 1 are the original numerical values for hydrophobicity, pI and RD. As can be seen from Table 1, the values of P_1^0 (Hydrophobicity) is in the range [−2.53~1.38], and the values of P_2^0 (isoelectric point) are in the range of 2.97~10.76, while P_3^0 (relative distance) varies between 1469 and 3355. Therefore, the normalization of these values is needed. Here we normalize them by the formulary below:

$$p'_n(AA_i) = P_n^0(AA_i) - \min_{j=1,\ldots,20}\left\{P_n^0(AA_j)\right\},$$

$$P_n^*(AA_i) = \frac{p'_n(AA_i)}{\max\limits_{j=1,\ldots,20}\left\{p'_n(AA_j)\right\}}.\ i = 1,2,\ldots,\ 20,\ n = 1,2,3. \quad (1)$$

Table 1. The original numerical values for the properties of the 20 native amino acids.

Amino Acid (AA)	Hydrophobicity [a] (P_1^0)	pI [b] (P_2^0)	RD [b] (P_3^0)
A	0.62	6.02	1889
C	0.29	5.02	3355
D	−0.90	2.97	2209
E	−0.74	3.22	1812
F	1.19	5.48	1916
G	0.48	5.97	2078
H	−0.40	7.59	1507
I	1.38	6.02	1765
K	−1.50	9.74	1797
L	1.06	5.98	1822
M	0.64	5.75	1689
N	−0.78	5.42	1943
P	0.12	6.30	1720
Q	−0.85	5.65	1538
R	−2.53	10.76	1697
S	−0.18	5.68	2000
T	−0.05	6.53	1469
V	1.08	5.97	1680
W	0.81	5.89	2317
Y	0.26	5.66	1787

[a] Taken from [41]; [b] Taken from [47–49].

Clearly, the normalized values for properties of the 20 native amino acids are in the interval [0,1]. The corresponding values are listed in Table 2. The last row in this table gives the average values.

Table 2. The normalized values for the properties of the 20 native amino acids.

AA	P_1^*	P_2^*	P_3^*
A	0.8056	0.3915	0.2227
C	0.7212	0.2632	1.0000
D	0.4169	0	0.3924
E	0.4578	0.0321	0.1819
F	0.9514	0.3222	0.2370
G	0.7698	0.3851	0.3229
H	0.5448	0.5931	0.0201
I	1.0000	0.3915	0.1569
K	0.2634	0.8691	0.1739
L	0.9182	0.3864	0.1872
M	0.8107	0.3569	0.1166
N	0.4476	0.3145	0.2513
P	0.6777	0.4275	0.1331
Q	0.4297	0.3440	0.0366
R	0	1.0000	0.1209
S	0.6010	0.3479	0.2815
T	0.6343	0.4570	0
V	0.9233	0.3851	0.1119
W	0.8542	0.3748	0.4496
Y	0.7136	0.3453	0.1686
$\overline{P_n}$	0.6471	0.3994	0.2283

For each amino acid (AA), we associate it with a triple ($t(1)$, $t(2)$, $t(3)$), where

$$t(n) = \begin{cases} +1 & \text{if } P_n^*(AA) \geq \overline{P_n} \\ -1 & \text{otherwise} \end{cases} \quad (n = 1, 2, 3) \tag{2}$$

All the amino acids with a same triple form a group. In this way, the 20 native amino acids can be classified into 6 groups:

G_{I} = {A, Y, V, M, L, I},
G_{II} = {C, W, G, F},
G_{III} = {D, S, N},
G_{IV} = {E, Q},
G_{V} = {H, T, R, K},
G_{VI} = {P}.

For each group, the first amino acid is selected to be the representative. That is, A, C, D, E, H and P are used to stand for the six groups, respectively. The value of the property of a group is defined as the average value of the property of amino acids belonging to the group. Listed in Table 3 are the corresponding values of the six groups.

Table 3. The values for properties of the six groups.

Group	Representative	P_1	P_2	P_3
G_{I}	A	0.8619	0.3761	0.1607
G_{II}	C	0.8242	0.3363	0.5024
G_{III}	D	0.4885	0.2208	0.3084
G_{IV}	E	0.4437	0.1881	0.1092
G_{V}	H	0.3606	0.7298	0.0787
G_{VI}	P	0.6777	0.4275	0.1331

At the same time, a protein primary sequence can be reduced into a six-letter sequence by replacing each element in the protein sequence with its representative letter. Suppose $S = S_1S_2 \ldots S_L$ is a given six-letter sequence, we inspect it by stepping one element at a time. For the step k ($k = 1, 2, \ldots, L$), a 3-D space point $q_k = (x_k, y_k, z_k)$ can be constructed as follows:

$$(x_k, y_k, z_k) = (x_{k-1}, y_{k-1}, z_{k-1}) + (P_1(S_k), P_2(S_k), P_3(S_k)), \tag{3}$$

where $(x_0, y_0, z_0) = (0, 0, 0)$. When k runs from 1 to L, we get L points $q_1, q_2, \ldots, q_L$. Connecting these points one by one sequentially with straight lines, a three-dimensional curve can be drawn. One can further associate the graph with some structural matrices. Here we adopt the *L/L* matrix and denote it by M, whose (i,j)-entry is defined as follows:

$$m_{ij} = \begin{cases} \frac{d(i,j)}{d(i,i+1)+d(i+1,i+2)+\ldots+d(j-1,j)} & \text{if } i < j \\ 0 & \text{if } i = j\,, \\ m_{ji} & \text{if } i > j \end{cases} \tag{4}$$

where $d(i, j)$ is the Euclidean distance between points q_i and q_j. It is not difficult to see that $\lim_{t\to+\infty} {}^tM$ is a (0,1) matrix; here ${}^tM$ stands for the product of Hadmammard multiplication of the matrix M by itself t-times. In this paper, we call the limit matrix as a generalized adjacency matrix (GAM) generated by points $q_1, q_2, \ldots, q_L$, and denote it by M_G. Obviously, $[M_G]_{ij} = 1$ if and only if q_i and q_j lie on a straight line in the graph.

As mentioned above, once a symmetric matrix is given, one can calculate its ALE-index by the following formula:

$$\chi = \frac{1}{2}\left(\frac{1}{L}\,||\cdot||_{m1} + \sqrt{\frac{L-1}{L}}\,||\cdot||_F\right), \tag{5}$$

where L is the order of the matrix, $||\cdot||_{m1}$ and $||\cdot||_F$ are the $m1$- and F-norms of a matrix, respectively. In order to reduce variations caused by comparison of matrices with different sizes, we consider instead of $\chi(M_G)$ a normalized ALE-index $\chi'(M_G) = \frac{\chi(M_G)}{\sqrt{6L}}$.

In addition, following the similar procedures in capturing the sequence-order information of a protein [33,34], for the six-letter sequence $S = S_1S_2 \cdots S_L$, we extract a set of new order-correlated factors as defined below:

$$\begin{array}{l} \theta_1 = \frac{1}{L-1} \times \frac{1}{3} \times \sum_{n=1}^{3} g_n(S,1), \\ \theta_2 = \frac{1}{L-2} \times \frac{1}{3} \times \sum_{n=1}^{3} g_n(S,2), \\ \qquad \cdots \\ \theta_\lambda = \frac{1}{L-\lambda} \times \frac{1}{3} \times \sum_{n=1}^{3} g_n(S,\lambda). \end{array} \quad (\lambda < L) \tag{6}$$

where $\theta_k (k = 1, 2, \ldots, \lambda)$ is called the k-th tier correlation factor, $g_n(S, k)$ represents the coupling mode function as given by

$$g_n(S,k) = \sqrt{\sum_{i=1}^{L-k} (P_n(S_i) - P_n(S_{i+k}))^2} \tag{7}$$

Factor θ_1 reflects the coupling mode between the most contiguous elements along a six-letter sequence (Figure 1a); θ_2 reflects the coupling mode between the second-most contiguous (Figure 1b); θ_3 reflects the coupling mode between the third-most contiguous (Figure 1c), and so on. λ is the highest rank of the coupling mode.

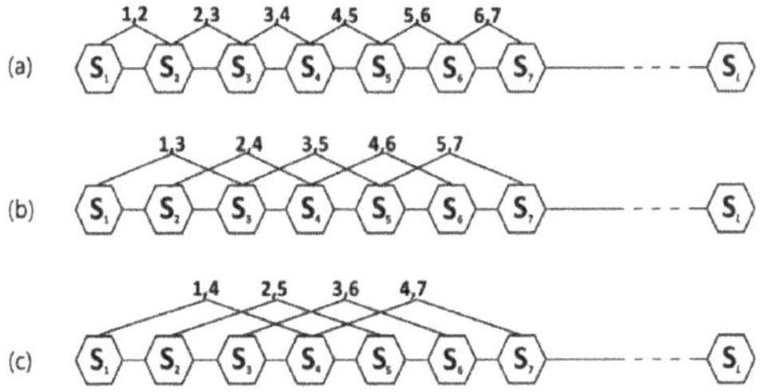

Figure 1. A schematic diagram to show: (**a**) the first-tier; (**b**) the second tier; and (**c**) the third-tier sequence order correlation mode along a sequence. Where the regular hexagon is used to show that each element of the sequence corresponds to one of the six amino acid groups.

Consequently, a protein sequence can be characterized by a $(21 + \lambda)$ dimensional vector V:

$$V = (v_1, v_2, \ldots, v_{20}, v_{20+1}, \ldots, v_{20+\lambda}, v_{20+\lambda+1}), \tag{8}$$

where

$$v_{\rm i} = \begin{cases} f_i & 1 \le i \le 20 \\ w_1 \theta_{\rm i-20} & 20+1 \le i \le 20+\lambda \\ w_2 \chi' & i = 20+\lambda+1 \end{cases} \tag{9}$$

Here w_1 and w_2 are weight factors. It is easy to see that the first 20 components reflect the effect of the amino acid composition, whereas the last $\lambda + 1$ components reflect the effect of sequence order and property of the residues. For convenience, a set of such $21 + \lambda$ components as formulated by Equations (8) and (9) is called the generalized pseudoamino acid composition of a protein sequence, and denoted by G-PseAAC.

3. Results

In this section, we will illustrate the use of the new quantitative characterization of protein sequences with two experiments. As we can see from Equations (8) and (9), there are three adjustable parameters for the G-PseAAC: λ, w_1, and w_2. It is not known beforehand which λ, w_1, and w_2 are best for a given problem. Three datasets are considered in this paper. The first one is used for determining these parameters and others for testing purpose.

3.1. Experiment I: Phylogenetic Analysis on Two Datasets

The first dataset used in this paper is composed of β-globin protein of 17 species (see Table 4). According to the method proposed, we associate each of the 17 protein sequences with a $\tau = 21 + \lambda$ dimensional vector. These vectors are then used to define a pair-wise evolutionary distance between any two protein sequences i and j:

$$D(i,j) = d(V_i, V_j) = \sqrt{\sum_{k=1}^{\tau} (v_{ik} - v_{jk})^2} \tag{10}$$

where $V_i = (v_{i1}, v_{i2}, \ldots, v_{i,\tau})$ and $V_j = (v_{j1}, v_{j2}, \ldots, v_{j,\tau})$ are the corresponding vectors for sequences i and j, respectively. Thus, a 17×17 real symmetric matrix D_{17} is obtained. On the basis of the achieved distance matrix D_{17}, a phylogenetic tree can be constructed using a UPGMA (Unweighted Pair Group Method with Arithmetic Mean) program included in the MEGA4 package. It is found that, when $\lambda = 7$ and $w_1 = w_2 = 1.6$, the non-mammals, including Guttata, Gallus and Muscovy duck, appear to cluster together and stay outside of the mammals, while Opossum is distinguished from the remaining mammals. In addition, Primate group {Human, Chimpanzee, Gorilla}, Cetartiodactyla group {Cattle, Banteng, Sheep, Goat}, Lagomorpha group {Rabbit, European hare}, and Rodentia group {House mouse, Western wild mouse, Spiny mouse, Norway rat} form separate branches, respectively (cf. Figure 2). This result is in accordance with the accepted taxonomy and the literature [1,12,30].

Table 4. The β-globin protein of 17 species.

No.	Species	Accession Number	Length (aa)
1	Human	ALU64020	147
2	Gorilla	P02024	147
3	Chimpanzee	P68873	147
4	Cattle	CAA25111	145
5	Banteng	BAJ05126	145
6	Goat	AAA30913	145
7	Sheep	ABC86525	145
8	European hare	CAA68429	147
9	Rabbit	CAA24251	147
10	House mouse	ADD52660	147
11	Western wild mouse	ACY03394	147
12	Spiny mouse	ACY03377	147
13	Norway rat	CAA29887	147
14	Opossum	AAA30976	147
15	Guttata	ACH46399	147
16	Gallus	CAA23700	147
17	Muscovy duck	CAA33756	147

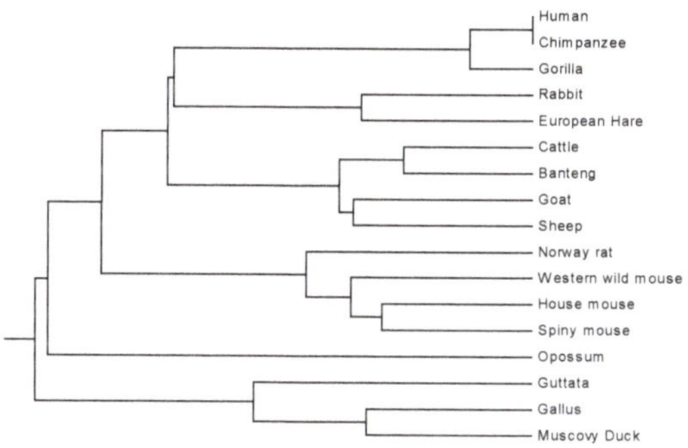

Figure 2. The relationship tree of 17 species.

Using the above-determined values for λ, w_1, and w_2, we infer the relationship of 72 coronavirus spike (S) proteins. The coronavirus, whose name is derived from its crown-like shape, is a positive-sense, single-stranded RNA virus in the family *Coronaviridae*. It was first identified in the 1960s from the nasal cavities of patients with the common cold. Most coronaviruses are not dangerous, but some strains could cause severe, sometimes fatal, diseases in humans and other animals. The MERS coronavirus (commonly shortened to MERS-CoV) is the virus that causes the Middle East respiratory syndrome (MERS). MERS was first reported in 2012 in Saudi Arabia and then in other countries in the Middle East, Africa, Asia, Europe and America. As of July 2016, 1769 laboratory-confirmed cases of MERS-CoV infection, including at least 630 related deaths (the case fatality rate is >30%), have been reported in over 27 countries (http://www.who.int/emergencies/mers-cov/en/). People also died from a severe acute respiratory syndrome (SARS), which first emerged in 2002 in Guangdong Province, China, and then spread globally. SARS resulted in more than 8000 infections with a case-fatality rate of ~10%. The virus that causes SARS is officially called SARS coronavirus (SARS-CoV). Both MERS-CoV and SARS-CoV are identified as members of the beta group of coronavirus, *Betacoronavirus*, while they are distinct from each other. The name, accession number, and abbreviation of the 72 sequences are listed in Table 5. According to the existing taxonomic groups, sequences 1–5 belong to group alpha (formerly known as Coronavirus group 1 (CoV-1)), sequences 6–8 are members of group gamma (formerly CoV-3), and the remaining belongs to group beta (formerly CoV-2). Refer to Table 5 for details.

Table 5. The accession number, name and abbreviation for 72 coronavirus spike proteins.

NO.	Accession Number	Virus Name/Strain	Abbreviation
1	CAB91145	Transmissible gastroenteritis virus, genomic RNA	TGEVG
2	NP_058424	Transmissible gastroenteritis virus	TGEV
3	AAK38656	Porcine epidemic diarrhea virus strain CV777	PEDVC
4	NP_598310	Porcine epidemic diarrhea virus	PEDV
5	BAL45637	Human coronavirus 229E	HCoV-229E
6	AAP92675	Avian infectious bronchitis virus isolate BJ	IBVBJ
7	AAS00080	Avian infectious bronchitis virus strain Ca199	IBVC
8	NP_040831	Avian infectious bronchitis virus	IBV
9	NP_937950	Human coronavirus OC43	HCoV-OC43
10	AAK83356	Bovine coronavirus isolate BCoV-ENT	BCoVE
11	AAL57308	Bovine coronavirus isolate BCoV-LUN	BCoVL
12	AAA66399	Bovine coronavirus strain Mebus	BCoVM
13	AAL40400	Bovine coronavirus strain Quebec	BCoVQ
14	NP_150077	Bovine coronavirus	BCoV
15	AAB86819	Mouse hepatitis virus strain MHV-A59C12 mutant	MHVA
16	YP_209233	Murine hepatitis virus strain JHM	MHVJHM
17	AAF69334	Mouse hepatitis virus strain Penn 97-1	MHVP
18	AAF69344	Mouse hepatitis virus strain ML-10	MHVM
19	NP_045300	Mouse hepatitis virus	MHV
20	AAU04646	SARS coronavirus civet007	civet007
21	AAU04649	SARS coronavirus civet010	civet010
22	AAU04664	SARS coronavirus civet020	civet020
23	AAV91631	SARS coronavirus A022	A022
24	AAV49730	SARS coronavirus B039	B039
25	AAP51227	SARS coronavirus GD01	GD01
26	AAS00003	SARS coronavirus GZ02	GZ02
27	AAP30030	SARS coronavirus BJ01	BJ01
28	AAP13567	SARS coronavirus CUHK-W1	CUHK-W1
29	AAP37017	SARS coronavirus TW1	TW1
30	AAR87523	SARS coronavirus TW2	TW2
31	BAC81348	SARS coronavirus TWH genomic RNA	TWH
32	BAC81362	SARS coronavirus TWJ genomic RNA	TWJ
33	AAQ01597	SARS coronavirus Taiwan TC1	TaiwanTC1
34	AAQ01609	SARS coronavirus Taiwan TC2	TaiwanTC2
35	AAP97882	SARS coronavirus Taiwan TC3	TaiwanTC3
36	AAP13441	SARS coronavirus Urbani	Urbani
37	AAP72986	SARS coronavirus HSR 1	HSR1
38	AAQ94060	SARS coronavirus AS	AS
39	AAP94737	SARS coronavirus CUHK-AG01	CUHK-AG01
40	AAP94748	SARS coronavirus CUHK-AG02	CUHK-AG02
41	AAP94759	SARS coronavirus CUHK-AG03	CUHK-AG03
42	AAP30713	SARS coronavirus CUHK-Su10	CUHK-Su10
43	AAP33697	SARS coronavirus Frankfurt 1	Frankfurt1
44	AAR14803	SARS coronavirus PUMC01	PUMC01
45	AAR14807	SARS coronavirus PUMC02	PUMC02
46	AAR14811	SARS coronavirus PUMC03	PUMC03
47	AAP41037	SARS coronavirus TOR2	TOR2
48	AAP50485	SARS coronavirus FRA	FRA
49	AAR23250	SARS coronavirus Sin01-11	Sino1-11
50	AHX00731	MERS coronavirus	KFU-HKU1
51	AHX00711	MERS coronavirus	KFU-HKU13
52	AHX00721	MERS coronavirus	KFU-HKU19Dam
53	AIY60578	MERS coronavirus	Abu-Dhabi_UAE_9
54	AIY60568	MERS coronavirus	Abu-Dhabi_UAE_33
55	AIZ74417	MERS coronavirus	Hu-France(UAE)-FRA1

Table 5. *Cont.*

NO.	Accession Number	Virus Name/Strain	Abbreviation
56	AIZ74433	MERS coronavirus	Hu-France-FRA2
57	ALJ54502	MERS coronavirus	Hu/Qunfidhah-KSA-Rs1338
58	AKN24821	MERS coronavirus	KFMC-1
59	AKN24830	MERS coronavirus	KFMC-7
60	ALJ76282	MERS coronavirus	Hu/Taif, KSA-2083
61	ALJ76281	MERS coronavirus	Hu/Taif, KSA-5920
62	ALJ54493	MERS coronavirus	Hu/Makkah-KSA-728
63	ALB08267	MERS coronavirus	KOREA/Seoul/014-1
64	ALB08278	MERS coronavirus	KOREA/Seoul/014-2
65	ALR69641	MERS coronavirus	D2731.3
66	AKQ21055	MERS coronavirus	ADFCA-HKU1
67	AKQ21064	MERS coronavirus	ADFCA-HKU2
68	AKQ21073	MERS coronavirus	ADFCA-HKU3
69	ALA50001	MERS coronavirus	camel/Taif/T68
70	ALA50012	MERS coronavirus	camel/Taif/T89
71	ALT66813	MERS coronavirus	Jordan_1
72	ALT66802	MERS coronavirus	Jordan_10

The corresponding phylogenetic tree constructed by our method is shown in Figure 3. Observing Figure 3, we find that TGEVG, TGEV, PEDVC, PEDV and HCoV-229E, which belong to group alpha, are clearly clustered together, and so do the three gamma coronaviruses IBV, IBVBJ, IBVC. In the subtree of the group beta, MERS-CoVs appear to cluster together, and SARS-CoVs are situated at an independent branch, while BCoV, BCoVM, BCoVQ, BCoVE, BCoVL, HCoV-OC43, MHV, MHVA, MHVM, MHVP and MHVJHM form a separate branch. The resulting cluster agrees well with the established taxonomic groups.

3.2. Experiment II: Identification of DNA-Binding Proteins

Numerous biological mechanisms depend on nucleic acid-protein interactions. The first step for understanding these mechanisms is to identify the interacting molecules. There are different strategies for determining DNA sequences that bind specifically to a known protein. However, it is difficult to accurately identify DNA-binding proteins [50]. Existing experimental techniques have low practical value due to time consumption and expensive costs [51]. Therefore, developing an efficient computational approach for identifying DNA-binding proteins is becoming increasingly important. In this section, we explore the application of the G-PseAAC to the identification of DNA-binding proteins. The parameters λ, w_1, and w_2 used here are the same as those determined in Section 3.1.

The dataset used here is taken from [51]. Itsoriginal version was created in 2009 by Kumar et al. [52], in which the DNA-binding proteins are extracted from the Pfam database [53] with keywords of "DNA-binding domain" and pairwise sequence identity cutoff of 25%, while the non DNA-binding domains are randomly selected from Pfam protein families that are unrelated to the DNA-binding protein family. Xu et al. [51] removed some sequences from the original dataset, and its current version is composed of 1585 protein sequences. This benchmark dataset contains 770 DNA-binding proteins and 815 non DNA-binding proteins, which form the positive sample set and negative sample set, respectively. We randomly divide the 770 DNA-binding proteins into two parts, one has 410 sequences and the other 360 sequences. Also, we randomly select 410 and 405 sequences from the 815 non DNA-binding proteins, respectively. We conduct two sets of data. Set I contains 410 DNA-binding proteins and 410 non DNA-binding proteins. This set serves as a training set. The remaining protein sequences (360 DNA-binding proteins and 405 non DNA-binding proteins) form Set II, which serves as a test set.

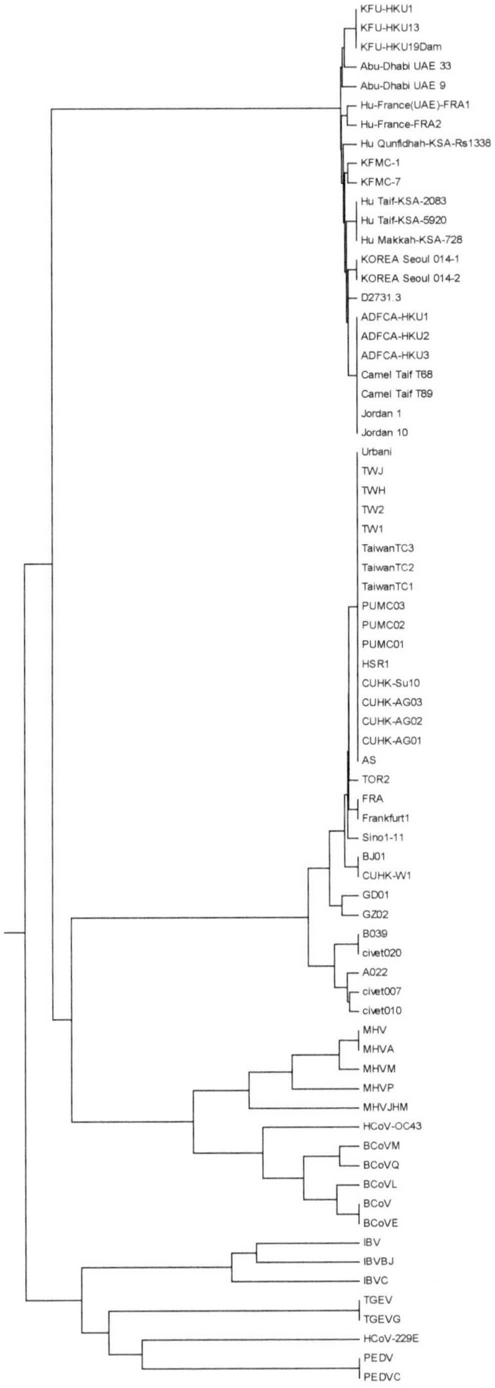

Figure 3. The relationship tree of 72 coronavirus spike proteins.

Support vector machine (SVM) is employed as the classifier, and its implementation is based on the package LIBSVM (a Library for Support Vector Machines) v3.17 [54], which is open sourced and can be freely downloaded from http://www.csie.ntu.edu.tw/~cjlin/libsvm. There are four types of kernel functions in LIBSVM: linear kernel; polynomial kernel; radial basis function (RBF) kernel; and sigmoid kernel. Among them, the RBF kernel is deemed a reasonable first choice [55]. The main reason is that, taking the form $K\left(V_i, V_j\right) = e^{-\gamma\|V_i - V_j\|^2}$, the RBF kernel can non-linearly map samples into a higher dimensional space so it can handle the non-linearly separable data. Accordingly, the RBF kernel is also adopted in this paper. The model selection of this kernel involves two parameters to be decided: the penalty parameter C and the kernel parameter γ. We first convert each of the 1585 protein sequences into a 28-D vector, and then the vectors belonging to Set I are scaled and fed to the SVM. With an optimization procedure using a grid search strategy in LIBSVM, the parameter pair (C, γ) is determined as (8, 0.5) (It should be pointed out that the optimal values for one round of cross-validation may not be the same for another.). In literature, a set of metrics are often used to measure the prediction quality. To make it intuitive and easy to understand for readers, here we adopt the definition and notations used in [40,41,56–60] to describe the corresponding evaluation metrics:

$$S_n = 1 - \frac{N_-^+}{N^+},$$

$$S_p = 1 - \frac{N_+^-}{N^-},$$

$$Acc = 1 - \frac{N_-^+ + N_+^-}{N^+ + N^-},$$

$$MCC = \frac{1 - \left(\frac{N_-^+}{N^+} + \frac{N_+^-}{N^-}\right)}{\sqrt{\left(1 + \frac{N_+^- - N_-^+}{N^+}\right)\left(1 + \frac{N_-^+ - N_+^-}{N^-}\right)}},$$

$$F1_M = 2 \times \frac{\text{Precision} \times \text{Recall}}{\text{Precision} + \text{Recall}},$$

where N^+ is the total number of DNA-binding proteins investigated, while N_-^+ the number of DNA-binding proteins incorrectly predicted to be of non DNA-binding proteins; N^- the total number of non DNA-binding proteins investigated, while N_+^- the number of non DNA-binding proteins incorrectly predicted as DNA-binding proteins. Precision $= \frac{N^+ - N_-^+}{N^+ - N_-^+ + N_+^-}$, Recall $= 1 - \frac{N_-^+}{N^+}$. It should be pointed out that the set of metrics above is valid only for the single-label system (such as the case at hand). For the multi-label systems whose existence has become more frequent in system biology [61–64] and system medicine [65], a completely different set of metrics as defined in [66] is needed.

With the best pair (C, γ) obtained in the training stage, Set II is fed to the SVM. We find that $N_-^+ = 17$ and $N_+^- = 22$. We thus have

$$S_n = 95.28\%,\ S_p = 94.57\%,\ Acc = 94.90\%,\ MCC = 0.8978,\ F1_M = 94.62\%.$$

Repeating the above random division procedure three times, we perform three cross-validation tests and list the results in Table 6. As can be seen, the accuracy (*Acc*), Matthew's correlation coefficient (*MCC*), and F1-measure (*F1_M*) in each cross-validation test are greater than 94.90%, 0.8977, and 94.59%, respectively. This result indicates that our method is promising in identifying DNA-binding proteins.

Table 6. The results of three different cross-validation tests.

Test	1	2	3	Average
S_n (%)	95.28	94.72	95.00	95.00
S_p (%)	94.57	95.06	95.06	94.90
Acc (%)	94.90	94.90	95.03	94.94
MCC	0.8978	0.8977	0.9004	0.8986
F1_M (%)	94.62	94.59	94.73	94.65

4. Discussion

4.1. Selection of Properties for Amino Acids

In addition to the three physical–chemical properties mentioned above, both hydrophilicity and molecular weight of amino acids can play important roles for characterization of proteins. Therefore, one can consider r-combinations of the five properties to describe a protein sequence. The purpose of this paper is to find an appropriate way for converting a protein sequence of 20 kinds of amino acids into a string over a "small" alphabet. If we take r to be 3, by the scheme described in Section 2, the triple $(t(1), t(2), t(3))$ has at most $2^3 = 8$ different forms. This means that the 20 native amino acids can thus be classified into no more than eight groups, whereas if the 5-combination or 4-combination is selected, by the similar scheme, $(t(1), t(2), \cdots, t(r))$ will have $2^5 = 32$ or $2^4 = 16$ possible forms. Compared with "20," the figure is not "small." Therefore, r is taken to be 3 in this paper. By means of each of the 3-combinations of the five properties, the same experiments are performed. As a result, we find that hydrophobicity, isoelectric point, and relative distance form the best 3-combination.

4.2. Feature Analysis

As we see from Equations (8) and (9), the 28-D feature vector consists of three parts: 20 amino acid compositions; 7 correlation factors; and 1 ALE-index. One may be interested in knowing whether or not the last two parts are significant. First and foremost, let us see what would happen if only the first part was used? Without loss of generality, suppose S is a protein sequence and the counts of 20 native amino acids are $n_1, n_2, \cdots, n_{20}$, respectively. Then we have a multi-set $M(S) = \{n_1 \cdot A, n_2 \cdot C, \cdots, n_{20} \cdot Y\}$. Based on the knowledge of combinatorics, it is not difficult to see that there are a total of $\frac{|S|!}{n_1! \cdot n_2! \cdots n_{20}!} = \frac{(n_1 + n_2 + \cdots + n_{20})!}{n_1! \cdot n_2! \cdots n_{20}!}$ different sequence/strings possessing the same amino acid compostion. This suggests that the amino acid composition alone is not sufficient to represent and compare protein sequences. What would happen if only the first two parts were used (i.e., without using the ALE-index)? By using the vector with the first 27 components, experiments I and II are performed. For the first dataset, there is no significant difference between the tree constructed with the 27-D vector and that with the 28-D vector. For the second dataset, the corresponding relationship tree of coronavirus spike proteins is shown in Figure 4. From Figure 4, it is easy to see that MERS-CoVs belonging to *Betacoronavirus* appear to cluster together with the three *Gammacoronaviruses*, instead of the other *Betacoronaviruses*. This phenomenon is disappointing. For the third dataset, we repeat the three cross-validation tests with the 27-D vector and list the corresponding results in Table 7. By comparing Table 7 with Table 6, we can find that the prediction quality diminished slightly. These results indicate that the ALE-index can make a very positive contribution to the performance of experiments.

Table 7. Results of the three cross-validation tests with the 27-D vector.

Test	1	2	3	Average
S_n (%)	95.00	93.61	94.44	94.35
S_p (%)	94.32	94.32	95.06	94.57
Acc (%)	94.64	93.99	94.78	94.47

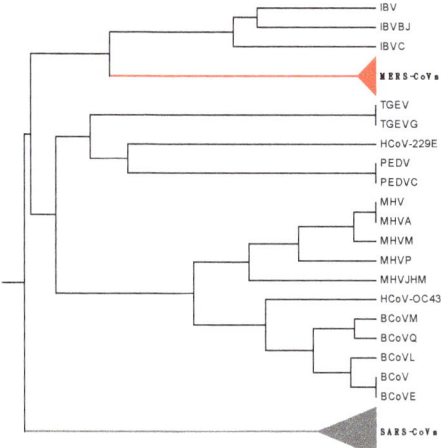

Figure 4. The relationship tree of the coronavirus spike proteins with the 27-D vector.

5. Conclusions

By means of three important physicochemical properties of amino acids, we first classify the 20 native amino acids into six groups, and assign to each group a representative symbol. Then, by substituting each letter with its representative letter, we convert a protein primary sequence into a six-letter sequence, which can be regarded as a coarse-grained description of the protein primary sequence. In comparison with the string composed of 20 kinds of amino acids, the reduced sequence not only makes the generalization from representations of DNA sequences to those of proteins easier, but also enables us to focus more on the information of our interest. On the basis of the six-letter sequence, we obtain a generalized adjacency matrix (GAM) and then its normalized ALE-index. Also, we extract λ order-correlated factors via the reduced sequence. Combining these elements with the frequencies of occurrenceof 20 native amino acids, we constructa $(21 + \lambda)$ dimensional vector to characterize a protein sequence. Our method is tested byphylogenetic analysis and identification of DNA-binding proteins. The feature analysis implies that the $\lambda + 1$ components beyond the amino acid composition play very important roles in the performance of the experiment. As shown in a series of recent publications (see, e.g., [58,67–72]) in demonstrating new methods or approaches, user-friendly and publicly accessible web-servers will significantly enhance their impacts [73]. We will make efforts in our future work to further improve our method and provide a web-server for the new method presented.

Acknowledgments: The authors wish to thank the four anonymous referees for their valuable suggestions and support. The authors also thank Ren Zhang, University of Wollongong, Australia, for helpful discussions. This work was partially supported by the National Natural Science Foundation of China (No. 11171042), the Program for Liaoning Innovative Research Team in University (LT2014024), the Natural Science Foundation of Liaoning Province (201602005), and the Liaoning Bai Qian Wan Talents Program (2012921060).

Author Contributions: Chun Li and Yan-Xia Lin conceived the study and wrote the paper. Xueqin Li participated in the design of the study and analysis of the results.

Conflicts of Interest: The authors declare no conflict of interest.

References

1. Randic, M.; Vracko, M.; Nandy, A.; Basak, S.C. On 3-D graphical representation of DNA primary sequences and their numerical characterization. *J. Chem. Inf. Comput. Sci.* **2000**, *40*, 1235–1244. [CrossRef] [PubMed]
2. Yao, Y.H.; Yan, S.; Han, J.; Dai, Q.; He, P.-A. A novel descriptor of protein sequences and its application. *J. Theor. Biol.* **2014**, *347*, 109–117. [CrossRef] [PubMed]
3. Hamori, E.; Ruskin, J. H curves, a novel method of representation of nucleotide series especially suited for long DNA sequences. *J. Biol. Chem.* **1983**, *258*, 1318–1327. [PubMed]
4. Hamori, E. Novel DNA sequencerepresentations. *Nature* **1985**, *314*, 585–586. [CrossRef] [PubMed]
5. Gates, M.A. Simpler DNA sequence representations. *Nature* **1985**, *316*, 219. [CrossRef] [PubMed]
6. Jeffrey, H.J. Chaos game representation of gene structure. *Nucleic Acids Res.* **1990**, *18*, 2163–2170. [CrossRef] [PubMed]
7. Nandy, A. A new graphical representation and analysis of DNA sequence structure: I Methodology and application to globin genes. *Curr. Sci.* **1994**, *66*, 309–314.
8. Nandy, A. Graphical representation of long DNA sequences. *Curr. Sci.* **1994**, *66*, 821.
9. Leong, P.M.; Morgenthaler, S. Random walk and gap plots of DNA sequences. *Comput. Appl. Biosci.* **1995**, *11*, 503–507. [CrossRef] [PubMed]
10. Zhang, R.; Zhang, C.T. Z curves, an intuitive tool for visualizing and analyzing DNA sequences. *J. Biomol. Str. Dyn.* **1994**, *11*, 767–782. [CrossRef] [PubMed]
11. Zhang, R.; Zhang, C.T. A brief review: The Z-curve theory and its application in genome analysis. *Curr. Genomics* **2014**, *15*, 78–94. [CrossRef] [PubMed]
12. Randic, M.; Vracko, M.; Lers, N.; Plavsic, D. Analysis ofsimilarity/dissimilarity of DNA sequences based on novel 2-Dgraphical representation. *Chem. Phys. Lett.* **2003**, *371*, 202–207. [CrossRef]
13. Randic, M.; Novic, M.; Plavsic, D. Milestones in graphical bioinformatics. *Int. J. Quantum Chem.* **2013**, *113*, 2413–2446. [CrossRef]
14. Li, C.; Fei, W.C.; Zhao, Y.; Yu, X.Q. Novel graphical representation and numerical characterization of DNA sequences. *Appl. Sci.* **2016**, *6*, 63. [CrossRef]
15. Sen, D.; Dasgupta, S.; Pal, I.; Manna, S.; Basak, S.C.; Nandy, A. Intercorrelation of major DNA/RNA sequence descriptors—A preliminary study. *Curr. Comput. Aided Drug Des.* **2016**, *12*, 216–228. [CrossRef] [PubMed]
16. Feng, Z.P.; Zhang, C.T. A graphic representation of protein sequence and predicting the subcellular locations of prokaryotic proteins. *Int. J. Biochem. Cell Biol.* **2002**, *34*, 298–307. [CrossRef]
17. Randic, M. 2-D Graphical representation of proteins based on virtual genetic code. *SAR QSAR Environ. Res.* **2004**, *15*, 147–157. [CrossRef] [PubMed]
18. Randic, M.; Zupan, J.; Balaban, A.T. Unique graphical representation of protein sequences based on nucleotide triplet codons. *Chem. Phys. Lett.* **2004**, *397*, 247–252. [CrossRef]
19. Randic, M.; Butina, D.; Zupan, J. Novel 2-D graphical representation of proteins. *Chem. Phys. Lett.* **2006**, *419*, 528–532. [CrossRef]
20. Randic, M.; Zupan, J.; Balaban, A.T.; Vikic-Topic, D.; Plavsic, D. Graphical representation of proteins. *Chem. Rev.* **2011**, *111*, 790–862. [CrossRef] [PubMed]
21. Novic, M.; Randic, M. Representation of proteins as walks in 20-D space. *SAR QSAR Environ. Res.* **2008**, *19*, 317–337.
22. Aguero-Chapin, G.; Gonzalez-Diaz, H.; Molina, R.; Varona-Santos, J.; Uriarte, E.; Gonzalez-Diaz, Y. Novel 2D maps and coupling numbers for protein sequences. The first QSAR study of polygalacturonases; isolation and prediction of a novel sequence from *Psidiumguajava* L. *FEBS Lett.* **2006**, *580*, 723–730. [CrossRef] [PubMed]
23. Li, C.; Xing, L.L.; Wang, X. 2-D graphical representation of protein sequences and its application to coronavirus phylogeny. *BMB Rep.* **2008**, *41*, 217–222. [CrossRef] [PubMed]
24. Nandy, A.; Ghosh, A.; Nandy, P. Numerical characterization of protein sequences and application to voltage-gated sodium channel α subunit phylogeny. *Silico Biol.* **2009**, *9*, 77–87.
25. Ghosh, A.; Nandy, A. Graphical representation and mathematical characterization of protein sequences and applications to viral proteins. *Adv. Protein Chem. Struct. Biol.* **2011**, *83*, 1–42. [PubMed]
26. Sun, D.D.; Xu, C.R.; Zhang, Y.S. A novel method of 2D graphical representation for proteins and its application. *MATCH Commun. Math. Comput. Chem.* **2016**, *75*, 431–446.

27. Qi, Z.H.; Jin, M.Z.; Li, S.L.; Feng, J. A protein mapping method based on physicochemical properties and dimension reduction. *Comput. Biol. Med.* **2015**, *57*, 1–7. [CrossRef] [PubMed]
28. Randic, M.; Balaban, A.T. On a four-dimensional representation of DNA primary sequences. *J. Chem. Inf. Comput. Sci.* **2003**, *43*, 532–539. [CrossRef] [PubMed]
29. Li, C.; Yang, Y.; Jia, M.D.; Zhang, Y.Y.; Yu, X.Q.; Wang, C.Z. Phylogenetic analysis of DNA sequences based on *k*-word and rough set theory. *Physica A* **2014**, *398*, 162–171. [CrossRef]
30. Randic, M.; Guo, X.F.; Basak, S.C. On the characterization of DNA primary sequences by triplet of nucleic acid bases. *J. Chem. Inf. Comput. Sci.* **2001**, *41*, 619–626. [CrossRef] [PubMed]
31. Randic, M.; Vracko, M. On the similarity of DNA primary sequences. *J. Chem. Inf. Comput. Sci.* **2000**, *40*, 599–606. [CrossRef] [PubMed]
32. Li, C.; Wang, J. New invariant of DNA sequences. *J. Chem. Inf. Model.* **2005**, *36*, 115–120. [CrossRef] [PubMed]
33. Chou, K.C. Prediction of protein cellular attributes using pseudo-amino acid composition. *Proteins Struct. Funct. Bioinform.* **2001**, *43*, 246–255. [CrossRef] [PubMed]
34. Chou, K.C. Using amphiphilic pseudo amino acid composition to predict enzyme subfamily classes. *Bioinformatics* **2005**, *21*, 10–19. [CrossRef] [PubMed]
35. Cao, D.S.; Xu, Q.S.; Liang, Y.Z. Propy: A tool to generate various modes of Chou's PseAAC. *Bioinformatics* **2013**, *29*, 960–962. [CrossRef] [PubMed]
36. Du, P.; Gu, S.; Jiao, Y. PseAAC-General: Fast building various modes of general form of Chou's pseudo amino acid composition for large-scale protein datasets. *Int. J. Mol. Sci.* **2014**, *15*, 3495–3506. [CrossRef] [PubMed]
37. Chou, K.C. Pseudo amino acid composition and its applications in bioinformatics, proteomics and system biology. *Curr. Proteom.* **2009**, *6*, 262–274. [CrossRef]
38. Kabir, M.; Hayat, M. iRSpot-GAEnsC: Identifying recombination spots via ensemble classifier and extending the concept of Chou's PseAAC to formulate DNA samples. *Mol. Genet. Genom.* **2016**, *291*, 285–296. [CrossRef] [PubMed]
39. Tahir, M.; Hayat, M. iNuc-STNC: A sequence-based predictor for identification of nucleosome positioning in genomes by extending the concept of SAAC and Chou's PseAAC. *Mol. Biosyst.* **2016**, *12*, 2587–2593. [CrossRef] [PubMed]
40. Chen, W.; Feng, P.M.; Lin, H.; Chou, K.C. iRSpot-PseDNC: Identify recombination spots with pseudo dinucleotide composition. *Nucleic Acids Res.* **2013**, *41*, e68. [CrossRef] [PubMed]
41. Qiu, W.R.; Xiao, X.; Chou, K.C. iRSpot-TNCPseAAC: Identify recombination spots with trinucleotide composition and pseudo amino acid components. *Int. J. Mol. Sci.* **2014**, *15*, 1746–1766. [CrossRef] [PubMed]
42. Li, L.Q.; Yu, S.J.; Xiao, W.D.; Li, Y.S.; Huang, L.; Zheng, X.Q.; Zhou, S.W.; Yang, H. Sequence-based identification of recombination spots using pseudo nucleic acid representation and recursive feature extraction by linear kernel SVM. *BMC Bioinform.* **2014**, *15*, 340. [CrossRef] [PubMed]
43. Chen, W.; Lei, T.Y.; Jin, D.C.; Chou, K.C. PseKNC: A flexible web-server for generating pseudo K-tuple nucleotide composition. *Anal. Biochem.* **2014**, *456*, 53–60. [CrossRef] [PubMed]
44. Chen, W.; Zhang, X.; Brooker, J.; Lin, H.; Zhang, L.Q.; Chou, K.C. PseKNC-General: A cross-platform package for generating various modes of pseudo nucleotide compositions. *Bioinformatics* **2015**, *31*, 119–120. [CrossRef] [PubMed]
45. Chen, W.; Lin, H.; Chou, K.C. Pseudo nucleotide composition or PseKNC: An effective formulation for analyzing genomic sequences. *Mol. Biosyst.* **2015**, *11*, 2620–2634. [CrossRef] [PubMed]
46. Liu, B.; Liu, F.; Wang, X.L.; Chen, J.; Fang, L.; Chou, K.C. Pse-in-One: A web server for generating various modes of pseudo components of DNA, RNA, and protein sequences. *Nucleic Acids Res.* **2015**, *43*, W65–W71. [CrossRef] [PubMed]
47. Grantham, R. Amino acid difference formula to help explain protein. *Science* **1974**, *185*, 862–864. [CrossRef] [PubMed]
48. Ma, F.; Wu, Y.T.; Xu, X.F. Correlation analysis of some physical chemistry properties among genetic codons and amino acids. *J. Anhui Agric. Univ.* **2003**, *30*, 439–445.
49. Li, C.; Wang, J.; Zhang, Y.; Wang, J. Similarity analysis of protein sequences based on the normalized relative entropy. *Comb. Chem. High Throughput Scr.* **2008**, *11*, 477–481. [CrossRef]

50. Hegarat, N.; Francois, J.C.; Praseuth, D. Modern tools for identification of nucleic acid-binding proteins. *Biochimie* **2008**, *90*, 1265–1272. [CrossRef] [PubMed]
51. Xu, R.F.; Zhou, J.Y.; Liu, B.; Yao, L.; He, Y.L.; Zou, Q.; Wang, X.L. enDNA-Prot: Identification of DNA-binding proteins by applying ensemble learning. *Biomed. Res. Int.* **2014**, *2014*, 294279. [CrossRef] [PubMed]
52. Kumar, K.K.; Pugalenthi, G.; Suganthan, P.N. DNA-Prot: Identification of DNA binding proteins from protein sequence information using random forest. *J. Biomol. Struct. Dyn.* **2009**, *26*, 679–686. [CrossRef] [PubMed]
53. Sonnhammer, E.L.; Eddy, S.R.; Durbin, R. Pfam: A comprehensive database of protein domain families based on seed alignments. *Proteins* **1997**, *28*, 405–420. [CrossRef]
54. Chang, C.C.; Lin, C.J. Libsvm: A library for support vector machines. *ACM Trans. Intell. Syst. Technol.* **2011**, *2*, 1–27. [CrossRef]
55. Hsu, C.W.; Chang, C.C.; Lin, C.J. A Practical Guide to Support Vector Classification. Available online: Https://www.csie.ntu.edu.tw/~cjlin/libsvm (accessed on 17 August 2014).
56. Lin, H.; Deng, E.Z.; Ding, H.; Chen, W.; Chou, K.C. iPro54-PseKNC: A sequence-based predictor for identifying sigma-54 promoters in prokaryote with pseudo k-tuple nucleotide composition. *Nucleic Acids Res.* **2014**, *42*, 12961–12972. [CrossRef] [PubMed]
57. Liu, B.; Fang, L.; Long, R.; Lan, X.; Chou, K.C. iEnhancer-2L: A two-layer predictor for identifying enhancers and their strength by pseudo k-tuple nucleotide composition. *Bioinformatics* **2016**, *32*, 362–369. [CrossRef] [PubMed]
58. Jia, J.; Zhang, L.; Liu, Z.; Xiao, X.; Chou, K.C. pSumo-CD: Predicting sumoylation sites in proteins with covariance discriminant algorithm by incorporating sequence-coupled effects into general PseAAC. *Bioinformatics* **2016**, *32*, 3133–3141. [CrossRef] [PubMed]
59. Chen, W.; Feng, P.; Ding, H.; Lin, H.; Chou, K.C. Using deformation energy to analyze nucleosome positioning in genomes. *Genomics* **2016**, *107*, 69–75. [CrossRef] [PubMed]
60. Chen, W.; Tang, H.; Ye, J.; Lin, H.; Chou, K.C. iRNA-PseU: Identifying RNA pseudouridine sites. *Mol. Ther. Nucleic Acids* **2016**, *5*, e332.
61. Chou, K.C.; Wu, Z.C.; Xiao, X. iLoc-Euk: A Multi-Label Classifier for Predicting the Subcellular Localization of Singleplex and Multiplex Eukaryotic Proteins. *PLoS ONE* **2011**, *6*, e18258. [CrossRef] [PubMed]
62. Chou, K.C.; Wu, Z.C.; Xiao, X. iLoc-Hum: Using accumulation-label scale to predict subcellular locations of human proteins with both single and multiple sites. *Mol. Biosyst.* **2012**, *8*, 629–641. [CrossRef] [PubMed]
63. Wu, Z.C.; Xiao, X.; Chou, K.C. iLoc-Plant: A multi-label classifier for predicting the subcellular localization of plant proteins with both single and multiple sites. *Mol. Biosyst.* **2011**, *7*, 3287–3297. [CrossRef] [PubMed]
64. Lin, W.Z.; Fang, J.A.; Xiao, X.; Chou, K.C. iLoc-Animal: A multi-label learning classifier for predicting subcellular localization of animal proteins. *Mol. Biosyst.* **2013**, *9*, 634–644. [CrossRef] [PubMed]
65. Xiao, X.; Wang, P.; Lin, W.Z.; Jia, J.H.; Chou, K.C. iAMP-2L: A two-level multi-label classifier for identifying antimicrobial peptides and their functional types. *Anal. Biochem.* **2013**, *436*, 168–177. [CrossRef] [PubMed]
66. Chou, K.C. Some remarks on predicting multi-label attributes in molecular biosystems. *Mol. Biosyst.* **2013**, *9*, 1092–1100. [CrossRef] [PubMed]
67. Qiu, W.R.; Sun, B.Q.; Xiao, X.; Xu, Z.C.; Chou, K.C. iPTM-mLys: Identifying multiple lysine PTM sites and their different types. *Bioinformatics* **2016**, *32*, 3116–3123. [CrossRef] [PubMed]
68. Qiu, W.R.; Sun, B.Q.; Xiao, X. iHyd-PseCp: Identify hydroxyproline and hydroxylysine in proteins by incorporating sequence-coupled effects into general PseAAC. *Oncotarget* **2016**, *7*, 44310–44321. [CrossRef] [PubMed]
69. Qiu, W.R.; Xiao, X.; Xu, Z.H.; Chou, K.C. iPhos-PseEn: Identifying phosphorylation sites in proteins by fusing different pseudo components into an ensemble classifier. *Oncotarget* **2016**, *7*, 51270–51283. [CrossRef] [PubMed]
70. Chen, W.; Ding, H.; Feng, P.; Lin, H.; Chou, K.C. iACP: A sequence-based tool for identifying anticancer peptides. *Oncotarget* **2016**, *7*, 16895–16909. [CrossRef] [PubMed]
71. Jia, J.; Liu, Z.; Xiao, X.; Liu, B.X.; Chou, K.C. iCar-PseCp: Identify carbonylation sites in proteins by Monto Carlo sampling and incorporating sequence coupled effects into general PseAAC. *Oncotarget* **2016**, *7*, 34558–34570. [CrossRef] [PubMed]

72. Xiao, X.; Ye, H.X.; Liu, Z.; Jia, J.H.; Chou, K.C. iROS-gPseKNC: Predicting replication origin sites in DNA by incorporating dinucleotide position-specific propensity into general pseudo nucleotide composition. *Oncotarget* **2016**, *7*, 34180–34189. [CrossRef] [PubMed]
73. Chou, K.C. Impacts of bioinformatics to medicinal chemistry. *Med. Chem.* **2015**, *11*, 218–234. [CrossRef] [PubMed]

applied sciences

MDPI

Article

A Liquid-Solid Coupling Hemodynamic Model with Microcirculation Load

Bai Li * and Xiaoyang Li

Biomechanical Research Laboratory, Center of Engineering Mechanics, Beijing University of Technology, No.100 Pingleyuan, Chaoyang District, Beijing 100124, China; lixy@bjut.edu.cn (X.L.)

* Correspondence: litaibai1987@emails.bjut.edu.cn; Tel.: +86-180-4652-8515

Academic Editor: Yang Kuang

Received: 17 November 2015; Accepted: 13 January 2016; Published: 20 January 2016

Abstract: From the aspect of human circulation system structure, a complete hemodynamic model requires consideration of the influence of microcirculation load effect. This paper selected the seepage in porous media as the simulant of microcirculation load. On the basis of a bi-directional liquid-solid coupling tube model, we built a liquid-solid-porous media seepage coupling model. The simulation parameters accorded with the physiological reality. Inlet condition was set as transient single-pulse velocity, and outlet as free outlet. The pressure in the tube was kept at the state of dynamic stability in the range of 80–120 mmHg. The model was able to simulate the entire propagating process of pulse wave. The pulse wave velocity simulated was 6.25 m/s, which accorded with the physiological reality. The complex pressure wave shape produced by reflections of pressure wave was also observed. After the model changed the cardiac cycle length, the pressure change according with actual human physiology was simulated successfully. The model in this paper is well-developed and reliable. It demonstrates the importance of microcirculation load in hemodynamic model. Moreover the properties of the model provide a possibility for the simulation of dynamic adjustment process of human circulation system, which indicates a promising prospect in clinical application.

Keywords: hemodynamic model; microcirculation load; liquid-solid-porous media seepage coupling

1. Introduction

A hemodynamic model is able to provide theoretical evidence for suitable selections of clinical treatment plan, and hence has important meanings. Take human physiology as an example: if the heart is regarded as a power source and the arteries at each level as transportation pipelines, the human microcirculation system can be seen as the system load. There is a complex coupling relationship between various components of the circulation system, which influence each other severely. A change in one factor can even cause a change in the environment of the entire circulation system. Complex phenomena in the human blood circulation system should be the properties produced by the coupling of various parts of the system, not only relying on the selection of boundary conditions. Therefore, a well-developed hemodynamic model must be able to show completely different components of the circulation system and their effects. Only in this way can it provide a correct simulation of the numerous phenomena of human blood flow, and provide evidence for research on producing and developing principles for dealing with some diseases.

Nowadays, there are many studies on the hemodynamics model, but none of them is able to show the properties of the human circulation system. Though the hemodynamic model of a rigid tube wall [1–4] has been widely utilized, it is not able to show the stress-strain properties of solids, which places severe limitations on clinical applications. The model of a flexible tube wall [5–7] is not only able to show the deformation of the solid, but also reflects the influence of solid deformation on flow field, which has great potential for application [8]. However, the model fails to show the complex pressure wave of physiology [9] and simulate the propagation of a pulse wave. In these studies, the outlet pressures were all set as a fixed value or a measured value, and it was assumed that these outlet conditions will not change with the change of the flow field. Since pressure influences the deformation of the tube wall, the change of outlet pressure conditions will have a profound influence on the flow field. Therefore, the results were not ideal in those studies relating to the adjustment process of the entire human circulation system [10]. In the study on myocardial bridge [11], Schwarz observed the influence of changes of downstream flow field on the front-end flow field. It follows that changeable load condition is one of the most important components neglected in most hemodynamic models. Through analysis of the characteristics of a capillary microcirculation system and the flow properties of the kidney and other organs, it was found that the seepage in porous media is a load model that accords relatively well with the form of physiological changes. Therefore, this paper needs to add a load of porous media seepage in the classical bi-directional liquid-solid coupling model to completely simulate the human circulation system.

Additionally, in order to make the load of porous media seepage accord with physiology, there are requirements for the tube length and other flow field factors. The human microcirculation system is located at the end of the human blood circulation system. Before arriving at the microcirculation system, the fluid will flow through the entire circulation system and experience severe changes. However in previous studies, the flow field was not long enough for the simulation of these changes, so the model requires a flow field that is equivalent to the actual physiology to simulate the physiological flow field correctly. In addition, the deformation of tube wall is an important factor influencing the flow field. Thus, in the calculating simulation of Dong [12], as the influence of liquid-solid coupling was not considered, his simulation results were not very satisfying, though the load of porous media seepage was also selected. It can be seen that a bi-directional liquid-solid coupling model with enough developing space for a flow field is the precondition for a load to produce correct results. Only when all the factors in this system can meet physiological needs can the results accord with physiology.

Accordingly, this paper built a bi-directional liquid-solid model with long straight tube to simulate the human blood vessel. A load of porous media seepage was arranged at the outlet of the liquid-solid coupling model to simulate the human microcirculation system. Using the most ideal single-pulse inlet condition and free outlet's boundary condition, we successfully simulated various complex phenomena in human blood vessels. The mechanisms of these phenomena were revealed through analysis. Moreover, the heart rate was changed for a comparison with human physiological reality, thus verifying the reliability of the model proposed.

2. Model and Methods

2.1. Hemodynamic Model

According to the discussions above, this paper built a model of a long straight tube with a flexible wall and a load of porous media seepage. As shown in Figure 1, the model mainly consists of three parts: the straight tube is used for simulating the flow field region of the blood vessel; the flexible tube wall is used for simulating deformation of the vascular wall and its influence on flow; and the tube with porous media seepage is used for simulating flow in the microcirculation system. The control equations of the various regions are as follows:

2.1.1. Fluid Region

Continuity equation:

$$\frac{\partial u_i}{\partial x_i} = 0 \tag{1}$$

Momentum equation:

$$\frac{\partial u_i}{\partial t} + u_j \frac{\partial u_i}{\partial x_j} = -\frac{1}{\rho_f} \frac{\partial p}{\partial x_i} + \frac{1}{\rho_f} \frac{\partial \tau_{ij}}{\partial x_j} \tag{2}$$

where u_i is fluid velocity, p is fluid pressure, ρ_f is fluid density, and τ_{ij} is the fluid stress tensor. The configuration of the fluid region is changeable.

2.1.2. Solid Region

Structural momentum equation:

$$\frac{\partial \sigma_{ij}}{\partial x_i} + f_i = \rho_p \frac{\partial u_i}{\partial t} \tag{3}$$

Solid constitutive equation:

$$\sigma_{ij} = D_{ijkl} \varepsilon_{kl} \tag{4}$$

where σ_{ij} is the solid stress tensor, D_{ijkl} is the Lagrange elasticity tensor, and ε_{kl} is the strain tensor.

2.1.3. Liquid-Solid Coupling Interface:

$$\sigma_f \cdot n_f = \sigma_p \cdot n_p \tag{5}$$

$$u_f = u_p \tag{6}$$

where σ is the stress tensor, n is the normal vector, and u is the velocity vector of the interface.

2.1.4. Region with Porous Media Seepage

Continuity equation:

$$\frac{\partial \varepsilon u_i}{\partial x_i} = 0 \tag{7}$$

Momentum equation:

$$\frac{\partial \varepsilon u_i}{\partial t} + u_j \frac{\partial \varepsilon u_i}{\partial x_j} = -\frac{1}{\rho_f} \frac{\partial \varepsilon p}{\partial x_i} + \frac{1}{\rho_f} \frac{\partial \varepsilon \tau_{ij}}{\partial x_j} - \frac{\varepsilon^2 \mu}{\rho_f \cdot k} u_i \tag{8}$$

where k is permeability, ε is porosity, and μ is the fluid viscosity factor.

2.1.5. Fluid-Porous Media Seepage Interface

$$I_{f,p} = -K_{fp} \left(V_f - V_m\right) \tag{9}$$

where $I_{f,p}$ is the momentum exchange capacity of the interface; K_{fp} is the interface conductivity, which is determined by the constitutive equation of porous media seepage; V is velocity.

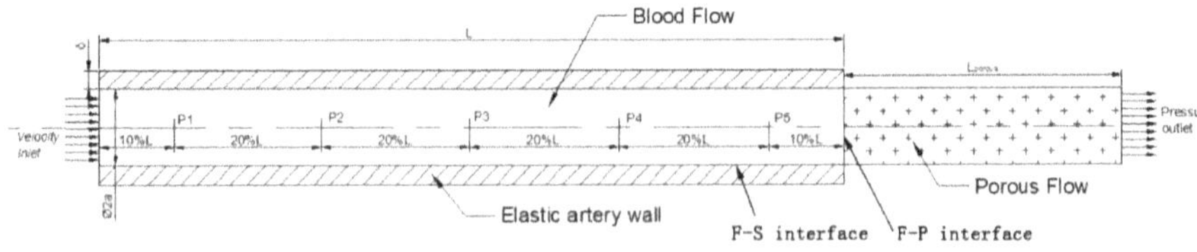

Figure 1. Calculation model and schematic of measuring point distribution.

2.2. Simulate Method

In order to comprehensively show the properties of the model, this paper selected the entire blood circulation system starting from the aorta as a study subject. The whole aorta has a straight tube of about 1 m. Because of the scale and strain rate of the aorta, the fluid was simplified as the Newtonian fluid [13]. The straight tube was set with a length of 1000 mm, diameter of 20 mm, and wall thickness of 2 mm, which accords with the aortal physiological parameters. The tube with porous media for seepage had a length of 300 mm, ensuring its internal flow developing completely. As the power source of the blood circulation system, the inlet took velocity as its boundary condition. Outlet was set as free outlet, located at the end of the tube with porous media for seepage.

The Computational methodology process of the hemodynamic model is shown in Figure 2. On the liquid-solid interface, fluid applied the calculated wall pressure on the solid; after the solid was deformed due to the pressure, the flow field grid was rebuilt. Then, after several iterations, a convergent result was obtained. Notably, due to the change in the flow field region, the fluid would be temporarily stored in the deformation; when the pressure in the tube dropped, the deformation would decrease and those stored fluids would reenter the flow field to flow. On the fluid-porous media seepage interface, fluid provided the pressure on interface. Under this pressure, the total fluid amount allowed to go through the seepage tube could be calculated. In this system, though solid and porous media seepage did not interact directly on an interface, they were connected together by the flow field. Therefore, the three parts operate synergistically, creating a complex multi-directional coupling relationship.

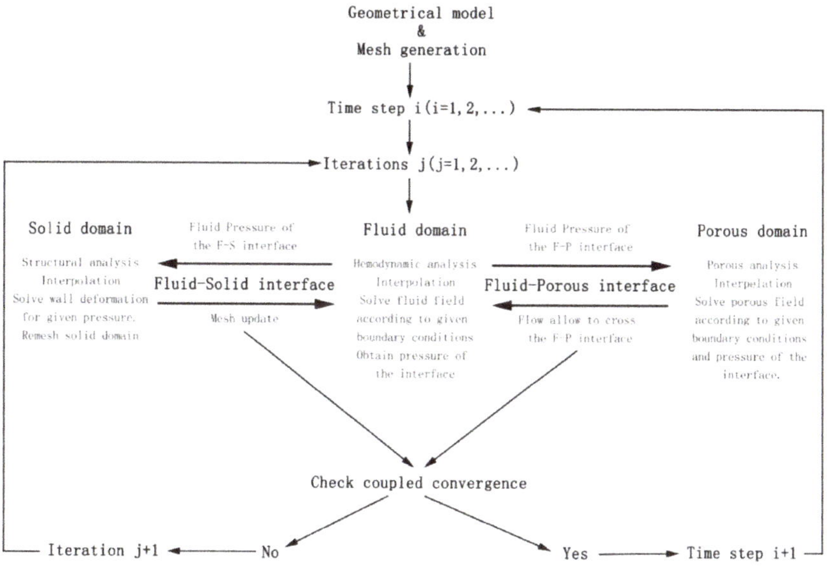

Figure 2. Computational methodology.

This paper divided the model into a structural grid. As shown in Figure 3, the tube cross-section was divided into butterfly grids. The model finally consisted of 84,728 cells, with 107,850 nodes. Therein, the fluid region had 52,668 cells, the solid region had 16,632 cells, and the tube with porous media for seepage had 15,428 cells. CFD-ACE SOLVER was chosen to carry out the entire calculation for this paper, and an arbitrary Lagrangian-Eulerian (ALE) method was employed. The first-order upwind difference scheme was used. A constant time step $\Delta t = 0.02$ s was employed in this study. When the program was running, corresponding physical quantities such as pressure and displacement transfer across fluid-structure interfaces through the coupling of the two sets of codes until a convergence criterion (10^{-4}) was reached for each time step. Repeated computations with a finer grid (253,800 nodes, 205,386 cells) and coarser grid (33,165 nodes, 24,220 cells) were carried out. Results showed that the alteration of maximum pressure at the same location of the fluid domain was 6% between the finer grid and the coarser grid, and alteration was 3% between the finer grid and the grid in this paper.

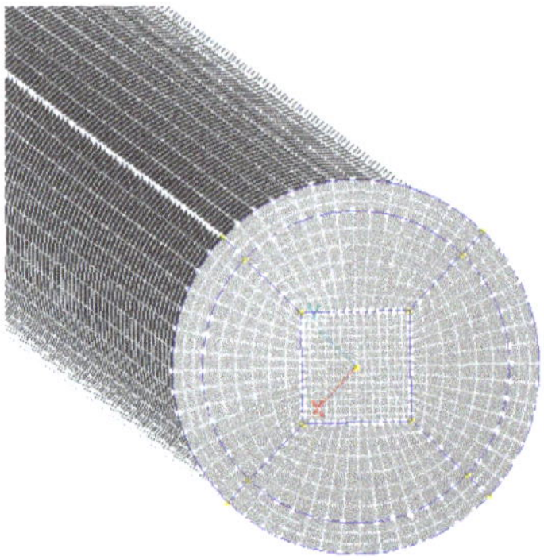

Figure 3. Grid division for the tube cross-section.

2.3. Calculation Examples and Parameters

In order to verify the reliability and superiority of the hemodynamic model proposed, five calculation examples (see Table 1) were taken through changing the tube length, load conditions, and cardiac cycle. All the calculation examples were based on the physiological parameters of the descending aorta, and the transient flow was adopted for simulating actual the working state of a blood vessel. The fluid in the tube was simplified as the Newtonian fluid of blood parameters, with a density of 1050 kg/m^3 and a dynamic viscosity coefficient of 0.0035 Pa·s [14]. The density of the flexible tube wall was 1120 kg/m^3, Young's modulus was 0.5 MPa [15], and Poisson's ratio is 0.49 [16]. The permeability of the porous media was $9.5 \times 10^{-9} m^2$, and the porosity was 100%.

Table 1. Five calculation examples and their parameters.

Calculation Examples	Permeability of Porous Media k	Length of Flow Field	Cardiac Cycle T
Example 1. Normal physiology	9.5×10^{-9}	1000 mm	0.8
Example 2. No seepage load	N/A	1000 mm	0.8
Example 3. Short tube	9.5×10^{-9}	200 mm	0.8
Example 4. Accelerated heart rate	9.5×10^{-9}	1000 mm	0.6
Example 5. Decreased heart rate	9.5×10^{-9}	1000 mm	1

All the calculation examples selected velocity inlet condition and free outlet, and the inlet condition and outlet condition are shown in Figure 4. The inlet condition was velocity inlet, which was used for simulating the process of cardiac impulse. Systole was 0–0.2 s, and the maximum systolic velocity was 0.8 m/s [17]. Diastole ranged from 0.2 s to the end of a cardiac cycle. Inlet velocity at this period was kept at 0 m/s to simulate the state of a closed heart valve without reflux. The average Reynolds number was about 661, which was based on the average velocity of the inlet. The flow regime was laminar flow. The outlet condition was pressure-free, *i.e.*, the pressure was always kept at 0 Pa. This not only eliminated the influence of disturbance from outer pressure factors, but also achieved similarity with the physiological reality in the vein.

For the two calculation examples where the heart rate was changed, this paper assumed the outflow after every heart pulse to be consistent. Thus one only has to change the length of the diastole to achieve a change in the cardiac cycle. As shown in Figure 4, for different cardiac cycle calculation examples, the systole was kept unchanged while the diastole changed. Accordingly, in the example of normal physiology, heart rate was 75/min and cardiac cycle was 0.8 s (0.2 s for systole, and 0.6 s for diastole); in the example of accelerated heart rate, heart rate was 100 min^{-1} and cardiac cycle was 0.6 s

(0.2 s for systole, and 0.4 s for diastole); in the example of decreased heart rate, heart rate was 60 min^{-1} and cardiac cycle was 1 s (0.2 s for systole, and 0.8 s for diastole).

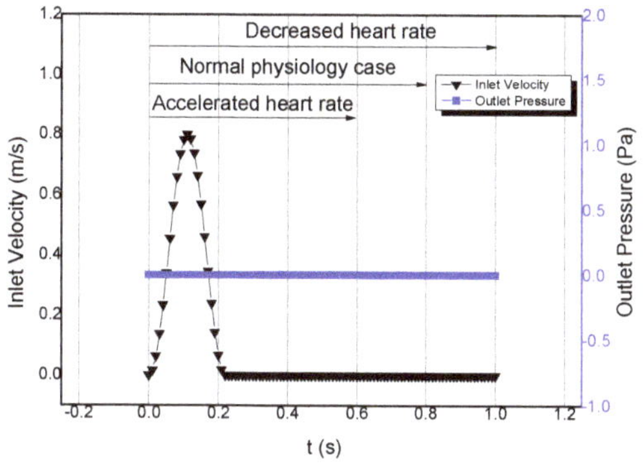

Figure 4. Boundary condition of the flow field.

3. Results and Discussion

This paper firstly compared the calculation example with no seepage load. Using the index of vascular pressure, the rationality of the liquid-solid coupling hemodynamic model with seepage load proposed in this paper was analyzed. Then, through simulation of pressure wave propagation, reflection, and other physiological phenomena in the blood vessel, the importance of the length of the straight tube was analyzed, which supports the reliability of the model proposed. Finally, the superiority of the model was discussed from the formation of secondary pressure wave in blood vessel and the influence of cardiac cycle.

In the process of data analysis, in order to show the different variation forms in the entire flow field at different positions, this paper selected five measuring points in the tube. All of them were located in the center of the tube cross-section. As shown in Figure 1, the distances of the five measuring points P1–P5 from the inlet were 0.1 m, 0.3 m, 0.5 m, 0.7 m, and 0.9 m, respectively. Under such a configuration, we could comprehensively master the similarities and differences at each position of the tube, through analysis of the pressure-time relationship at various measuring points of the flow field. In addition, we could analyze phase differences of various feature points in one cycle.

3.1. Physiological Pressure Level

Calculation results showed that the influence of load on the flow field was tremendous. Thus, between calculation examples with load and without load, the properties of flow field exhibited major differences. In Figure 5, there is a comparison between the pressure results of calculation examples with normal physiology and no seepage load. Therein, A is the systolic pressure of various positions in the tube, B is the diastolic pressure, and C is the amplitude of the pressure in the tube. In the example of normal physiology, the systolic pressure of all positions in the tube was maintained at around 120 mmHg; the diastolic pressure was kept at 80 mmHg, and the amplitude of pressure at 30–40 mmHg. These observations accorded with the physiological reality. In the example without seepage load, the systolic pressure near the inlet was 29 mmHg, while that near the outlet was 6 mmHg. The diastolic pressure near the inlet was −32 mmHg, while that near the outlet was −6 mmHg. With the approach to the outlet, the amplitude decreased continuously. It can be seen that the use of pressure alone as the outlet condition led to a larger negative pressure, and the pressure amplitude in the tube was changing. These results do not accord with the physiological conditions. The seepage load played a role in impeding flow in the model, which had a great effect on whether the fluid would be stored in the deformation of the tube wall. Thus the pressure in the tube was redistributed and reached the physiological pressure level.

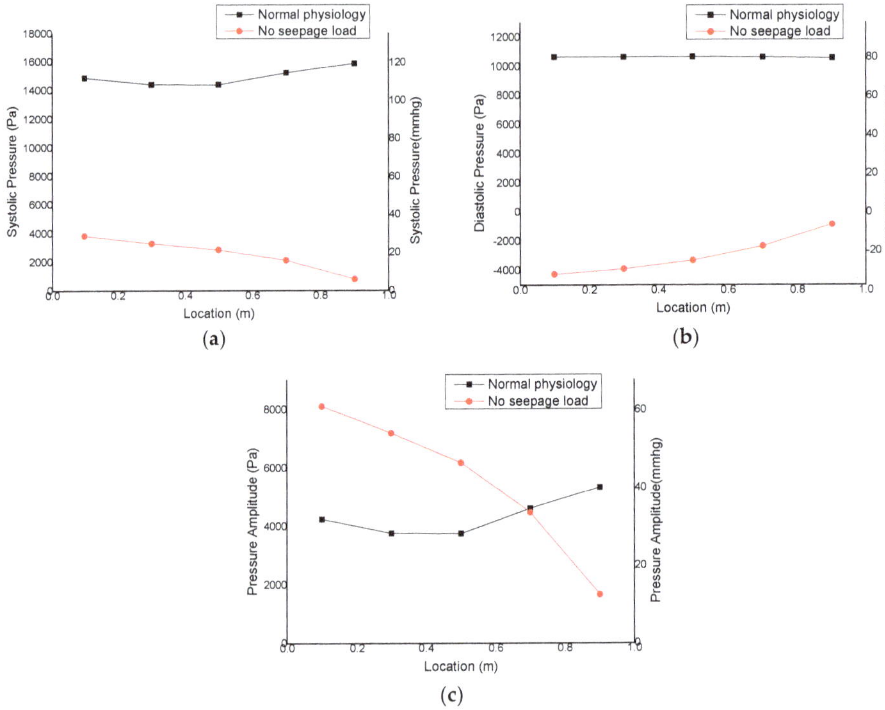

Figure 5. Comparisons of pressure in the examples with seepage load and pressure-free outlet: (**a**) maximum vascular pressure; (**b**) minimum vascular pressure; (**c**) pressure amplitude.

Through studying the process of the flow field developing in the tube, the internal mechanism of how the normal physiology example maintained the physiological pressure level can be revealed. When the load had a very strong hindering effect, the fluid entering the flow field was not able to flow out completely. Part of the fluid was stored in the deformation of tube wall and participated in the next cardiac cycle. Thus, the initial pressure of the second cycle would be increased. After several cardiac cycles, when the initial pressure of the flow field increased to a certain level, the newly input fluid was able to pass through the load and flow out of the flow field in one cycle. Accordingly, the fluid in the tube resumed a stable state, and the initial pressure of flow field also reached a dynamic balance.

Figure 6 is the curve of pressure change with time in the flow field of the normal physiology example, which shows the whole process of flow field developing from completely static to stable. The five lines in the figure represent the five positions from inlet to outlet in the tube. The time ranged from 0 s to 16 s, with 20 cycles in total. It can be seen from the figure that the pressure in flow field increased rapidly from 0 s to 7 s, and gradually became stable after 7 s. The final pressure at all measuring points in the tube was kept in the range of physiological pressure of 80–120 mmHg. This indicates that the load influenced the total fluid quantity stored through the elastic deformation of tube wall, which further influenced the systolic pressure, diastolic pressure, and pressure amplitude in the tube.

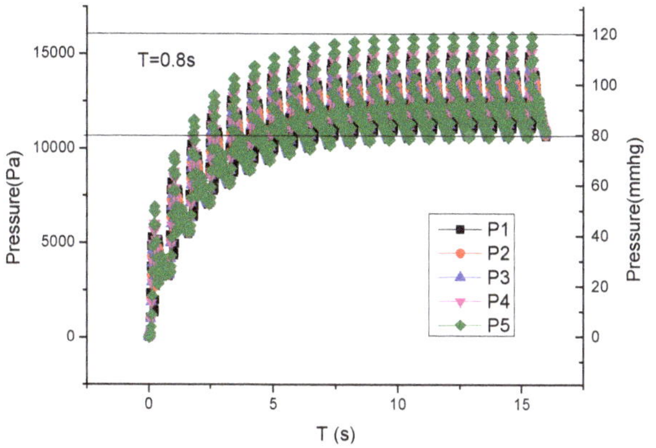

Figure 6. Changing process of pressure in the tube from static flow field (P1–P5 are five measuring points distributing axially along the tube).

3.2. Pulse Wave Propagation

This model showed very complex phenomena in time and in space, which included the propagation of pulse wave. When fluid entered the tube, fluid would be stored in the increased flow field from tube wall deformation, which would not cause influence for downstream flow field temporarily. Due to such property, the moments when the pressure wave peak occurred at various positions of the tube were different. Figure 7 is the pressure nephogram of the tube cross-section, wherein A, B, C, and D represent four moments (0 s, 0.1 s, 0.16 s, and 0.2 s, respectively). From the nephogram, we can clearly observe the propagation of the pressure wave peak, that is, the propagation of the pulse wave. At the initial moment, the pressure in the tube was maintained at a stable value. At 0.1 s, inlet velocity just reached the wave peak, and the inlet pressure reached the maximum. At 0.16 s, the pressure wave peak started to move forward, and the pressure decreased on both sides of the wave peak. Until0.2 s, the pressure wave peak was near the outlet. Comparing the flow field of these moments, the whole process of the pressure wave peak moving from inlet to outlet can be seen distinctly. This is the same as the result of Olson's research [18]. According to the academic monograph of Fung [13], propagation of the pulse wave exists widely. Therefore, it is necessary to take the effect of the pulse wave into account in hemodynamic simulations.

Results of the short tube example show that the difference in peak occurring moment was small. As shown in Figure 8a, the pressure peak values in tube all occurred at 0.18 s. The tube with a length of only 200 mm failed to show the phenomenon of pressure wave propagation. Comparatively, in the model of this paper, the tube length was the same as the actual aorta length. This provided the flow field with enough developing space. It follows that tube length is one of the essential conditions for simulation results to accord with the physiological reality.

Through data treatment, the moments when pressure peaks occurred at five measuring points were obtained and subjected to linear fitting, as seen in Figure 8b. It can be seen that the measuring point P1 at 0.1 m away from the inlet reached the pressure maximum first. Other measuring points showed the pressure peak value successively; the closer to the inlet, the earlier the peak occurred. The two measuring points at 0.1 m and 0.9 m away from the inlet had a time difference of 0.12 s in the occurrence of pressure peak. Through the linear fitting for peak-occurring moments, we can obtain the relationship between peak-occurring moment and the distance from inlet. According to the fitting result, the velocity of the pulse wave in the tube was 6.25 m/s, which is close to the actual pulse wave velocity [19]. Also, it is the same as the one-dimensional pressure wave velocity [20]. This confirms that this model was able to correctly simulate the pulse wave propagation in human aortas.

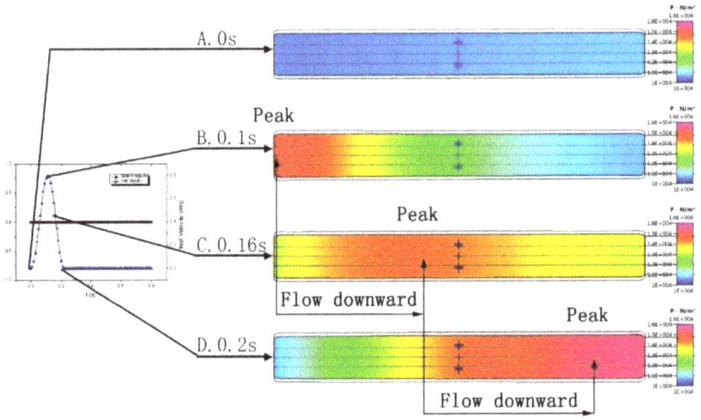

Figure 7. Pressure nephograms (A. t = 0 s; B. t = 0.1 s; C. t = 0.16 s; D. t = 0.2 s).

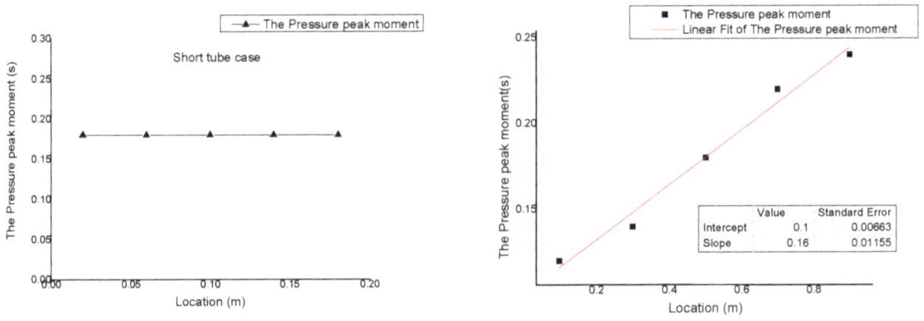

(**a**) (**b**)

Figure 8. Moments corresponding to the pressure peak and their fitting results: (**a**) peak-occurring moments in short tube example; (**b**) peak-occurring moments in a normal physiology example.

3.3. Pulse Wave Reflection

Another phenomenon shown in the model is the reflection of the pulse wave. When pressure propagated to the load interface, seepage load played a hindering effect. Accordingly, not all the fluid was able to flow out of the flow field smoothly; part of it would be held in the flow field. At this time the tube wall near the outlet would expand, and the pressure would increase accordingly. When the pressure downstream was larger than that upstream, it would also propagate back towards the inlet, producing a reflection wave. If the pulse wave is taken as a major pressure wave, this reflection wave is typically called a secondary pressure wave. By overlapping this reflection wave with the incoming major pressure wave, a pressure-time curve with double wave peaks can be produced. In previous studies, such a curve failed to be simulated. However, in the results of physiological measurement and of simulation by our model, such a curve shape with a double wave peak was visible.

Figure 9 shows a pressure-time curve, plotted with the observations of two measuring points at 0.3 m and 0.7 m away from the inlet. We can clearly see the double peak structure from the figure. The pressure-time curve was formed through the overlapping of a major pressure wave and a reflection wave; the reflection wave was obviously smaller than the major pressure wave. Comparing the curves of the two measuring points, it can be seen that though the difference between major pressure waves was not large, there was a significant difference between reflection waves.

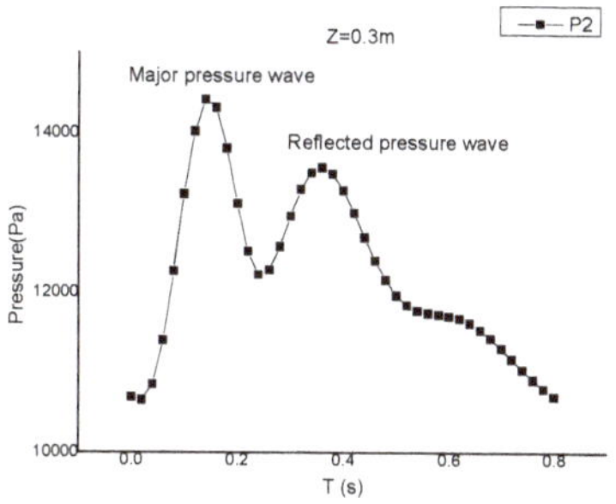

(a)

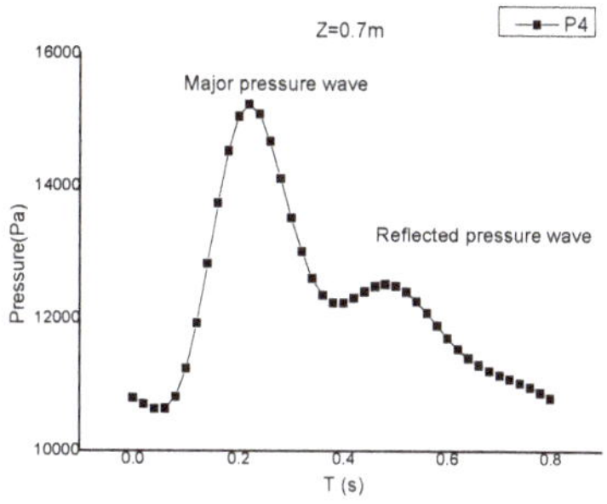

(b)

Figure 9. Pressure-time curves of measuring points: (**a**) P2 (0.3 m); (**b**) P4 (0.7 m).

This paper studied the occurring moments of the major pressure wave peak and reflection wave peak, as shown in Figure 10. It can be seen that the occurring moments of major pressure peak were in linear relationship with distance. However, the reflection wave peak presented a non-linear feature, e.g., the peak-occurring time was earlier at the position of 0.3 m than at the position of 0.1 m, while it was earlier at the position of 0.7 m than at the position of 0.9 m. This exactly suggests that secondary pressure wave is also propagated in the form of a wave. As shown by marks of the figure, the major pressure wave was a non-reflected primary pressure wave, propagating towards the outlet. However, at the positions of 0.3 m and 0.1 m, the pressure waves were secondary pressure waves that have been reflected once by the load and hence propagated towards the inlet. Similarly, the pressure waves at 0.7 m and 0.9 m were the third pressure waves that experienced another reflection by the inlet, and propagated towards the outlet again. Conclusively, in Figure 9, the reflection wave at 0.3 m was the secondary pressure wave undergoing one reflection, and that at 0.7 m was the third pressure wave undergoing two reflections. That is why there was a large difference between the magnitudes of reflection waves at these two positions.

Through comparing with physiological measurements [9], it can be seen that the secondary pressure wave simulated by the model of this paper accords with the physiological reality. Without the coupling of liquid-solid-porous media seepage, it is impossible to simulate this physiological blood pressure, which is produced by repeated reflections and overlapping of pressure waves. In addition, this complex wave shape produced a profound influence on the flow field by influencing the tube wall deformation. The wall shear stress and other flow field parameters would change accordingly. Therefore, the simulation of this paper accords with the physiological reality, and this complex wave structure is also an important guarantee for the correctness of the study later.

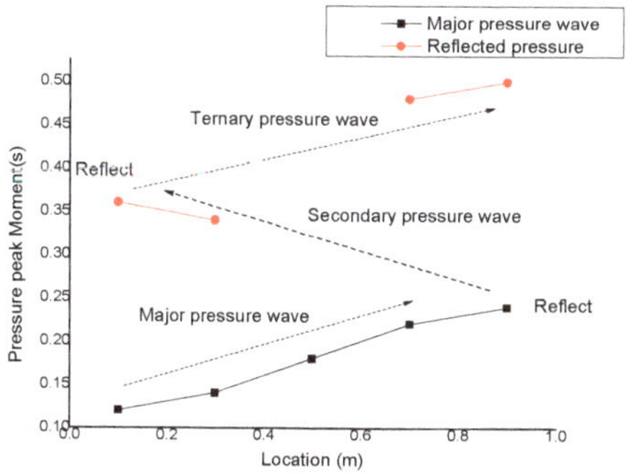

Figure 10. Moments corresponding to major pressure peak and reflection wave peak.

3.4. Influence of Cardiac Cycle

In order to verify the reliability of the model in terms of physiological parameter change, this paper used the cardiac cycle as the study subject. In human physiological phenomena, a change in the cardiac cycle has a great influence on the pressure in a blood vessel. The blood flux output after every cardiac pulse is fixed. However, with a change in the cardiac cycle, the time of fluid flowing out of the flow field will change, and the pressure required by a fluid to pass through the load will increase, finally influencing the pressure in the tube. Therefore, the pressure change caused by heart rate is not only related to flux but also has a close connection with vascular wall deformation and load. It can be seen that the influence of the cardiac cycle on vascular pressure is produced by a coupling of multiple factors, and all factors in the system are required to accord with the physiological reality. In a numerical simulation, a short straight tube could not meet the needs of storing fluid, and normal pressure outlet conditions could not change with the tube's flow field. Hence, both could not simulate the pressure change caused by cardiac cycle change. Comparatively, the model in this paper provides fluid with enough developing space and storing ability, as well as the load condition changing with the flow field. Therefore it can complete the task the numerical simulation cannot. This paper selected two cardiac cycles (0.6 s and 1 s), and compared them with the normal physiology example.

First, this paper analyzed the systolic pressure and diastolic pressure of various positions under the circumstances of different cardiac cycles. As shown in Figure 11, three lines represent the results of the three calculation examples with cardiac cycles of 0.6 s, 0.8 s, and 1 s, respectively. It can be seen that the shortening of the cardiac cycle brought simultaneous increases in diastolic pressure and systolic pressure. When the cardiac cycle was 0.6 s, the systolic pressure was 140–150 mmHg and diastolic pressure was 110–115 mmHg. When the cardiac cycle was 1 s, systolic pressure was 90–100 mmHg and diastolic pressure was about 60 mmHg. The above described pressure change was the same as the actual changing principle of human blood pressure under rest and motion states. Thus the change of vascular pressure in our model accords with the physiological reality.

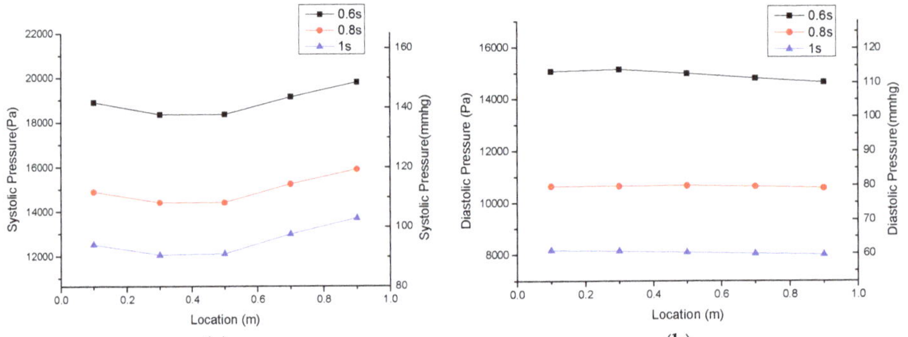

Figure 11. Comparison of pressure in tube under different cardiac cycle (**a**) systolic pressure; (**b**) diastolic pressure.

Secondly, this paper analyzed the changes of tube displacement under different cardiac cycles. In Figure 12, three lines represent the tube displacement circumstances in the three examples, with cardiac cycles of 0.6 s, 0.8 s, and 1 s. Figure 12a shows the initial displacement circumstance of tube at 0 s. Figure 12b shows the circumstance of tube displacement at 0.1 s, when the peak occurred at the inlet. Figure 12c shows the circumstance of tube displacement at 0.16 s, when the peak had not reached the outlet. Figure 12d shows the displacement circumstance at 0.2 s, when the peak reached the outlet. It can be seen that when the cardiac cycle decreased, tube wall deformation increased markedly, and *vice versa*. The inlet pressure reached its peak value at 0.1 s. The pressure peak moved forward at 0.16 s and arrived at the outlet at 0.2 s. However, like the pressure distribution, the curve shape changed little under different frequencies. This illustrates that the cardiac cycle has limited influence on the change of flow in the tube, and the pressure change is mainly caused by a change in the fluid amount stored through tube wall deformation.

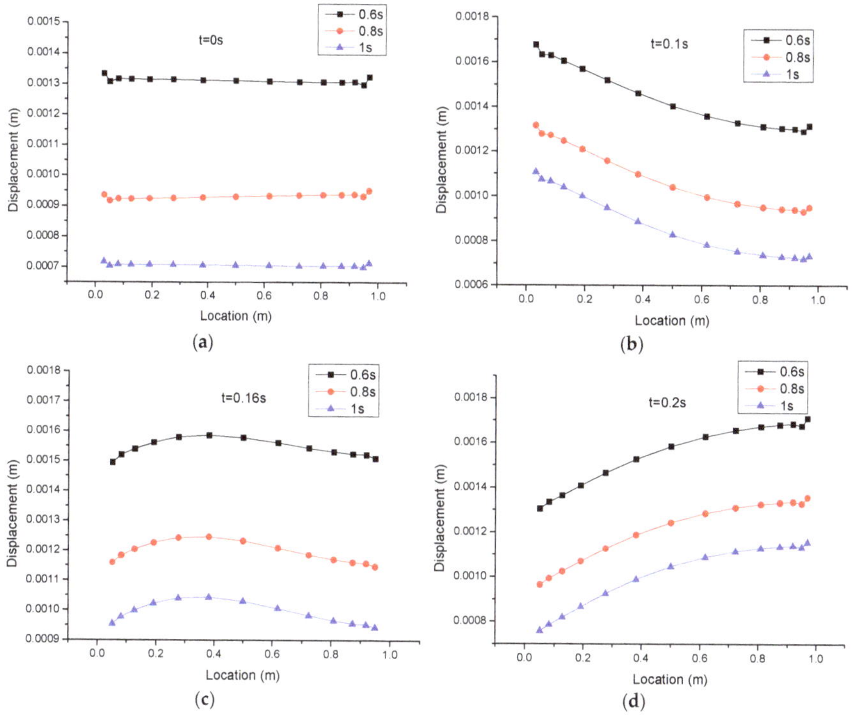

Figure 12. Comparison of tube wall deformation under different cardiac cycle lengths: (**a**) 0 s; (**b**) 0.1 s; (**c**) 0.16 s; (**d**) 0.2 s.

With the coupling effect of three major systems, the proposed model successfully simulated the pressure changes in the tube only by adjusting cardiac cycle length. The simulated pressure changes have been proved to accord with the actual physiological features. This illustrates that the model of this paper is a well-developed and reliable system, which is able to show comprehensively the transition process of the circulation system under different states.

4. Conclusions

Conclusively, the liquid-solid coupling hemodynamic model built in this paper considering microcirculation load effect is able to effectively simulate pulse wave propagation and reflection. It can also reflect the physiological features of human aortas and circulation system. The model is a hemodynamic model according with actual physiological process. Through data analysis, it was revealed that in the human blood circulation system, the coupling relationship of liquid, solid, and porous media seepage is exactly the deep reason why the various complex flowing phenomena of human blood occur in the circulation process. The effects of the propagation and reflection of the pulse wave should not be ignored in hemodynamic simulations. Meanwhile, through this model, we were able to simulate the motion and physiological features of flow in a blood vessel more rationally. This study thus provides an important theoretical foundation and technical methods for analyzing cause, development, and clinical treatment of atherosclerosis, aneurysms, high blood pressure, and other cardiovascular and cerebrovascular diseases in the future.

The constitutive equation of microcirculation load has a great influence on the bloodstream, and has decisive influence on various phenomena in the flow field. Therefore, in hemodynamic study and simulation, the influence of microcirculation load must be considered. The simulations will achieve a better correspondence with human physiology through deep study of microcirculation load properties and improving the constitutive equation of the model.

The model is able to realize the transition of the circulation system under different flow field conditions, providing the possibility of simulating the dynamic condition of the human circulation system. Through studying dynamic properties, we cannot only analyze the properties of the blood circulation system under different physiological states, but also simulate the transition process of the human body between different physiological states. These have important meanings for research on aneurysms and other blood diseases.

However, the present work is limited by physiological inaccuracies in the geometrical shape of the models. Further work will be conducted on the simulation in actual arterial geometries extracted from CT angiography or MRI. High-precision grids will be employed. In addition, the effect of non-Newtonian fluid on the small arteries will be considered.

Acknowledgments: Acknowledgments: This study was supported by "Specialized Research Fund for the Doctoral Program of Higher Education" (20101103110001). The authors are very grateful to the reviewers for reviewing this manuscript and giving very constructive suggestions.

Author Contributions: Author Contributions: Bai Li and Xiaoyang Li conceived and designed the study. Bai Li performed the simulation and wrote the paper. All authors read and approved the manuscript.

Conflicts of Interest: Conflicts of Interest: The authors declare no conflict of interest.

References

1. Rojas, H.A.G. Numerical implementation of viscoelastic blood flow in a simplified arterial geometry. *Med. Eng. Phys.* **2007**, *29*, 491–496. [CrossRef] [PubMed]
2. Shojima, M.; Oshima, M.; Takagi, K.; Torii, R.; Nagata, K.; Shirouzu, I.; Morita, A.; Kirino, T. Role of the bloodstream impacting force and the local pressure elevation in the rupture of cerebral aneurysms. *Stroke* **2005**, *36*, 1933–1938. [CrossRef] [PubMed]
3. Aenis, M.; Stancampiano, A.P.; Wakhloo, A.K.; Lieber, B.B. Modeling of flow in a straight stented and nonstented side wall aneurysm model. *J. Biomech. Eng. ASME* **1997**, *119*, 206–212. [CrossRef]
4. Milner, J.S.; Moore, J.A.; Rutt, B.K.; Steinman, D.A. Hemodynamics of human carotid artery bifurcations: Computational studies with models reconstructed from magnetic resonance imaging of normal subjects. *J. Vasc. Surg.* **1998**, *28*, 143–156. [CrossRef]
5. Molony, D.S.; Callanan, A.; Kavanagh, E.G.; Walsh, M.T.; McGloughlin, T.M. Fluid-structure interaction of a patient-specific abdominal aortic aneurysm treated with an endovascular stent-graft. *Biomed. Eng. Online* **2009**, *8*. [CrossRef] [PubMed]
6. Wang, X.H.; Li, X.Y. The influence of wall compliance on flow pattern in a curved artery exposed to a dynamic physiological environment: An elastic wall model versus a rigid wall model. *J. Mech. Med. Biol.* **2012**, *12*. [CrossRef]
7. Wang, X.; Li, X. Biomechanical behaviour of cerebral aneurysm and its relation with the formation of intraluminal thrombus: A patient-specific modelling study. *Comput. Methods Biomech.* **2013**, *16*, 1127–1134. [CrossRef] [PubMed]
8. Le, T.; Borazjani, I.; Sotiropoulos, F. A Computational Fluid Dynamic (CFD) Tool for Optimization and Guided Implantation of Biomedical Devices. *J. Med. Devices* **2009**, *3*, 27553. [CrossRef]
9. Nichols, W.; O'Rourke, M.; Vlachopoulos, C. *McDonald's Blood Flow in Arteries*; Hodder Arnold: London, UK, 2011; pp. 225–227.
10. Wang, X.; Li, X. Computer-based mechanical analysis of stenosed artery with thrombotic plaque: The influences of important physiological parameters. *J. Mech. Med. Biol.* **2012**, *12*. [CrossRef]

11. Schwarz, E.R.; Klues, H.G.; VomDahl, J.; Klein, I.; Krebs, W.; Hanrath, P. Functional characteristics of myocardial bridging—A combined angiographic and intracoronary Doppler flow study. *Eur. Heart J.* **1997**, *18*, 434–442. [CrossRef] [PubMed]
12. Dong, J.; Wong, K.K.L.; Tu, J. Hemodynamics analysis of patient-specific carotid bifurcation: A CFD model of downstream peripheral vascular impedance. *Int. J. Numer. Method Biomed. Eng.* **2013**, *29*, 476–491. [CrossRef] [PubMed]
13. Fung, Y.C. *Biomechanics, Mechanical Properties of Living Tissues*; Springer: Berlin, Germany, 2005.
14. Molony, D.S.; Callanan, A.; Morris, L.G.; Doyle, B.J.; Walsh, M.T.; McGloughlin, T.M. Geometrical enhancements for abdominal aortic stent-grafts. *J. Endovasc. Ther.* **2008**, *15*, 518–529. [CrossRef] [PubMed]
15. Patel, D.J.; Vaishnav, R.N. The rheology of large blood vessels. *Cardiovasc. Fluid Dyn.* **1972**, *2*, 2–65.
16. Di Martino, E.S.; Guadagni, G.; Fumero, A.; Ballerini, G.; Spirito, R.; Biglioli, P.; Redaelli, A. Fluid-structure interaction within realistic three-dimensional models of the aneurysmatic aorta as a guidance to assess the risk of rupture of the aneurysm. *Med. Eng. Phys.* **2001**, *23*, 647–655. [CrossRef]
17. Lam, S.K.; Fung, G.S.; Cheng, S.W.; Chow, K.W. A computational study on the biomechanical factors related to stent-graft models in the thoracic aorta. *Med. Biol. Eng. Comput.* **2008**, *46*, 1129–1138. [CrossRef] [PubMed]
18. Olson, R.M. Aortic blood pressure and velocity as a function of time and position. *J. Appl. Physiol.* **1968**, *4*, 563–569.
19. Asmar, R.; O'Rourke, M.; Safar, M. *Arterial Stiffness and Pulse Wave Velocity: Clinical Applications*; Elsevier: Paris, France, 1999.
20. Young, T. The Croonian Lecture: On the Functions of the Heart and Arteries. *Philos. Trans. R. Soc. Lond.* **1809**, *99*, 1–31. [CrossRef]

MDPI
St. Alban-Anlage 66
4052 Basel
Switzerland
Tel. +41 61 683 77 34
Fax +41 61 302 89 18
www.mdpi.com

Applied Sciences Editorial Office
E-mail: applsci@mdpi.com
www.mdpi.com/journal/applsci

www.ingramcontent.com/pod-product-compliance
Lightning Source LLC
LaVergne TN
LVHW071939080526
838202LV00064B/6637
* 9 7 8 3 0 3 9 2 1 2 1 7 0 *